Geometric Formulas

Conversion between Radians and Degrees: π radians $= 180°$

Triangle

$A = \frac{1}{2}bh$

$\quad = \frac{1}{2}ab\sin\theta$

Circle

$A = \pi r^2$

$C = 2\pi r$

Sector of Circle

$A = \frac{1}{2}r^2\theta \quad (\theta \text{ in radians})$

$s = r\theta \quad (\theta \text{ in radians})$

Sphere

$V = \frac{4}{3}\pi r^3 \quad A = 4\pi r^2$

Cylinder

$V = \pi r^2 h$

Cone

$V = \frac{1}{3}\pi r^2 h$

Trigonometric Functions

$\sin\theta = \dfrac{y}{r}$

$\cos\theta = \dfrac{x}{r}$

$\tan\theta = \dfrac{y}{x}$

$\tan\theta = \dfrac{\sin\theta}{\cos\theta}$

$\cos^2\theta + \sin^2\theta = 1$

$\sin(A\pm B) = \sin A\cos B \pm \cos A\sin B$

$\cos(A\pm B) = \cos A\cos B \mp \sin A\sin B$

$\sin(2A) = 2\sin A\cos A$

$\cos(2A) = 2\cos^2 A - 1 = 1 - 2\sin^2 A$

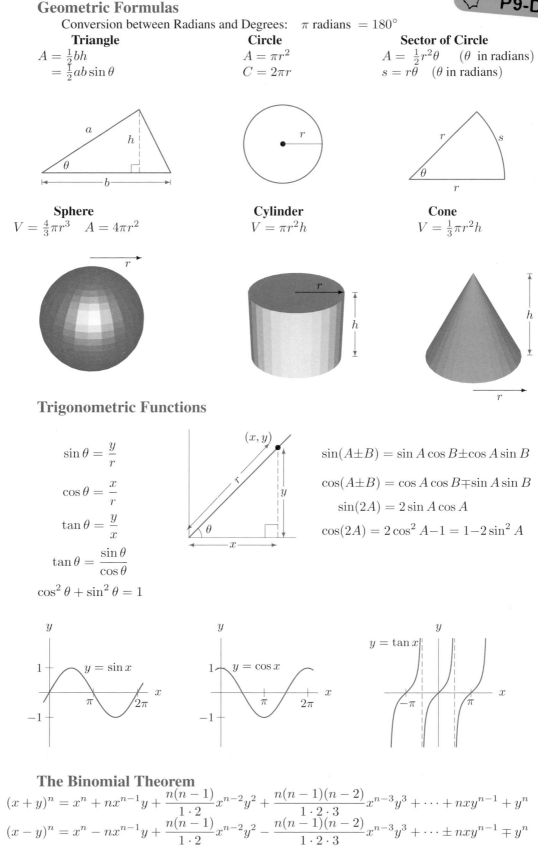

The Binomial Theorem

$$(x+y)^n = x^n + nx^{n-1}y + \frac{n(n-1)}{1\cdot 2}x^{n-2}y^2 + \frac{n(n-1)(n-2)}{1\cdot 2\cdot 3}x^{n-3}y^3 + \cdots + nxy^{n-1} + y^n$$

$$(x-y)^n = x^n - nx^{n-1}y + \frac{n(n-1)}{1\cdot 2}x^{n-2}y^2 - \frac{n(n-1)(n-2)}{1\cdot 2\cdot 3}x^{n-3}y^3 + \cdots \pm nxy^{n-1} \mp y^n$$

MULTIVARIABLE CALCULUS

Fourth Edition

MULTIVARIABLE CALCULUS

Fourth Edition

Produced by the Calculus Consortium and initially funded by a National Science Foundation Grant.

William G. McCallum
University of Arizona

Deborah Hughes-Hallett
University of Arizona

Daniel Flath
Macalester College

Brad G. Osgood
Stanford University

Andrew M. Gleason
Harvard University

Douglas Quinney
University of Keele

Patti Frazer Lock
St. Lawrence University

Jeff Tecosky-Feldman
Haverford College

David Mumford
Brown University

Thomas W. Tucker
Colgate University

with the assistance of

Sheldon P. Gordon
SUNY at Farmingdale

Adrian Iovita
University of Washington

Wayne Raskind
University of Southern California

Joe B. Thrash
University of Southern Mississippi

Coordinated by
Elliot J. Marks

WILEY

John Wiley & Sons, Inc.

ACQUISITIONS EDITOR	Michael Boezi
ASSOCIATE PUBLISHER	Laurie Rosatone
DEVELOPMENTAL EDITOR	Anne Scanlan-Rohrer/Two Ravens Editorial
ASSISTANT EDITOR	Kelly Boyle
EXECUTIVE MARKETING MANAGER	Julie Z. Lindstrom
MEDIA EDITOR	Lisa Schnettler
SENIOR PRODUCTION EDITOR	Ken Santor
COVER DESIGNER	Madelyn Lesure
COVER AND CHAPTER OPENING PHOTO	©Pete Turner/The Image Bank/Getty Images

Problems from Calculus: The Analysis of Functions, by Peter D. Taylor (Toronto: Wall & Emerson, Inc., 1992). Reprinted with permission of the publisher.

This book was set in Times Roman by the Consortium using TeX, Mathematica, and the package AsTeX, which was written by Alex Kasman. It was printed and bound by Von Hoffmann Press. The cover was printed by Von Hoffmann Press.

This book is printed on acid-free paper.

The paper in this book was manufactured by a mill whose forest management programs include sustained yield harvesting of its timberlands. Sustained yield harvesting principles ensure that the numbers of trees cut each year does not exceed the amount of new growth.

To order books or for customer service please, call 1(800)-CALL-WILEY (225-5945).

This material is based upon work supported by the National Science Foundation under Grant No. DUE-9352905. Opinions expressed are those of the authors and not necessarily those of the Foundation.

ISBN 0-471-48480-6

Printed in the United States of America

10 9 8 7 6 5 4 3 2

PREFACE

Calculus is one of the greatest achievements of the human intellect. Inspired by problems in astronomy, Newton and Leibniz developed the ideas of calculus 300 years ago. Since then, each century has demonstrated the power of calculus to illuminate questions in mathematics, the physical sciences, engineering, and the social and biological sciences.

Calculus has been so successful because of its extraordinary power to reduce complicated problems to simple rules and procedures. Therein lies the danger in teaching calculus: it is possible to teach the subject as nothing but the rules and procedures—thereby losing sight of both the mathematics and of its practical value. This edition of *Calculus* continues our effort to refocus the teaching of calculus on concepts as well as procedures.

A Focused Vision: Conceptual Understanding

Our goal is to provide students with a clear understanding of the ideas of calculus as a solid foundation for subsequent courses in mathematics and other disciplines. When we designed this curriculum we started with a clean slate. We increased the emphasis on some topics, such as differential equations, and decreased the emphasis on others after discussions with mathematicians, engineers, physicists, chemists, biologists, and economists. We focused on key concepts, emphasizing depth of understanding rather than breadth of coverage.

The Fourth Edition: Flexible and Focused

Since our First Edition struck a new balance between concepts, modeling, and skills in 1992, we have helped widen the range of choices for calculus instructors. Under our approach, which we called the "Rule of Four," ideas are presented graphically, numerically, symbolically, and verbally. Our problems probe student understanding in areas often taken for granted. The influence of these problems, praised for their creativity and variety, has extended far beyond the users of our textbook.

The text retains the hallmarks of our earlier editions: The Rule of Four, an emphasis on modeling, exposition that students can read, and a flexible approach to technology. Working in collaboration, consortium members and colleagues in mathematics and client disciplines have shaped a flexible curriculum that balances new and traditional topics.

In adding new topics, we have kept in mind two competing considerations: a textbook's role as a reference suitable for differing audiences, and the danger of encyclopedic clutter that overwhelms students and obscures the main vista. Our treatment of new topics is guided by the Rule of Four and our focus on conceptual understanding.

The Development of Mathematical Thinking

The first stage in the development of mathematical thinking is the acquisition of a clear intuitive picture of the central ideas. In the next stage, the student learns to reason with the intuitive ideas and explain the reasoning clearly in plain English. After this foundation has been laid, there is a choice of direction. All students benefit from both theory and modeling, but the balance may differ for different groups. Some students, such as mathematics majors, may prefer more theory, while others, such as science and engineering majors, may prefer more modeling.

The Development of Mathematical Skills

To use calculus effectively, students need skill in both symbolic manipulation and the use of technology. The exact proportions of each may vary widely, depending on the preparation of the student and the wishes of the instructor. The book is adaptable to many different combinations.

The book does not require any specific software or technology. It has been used with graphing calculators, graphing software, and computer algebra systems. Any technology with the ability to graph functions and perform numerical integration will suffice. Students are expected to use their own judgment to determine where technology is useful.

Content

This content represents our vision of how calculus can be taught. It is flexible enough to accommodate individual course needs and requirements. Topics can easily be added or deleted, or the order changed.

Chapter 12: Functions of Several Variables

This chapter introduces functions of many variables from several points of view, using surface graphs, contour diagrams, and tables. We assume throughout that functions of two or more variables are defined on regions with piecewise smooth boundaries. We conclude with a section on continuity.

Chapter 13: A Fundamental Tool: Vectors

This chapter introduces vectors geometrically and algebraically and discusses the dot and cross product.

Chapter 14: Differentiating Functions of Several Variables

Partial derivatives, directional derivatives, gradients, and local linearity are introduced. The chapter also discusses higher order partial derivatives, quadratic Taylor approximations, and differentiability.

Chapter 15: Optimization

The ideas of the previous chapter are applied to optimization problems, both constrained and unconstrained.

Chapter 16: Integrating Functions of Several Variables

This chapter discusses double and triple integrals in Cartesian, polar, cylindrical, and spherical coordinates, as well as the formula for a general change of variable.

Chapter 17: Parameterization and Vector Fields

This chapter discusses parameterized curves and motion, vector fields and flowlines, and parameterized surfaces.

Chapter 18: Line Integrals

This chapter introduces line integrals and shows how to calculate them using parameterizations. Conservative fields, gradient fields, the Fundamental Theorem of Calculus for Line Integrals, and Green's Theorem are discussed.

Chapter 19: Flux Integrals

This chapter introduces flux integrals and shows how to calculate them over surface graphs, portions of cylinders, portions of spheres, and arbitrary parameterized surfaces.

Chapter 20: Calculus of Vector Fields

The purpose of this chapter is to give students a practical understanding of the divergence and curl, as well as of the Divergence and Stokes' Theorems.

Appendices

There are appendices on roots, accuracy, and bounds; complex numbers; Newton's Method; and determinants.

Supplementary Materials and Additional Resources

Supplements for the instructor can be obtained by sending a request on your institutional letterhead to Mathematics Marketing Manager, John Wiley & Sons, Inc., 111 River Street, Hoboken, NJ 07030, or by contacting your Wiley representative. The following supplementary materials are available for this edition:

- **Instructor's Manual** containing teaching tips, calculator programs, overhead transparency masters.
- **Test Bank** containing test questions and solutions arranged by section.
- **Instructor's Solution Manual** with complete solutions to all problems.
- **Student's Solution Manual** with complete solutions to half the odd-numbered problems.
- **Additional Material**, elaborating specially marked points in the text, as well as password protected electronic versions of the instructor ancillaries, can be found on the web at www.wiley.com/college/hugheshallett.

ConcepTests

ConcepTests, modeled on the pioneering work of Harvard physicist Eric Mazur, are questions designed to promote active learning during class, particularly (but not exclusively) in large lectures. Our evaluation data show students taught with ConcepTest outperformed students taught by traditional lecture methods 73% versus 17% on conceptual questions, and 63% versus 54% on computational problems. ConcepTests arranged by section are available from your Wiley representative.

Faculty Resource Network

A peer-to-peer network of academic faculty dedicated to the effective use of technology in the classroom, this group can help you apply innovative classroom techniques and implement specific software packages. Visit www.facultyresourcenetwork.com or speak to your Wiley representative.

eGrade Plus

eGrade Plus is a powerful online tool that provides a completely integrated suite of teaching and learning resources in one easy-to-use Web site. *eGrade Plus* includes an online version of the text, with fully integrated electronic versions of all student supplements, including the Student Solutions Manual, the Student Study Guide, and an Algebra and Trigonometry Refresher. Instructors will have additional access to electronic versions of the Instructor's Manual, the Instructor's Solutions Manual, additional projects, as well as other valuable resources. *eGrade Plus* also offers an online assessment system with full gradebook capabilities, which contains over 1000 skill-building questions from the Exercise sections in each chapter. Please view our online demo at www.wiley.com/college/egradeplus. Here you will find additional information about the features and benefits of eGrade Plus, how to request a "test drive" of eGrade Plus, and how to adopt it for class use.

Acknowledgements

First and foremost, we want to express our appreciation to the National Science Foundation for their faith in our ability to produce a revitalized calculus curriculum and, in particular, to our program officers, Louise Raphael, John Kenelly, John Bradley, and James Lightbourne. We also want to thank the members of our Advisory Board, Benita Albert, Lida Barrett, Simon Bernau, Robert Davis, M. Lavinia DeConge-Watson, John Dossey, Ron Douglas, Eli Fromm, William Haver, Seymour Parter, John Prados, and Stephen Rodi.

In addition, a host of other people around the country and abroad deserve our thanks for their contributions to shaping this edition. They include: Huriye Arikan, Ralph Baierlein, Ruth Baruth, Frank Beatrous, Alon Ben-David, Ebo Bentil, Paul Blanchard, Michael Boezi, Kelly Boyle, Melkana Brakalova, David Bressoud, Stephen Boyd, Morton Brown, Lucille Buonocore, Rick Cangelosi, Jo Cannon, Ray Cannon, Marilyn Carlson, Phil Cheifetz, C.K. Cheung, Scott Clark, Eric Connally, Caspar Curjel, Ehud de Shalit, Ann Davidian, Bob Decker, Dennis De Turck, Tom Dick, Srdjan Divac, Tevian Dray, Jeff Edmunds, Wade Ellis, Johann Engelbrecht, Bill Faris, Paul Feehan, Hermann Flaschka, Sol Friedberg, Paul Froeschl, Amanda Galtman, Lynn Garner, Tom Gearhart, Chris Goff, Joshua Green, Cristi Guevara, John Hagood, Gregory Hartman,

Dudley Herschbach, Rob Indik, Adrian Iovita, Pallavi Jayawant, Millie Johnson, Alex Kasman, Matthias Kawski, Mike Klucznik, Donna Krawczyk, Mark Kunka, Ted Laetsch, Stephane Lafortune, Brigitte Lahme, Carl Leinbach, Carl Leinert, David Levermore, Don Lewis, Jim Lewis, Julie Lindstrom, Guadalupe Lozano, Dili Lu, Chris Lucas, John Lucas, Dan Madden, Alex Mallozzi, Corinne Manogue, Lisa A. Mantini, Nancy Marcus, Eric Mazur, Dan McGee, Tom McMahon, Ansie Meiring, Dave Meredith, Andrew Metrick, Lang Moore, Jerry Morris, Alan Newell, Steve Olson, John Orr, Arnie Ostebee, Igor Padure, Howard Penn, Amy Radunskaya, Wayne Raskind, Jeffrey Rauch, Janet Ray, Maria Robinson, Laurie Rosatone, Tenzin Chokey Rubling, Ann Ryu, Ayse Sahin, Nataliya Sandler, Ken Santor, Anne Scanlan-Rohrer, Pat Shure, Scott Pilzer, Lynne Small, David Smith, Seung Hye Song, Bob Speiser, Howard Stone, Steve Strogatz, "Suds" Sudholz, Al Taylor, Peter Taylor, Ralph Teixeira, Sally Thomas, Alan Tucker, Jerry Uhl, Doug Ulmer, Ignatios Vakalis, Bill Vélez, Aaron Wootton, Xianbao Xu, Kathy Yoshiwara, Lee Zia, and Paul Zorn.

Reports from the following reviewers were most helpful for the third edition:

Jennifer Aikens, Dean Allison, Cathy Anderson, Richard I. Avery, Janet Beery, Hamid Behmard, Victor Belfi, Tom Bengtson , David Borthwick, Sean Bradley, Carl T. Brezovec, Paul Wayne Britt, Denise Brown, Dave Bush, Don A. Campbell, Jack Carter, Stan Chadick, Ruth Charney, Jon Clauss, Ann Commito, Lawerence J. Cohen, Mary V. Connolly, Randall Crist, Rick Dalrymple, John Davis, Mike Dellens, Matt DeLong, Sylvia Desjardins, Meighan I. Dillon, Dawn Draus, Steven R. Dunbar, Jim Edmondson, Karen Edwards, Murray Eisenberg, Sue Esch, Phil Farmer, Richard E. Faulkenberry, George F. Feissner, Nelson Fong, Paula Castagna, Ben Fusaro, Kris Green, Michael Green, Leah Griffith, Cheryl Grood, David L. Gross, Brenda Guthrie, John Hagood, William Hamill, John Haverhals, Elizabeth Hentges, Curtis Herink, Jim Holmes, John Hoover, Rick Hough, Thomas Hunter, Kathy Ivey, Craig A. Jensen, Sue Jensen, Jerry Johnson, Nancy L. Johnson, Guarav Khanna, Helmut Knaust, Robert Kowalczyk, Anne Landry, James W. Lea, Phyllis Leonard, Tom Linton, Harold Lund, Juan Manfredi, William O. Martin, Alex Martsinkovsky, Anita Maxwell, Barbara McGee, Mary Ann McLoughlin, Douglas B. Meade, Richard Mercer, Raquel Mesa, C.R. Messer, Mark A. Mills, Yair Minsky, Derek Mitchem, Jeff Mock, Doug Mupasiri, Douglas Nelson, Dave Nolan, Sandy Norman, David A. Olson, Amos Ong, Grace Orzech, Wes Ostertag, Jeffrey S. Palmer, Barry Perratt, Peter Philliou, Shirley Pomeranz, G.J. Porter, Rebecca Porter, Vrunda Prabhu, Harald Proppe, Margie Reinhart, Mike Robillard, Donna Fields Rochon, Peter Rojas, Ned I. Rosen, Peter Rosnick, Tom Rousseau, Brenda Rudin, Mark D. Schlatter, Howard Schwesky, Patrick Shanahan, Michael Sheard, Angela B. Shiflet, Mehrdad Simkani, Lynne Small, Katy Snyder, Edward J. Soares, Ernie Solheid, Richard Stout, Robert Styer, James Sullivan, Ted Sundstrom, Stuart Swain, Ray Tennant, Buren Thomas, George Tintera, Dave Trautman, Mark Turner, Marian Van Vleet, Mike Veatch, Rachel Westlake, Bill Waltmann, Michael Ward, Steve Waters, Jack Winn, John Wood, Lisa Woodrum, William Wunderlich, Christos Xenophontos, W. Xie, Deane Yang, Kathie Yoder, Bruce Yoshiwara, Andrzej Zarach, Ed Zeidman, and Peter Zekert.

Reports from the following reviewers were most helpful for the fourth edition:

Patricia Beaulieu, Phil Blau, Russell D. Blyth, Ray Cannon, Paula Castagna, Annette Cooper, Randall Crist, James D'Archangelo, Michael Divinia, Diane Doyle, Jim Edmondson, Leslie O. Foged, Scott R. Fulton, Lynn E. Garner, Sandy Grabiner, Steve Harris, Jeff Hirst, Vlajko Kocic, Kenneth Levasseur, Greg Marks, Kevin McDougal, Wilbur Mellema, Dan Oberlin, Tom Rousseau, Jeffrey Scroggs, Ernie S. Solheid, Barry S. Spieler, Diana Staats, Steve Szydlik, Mark Turner, Tony Vance, Gordon Woodward, and William Wunderlich.

William G. McCallum	Patti Frazer Lock	Douglas Quinney
Deborah Hughes-Hallett	David Mumford	Jeff Tecosky-Feldman
Andrew M. Gleason	Brad G. Osgood	Thomas W. Tucker
Daniel E. Flath		

To Students: How to Learn from this Book

- This book may be different from other math textbooks that you have used, so it may be helpful to know about some of the differences in advance. This book emphasizes at every stage the *meaning* (in practical, graphical or numerical terms) of the symbols you are using. There is much less emphasis on "plug-and-chug" and using formulas, and much more emphasis on the interpretation of these formulas than you may expect. You will often be asked to explain your ideas in words or to explain an answer using graphs.

- The book contains the main ideas of multivariable calculus in plain English. Your success in using this book will depend on your reading, questioning, and thinking hard about the ideas presented. Although you may not have done this with other books, you should plan on reading the text in detail, not just the worked examples.

- There are very few examples in the text that are exactly like the homework problems. This means that you can't just look at a homework problem and search for a similar–looking "worked out" example. Success with the homework will come by grappling with the ideas of calculus.

- Many of the problems that we have included in the book are open-ended. This means that there may be more than one approach and more than one solution, depending on your analysis. Many times, solving a problem relies on common sense ideas that are not stated in the problem but which you will know from everyday life.

- Some problems in this book assume that you have access to a graphing calculator or computer; preferably one that can draw surface graphs, contour diagrams, and vector fields, and can compute multivariable integrals and line integrals numerically. There are many situations where you may not be able to find an exact solution to a problem, but you can use a calculator or computer to get a reasonable approximation.

- This book attempts to give equal weight to three methods for describing functions: graphical (a picture), numerical (a table of values) and algebraic (a formula). Sometimes you may find it easier to translate a problem given in one form into another. For example, if you have to find the maximum of a function, you might use a contour diagram to estimate its approximate position, use its formula to find equations that give the exact position, then use a numerical method to solve the equations. The best idea is to be flexible about your approach: if one way of looking at a problem doesn't work, try another.

- Students using this book have found discussing these problems in small groups very helpful. There are a great many problems which are not cut-and-dried; it can help to attack them with the other perspectives your colleagues can provide. If group work is not feasible, see if your instructor can organize a discussion session in which additional problems can be worked on.

- You are probably wondering what you'll get from the book. The answer is, if you put in a solid effort, you will get a real understanding of one of the most important accomplishments of the millennium – calculus – as well as a real sense of the power of mathematics in the age of technology.

CONTENTS

Chapter Twelve

FUNCTIONS OF SEVERAL VARIABLES

Many quantities depend on more than one variable: the amount of food grown depends on the amount of rain and the amount of fertilizer used; the rate of a chemical reaction depends on the temperature and the pressure of the environment in which it proceeds; the strength of the gravitational attraction between two bodies depends on their masses and their distance apart; and the rate of fallout from a volcanic explosion depends on the distance from the volcano and the time since the explosion. Each example involves a function of two or more variables. In this chapter, we will see many different ways of looking at functions of several variables.

12.1 FUNCTIONS OF TWO VARIABLES

Function Notation

Suppose you want to calculate your monthly payment on a five-year car loan; this depends on both the amount of money you borrow and the interest rate. These quantities can vary separately: the loan amount can change while the interest rate remains the same, or the interest rate can change while the loan amount remains the same. To calculate your monthly payment you need to know both. If the monthly payment is m, the loan amount is L, and the interest rate is $r\%$, then we express the fact that m is a function of L and r by writing:

$$m = f(L, r).$$

This is just like the function notation of one-variable calculus. The variable m is called the *dependent variable*, and the variables L and r are called the independent variables. The letter f stands for the *function* or rule that gives the value of m corresponding to given values of L and r.

A function of two variables can be represented graphically, numerically by a table of values, or algebraically by a formula. In this section, we give examples of each.

Graphical Example: A Weather Map

Figure 12.1 shows a weather map from a newspaper. What information does it convey? It displays the predicted high temperature, T, in degrees Fahrenheit (°F), throughout the US on that day. The curves on the map, called *isotherms*, separate the country into zones, according to whether T is in the 60s, 70s, 80s, 90s, or 100s. (*Iso* means same and *therm* means heat.) Notice that the isotherm separating the 80s and 90s zones connects all the points where the temperature is exactly 90°F.

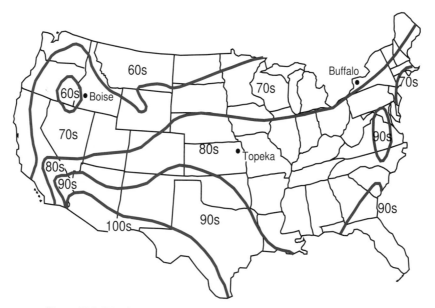

Figure 12.1: Weather map showing predicted high temperatures, T, on a summer day

Example 1   Estimate the predicted value of T in Boise, Idaho; Topeka, Kansas; and Buffalo, New York.

Solution   Boise and Buffalo are in the 70s region, and Topeka is in the 80s region. Thus, the predicted temperature in Boise and Buffalo is between 70 and 80 while the predicted temperature in Topeka is between 80 and 90. In fact, we can say more. Although both Boise and Buffalo are in the 70s, Boise is quite close to the $T = 70$ isotherm, whereas Buffalo is quite close to the $T = 80$ isotherm. So we estimate the temperature to be in the low 70s in Boise and in the high 70s in Buffalo. Topeka is about halfway between the $T = 80$ isotherm and the $T = 90$ isotherm. Thus, we guess the temperature in Topeka to be in the mid 80s. In fact, the actual high temperatures for that day were 71°F for Boise, 79°F for Buffalo, and 86°F for Topeka.

The predicted high temperature, T, illustrated by the weather map is a function of (that is, depends on) two variables, often longitude and latitude, or miles east-west and miles north-south of a fixed point, say, Topeka. The weather map in Figure 12.1 is called a *contour map* or *contour diagram* of that function. Section 12.2 shows another way of visualizing functions of two variables using surfaces; Section 12.3 looks at contour maps in detail.

Numerical Example: Beef Consumption

Suppose you are a beef producer and you want to know how much beef people will buy. This depends on how much money people have and on the price of beef. The consumption of beef, C (in pounds per week per household) is a function of household income, I (in thousands of dollars per year), and the price of beef, p (in dollars per pound). In function notation, we write:

$$C = f(I, p).$$

Table 12.1 contains values of this function. Values of p are shown across the top, values of I are down the left side, and corresponding values of $f(I, p)$ are given in the table.[1] For example, to find the value of $f(40, 3.50)$, we look in the row corresponding to $I = 40$ under $p = 3.50$, where we find the number 4.05. Thus,

$$f(40, 3.50) = 4.05.$$

This means that, on average, if a household's income is \$40,000 a year and the price of beef is \$3.50/lb, the family will buy 4.05 lbs of beef per week.

Table 12.1   *Quantity of beef bought (pounds/household/week)*

		Price of beef, ($/lb)			
		3.00	3.50	4.00	4.50
Household	20	2.65	2.59	2.51	2.43
income	40	4.14	4.05	3.94	3.88
per year,	60	5.11	5.00	4.97	4.84
I	80	5.35	5.29	5.19	5.07
(1000)	100	5.79	5.77	5.60	5.53

Notice how this differs from the table of values of a one-variable function, where one row or one column is enough to list the values of the function. Here many rows and columns are needed because the function has a value for every *pair* of values of the independent variables.

Algebraic Examples: Formulas

In both the weather map and beef consumption examples, there is no formula for the underlying function. That is usually the case for functions representing real-life data. On the other hand, for many idealized models in physics, engineering, or economics, there are exact formulas.

---

[1] Adapted from Richard G. Lipsey, *An Introduction to Positive Economics*, 3rd ed., (London: Weidenfeld and Nicolson, 1971).

Example 2   Give a formula for the function $M = f(B, t)$ where M is the amount of money in a bank account t years after an initial investment of B dollars, if interest is accrued at a rate of 5% per year compounded annually.

Solution   Annual compounding means that M increases by a factor of 1.05 every year, so

$$M = f(B, t) = B(1.05)^t.$$

Example 3   A cylinder with closed ends has radius r and height h. If its volume is V and its surface area is A, find formulas for the functions $V = f(r, h)$ and $A = g(r, h)$.

Solution   Since the area of the circular base is πr^2, we have

$$V = f(r, h) = \text{Area of base} \cdot \text{Height} = \pi r^2 h.$$

The surface area of the side is the circumference of the bottom, $2\pi r$, times the height h, giving $2\pi rh$. Thus,

$$A = g(r, h) = 2 \cdot \text{Area of base} + \text{Area of side} = 2\pi r^2 + 2\pi rh.$$

A Tour of 3-Space

In Section 12.2 we see how to visualize a function of two variables as a surface in space. Now we see how to locate points in three-dimensional space (3-space).

Imagine three coordinate axes meeting at the *origin*: a vertical axis, and two horizontal axes at right angles to each other. (See Figure 12.2.) Think of the xy-plane as being horizontal, while the z-axis extends vertically above and below the plane. The labels x, y, and z show which part of each axis is positive; the other side is negative. We generally use *right-handed axes* in which looking down the positive z-axis gives the usual view of the xy-plane. We specify a point in 3-space by giving its coordinates (x, y, z) with respect to these axes. Think of the coordinates as instructions telling you how to get to the point; start at the origin, go x units along the x-axis, then y units in the direction parallel to the y-axis and finally z units in the direction parallel to the z-axis. The coordinates can be positive, zero or negative; a zero coordinate means "don't move in this direction," and a negative coordinate means "go in the negative direction parallel to this axis." For example, the origin has coordinates $(0, 0, 0)$, since we get there from the origin by doing nothing at all.

Figure 12.2: Coordinate axes in three-dimensional space   Figure 12.3: The point $(1, 2, 3)$ in 3-space   Figure 12.4: The point $(0, 0, -1)$ in 3-space

Example 4   Describe the position of the points with coordinates $(1, 2, 3)$ and $(0, 0, -1)$.

Solution   We get to the point $(1, 2, 3)$ by starting at the origin, going 1 unit along the x-axis, 2 units in the direction parallel to the y-axis, and 3 units up in the direction parallel to the z-axis. (See Figure 12.3.)

To get to $(0, 0, -1)$, we don't move at all in the x and y directions, but move 1 unit in the negative z direction. So the point is on the negative z-axis. (See Figure 12.4.) You can check that the position of the point is independent of the order of the x, y, and z displacements.

Example 5   You start at the origin, go along the y-axis a distance of 2 units in the positive direction, and then move vertically upward a distance of 1 unit. What are the coordinates of your final position?

Solution   You started at the point $(0, 0, 0)$. When you went along the y-axis, your y-coordinate increased to 2. Moving vertically increased your z-coordinate to 1; your x-coordinate did not change because you did not move in the x direction. So your final coordinates are $(0, 2, 1)$. (See Figure 12.5.)

Figure 12.5: The point $(0, 2, 1)$ is reached by moving 2 along the y-axis and 1 upward

It is often helpful to picture a three dimensional coordinate system in terms of a room. The origin is a corner at floor level where two walls meet the floor. The z-axis is the vertical intersection of the two walls; the x- and y-axes are the intersections of each wall with the floor. Points with negative coordinates lie behind a wall in the next room or below the floor.

Graphing Equations in 3-Space

We can graph an equation involving the variables x, y, and z in 3-space; such a graph is a picture of all points (x, y, z) that satisfy the equation.

Example 6   What do the graphs of the equations $z = 0$, $z = 3$, and $z = -1$ look like?

Solution   To graph $z = 0$, we visualize the set of points whose z-coordinate is zero. If the z-coordinate is 0, then we must be at the same vertical level as the origin, that is, we are in the horizontal plane containing the origin. So the graph of $z = 0$ is the middle plane in Figure 12.6. The graph of $z = 3$ is a plane parallel to the graph of $z = 0$, but three units above it. The graph of $z = -1$ is a plane parallel to the graph of $z = 0$, but one unit below it.

Figure 12.6: The planes $z = -1$, $z = 0$, and $z = 3$

The plane $z = 0$ contains the x- and y-coordinate axes, and is called the xy-plane. There are two other coordinate planes. The yz-plane contains both the y- and the z-axes, and the xz-plane contains the x- and z-axes. (See Figure 12.7.)

xy-plane xz-plane yz-plane

Figure 12.7: The three coordinate planes

Example 7   Which of the points $A = (1, -1, 0)$, $B = (0, 3, 4)$, $C = (2, 2, 1)$, and $D = (0, -4, 0)$ lies closest to the xz-plane? Which point lies on the y-axis?

Solution   The magnitude of the y-coordinate gives the distance to the xz-plane. The point A lies closest to that plane, because it has the smallest y-coordinate in magnitude. To get to a point on the y-axis, we move along the y-axis, but we don't move at all in the x or z directions. Thus, a point on the y-axis has both its x- and z-coordinates equal to zero. The only point of the four that satisfies this is D. (See Figure 12.8.)

In general, if a point has one of its coordinates equal to zero, it lies in one of the coordinate planes. If a point has two of its coordinates equal to zero, it lies on one of the coordinate axes.

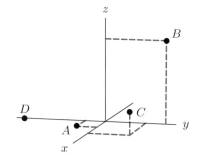

Figure 12.8: Which point lies closest to the xz-plane? Which point lies on the y-axis?

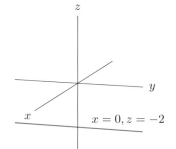

Figure 12.9: The line $x = 0$, $z = -2$

Example 8   You are 2 units below the xy-plane and in the yz-plane. What are your coordinates?

Solution   Since you are 2 units below the xy-plane, your z-coordinate is -2. Since you are in the yz-plane, your x-coordinate is 0; your y-coordinate can be anything. Thus, you are at the point $(0, y, -2)$. The set of all such points forms a line parallel to the y-axis, 2 units below the xy-plane, and in the yz-plane. (See Figure 12.9.)

Example 9   You are standing at the point $(4, 5, 2)$, looking at the point $(0.5, 0, 3)$. Are you looking up or down?

Solution   The point you are standing at has z-coordinate 2, whereas the point you are looking at has z-coordinate 3; hence you are looking up.

Example 10   Imagine that the yz-plane in Figure 12.7 is a page of this book. Describe the region behind the page algebraically.

Solution   The positive part of the x-axis pokes out of the page; moving in the positive x direction brings you out in front of the page. The region behind the page corresponds to negative values of x, so it is the set of all points in 3-space satisfying the inequality $x < 0$.

Distance Between Two Points

In 2-space, the formula for the distance between two points (x, y) and (a, b) is given by

$$\text{Distance} = \sqrt{(x - a)^2 + (y - b)^2}.$$

The distance between two points (x, y, z) and (a, b, c) in 3-space is represented by PG in Figure 12.10. The side PE is parallel to the x-axis, EF is parallel to the y-axis, and FG is parallel to the z-axis.

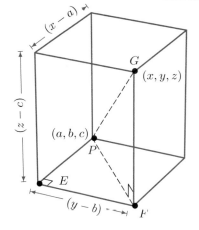

Figure 12.10: The diagonal PG gives the distance between the points (x, y, z) and (a, b, c)

Using Pythagoras' theorem twice gives

$$(PG)^2 = (PF)^2 + (FG)^2 = (PE)^2 + (EF)^2 + (FG)^2 = (x-a)^2 + (y-b)^2 + (z-c)^2.$$

Thus, a formula for the distance between the points (x, y, z) and (a, b, c) in 3-space is

$$\text{Distance} = \sqrt{(x-a)^2 + (y-b)^2 + (z-c)^2}.$$

Example 11   Find the distance between $(1, 2, 1)$ and $(-3, 1, 2)$.

Solution $\text{Distance} = \sqrt{(-3-1)^2 + (1-2)^2 + (2-1)^2} = \sqrt{18} = 4.24.$

Example 12   Find an expression for the distance from the origin to the point (x, y, z).

Solution   The origin has coordinates $(0, 0, 0)$, so the distance from the origin to (x, y, z) is given by

$$\text{Distance} = \sqrt{(x-0)^2 + (y-0)^2 + (z-0)^2} = \sqrt{x^2 + y^2 + z^2}.$$

Example 13   Find an equation for a sphere of radius 1 with center at the origin.

Solution   The sphere consists of all points (x, y, z) whose distance from the origin is 1, that is, which satisfy the equation

$$\sqrt{x^2 + y^2 + z^2} = 1.$$

This is an equation for the sphere. If we square both sides we get the equation in the form

$$x^2 + y^2 + z^2 = 1.$$

Note that this equation represents the *surface* of the sphere. The solid ball enclosed by the sphere is represented by the inequality $x^2 + y^2 + z^2 \leq 1$.

Exercises and Problems for Section 12.1

Exercises

1. Which of the points $A = (1.3, -2.7, 0)$, $B = (0.9, 0, 3.2)$, $C = (2.5, 0.1, -0.3)$ is closest to the yz-plane? Which one lies on the xz-plane? Which one is farthest from the xy-plane?

2. Which of the points $A = (23, 92, 48)$, $B = (-60, 0, 0)$, $C = (60, 1, -92)$ is closest to the yz-plane? Which lies on the xz-plane? Which is farthest from the xy-plane?

3. You are at the point $(-1, -3, -3)$, standing upright and facing the yz-plane. You walk 2 units forward, turn left, and walk for another 2 units. What is your final position? From the point of view of an observer looking at the coordinate system in Figure 12.2 on page 606, are you in front of or behind the yz-plane? To the left or to the right of the xz-plane? Above or below the xy-plane?

4. You are at the point $(3, 1, 1)$, standing upright and facing the yz-plane. You walk 2 units forward, turn left, and walk another 2 units. What is your final position? From the point of view of an observer looking at the coordinate system in Figure 12.2 on page 606, are you in front of or behind the yz-plane? To the left or to the right of the xz-plane? Above or below the xy-plane?

5. Which of the points $P = (1, 2, 1)$ and $Q = (2, 0, 0)$ is closest to the origin?

6. Which two of the three points $P_1 = (1, 2, 3)$, $P_2 = (3, 2, 1)$ and $P_3 = (1, 1, 0)$ are closest to each other?

7. Find the equation of the sphere of radius 5 centered at the origin.

8. Find the equation of the sphere of radius 5 centered at $(1, 2, 3)$.

Sketch graphs of the equations in Exercises 9–11 in 3-space.

9. $x = -3$ **10.** $y = 1$

11. $z = 2$ and $y = 4$

Exercises 12–14 refer to the map in Figure 12.1 on page 604.

12. Give the range of daily high temperatures for

   (a) Pennsylvania   **(b)** North Dakota

   (c) California.

13. Sketch a possible graph of the predicted high temperature T on a line north-south through Topeka.

14. Sketch possible graphs of the predicted high temperature on a north-south line and an east-west line through Boise.

For Exercises 15–17, refer to Table 12.1 on page 605, where p is the price of beef and I is annual household income.

15. Make a table showing the amount of money, M, that the average household spends on beef (in dollars per household per week) as a function of the price of beef and household income.

16. Give tables for beef consumption as a function of p, with I fixed at $I = 20$ and $I = 100$. Give tables for beef consumption as a function of I, with p fixed at $p = 3.00$ and $p = 4.00$. Comment on what you see in the tables.

17. How does beef consumption vary as a function of household income if the price of beef is held constant?

Problems

18. The gravitational force, F newtons, exerted on an object by the earth depends on its mass, m kilograms, and its distance, r meters, from the center of the earth, so $F = f(m, r)$. Interpret the following statement in terms of gravitation: $f(100, 7000000) \approx 820$.

19. The balance, B, in dollars, in a bank account depends on the amount deposited, A dollars, the annual interest rate, $r\%$, and the time, t, in months since the deposit, so $B = f(A, r, t)$.

   (a) Is f an increasing or decreasing function of A? Of r? Of t?

   (b) Interpret the statement $f(1250, 1, 25) \approx 1276$. Give units.

20. The monthly payments, P dollars, on a mortgage in which A dollars were borrowed at an annual interest rate of $r\%$ for t years is given by $P = f(A, r, t)$. Is f an increasing or decreasing function of A? Of r? Of t?

21. Consider the acceleration due to gravity, g, at a height h above the surface of a planet of mass m.

   (a) If m is held constant, is g an increasing or decreasing function of h? Why?

   (b) If h is held constant, is g an increasing or decreasing function of m? Why?

22. A car rental company charges $40 a day and 15 cents a mile for its cars.

   (a) Write a formula for the cost, C, of renting a car as a function, f, of the number of days, d, and the number of miles driven, m.

   (b) If $C = f(d, m)$, find $f(5, 300)$ and interpret it.

23. The temperature adjusted for wind-chill is a temperature which tells you how cold it feels, as a result of the combination of wind and temperature.[2] See Table 12.2.

   (a) If the temperature is $0°F$ and the wind speed is 15 mph, how cold does it feel?

---

[2]Data from www.nws.noaa.gov/om/windchill, accessed on May 30, 2004.

(b) If the temperature is $35°$F, what wind speed makes it feel like $24°$F?

(c) If the temperature is $25°$F, what wind speed makes it feel like $12°$F?

(d) If the wind is blowing at 20 mph, what temperature feels like $0°$F?

Table 12.2 *Temperature adjusted for wind-chill ($°F$) as a function of wind speed and temperature*

Temperature (°F)

	35	30	25	20	15	10	5	0
5	31	25	19	13	7	1	−5	−11
10	27	21	15	9	3	−4	−10	−16
15	25	19	13	6	0	−7	−13	−19
20	24	17	11	4	−2	−9	−15	−22
25	23	16	9	3	−4	−11	−17	−24

Wind Speed (mph) labels the rows.

24. Using Table 12.2, make tables of the temperature adjusted for wind-chill as a function of wind speed for temperatures of $20°$F and $0°$F.

25. Find a possible formula for the function $f(x, y)$ whose values are in Table 12.3.

Table 12.3

y

x	−2	−1	0	1	2
−2	4	1	0	1	4
−1	4	1	0	1	4
0	4	1	0	1	4
1	4	1	0	1	4
2	4	1	0	1	4

26. A cube is located such that its top four corners have the coordinates $(-1, -2, 2)$, $(-1, 3, 2)$, $(4, -2, 2)$ and $(4, 3, 2)$. Give the coordinates of the center of the cube.

27. Describe the set of points whose distance from the x-axis is 2.

28. Describe the set of points whose distance from the x-axis equals the distance from the yz-plane.

29. Find a formula for the shortest distance between a point (a, b, c) and the y-axis.

30. Find the equations of planes that just touch the sphere $(x - 2)^2 + (y - 3)^2 + (z - 3)^2 = 16$ and are parallel to

    (a) The xy-plane   **(b)** The yz-plane

    (c) The xz-plane

31. Find an equation of the largest sphere contained in the cube determined by the planes $x = 2, x = 6$; $y = 5, y = 9$; and $z = -1, z = 3$.

32. Let $f(x, y, z) = x^2 + y^2 + z^2 - 2x + 3y + z$.

    (a) Find a point (a, b, c) so that f can be expressed in terms of distance from (a, b, c). [Hint: Complete the square.]

    (b) Express f in terms of the distance d from the point you found in part (a).

33. Which of the following functions can be expressed in terms of the distance from the y-axis?

    (a) $f(x, y, z) = 10e^y$

    (b) $g(x, y, z) = x^2 z^2$

    (c) $h(x, y, z) = 1/\sqrt{x^2 + b^2 + z^2}$, for constant b.

12.2   GRAPHS OF FUNCTIONS OF TWO VARIABLES

The weather map on page 604 is one way of visualizing a function of two variables. In this section we see how to visualize a function of two variables in another way, using a surface in 3-space.

Visualizing a Function of Two Variables Using a Graph

For a function of one variable, $y = f(x)$, the graph of f is the set of all points (x, y) in 2-space such that $y = f(x)$. In general, these points lie on a curve in the plane. When a computer or calculator graphs f, it approximates by plotting points in the xy-plane and joining consecutive points by line segments. The more points, the better the approximation.

Now consider a function of two variables.

> The **graph** of a function of two variables, f, is the set of all points (x, y, z) such that $z = f(x, y)$. In general, the graph of a function of two variables is a surface in 3-space.

Plotting the Graph of the Function $f(x, y) = x^2 + y^2$

To sketch the graph of f we connect points as for a function of one variable. We first make a table of values of f, such as in Table 12.4.

Table 12.4 *Table of values of $f(x, y) = x^2 + y^2$*

		-3	-2	-1	0	1	2	3
	-3	18	13	10	9	10	13	18
	-2	13	8	5	4	5	8	13
	-1	10	5	2	1	2	5	10
x	0	9	4	1	0	1	4	9
	1	10	5	2	1	2	5	10
	2	13	8	5	4	5	8	13
	3	18	13	10	9	10	13	18

Now we plot points. For example, we plot $(1, 2, 5)$ because $f(1, 2) = 5$ and we plot $(0, 2, 4)$ because $f(0, 2) = 4$. Then, we connect the points corresponding to the rows and columns in the table. The result is called a *wire-frame* picture of the graph. Filling in between the wires gives a surface. That is the way a computer drew the graphs in Figure 12.11 and 12.12. As more points are plotted, we get the surface in Figure 12.13, called a *paraboloid*.

You should check to see if the sketches make sense. Notice that the graph goes through the origin since $(x, y, z) = (0, 0, 0)$ satisfies $z = x^2 + y^2$. Observe that if x is held fixed and y is allowed to vary, the graph dips down and then goes back up, just like the entries in the rows of Table 12.4. Similarly, if y is held fixed and x is allowed to vary, the graph dips down and then goes back up, just like the columns of Table 12.4.

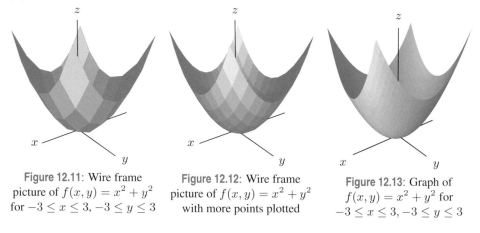

Figure 12.11: Wire frame picture of $f(x, y) = x^2 + y^2$ for $-3 \le x \le 3, -3 \le y \le 3$

Figure 12.12: Wire frame picture of $f(x, y) = x^2 + y^2$ with more points plotted

Figure 12.13: Graph of $f(x, y) = x^2 + y^2$ for $-3 \le x \le 3, -3 \le y \le 3$

New Graphs from Old

We can use the graph of a function to visualize the graphs of related functions.

Example 1   Let $f(x, y) = x^2 + y^2$. Describe in words the graphs of the following functions:
(a) $g(x, y) = x^2 + y^2 + 3$,   (b) $h(x, y) = 5 - x^2 - y^2$,   (c) $k(x, y) = x^2 + (y - 1)^2$.

Solution   We know from Figure 12.13 that the graph of f is a paraboloid, or bowl with its vertex at the origin. From this we can work out what the graphs of g, h, and k will look like.
(a) The function $g(x, y) = x^2 + y^2 + 3 = f(x, y) + 3$, so the graph of g is the graph of f, but raised by 3 units. See Figure 12.14.
(b) Since $-x^2 - y^2$ is the negative of $x^2 + y^2$, the graph of $-x^2 - y^2$ is an upside down paraboloid. Thus, the graph of $h(x, y) = 5 - x^2 - y^2 = 5 - f(x, y)$ looks like an upside down paraboloid with vertex at $(0, 0, 5)$, as in Figure 12.15.
(c) The graph of $k(x, y) = x^2 + (y - 1)^2 = f(x, y - 1)$ is a paraboloid with vertex at $x = 0, y = 1$, since that is where $k(x, y) = 0$, as in Figure 12.16.

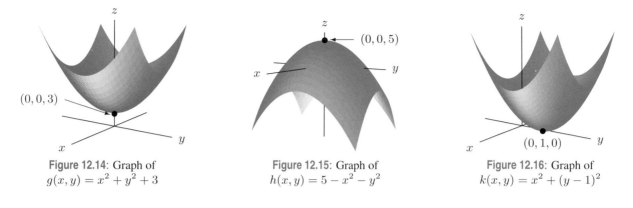

Figure 12.14: Graph of $g(x, y) = x^2 + y^2 + 3$

Figure 12.15: Graph of $h(x, y) = 5 - x^2 - y^2$

Figure 12.16: Graph of $k(x, y) = x^2 + (y - 1)^2$

Example 2   Describe the graph of $G(x, y) = e^{-(x^2 + y^2)}$. What symmetry does it have?

Solution   Since the exponential function is always positive, the graph lies entirely above the xy-plane. From the graph of $x^2 + y^2$ we see that $x^2 + y^2$ is zero at the origin and gets larger as we move farther from the origin in any direction. Thus, $e^{-(x^2 + y^2)}$ is 1 at the origin, and gets smaller as we move away from the origin in any direction. It can't go below the xy-plane; instead it flattens out, getting closer and closer to the plane. We say the surface is *asymptotic* to the xy-plane. (See Figure 12.17.)

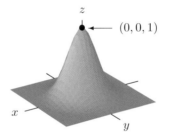

Figure 12.17: Graph of $G(x, y) = e^{-(x^2 + y^2)}$

Now consider a point (x, y) on the circle $x^2 + y^2 = r^2$. Since

$$G(x, y) = e^{-(x^2 + y^2)} = e^{-r^2},$$

the value of the function G is the same at all points on this circle. Thus, we say the graph of G has *circular symmetry*.

Cross-Sections and the Graph of a Function

We have seen that a good way to analyze a function of two variables is to let one variable vary at a time while the other is kept fixed.

For a function $f(x, y)$, the function we get by holding x fixed and letting y vary is called a **cross-section** of f with x fixed. The graph of the cross-section of $f(x, y)$ with $x = c$ is the curve, or cross-section, we get by intersecting the graph of f with the plane $x = c$. We define a cross-section of f with y fixed similarly.

For example, the cross-section of $f(x, y) = x^2 + y^2$ with $x = 2$ is $f(2, y) = 4 + y^2$. The graph of this cross-section is the curve we get by intersecting the graph of f with the plane perpendicular to the x-axis at $x = 2$. (See Figure 12.18.)

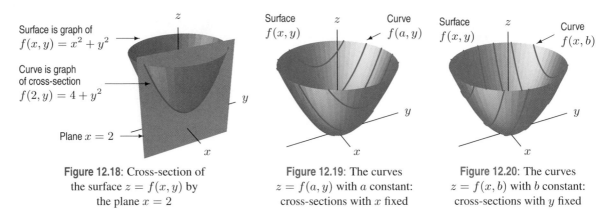

Figure 12.18: Cross-section of the surface $z = f(x, y)$ by the plane $x = 2$

Figure 12.19: The curves $z = f(a, y)$ with a constant: cross-sections with x fixed

Figure 12.20: The curves $z = f(x, b)$ with b constant: cross-sections with y fixed

Figure 12.19 shows graphs of other cross-sections of f with x fixed; Figure 12.20 shows graphs of cross-sections with y fixed.

Example 3   Describe the cross-sections of the function $g(x, y) = x^2 - y^2$ with y fixed and then with x fixed. Use these cross-sections to describe the shape of the graph of g.

Solution   The cross-sections with y fixed at $y = b$ are given by

$$z = g(x, b) = x^2 - b^2.$$

Thus, each cross-section with y fixed gives a parabola opening upward, with minimum $z = -b^2$. The cross-sections with x fixed are of the form

$$z = g(a, y) = a^2 - y^2,$$

which are parabolas opening downward with a maximum of $z = a^2$. (See Figures 12.21 and 12.22.) The graph of g is shown in Figure 12.23. Notice the upward opening parabolas in the x-direction and the downward opening parabolas in the y-direction. We say that the surface is *saddle-shaped*.

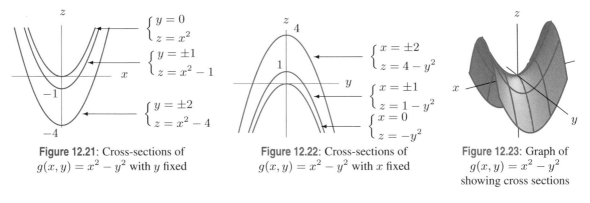

Figure 12.21: Cross-sections of $g(x, y) = x^2 - y^2$ with y fixed

Figure 12.22: Cross-sections of $g(x, y) = x^2 - y^2$ with x fixed

Figure 12.23: Graph of $g(x, y) = x^2 - y^2$ showing cross sections

Linear Functions

Linear functions are central to single variable calculus; they are equally important in multivariable calculus. You may be able to guess the shape of the graph of a linear function of two variables. (It's a plane.) Let's look at an example.

Example 4   Describe the graph of $f(x, y) = 1 + x - y$.

Solution   The plane $x = a$ is vertical and parallel to the yz-plane. Thus, the cross-section with $x = a$ is the line $z = 1 + a - y$ which slopes downward in the y-direction. Similarly, the plane $y = b$ is parallel to the xz-plane. Thus, the cross-section with $y = b$ is the line $z = 1 + x - b$ which slopes upward in the x-direction. Since all the cross-sections are lines, you might expect the graph to be a flat plane, sloping down in the y-direction and up in the x-direction. This is indeed the case. (See Figure 12.24.)

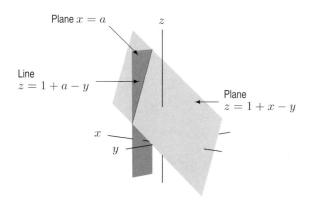

Figure 12.24: Graph of the plane $z = 1 + x - y$ showing cross-section with $x = a$

When One Variable is Missing: Cylinders

Suppose we graph an equation like $z = x^2$ which has one variable missing. What does the surface look like? Since y is missing from the equation, the cross-sections with y fixed are all the same parabola, $z = x^2$. Letting y vary up and down the y-axis, this parabola sweeps out the trough-shaped surface shown in Figure 12.25. The cross-sections with x fixed are horizontal lines obtained by cutting the surface by a plane perpendicular to the x-axis. This surface is called a *parabolic cylinder*, because it is formed from a parabola in the same way that an ordinary cylinder is formed from a circle; it has a parabolic cross-section instead of a circular one.

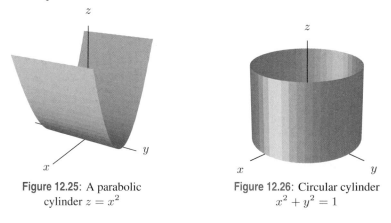

Figure 12.25: A parabolic cylinder $z = x^2$

Figure 12.26: Circular cylinder $x^2 + y^2 = 1$

Example 5   Graph the equation $x^2 + y^2 = 1$ in 3-space.

Solution   Although the equation $x^2 + y^2 = 1$ does not represent a function, the surface representing it can be graphed by the method used for $z = x^2$. The graph of $x^2 + y^2 = 1$ in the xy-plane is a circle. Since z does not appear in the equation, the intersection of the surface with any horizontal plane will be the same circle $x^2 + y^2 = 1$. Thus, the surface is the cylinder shown in Figure 12.26.

Exercises and Problems for Section 12.2

Exercises

1. Without a calculator or computer, match the functions with their graphs in Figure 12.27.

   (a) $z = 2 + x^2 + y^2$ (b) $z = 2 - x^2 - y^2$
   (c) $z = 2(x^2 + y^2)$ (d) $z = 2 + 2x - y$
   (e) $z = 2$

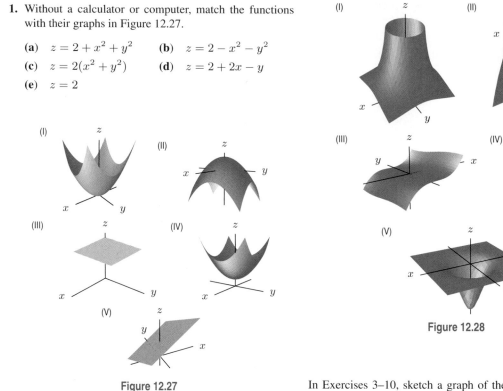

Figure 12.27

Figure 12.28

2. Without a calculator or computer, match the functions with their graphs in Figure 12.28.

   (a) $z = \dfrac{1}{x^2 + y^2}$ (b) $z = -e^{-x^2 - y^2}$
   (c) $z = x + 2y + 3$ (d) $z = -y^2$
   (e) $z = x^3 - \sin y$.

In Exercises 3–10, sketch a graph of the surface and briefly describe it in words.

3. $z = 3$ 4. $x^2 + y^2 + z^2 = 9$
5. $z = x^2 + y^2 + 4$ 6. $z = 5 - x^2 - y^2$
7. $z = y^2$ 8. $2x + 4y + 3z = 12$
9. $x^2 + y^2 = 4$ 10. $x^2 + z^2 = 4$

Problems

11. For each of the following functions, decide whether it could be a bowl, a plate, or neither. Consider a plate to be any fairly flat surface and a bowl to be anything that could hold water, assuming the positive z-axis is up.

    (a) $z = x^2 + y^2$ (b) $z = 1 - x^2 - y^2$
    (c) $x + y + z = 1$ (d) $z = -\sqrt{5 - x^2 - y^2}$
    (e) $z = 3$

12. For each function in Problem 11 sketch cross-sections.

13. Consider the function f given by $f(x, y) = y^3 + xy$. Draw graphs of cross-sections with:

    (a) x fixed at $x = -1$, $x = 0$, and $x = 1$.

    (b) y fixed at $y = -1$, $y = 0$, and $y = 1$.

14. Without a computer or calculator, match the equations (a)–(i) with the graphs (I)–(IX).

    (a) $z = xye^{-(x^2 + y^2)}$ (b) $z = \cos(\sqrt{x^2 + y^2})$
    (c) $z = \sin y$ (d) $z = -\dfrac{1}{x^2 + y^2}$
    (e) $z = \cos^2 x \cos^2 y$ (f) $z = \dfrac{\sin(x^2 + y^2)}{x^2 + y^2}$
    (g) $z = \cos(xy)$ (h) $z = |x||y|$
    (i) $z = (2x^2 + y^2)e^{1 - x^2 - y^2}$

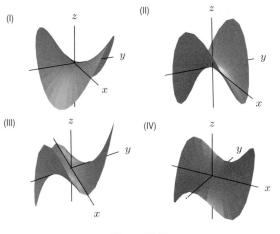

Figure 12.30

Problems 16–18 concern the concentration, C, in mg per liter, of a drug in the blood as a function of x, the amount, in mg, of the drug given and t, the time in hours since the injection. For $0 \le x \le 4$ and $t \ge 0$, we have $C = f(x, t) = te^{-t(5-x)}$.

16. Find $f(3, 2)$. Give units and interpret in terms of drug concentration.

17. Graph the following two single variable functions and explain their significance in terms of drug concentration.

    (a) $f(4, t)$ **(b)** $f(x, 1)$

18. Graph $f(a, t)$ for $a = 1, 2, 3, 4$ on the same axes. Describe how the graph changes as a increases and explain what this means in terms of drug concentration.

19. You like pizza and you like cola. Which of the graphs in Figure 12.31 represents your happiness as a function of how many pizzas and how much cola you have if

    (a) There is no such thing as too many pizzas and too much cola?
    (b) There is such a thing as too many pizzas or too much cola?
    (c) There is such a thing as too much cola but no such thing as too many pizzas?

15. Figure 12.29 contains graphs of the parabolas $z = f(x, b)$ for $b = -2, -1, 0, 1, 2$. Which of the graphs of $z = f(x, y)$ in Figure 12.30 best fits this information?

Figure 12.29

Figure 12.31

20. For each of the graphs I-IV in Problem 19 draw:

   (a) Two cross-sections with pizza fixed
   (b) Two cross-sections with cola fixed.

21. By setting one variable constant, find a plane that intersects the graph of $z = 4x^2 - y^2 + 1$ in a:

   (a) Parabola opening upward
   (b) Parabola opening downward
   (c) Pair of intersecting straight lines

22. By setting one variable constant, find a plane that intersects the graph of $z = (x^2 + 1)\sin y + xy^2$ in a:

   (a) Parabola
   (b) Straight line
   (c) Sine curve

23. A wave travels along a canal. Let x be the distance along the canal from the middle, t be the time, and z be the height of the water above the equilibrium level. The graph of z as a function of x and t is in Figure 12.32.

   (a) Draw the profile of the wave for $t = -1, 0, 1, 2$. (Put the x-axis to the right and the z-axis vertical.)
   (b) Is the wave traveling in the direction of increasing or decreasing x?
   (c) Sketch a surface representing a wave traveling in the opposite direction.

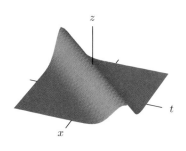

Figure 12.32

24. A swinging pendulum consists of a mass at the end of a string. At one moment the string makes an angle x with the vertical and the mass has speed y. At that time, the energy, E, of the pendulum is given by the expression[3]

$$E = 1 - \cos x + \frac{y^2}{2}.$$

   (a) Consider the surface representing the energy. Sketch a cross-section of the surface:

      (i) Perpendicular to the x-axis at $x = c$.
      (ii) Perpendicular to the y-axis at $y = c$.

   (b) For each of the graphs in Figures 12.33 and 12.34 use your answer to part (a) to decide which is the x-axis and which is the y-axis and to put reasonable units on each one.

Figure 12.33

Figure 12.34

12.3 CONTOUR DIAGRAMS

The surface which represents a function of two variables often gives a good idea of the function's general behavior—for example, whether it is increasing or decreasing as one of the variables increases. However it is difficult to read numerical values off a surface and it can be hard to see all of the function's behavior from a surface. Thus, functions of two variables are often represented by contour diagrams like the weather map on page 604. Contour diagrams have the additional advantage that they can be extended to functions of three variables.

Topographical Maps

One of the most common examples of a contour diagram is a topographical map like that shown in Figure 12.35. It gives the elevation in the region and is a good way of getting an overall picture of the terrain: where the mountains are, where the flat areas are. Such topographical maps are frequently colored green at the lower elevations and brown, red, or white at the higher elevations.

[3] Adapted from *Calculus in Context*, by James Callahan, Kenneth Hoffman, (New York: W.H. Freeman, 1995).

Figure 12.35: A topographical map showing the region around South Hamilton, NY

The curves on a topographical map that separate lower elevations from higher elevations are called *contour lines* because they outline the contour or shape of the land.[4] Because every point along the same contour has the same elevation, contour lines are also called *level curves* or *level sets*. The more closely spaced the contours, the steeper the terrain; the more widely spaced the contours, the flatter the terrain (provided, of course, that the elevation between contours varies by a constant amount). Certain features have distinctive characteristics. A mountain peak is typically surrounded by contour lines like those in Figure 12.36. A pass in a range of mountains may have contours that look like Figure 12.37. A long valley has parallel contour lines indicating the rising elevations on both sides of the valley (see Figure 12.38); a long ridge of mountains has the same type of contour lines, only the elevations decrease on both sides of the ridge. Notice that the elevation numbers on the contour lines are as important as the curves themselves. We usually draw contours for equally spaced values of z.

Figure 12.36: Mountain peak

Figure 12.37: Pass between two mountains

Figure 12.38: Long valley

Figure 12.39: Impossible contour lines

Notice that two contours corresponding to different elevations cannot cross each other as shown in Figure 12.39. If they did, the point of intersection of the two curves would have two different elevations, which is impossible (assuming the terrain has no overhangs).

Corn Production

Contour maps can display information about a function of two variables without reference to a surface. Consider the effect of weather conditions on US corn production. Figure 12.40 gives corn production $C = f(R, T)$ as a function of the total rainfall, R, in inches, and average temperature, T, in degrees Fahrenheit, during the growing season.[5] At the present time, $R = 15$ inches and

---

[4]In fact they are usually not straight lines, but curves. They may also be in disconnected pieces.

[5]Adapted from S. Beaty and R. Healy, "The Future of American Agriculture," *Scientific American* 248, No.2, February 1983.

$T = 76°F$. Production is measured as a percentage of the present production; thus, the contour through $R = 15, T = 76$, has value 100, that is, $C = f(15, 76) = 100$.

Example 1   Use Figure 12.40 to estimate $f(18, 78)$ and $f(12, 76)$ and interpret in terms of corn production.

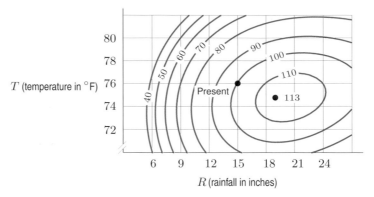

Figure 12.40: Corn production, C, as a function of rainfall and temperature

Solution   The point with R-coordinate 18 and T-coordinate 78 is on the contour $C = 100$, so $f(18, 78) = 100$. This means that if the annual rainfall were 18 inches and the temperature were 78°F, the country would produce about the same amount of corn as at present, although it would be wetter and warmer than it is now.

The point with R-coordinate 12 and T-coordinate 76 is about halfway between the $C = 80$ and the $C = 90$ contours, so $f(12, 76) \approx 85$. This means that if the rainfall fell to 12 inches and the temperature stayed at 76°, then corn production would drop to about 85% of what it is now.

Example 2   Use Figure 12.40 to describe in words the cross-sections with T and R constant through the point representing present conditions. Give a common sense explanation of your answer.

Solution   To see what happens to corn production if the temperature stays fixed at 76°F but the rainfall changes, look along the horizontal line $T = 76$. Starting from the present and moving left along the line $T = 76$, the values on the contours decrease. In other words, if there is a drought, corn production decreases. Conversely, as rainfall increases, that is, as we move from the present to the right along the line $T = 76$, corn production increases, reaching a maximum of more than 110% when $R = 21$, and then decreases (too much rainfall floods the fields).

If, instead, rainfall remains at the present value and temperature increases, we move up the vertical line $R = 15$. Under these circumstances corn production decreases; a 2° increase causes a 10% drop in production. This makes sense since hotter temperatures lead to greater evaporation and hence drier conditions, even with rainfall constant at 15 inches. Similarly, a decrease in temperature leads to a very slight increase in production, reaching a maximum of around 102% when $T = 74$, followed by a decrease (the corn won't grow if it is too cold).

Contour Diagrams and Graphs

Contour diagrams and graphs are two different ways of representing a function of two variables. How do we go from one to the other? In the case of the topographical map, the contour diagram was created by joining all the points at the same height on the surface and dropping the curve into the xy-plane.

How do we go the other way? Suppose we wanted to plot the surface representing the corn production function $C = f(R, T)$ given by the contour diagram in Figure 12.40. Along each contour the function has a constant value; if we take each contour and lift it above the plane to a height equal to this value, we get the surface in Figure 12.41.

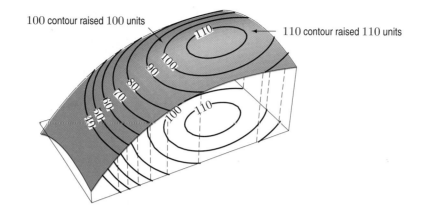

Figure 12.41: Getting the graph of the corn yield function from the contour diagram

Notice that the raised contours are the curves we get by slicing the surface horizontally. In general, we have the following result:

Contour lines, or level curves, are obtained from a surface by slicing it with horizontal planes.

Finding Contours Algebraically

Algebraic equations for the contours of a function f are easy to find if we have a formula for $f(x, y)$. Suppose the surface has equation

$$z = f(x, y).$$

A contour is obtained by slicing the surface with a horizontal plane with equation $z = c$. Thus, the equation for the contour at height c is given by:

$$f(x, y) = c.$$

Example 3   Find equations for the contours of $f(x, y) = x^2 + y^2$ and draw a contour diagram for f. Relate the contour diagram to the graph of f.

Solution   The contour at height c is given by

$$f(x, y) = x^2 + y^2 = c.$$

This is a contour only for $c \geq 0$, For $c > 0$ it is a circle of radius $\sqrt{c}$. For $c = 0$, it is a single point (the origin). Thus, the contours at an elevation of $c = 1, 2, 3, 4, \ldots$ are all circles centered at the origin of radius $1, \sqrt{2}, \sqrt{3}, 2, \ldots$. The contour diagram is shown in Figure 12.42. The bowl–shaped graph of f is shown in Figure 12.43. Notice that the graph of f gets steeper as we move further away from the origin. This is reflected in the fact that the contours become more closely packed as we move further from the origin; for example, the contours for $c = 6$ and $c = 8$ are closer together than the contours for $c = 2$ and $c = 4$.

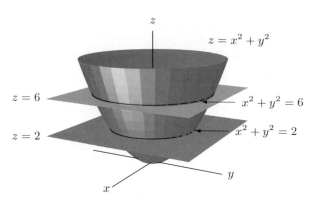

Figure 12.42: Contour diagram for $f(x, y) = x^2 + y^2$ (even values of c only)

Figure 12.43: The graph of $f(x, y) = x^2 + y^2$

Example 4   Draw a contour diagram for $f(x, y) = \sqrt{x^2 + y^2}$ and relate it to the graph of f.

Solution   The contour at level c is given by

$$f(x, y) = \sqrt{x^2 + y^2} = c.$$

For $c > 0$ this is a circle, just as in the previous example, but here the radius is c instead of $\sqrt{c}$. For $c = 0$, it is the origin. Thus, if the level c increases by 1, the radius of the contour increases by 1. This means the contours are equally spaced concentric circles (see Figure 12.44) which do not become more closely packed further from the origin. Thus, the graph of f has the same constant slope as we move away from the origin (see Figure 12.45), making it a cone rather than a bowl.

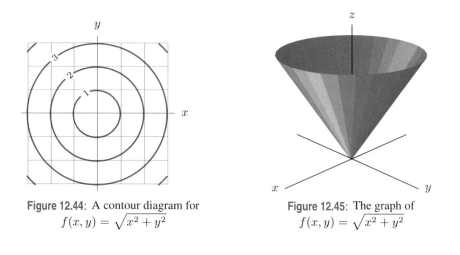

Figure 12.44: A contour diagram for $f(x, y) = \sqrt{x^2 + y^2}$

Figure 12.45: The graph of $f(x, y) = \sqrt{x^2 + y^2}$

In both of the previous examples the level curves are concentric circles because the surfaces have circular symmetry. Any function of two variables which depends only on the quantity $(x^2 + y^2)$ has such symmetry: for example, $G(x, y) = e^{-(x^2+y^2)}$ or $H(x, y) = \sin(\sqrt{x^2 + y^2})$.

Example 5   Draw a contour diagram for $f(x, y) = 2x + 3y + 1$.

Solution   The contour at level c has equation $2x + 3y + 1 = c$. Rewriting this as $y = -(2/3)x + (c - 1)/3$, we see that the contours are parallel lines with slope $-2/3$. The y-intercept for the contour at level c is $(c - 1)/3$; each time c increases by 3, the y-intercept moves up by 1. The contour diagram is shown in Figure 12.46.

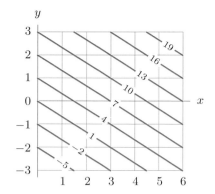

Figure 12.46: A contour diagram for $f(x, y) = 2x + 3y + 1$

Contour Diagrams and Tables

Sometimes we can get an idea of what the contour diagram of a function looks like from its table.

Example 6   Relate the values of $f(x, y) = x^2 - y^2$ in Table 12.5 to its contour diagram in Figure 12.47.

Table 12.5   *Table of values of $f(x, y) = x^2 - y^2$*

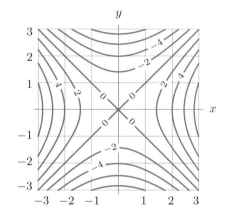

	3	0	−5	−8	−9	−8	−5	0
	2	5	0	−3	−4	−3	0	5
	1	8	3	0	−1	0	3	8
y	0	9	4	1	0	1	4	9
	−1	8	3	0	−1	0	3	8
	−2	5	0	−3	−4	−3	0	5
	−3	0	−5	−8	−9	−8	−5	0
		−3	−2	−1	0	1	2	3
					x			

Figure 12.47: Contour map of $f(x, y) = x^2 - y^2$

Solution   One striking feature of the values in Table 12.5 is the zeros along the diagonals. This occurs because $x^2 - y^2 = 0$ along the lines $y = x$ and $y = -x$. So the $z = 0$ contour consists of these two lines. In the triangular region of the table that lies to the right of both diagonals, the entries are positive. To the left of both diagonals, the entries are also positive. Thus, in the contour diagram, the positive contours lie in the triangular regions to the right and left of the lines $y = x$ and $y = -x$. Further, the table shows that the numbers on the left are the same as the numbers on the right; thus, each contour has two pieces, one on the left and one on the right. See Figure 12.47. As we move away from the origin along the x-axis, we cross contours corresponding to successively larger values. On the saddle-shaped graph of $f(x, y) = x^2 - y^2$ shown in Figure 12.48, this corresponds to climbing out of the saddle along one of the ridges. Similarly, the negative contours occur in pairs in the top and bottom triangular regions; the values get more and more negative as we go out along the y-axis. This corresponds to descending from the saddle along the valleys that are submerged below the xy-plane in Figure 12.48. Notice that we could also get the contour diagram by graphing the family of hyperbolas $x^2 - y^2 = 0, \pm 2, \pm 4, \ldots$.

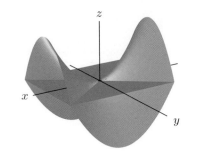

Figure 12.48: Graph of $f(x, y) = x^2 - y^2$ showing plane $z = 0$

Using Contour Diagrams: The Cobb-Douglas Production Function

Suppose you decide to expand your small printing business. How should you expand? Should you start a night-shift and hire more workers? Should you buy more expensive but faster computers which will enable the current staff to keep up with the work? Or should you do some combination of the two?

Obviously, the way such a decision is made in practice involves many other considerations—such as whether you could get a suitably trained night shift, or whether there are any faster computers available. Nevertheless, you might model the quantity, P, of work produced by your business as a function of two variables: your total number, N, of workers, and the total value, V, of your equipment.

How would you expect such a production function to behave? In general, having more equipment and more workers enables you to produce more. However, increasing equipment without increasing the number of workers will increase production a bit, but not beyond a point. (If equipment is already lying idle, having more of it won't help.) Similarly, increasing the number of workers without increasing equipment will increase production, but not past the point where the equipment is fully utilized, as any new workers would have no equipment available to them.

Example 7   Explain why the contour diagram in Figure 12.49 does not model the behavior expected of the production function, whereas the contour diagram in Figure 12.50 does.

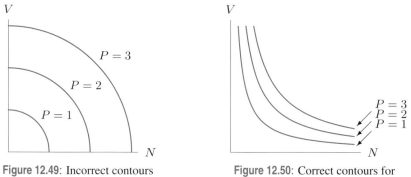

Figure 12.49: Incorrect contours for printing production

Figure 12.50: Correct contours for printing production

Solution   Look at Figure 12.49. Fixing V and letting N increase corresponds to moving to the right on the contour diagram. As you do so, you cross contours with larger and larger P values, meaning that production increases indefinitely. On the other hand, in Figure 12.50, as you move in the same direction you move nearly parallel to the contours, crossing them less and less frequently. Therefore, production increases more and more slowly as N increases with V fixed. Similarly, if you fix N and let V increase, the contour diagram in Figure 12.49 shows production increasing at a steady rate, whereas Figure 12.50 shows production increasing, but at a decreasing rate. Thus, Figure 12.50 fits the expected behavior of the production function best.

Formula for a Production Function

Production functions are often approximated by formulas of the form

$$P = f(N, V) = cN^\alpha V^\beta$$

where P is the quantity produced and c, α, and β are positive constants, $0 < \alpha < 1$ and $0 < \beta < 1$.

Example 8   Show that the contours of the function $P = cN^\alpha V^\beta$ have approximately the shape of the contours in Figure 12.50.

Solution   The contours are the curves where P is equal to a constant value, say P_0, that is, where

$$cN^\alpha V^\beta = P_0.$$

Solving for V we get

$$V = \left(\frac{P_0}{c}\right)^{1/\beta} N^{-\alpha/\beta}.$$

Thus, V is a power function of N with a negative exponent, so its graph has the shape shown in Figure 12.50.

The Cobb-Douglas Production Model

In 1928, Cobb and Douglas used a similar function to model the production of the entire US economy in the first quarter of this century. Using government estimates of P, the total yearly production between 1899 and 1922, of K, the total capital investment over the same period, and of L, the total labor force, they found that P was well approximated by the *Cobb-Douglas production function*

$$P = 1.01L^{0.75}K^{0.25}.$$

This function turned out to model the US economy surprisingly well, both for the period on which it was based, and for some time afterward.

Exercises and Problems for Section 12.3

Exercises

In Exercises 1–9, sketch a contour diagram for the function with at least four labeled contours. Describe in words the contours and how they are spaced.

1. $f(x, y) = x + y$

2. $f(x, y) = 3x + 3y$

3. $f(x, y) = x^2 + y^2$

4. $f(x, y) = -x^2 - y^2 + 1$

5. $f(x, y) = xy$

6. $f(x, y) = y - x^2$

7. $f(x, y) = x^2 + 2y^2$

8. $f(x, y) = \sqrt{x^2 + 2y^2}$

9. $f(x, y) = \cos\sqrt{x^2 + y^2}$

10. (a) For $z = f(x, y) = xy$, sketch and label the level curves $z = \pm 1, z = \pm 2$.
 (b) Sketch and label cross-sections of f with $x = \pm 1$, $x = \pm 2$.
 (c) The surface $z = xy$ is cut by a vertical plane containing the line $y = x$. Sketch the cross-section.

11. Total sales, Q, of a product is a function of its price and the amount spent on advertising. Figure 12.51 shows a contour diagram for total sales. Which axis corresponds to the price of the product and which to the amount spent on advertising? Explain.

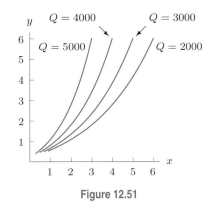

Figure 12.51

For each of the surfaces in Exercises 12–15, sketch a possible contour diagram, marked with reasonable z-values. (Note: There are many possible answers.)

12.

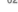

13.

14.

15.

16. Match Tables 12.6–12.9 with the contour diagrams (I)–(IV) in Figure 12.52.

Table 12.6

$y \backslash x$	-1	0	1
-1	2	1	2
0	1	0	1
1	2	1	2

Table 12.7

$y \backslash x$	-1	0	1
-1	0	1	0
0	1	2	1
1	0	1	0

Table 12.8

$y \backslash x$	-1	0	1
-1	2	0	2
0	2	0	2
1	2	0	2

Table 12.9

$y \backslash x$	-1	0	1
-1	2	2	2
0	0	0	0
1	2	2	2

Figure 12.52

17. Match the surfaces (a)–(e) in Figure 12.53 with the contour diagrams (I)–(V) in Figure 12.54.

Figure 12.53

Figure 12.54

Problems

18. Match the contour diagrams (a)–(d) with the surfaces (I)–(IV). Give reasons for your choice.

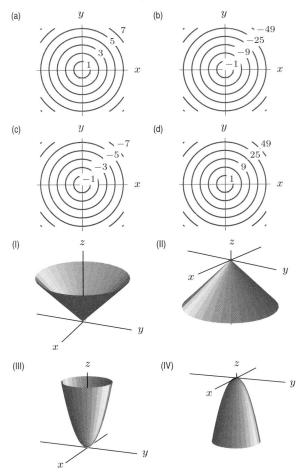

19. Match the pairs of functions (a)–(d) with the contour diagrams (I)–(IV). In each case, show which contours represent f and which represent g. (The x- and y-scales are equal.)

(a) $f(x, y) = x + y, g(x, y) = x - y$
(b) $f(x, y) = 2x + 3y, g(x, y) = 2x - 3y$
(c) $f(x, y) = x^2 - y, g(x, y) = 2y + \ln |x|$
(d) $f(x, y) = x^2 - y^2, g(x, y) = xy$

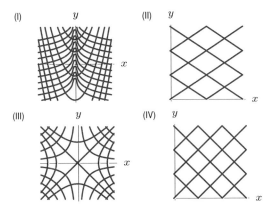

20. Figure 12.55 is a contour diagram of the monthly payment on a 5-year car loan as a function of the interest rate and the amount you borrow. The interest rate is 13% and you borrow $6000.

(a) What is your monthly payment?
(b) If interest rates drop to 11%, how much more can you borrow without increasing your monthly payment?
(c) Make a table of how much you can borrow, without increasing your monthly payment, as a function of the interest rate.

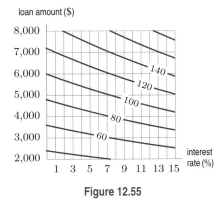

Figure 12.55

21. Figure 12.56 shows a contour map of a hill with two paths, A and B.

(a) On which path, A or B, will you have to climb more steeply?
(b) On which path, A or B, will you probably have a better view of the surrounding countryside? (Assuming trees do not block your view.)
(c) Alongside which path is there more likely to be a stream?

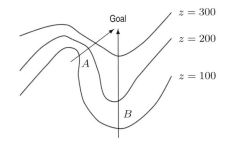

Figure 12.56

22. Match the functions (a)–(d) with the shapes of their level curves (I)–(IV). Sketch each contour diagram.

(a) $f(x, y) = x^2$ **(b)** $f(x, y) = x^2 + 2y^2$
(c) $f(x, y) = y - x^2$ **(d)** $f(x, y) = x^2 - y^2$

I.   Lines   II. Parabolas
III. Hyperbolas   IV. Ellipses

23. Figure 12.57 shows the density of the fox population P (in foxes per square kilometer) for southern England. Draw two different cross-sections along a north-south line and two different cross-sections along an east-west line of the population density P.

kilometers north

Figure 12.57

24. A manufacturer sells two goods, one at a price of $3000 a unit and the other at a price of $12,000 a unit. A quantity q_1 of the first good and q_2 of the second good are sold at a total cost of $4000 to the manufacturer.

   (a) Express the manufacturer's profit, π, as a function of q_1 and q_2.

   (b) Sketch curves of constant profit in the q_1q_2-plane for $\pi = 10,000$, $\pi = 20,000$, and $\pi = 30,000$ and the break-even curve $\pi = 0$.

25. Consider the Cobb-Douglas production function $P = f(L, K) = 1.01L^{0.75}K^{0.25}$. What is the effect on production of doubling both labor and capital?

26. Match each Cobb-Douglas production function (a)–(c) with a graph in Figure 12.58 and a statement (D)–(G).

   (a) $F(L, K) = L^{0.25}K^{0.25}$

   (b) $F(L, K) = L^{0.5}K^{0.5}$

   (c) $F(L, K) = L^{0.75}K^{0.75}$

   (D) Tripling each input triples output.

   (E) Quadrupling each input doubles output.

   (G) Doubling each input almost triples output.

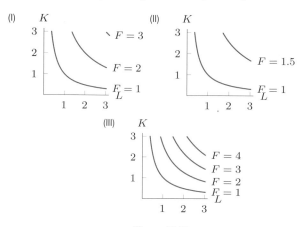

Figure 12.58

27. A general Cobb-Douglas production function has the form
$$P = cL^\alpha K^\beta.$$

What happens to production if labor and capital are both scaled up? For example, does production double if both labor and capital are doubled? Economists talk about

- *increasing returns to scale* if doubling L and K more than doubles P,
- *constant returns to scale* if doubling L and K exactly doubles P,
- *decreasing returns to scale* if doubling L and K less than doubles P.

What conditions on α and β lead to increasing, constant, or decreasing returns to scale?

28. **(a)** Sketch level curves of $f(x, y) = \sqrt{x^2 + y^2} + x$ for $f = 1, 2, 3$.

   (b) For what values of c can level curves $f = c$ be drawn?

29. Figure 12.59 is the contour diagram of $f(x, y)$. Sketch the contour diagram of each of the following functions.

   (a) $3f(x, y)$ **(b)** $f(x, y) - 10$

   (c) $f(x - 2, y - 2)$ **(d)** $f(-x, y)$

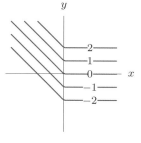

Figure 12.59

30. Figure 12.60 shows part of the contour diagram of $f(x, y)$. Complete the diagram for $x < 0$ if

   (a) $f(-x, y) = f(x, y)$ **(b)** $f(-x, y) = -f(x, y)$

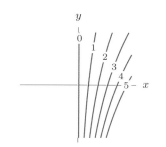

Figure 12.60

31. (a) Draw the contour diagram of $f(x, y) = g(y - x)$ if
  (i) $g(t) = t^2$ (ii) $g(t) = \sin t$

(b) What can you say about the level curves of a function of the form $f(x, y) = g(y - x)$ where $g(t)$ is a one-variable function?

32. Use the factored form of $f(x, y) = x^2 - y^2 = (x - y)(x + y)$ to sketch the contour $f(x, y) = 0$ and to find the regions in the xy-plane where $f(x, y) > 0$ and the regions where $f(x, y) < 0$. Explain how this sketch shows that the graph of $f(x, y)$ is saddle-shaped at the origin.

33. Use Problem 32 to find a formula for a "monkey-saddle" surface $z = g(x, y)$ which has three regions with $g(x, y) > 0$ and three with $g(x, y) < 0$.

12.4 LINEAR FUNCTIONS

What is a Linear Function of Two Variables?

Linear functions played a central role in one-variable calculus because many one-variable functions have graphs that look like a line when we zoom in. In two-variable calculus, a *linear function* is one whose graph is a plane. In Chapter 14, we see that many two-variable functions have graphs which look like planes when we zoom in.

What Makes a Plane Flat?

What makes the graph of the function $z = f(x, y)$ a plane? Linear functions of *one* variable have straight line graphs because they have constant slope. On a plane, the situation is a bit more complicated. If we walk around on a tilted plane, the slope is not always the same: it depends on the direction in which we walk. However, at every point on the plane, the slope is the same as long as we choose the same direction. If we walk parallel to the x-axis, we always find ourselves walking up or down with the same slope; the same is true if we walk parallel to the y-axis. In other words, the slope ratios $\Delta z / \Delta x$ (with y fixed) and $\Delta z / \Delta y$ (with x fixed) are each constant.

Example 1   A plane cuts the z-axis at $z = 5$, has slope 2 in the x direction and slope -1 in the y direction. What is the equation of the plane?

Solution   Finding the equation of the plane means constructing a formula for the z-coordinate of the point on the plane directly above the point (x, y) in the xy-plane. To get to that point start from the point above the origin, where $z = 5$. Then walk x units in the x direction. Since the slope in the x direction is 2, the height increases by $2x$. Then walk y units in the y direction; since the slope in the y direction is -1, the height decreases by y units. Since the height has changed by $2x - y$ units, the z-coordinate is $5 + 2x - y$. Thus, the equation for the plane is

$$z = 5 + 2x - y.$$

For any linear function, if we know its value at a point (x_0, y_0), its slope in the x direction, and its slope in the y direction, then we can write the equation of the function. This is just like the equation of a line in the one-variable case, except that there are two slopes instead of one.

If a **plane** has slope m in the x direction, slope n in the y direction, and passes through the point (x_0, y_0, z_0), then its equation is

$$z = z_0 + m(x - x_0) + n(y - y_0).$$

This plane is the graph of the **linear function**

$$f(x, y) = z_0 + m(x - x_0) + n(y - y_0).$$

If we write $c = z_0 - mx_0 - ny_0$, then we can write $f(x, y)$ in the equivalent form

$$f(x, y) = c + mx + ny.$$

Just as in 2-space a line is determined by two points, so in 3-space a plane is determined by three points, provided they do not lie on a line.

Example 2   Find the equation of the plane passing through the points $(1, 0, 1)$, $(1, -1, 3)$, and $(3, 0, -1)$.

Solution   The first two points have the same x-coordinate, so we use them to find the slope of the plane in the y-direction. As the y-coordinate changes from 0 to -1, the z-coordinate changes from 1 to 3, so the slope in the y-direction is $n = \Delta z / \Delta y = (3 - 1)/(-1 - 0) = -2$. The first and third points have the same y-coordinate, so we use them to find the slope in the x-direction; it is $m = \Delta z / \Delta x = (-1 - 1)/(3 - 1) = -1$. Because the plane passes through $(1, 0, 1)$, its equation is

$$z = 1 - (x - 1) - 2(y - 0) \quad \text{or} \quad z = 2 - x - 2y.$$

You should check that this equation is also satisfied by the points $(1, -1, 3)$ and $(3, 0, -1)$.

Example 2 was made easier by the fact that two of the points had the same x-coordinate and two had the same y-coordinate. An alternative method, which works for any three points, is to substitute the x, y, and z-values of each of the three points into the equation $z = c + mx + ny$. The resulting three equations in c, m, n are then solved simultaneously.

Linear Functions from a Numerical Point of View

To avoid flying planes with empty seats, airlines sell some tickets at full price and some at a discount. Table 12.10 shows an airline's revenue in dollars from tickets sold on a particular route, as a function of the number of full-price tickets sold, f, and the number of discount tickets sold, d.

Table 12.10   *Revenue from ticket sales (dollars)*

		Full-price tickets (f)			
		100	200	300	400
	200	39,700	63,600	87,500	111,400
	400	55,500	79,400	103,300	127,200
Discount tickets (d)	600	71,300	95,200	119,100	143,000
	800	87,100	111,000	134,900	158,800
	1000	102,900	126,800	150,700	174,600

In every column, the revenue jumps by \$15,800 for each extra 200 discount tickets. Thus, each column is a linear function of the number of discount tickets sold. In addition, every column has the same slope, $15,800/200 = 79$ dollars/ticket. This is the price of a discount ticket. Similarly, each row is a linear function and all the rows have the same slope, 239, which is the price in dollars of a full-fare ticket. Thus, R is a linear function of f and d, given by:

$$R = 239f + 79d.$$

We have the following general result:

A **linear function** can be recognized from its table by the following features:
- Each row and each column is linear.
- All the rows have the same slope.
- All the columns have the same slope (although the slope of the rows and the slope of the columns are generally different).

Example 3   The table contains values of a linear function. Fill in the blank and give a formula for the function.

$x \backslash y$	1.5	2.0
2	0.5	1.5
3	−0.5	?

Solution   In the first column the function decreases by 1 (from 0.5 to −0.5) as x goes from 2 to 3. Since the function is linear, it must decrease by the same amount in the second column. So the missing entry must be $1.5 - 1 = 0.5$. The slope of the function in the x-direction is −1. The slope in the y-direction is 2, since in each row the function increases by 1 when y increases by 0.5. From the table we get $f(2, 1.5) = 0.5$. Therefore, the formula is

$$f(x, y) = 0.5 - (x - 2) + 2(y - 1.5) = -0.5 - x + 2y.$$

What Does the Contour Diagram of a Linear Function Look Like?

The formula for the airline revenue function in Table 12.10 is $R = 239f + 79d$, where f is the number of full-fares and d is the number of discount fares sold.

Notice that the contours of this function in Figure 12.61 are parallel straight lines. What is the practical significance of the slope of these contour lines? Consider the contour $R = 100,000$; that means we are looking at combinations of ticket sales that yield \$100,000 in revenue. If we move down and to the right on the contour, the f-coordinate increases and the d-coordinate decreases, so we sell more full-fares and fewer discount fares. This is because to receive a fixed revenue of \$100,000, we must sell more full-fares if we sell fewer discount fares. The exact trade-off depends on the slope of the contour; the diagram shows that each contour has a slope of about −3. This means that for a fixed revenue, we must sell three discount fares to replace one full-fare. This can also be seen by comparing prices. Each full fare brings in \$239; to earn the same amount in discount fares we need to sell $239/79 \approx 3.03 \approx 3$ fares. Since the price ratio is independent of how many of each type of fare we sell, this slope remains constant over the whole contour map; thus, the contours are all parallel straight lines.

Notice also that the contours are evenly spaced. Thus, no matter which contour we are on, a fixed increase in one of the variables causes the same increase in the value of the function. In terms of revenue, no matter how many fares we have sold, an extra fare, whether full or discount, brings the same revenue as before.

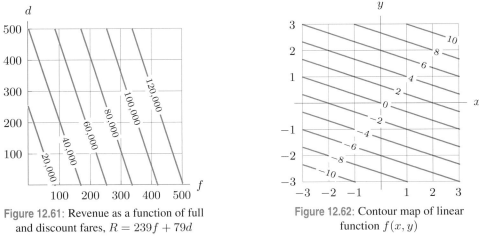

Figure 12.61: Revenue as a function of full
and discount fares, $R = 239f + 79d$

Figure 12.62: Contour map of linear
function $f(x, y)$

Example 4   Find the equation of the linear function whose contour diagram is in Figure 12.62.

Solution   Suppose we start at the origin on the $z = 0$ contour. Moving 2 units in the y direction takes us to the $z = 6$ contour; so the slope in the y direction is $\Delta z/\Delta y = 6/2 = 3$. Similarly, a move of 2 in the x-direction from the origin takes us to the $z = 2$ contour, so the slope in the x direction is $\Delta z/\Delta x = 2/2 = 1$. Since $f(0, 0) = 0$, we have $f(x, y) = x + 3y$.

Exercises and Problems for Section 12.4

Exercises

1. The charge, C, in dollars, to use an Internet service is a function of m, the number of months of use, and t, the total number of minutes on-line:

$$C = f(m, t) = 35 + 15m + 0.05t.$$

   (a) Is f a linear function?
   (b) Give units for the coefficients of m and t, and interpret them as charges.
   (c) Interpret the intercept 35 as a charge.
   (d) Find $f(3, 800)$ and interpret your answer.

2. Suppose that z is a linear function of x and y with slope 2 in the x direction and slope 3 in the y direction.

   (a) A change of 0.5 in x and -0.2 in y produces what change in z?
   (b) If $z = 2$ when $x = 5$ and $y = 7$, what is the value of z when $x = 4.9$ and $y = 7.2$?

3. (a) Find a formula for the linear function whose graph is a plane passing through point $(4, 3, -2)$ with slope 5 in the x-direction and slope -3 in the y-direction.
   (b) Sketch the contour diagram for this function.

Which of the tables of values in Exercises 4–7 could represent linear functions?

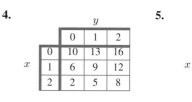

4.

		y	
	0	1	2
0	10	13	16
x 1	6	9	12
2	2	5	8

5.

		y	
	0	1	2
0	0	1	4
x 1	1	0	1
2	4	1	0

6.

		y	
	0	1	2
0	0	5	10
x 1	2	7	12
2	4	9	14

7.

		y	
	0	1	2
0	5	7	9
x 1	6	9	12
2	7	11	15

8. Find the equation of the linear function $z = c + mx + ny$ whose graph contains the points $(0, 0, 0)$, $(0, 2, -1)$, and $(-3, 0, -4)$.

9. Find the linear function whose graph is the plane through the points $(4, 0, 0)$, $(0, 3, 0)$ and $(0, 0, 2)$.

10. Find an equation for the plane containing the line in the xy-plane where $y = 1$, and the line in the xz-plane where $z = 2$.

11. Find the equation of the linear function $z = c + mx + ny$ whose graph intersects the xz-plane in the line $z = 3x + 4$ and intersects the yz-plane in the line $z = y + 4$.

Which of the contour diagrams in Exercises 12–15 could represent linear functions?

12.

13.

14.

15.

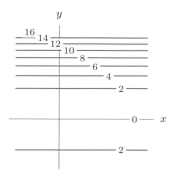

Problems

For Problems 16–17, find possible equations for linear functions with the given contour diagrams.

16.

17.

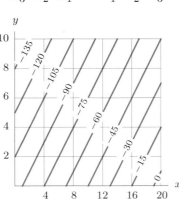

For Problems 18–19, find equations for linear functions with the given values.

18.

$x \backslash y$	−1	0	1	2
0	1.5	1	0.5	0
1	3.5	3	2.5	2
2	5.5	5	4.5	4
3	7.5	7	6.5	6

19.

$x \backslash y$	10	20	30	40
100	3	6	9	12
200	2	5	8	11
300	1	4	7	10
400	0	3	6	9

It is difficult to graph a linear function by hand. One method that works if the x, y, and z-intercepts are positive is to plot the intercepts and join them by a triangle as shown in Figure 12.63; this shows the part of the plane in the octant where $x \geq 0$, $y \geq 0$, $z \geq 0$. If the intercepts are not all positive, the same method works if the x, y, and z-axes are drawn from a different perspective. Use this method to graph the linear functions in Problems 20–23.

Figure 12.63

[6]From the August 28, 1994, issue of *Parade Magazine*.

20. $z = 2 - x - 2y$

21. $z = 6 - 2x - 3y$

22. $z = 4 + x - 2y$

23. $z = 2 - 2x + y$

Problems 24–26 concern Table 12.11, which gives the number of calories burned per minute for someone roller-blading, as a function of the person's weight and speed.[6]

Table 12.11

Calories burned per minute				
Weight	8 mph	9 mph	10 mph	11 mph
120 lbs	4.2	5.8	7.4	8.9
140 lbs	5.1	6.7	8.3	9.9
160 lbs	6.1	7.7	9.2	10.8
180 lbs	7.0	8.6	10.2	11.7
200 lbs	7.9	9.5	11.1	12.6

24. Does the data in Table 12.11 look approximately linear? Give a formula for B, the number of calories burned per minute in terms of the weight, w, and the speed, s. Does the formula make sense for all weights or speeds?

25. Who burns more total calories to go 10 miles: A 120 lb person going 10 mph or a 180 lb person going 8 mph? Which of these two people burns more calories per pound for the 10-mile trip?

26. Use Problem 24 to give a formula for P, the number of calories burned per pound, in terms of w and s, for a person weighing w lbs roller-blading 10 miles at s mph.

27. A manufacturer makes two products out of two raw materials. Let q_1, q_2 be the quantities sold of the two products, p_1, p_2 their prices, and m_1, m_2 the quantities purchased of the two raw materials. Which of the following functions do you expect to be linear, and why? In each case, assume that all variables except the ones mentioned are held fixed.

(a) Expenditure on raw materials as a function of m_1 and m_2.
(b) Revenue as a function of q_1 and q_2.
(c) Revenue as a function of p_1 and q_1.

28. Let f be the linear function $f(x, y) = c + mx + ny$, where c, m, n are constants and $n \neq 0$.

(a) Show that all the contours of f are lines of slope $-m/n$.
(b) For all x and y, show $f(x + n, y - m) = f(x, y)$.
(c) Explain the relation between parts (a) and (b).

Problems 29–30 refer to the linear function $z = f(x, y)$ whose values are in Table 12.12.

Table 12.12

	y				
	4	6	8	10	12
5	3	6	9	12	15
10	7	10	13	16	19
x 15	11	14	17	20	23
20	15	18	21	24	27
25	19	22	25	28	31

29. Each column of Table 12.12 is linear with the same slope, $m = \Delta z/\Delta x = 4/5$. Each row is linear with the same slope, $n = \Delta z/\Delta y = 3/2$. We now investigate the slope obtained by moving through the table along lines that are neither rows nor columns.

(a) Move down the diagonal of the table from the upper left corner ($z = 3$) to the lower right corner ($z = 31$). What do you notice about the changes in z? Now move diagonally from $z = 6$ to $z = 27$. What do you notice about the changes in z now?

(b) Move in the table along a line right one step, up two steps from $z = 19$ to $z = 9$. Then move in the same direction from $z = 22$ to $z = 12$. What do you notice about the changes in z?

(c) Show that $\Delta z = m\Delta x + n\Delta y$. Use this to explain what you observed in parts (a) and (b).

30. If we hold y fixed, that is we keep $\Delta y = 0$, and step in the positive x-direction, we get the x-slope, m. If instead we keep $\Delta x = 0$ and step in the positive y-direction, we get the y-slope, n. Fix a step in which neither $\Delta x = 0$ nor $\Delta y = 0$. The slope in the $\Delta x, \Delta y$ direction is

$$\text{Slope} = \frac{\text{Rise}}{\text{Run}} = \frac{\Delta z}{\text{Length of step}}$$
$$= \frac{\Delta z}{\sqrt{(\Delta x)^2 + (\Delta y)^2}}.$$

(a) Compute the slopes for the linear function in Table 12.12 in the direction of $\Delta x = 5, \Delta y = 2$.

(b) Compute the slopes for the linear function in Table 12.12 in the direction of $\Delta x = -10, \Delta y = 2$.

12.5 FUNCTIONS OF THREE VARIABLES

In applications of calculus, functions of any number of variables can arise. The density of matter in the universe is a function of three variables, since it takes three numbers to specify a point in space. Models of the US economy often use functions of ten or more variables. We need to be able to apply calculus to functions of arbitrarily many variables.

One difficulty with functions of more than two variables is that it is hard to visualize them. The graph of a function of one variable is a curve in 2-space, the graph of a function of two variables is a surface in 3-space, so the graph of a function of three variables would be a solid in 4-space. Since we can't easily visualize 4-space, we won't use the graphs of functions of three variables. On the other hand, it is possible to draw contour diagrams for functions of three variables, only now the contours are surfaces in 3-space.

Representing a Function of Three Variables Using a Family of Level Surfaces

A function of two variables, $f(x, y)$, can be represented by a family of level curves of the form $f(x, y) = c$ for various values of the constant, c.

A **level surface**, or **level set** of a function of three variables, $f(x, y, z)$, is a surface of the form $f(x, y, z) = c$, where c is a constant. The function f can be represented by the family of level surfaces obtained by allowing c to vary.

The value of the function, f, is constant on each level surface.

Example 1   The temperature, in °C, at a point (x, y, z) is given by $T = f(x, y, z) = x^2 + y^2 + z^2$. What do the level surfaces of the function f look like and what do they mean in terms of temperature?

Solution   The level surface corresponding to $T = 100$ is the set of all points where the temperature is $100°C$. That is, where $f(x, y, z) = 100$, so

$$x^2 + y^2 + z^2 = 100.$$

This is the equation of a sphere of radius 10, with center at the origin. Similarly, the level surface corresponding to $T = 200$ is the sphere with radius $\sqrt{200}$. The other level surfaces are concentric spheres. The temperature is constant on each sphere. We may view the temperature distribution as a set of nested spheres, like concentric layers of an onion, each one labeled with a different temperature, starting from low temperatures in the middle and getting hotter as we go out from the center. (See Figure 12.64.) The level surfaces become more closely spaced as we move farther from the origin because the temperature increases more rapidly the farther we get from the origin.

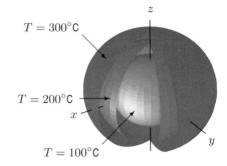

Figure 12.64: Level surfaces of $T = f(x, y, z) = x^2 + y^2 + z^2$, each one having a constant temperature

Example 2   What do the level surfaces of $f(x, y, z) = x^2 + y^2$ and $g(x, y, z) = z - y$ look like?

Solution   The level surface of f corresponding to the constant c is the surface consisting of all points satisfying the equation

$$x^2 + y^2 = c.$$

Since there is no z-coordinate in the equation, z can take any value. For $c > 0$, this is a circular cylinder of radius $\sqrt{c}$ around the z-axis. The level surfaces are concentric cylinders; on the narrow ones near the z-axis, f has small values; on the wider ones, f has larger values. See Figure 12.65.

The level surface of g corresponding to the constant c is the surface

$$z - y = c.$$

This time there is no x variable, so this surface is the one we get by taking each point on the straight line $z - y = c$ in the yz-plane and letting x vary. We get a plane which cuts the yz-plane diagonally; the x-axis is parallel to this plane. See Figure 12.66.

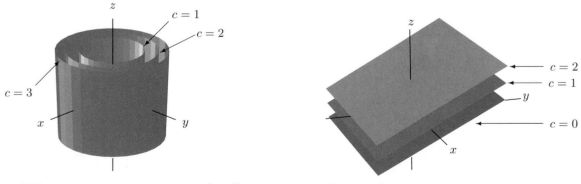

Figure 12.65: Level surfaces of $f(x, y, z) = x^2 + y^2$

Figure 12.66: Level surfaces of $g(x, y, z) = z - y$

Example 3   What do the level surfaces of $f(x, y, z) = x^2 + y^2 - z^2$ look like?

Solution   In Section 12.3, we saw that the two-variable quadratic function $g(x, y) = x^2 - y^2$ has a saddle-shaped graph and three types of contours. The contour equation $x^2 - y^2 = c$ gives a hyperbola opening right-left when $c > 0$, a hyperbola opening up-down when $c < 0$, and a pair of intersecting lines when $c = 0$. Similarly, the three-variable quadratic function $f(x, y, z) = x^2 + y^2 - z^2$ has three types of level surfaces depending on the value of c in the equation $x^2 + y^2 - z^2 = c$.

Suppose that $c > 0$, say $c = 1$. Rewrite the equation as $x^2 + y^2 = z^2 + 1$ and think of what happens as we cut the surface perpendicular to the z-axis by holding z fixed. The result is a circle, $x^2 + y^2 = $ constant, of radius at least 1 (since the constant $z^2 + 1 \geq 1$). The circles get larger as z gets larger. If we take the $x = 0$ cross-section instead, we get the hyperbola $y^2 - z^2 = 1$. The result is shown in Figure 12.70, with $a = b = c = 1$.

Suppose instead $c < 0$, say $c = -1$. Then the horizontal cross-sections of $x^2 + y^2 = z^2 - 1$ are again circles except that the radii shrink to 0 at $z = \pm 1$ and between $z = -1$ and $z = 1$ there are no cross-sections at all. The result is shown in Figure 12.71 with $a = b = c = 1$.

When $c = 0$, we get the equation $x^2 + y^2 = z^2$. Again the horizontal cross-sections are circles, this time with the radius shrinking down to exactly 0 when $z = 0$. The resulting surface, shown in Figure 12.72 with $a = b = c = 1$, is the cone $z = \sqrt{x^2 + y^2}$ studied in Section 12.3, together with the lower cone $z = -\sqrt{x^2 + y^2}$.

A Catalog of Surfaces

For later reference, here is a small catalog of the surfaces we have encountered.

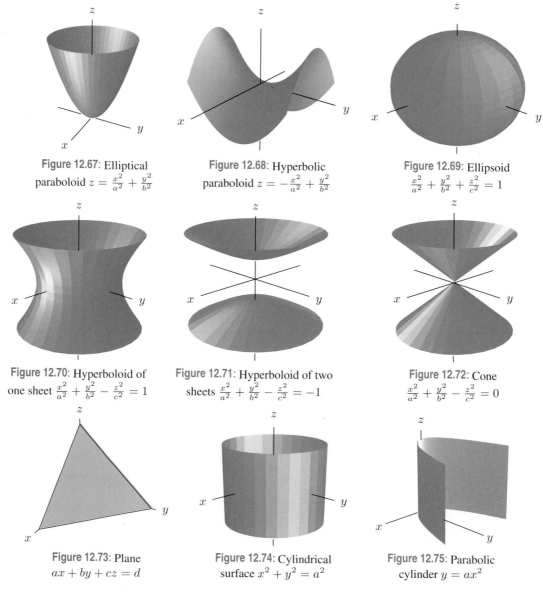

Figure 12.67: Elliptical paraboloid $z = \frac{x^2}{a^2} + \frac{y^2}{b^2}$

Figure 12.68: Hyperbolic paraboloid $z = -\frac{x^2}{a^2} + \frac{y^2}{b^2}$

Figure 12.69: Ellipsoid $\frac{x^2}{a^2} + \frac{y^2}{b^2} + \frac{z^2}{c^2} = 1$

Figure 12.70: Hyperboloid of one sheet $\frac{x^2}{a^2} + \frac{y^2}{b^2} - \frac{z^2}{c^2} = 1$

Figure 12.71: Hyperboloid of two sheets $\frac{x^2}{a^2} + \frac{y^2}{b^2} - \frac{z^2}{c^2} = -1$

Figure 12.72: Cone $\frac{x^2}{a^2} + \frac{y^2}{b^2} - \frac{z^2}{c^2} = 0$

Figure 12.73: Plane $ax + by + cz = d$

Figure 12.74: Cylindrical surface $x^2 + y^2 = a^2$

Figure 12.75: Parabolic cylinder $y = ax^2$

(These are viewed as equations in three variables x, y, and z)

How Surfaces Can Represent Functions of Two Variables and Functions of Three Variables

You may have noticed that we have used surfaces to represent functions in two different ways. First, we used a *single* surface to represent a two-variable function $f(x, y)$. Second, we used a *family* of level surfaces to represent a three-variable function $g(x, y, z)$. These level surfaces have equation $g(x, y, z) = c$.

What is the relation between these two uses of surfaces? For example, consider the function

$$f(x, y) = x^2 + y^2 + 3.$$

Define

$$g(x, y, z) = x^2 + y^2 + 3 - z$$

The points on the graph of f satisfy $z = x^2 + y^2 + 3$, so they also satisfy $x^2 + y^2 + 3 - z = 0$. Thus the graph of f is the same as the level surface

$$g(x, y, z) = x^2 + y^2 + 3 - z = 0.$$

In general, we have the following result:

A single surface that is the graph of a two-variable function $f(x, y)$ can be thought of as one member of the family of level surfaces representing the three-variable function

$$g(x, y, z) = f(x, y) - z.$$

The graph of f is the level surface $g = 0$.

Conversely, a single level surface $g(x, y, z) = c$ can be regarded as the graph of a function $f(x, y)$ if it is possible to solve for z. Sometimes the level surface is pieced together from the graphs of two or more two-variable functions. For example, if $g(x, y, z) = x^2 + y^2 + z^2$, then one member of the family of level surfaces is the sphere

$$x^2 + y^2 + z^2 = 1.$$

This equation defines z implicitly as a function of x and y. Solving it gives two functions

$$z = \sqrt{1 - x^2 - y^2} \qquad \text{and} \qquad z = -\sqrt{1 - x^2 - y^2}.$$

The graph of the first function is the top half of the sphere and the graph of the second function is the bottom half.

Exercises and Problems for Section 12.5

Exercises

1. Match the following functions with the level surfaces in Figure 12.76.

    (a) $f(x, y, z) = y^2 + z^2$ **(b)** $h(x, y, z) = x^2 + z^2$.

Figure 12.76

2. Find a formula for a function $f(x, y, z)$ whose level surfaces look like those in Figure 12.77.

Figure 12.77

In Exercises 3–5, represent the surface whose equation is given as the graph of a two-variable function, $f(x, y)$, and as the level surface of a three-variable function, $g(x, y, z) = c$. There are many possible answers.

3. The plane $4x - y - 2z = 6$

4. The top half of the sphere $x^2 + y^2 + z^2 - 10 = 0$

5. The bottom half of the ellipsoid $x^2 + y^2 + z^2/2 = 1$

In Exercises 6–9, decide if the given level surface can be expressed as the graph of a function, $f(x, y)$.

6. $2x + 3y - 5z - 10 = 0$ **7.** $x^2 + y^2 + z^2 - 1 = 0$

8. $z - x^2 - 3y^2 = 0$ **9.** $z^2 = x^2 + 3y^2$

10. Find a formula for a function $f(x, y, z)$ whose level surface $f = 4$ is a sphere of radius 2, centered at the origin.

11. Find a formula for a function $f(x, y, z)$ whose level surfaces are spheres centered at the point (a, b, c).

12. Write the level surface $x + 2y + 3z = 5$ as the graph of a function $f(x, y)$.

13. Write the level surface $x^2 + y + \sqrt{z} = 1$ as the graph of a function $f(x, y)$.

14. Which of the graphs in the catalog of surfaces on page 636 is the graph of a function of x and y?

Use the catalog on page 636 to identify the surfaces in Exercises 15–18.

15. $-x^2 - y^2 + z^2 = 1$ **16.** $-x^2 + y^2 - z^2 = 0$

17. $x^2 + y^2 - z = 0$ **18.** $x^2 + z^2 = 1$

Problems

19. Find a function $f(x, y, z)$ whose level surface $f = 1$ is the graph of the function $g(x, y) = x + 2y$.

20. Find two functions $f(x, y)$ and $g(x, y)$ so that the graphs of both together form the ellipsoid $x^2 + y^2/4 + z^2/9 = 1$.

21. Describe, in words, the level surface $f(x, y, z) = x^2/4 + z^2 = 1$.

22. Describe, in words, the level surface $g(x, y, z) = x^2 + y^2/4 + z^2 = 1$. [Hint: Look at cross-sections with constant x, y and z values.]

23. Describe in words the level surfaces of the function $g(x, y, z) = x + y + z$.

24. Find a formula for a function $g(x, y, z)$ whose level surfaces are planes parallel to the plane $z = 2x + 3y - 5$.

25. Describe in words the level surfaces of $f(x, y, z) = \sin(x + y + z)$.

26. Describe in words the level surfaces of $g(x, y, z) = e^{-(x^2+y^2+z^2)}$.

27. The surface S is the graph of $f(x, y) = \sqrt{1 - x^2 - y^2}$.

   (a) Explain why S is the upper hemisphere of radius 1, with equator in the xy-plane, centered at the origin.

   (b) Find a level surface $g(x, y, z) = c$ representing S.

28. The surface S is the graph of $f(x, y) = \sqrt{1 - y^2}$.

   (a) Explain why S is the upper half of a circular cylinder of radius 1, centered along the x-axis.

   (b) Find a level surface $g(x, y, z) = c$ representing S.

29. A cone C, with height 1 and radius 1, has its base in the xz-plane and its vertex on the positive y-axis. Find a function $g(x, y, z)$ such that C is part of the level surface $g(x, y, z) = 0$. [Hint: The graph of $f(x, y) = \sqrt{x^2 + y^2}$ is a cone which opens up and has vertex at the origin.]

30. What do the level surfaces of $f(x, y, z) = x^2 - y^2 + z^2$ look like? [Hint: Use cross-sections with y constant instead of cross-sections with z constant.]

31. Describe the surface $x^2 + y^2 = (2 + \sin z)^2$. In general, if $f(z) \geq 0$ for all z, describe the surface $x^2 + y^2 = (f(z))^2$.

12.6   LIMITS AND CONTINUITY

The sheer vertical face of Half Dome, in Yosemite National Park in California, was caused by glacial activity during the Ice Age. (See Figure 12.78.) The height of the terrain rises abruptly by nearly 1000 feet as we scale the rock from the west, whereas it is possible to make a gradual climb to the top from the east.

     If we consider the function h giving the height of the terrain above sea level in terms of longitude and latitude, then h has a *discontinuity* along the path at the base of the cliff of Half Dome. Looking at the contour map of the region in Figure 12.79, we see that in most places a small change in position results in a small change in height, except near the cliff. There, no matter how small a step we take, we get a large change in height. (You can see how crowded the contours get near the cliff; some end abruptly along the discontinuity.)

     This geological feature illustrates the ideas of continuity and discontinuity. Roughly speaking, a function is said to be *continuous* at a point if its values at places near the point are close to the value at the point. If this is not the case, the function is said to be *discontinuous*.

Figure 12.78: Half Dome in Yosemite National Park

Figure 12.79: A contour map of Half Dome

The property of continuity is one that, practically speaking, we usually assume of the functions we are studying. Informally, we expect (except under special circumstances) that values of a function do not change drastically when making small changes to the input variables. Whenever we model a one-variable function by an unbroken curve, we are making this assumption. Even when functions come to us as tables of data, we usually make the assumption that the missing function values between data points are close to the measured ones.

In this section we study limits and continuity a bit more formally in the context of functions of several variables. For simplicity we study these concepts for functions of two variables, but our discussion can be adapted to functions of three or more variables.

One can show that sums, products, and compositions of continuous functions are continuous, while the quotient of two continuous functions is continuous everywhere the denominator function is nonzero. Thus, each of the functions

$$\cos(x^2 y), \qquad \ln(x^2 + y^2), \qquad \frac{e^{x+y}}{x+y}, \qquad \ln(\sin(x^2 + y^2))$$

is continuous at all points (x, y) where it is defined. As for functions of one variable, the graph of a continuous function over an unbroken domain is unbroken—that is, the surface has no holes or rips in it.

Example 1 From Figures 12.80–12.83, which of the following functions appear to be continuous at $(0, 0)$?

(a) $f(x, y) = \begin{cases} \dfrac{x^2 y}{x^2 + y^2}, & (x, y) \neq (0, 0), \\ 0, & (x, y) = (0, 0). \end{cases}$

(b) $g(x, y) = \begin{cases} \dfrac{x^2}{x^2 + y^2}, & (x, y) \neq (0, 0), \\ 0, & (x, y) = (0, 0). \end{cases}$

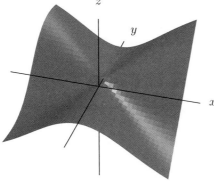

Figure 12.80: Graph of $z = x^2 y/(x^2 + y^2)$

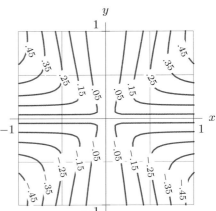

Figure 12.81: Contour diagram of $z = x^2 y/(x^2 + y^2)$

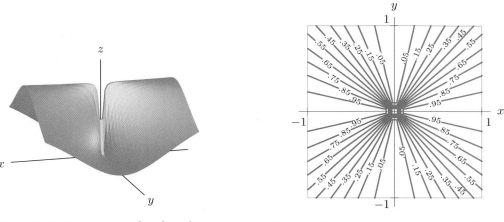

Figure 12.82: Graph of $z = x^2/(x^2 + y^2)$

Figure 12.83: Contour diagram of $z = x^2/(x^2 + y^2)$

Solution

(a) The graph and contour diagram of f in Figures 12.80 and 12.81 suggest that f is close to 0 when (x, y) is close to $(0, 0)$. That is, the figures suggest that f is continuous at the point $(0, 0)$; the graph appears to have no rips or holes there.

However, the figures cannot tell us for sure whether f is continuous. To be certain we must investigate the limit analytically, as is done in Example 2(a) on page 641.

(b) The graph of g and its contours near $(0, 0)$ in Figure 12.82 and 12.83 suggest that g behaves differently from f: The contours of g seem to "crash" at the origin and the graph rises rapidly from 0 to 1 near $(0, 0)$. Small changes in (x, y) near $(0, 0)$ can yield large changes in g, so we expect that g is not continuous at the point $(0, 0)$. Again, a more precise analysis is given in Example 2(b) on page 641.

The previous example suggests that continuity *at* a point depends on a function's behavior *near* the point. To study behavior near a point more carefully we need the idea of a limit of a function of two variables. Suppose that $f(x, y)$ is a function defined on a set in 2-space, not necessarily containing the point (a, b), but containing points (x, y) arbitrarily close to (a, b); suppose that L is a number.

The function f has a **limit** L at the point (a, b), written

$$\lim_{(x,y) \to (a,b)} f(x, y) = L,$$

if $f(x, y)$ is as close to L as we please whenever the distance from the point (x, y) to the point (a, b) is sufficiently small, but not zero.

We define continuity for functions of two variables in the same way as for functions of one variable:

A function f is **continuous at the point** (a, b) if

$$\lim_{(x,y) \to (a,b)} f(x, y) = f(a, b).$$

A function is **continuous on a region** R in the xy-plane if it is continuous at each point in R.

Thus, if f is continuous at the point (a, b), then f must be defined at (a, b) and the limit, $\lim_{(x,y)\to(a,b)} f(x, y)$, must exist and be equal to the value $f(a, b)$. If a function is defined at a point (a, b) but is not continuous there, then we say that f is *discontinuous* at (a, b).

We now apply the definition of continuity to the functions in Example 1, showing that f is continuous at $(0, 0)$ and that g is discontinuous at $(0, 0)$.

Example 2   Let f and g be the functions in Example 1. Use the definition of the limit to show that:

(a) $\displaystyle\lim_{(x,y)\to(0,0)} f(x, y) = 0$ (b) $\displaystyle\lim_{(x,y)\to(0,0)} g(x, y)$ does not exist.

Solution   To investigate these limits of f and g, we consider values of these functions near, but not at, the origin, where they are given by the formulas

$$f(x, y) = \frac{x^2 y}{x^2 + y^2} \qquad g(x, y) = \frac{x^2}{x^2 + y^2}.$$

(a) The graph and contour diagram of f both suggest that $\lim_{(x,y)\to(0,0)} f(x, y) = 0$. To use the definition of the limit, we estimate $|f(x, y) - L|$ with $L = 0$:

$$|f(x, y) - L| = \left| \frac{x^2 y}{x^2 + y^2} - 0 \right| = \left| \frac{x^2}{x^2 + y^2} \right| |y| \le |y| \le \sqrt{x^2 + y^2},$$

Now $\sqrt{x^2 + y^2}$ is the distance from (x, y) to $(0, 0)$. Thus, to make $|f(x, y) - 0| < 0.001$, for example, we need only require (x, y) be within 0.001 of $(0, 0)$. More generally, for any positive number u, no matter how small, we are sure that $|f(x, y) - 0| < u$ whenever (x, y) is no farther than u from $(0, 0)$. This is what we mean by saying that the difference $|f(x, y) - 0|$ can be made as small as we wish by choosing the distance to be sufficiently small. Thus, we conclude that

$$\lim_{(x,y)\to(0,0)} f(x, y) = \lim_{(x,y)\to(0,0)} \frac{x^2 y}{x^2 + y^2} = 0.$$

Notice that since this limit equals $f(0, 0)$, the function f is continuous at $(0, 0)$.

(b) Although the formula defining the function g looks similar to that of f, we saw in Example 1 that g's behavior near the origin is quite different. If we consider points $(x, 0)$ lying along the x-axis near $(0, 0)$, then the values $g(x, 0)$ are equal to 1, while if we consider points $(0, y)$ lying along the y-axis near $(0, 0)$, then the values $g(0, y)$ are equal to 0. Thus, within any distance (no matter how small) from the origin, there are points where $g = 0$ and points where $g = 1$. Therefore the limit $\lim_{(x,y)\to(0,0)} g(x, y)$ does not exist, and thus g is not continuous at $(0, 0)$.

While the notions of limit and continuity look formally the same for one- and two-variable functions, they are somewhat more subtle in the multivariable case. The reason for this is that on the line (1-space), we can approach a point from just two directions (left or right) but in 2-space there are an infinite number of ways to approach a given point.

Exercises and Problems for Section 12.6

Exercises

Are the functions in Exercises 1-6 continuous at all points in the given regions?

1. $\dfrac{1}{x^2 + y^2}$ on the square $-1 \le x \le 1, -1 \le y \le 1$

2. $\dfrac{1}{x^2 + y^2}$ on the square $1 \le x \le 2, 1 \le y \le 2$

3. $\dfrac{y}{x^2 + 2}$ on the disk $x^2 + y^2 \le 1$

4. $\dfrac{e^{\sin x}}{\cos y}$ on the rectangle $-\frac{\pi}{2} \le x \le \frac{\pi}{2}, 0 \le y \le \frac{\pi}{4}$

5. $\tan(xy)$ on the square $-2 \le x \le 2, -2 \le y \le 2$

6. $\sqrt{2x - y}$ on the disk $x^2 + y^2 \le 4$

In Exercises 7–11, find the limits of the functions $f(x, y)$ as $(x, y) \to (0, 0)$. Assume that polynomials, exponentials, logarithmic, and trigonometric functions are continuous.

7. $f(x, y) = e^{-x-y}$

8. $f(x, y) = x^2 + y^2$

9. $f(x, y) = \dfrac{x}{x^2 + 1}$

10. $f(x, y) = \dfrac{x + y}{(\sin y) + 2}$

11. $f(x, y) = \dfrac{\sin(x^2 + y^2)}{x^2 + y^2}$ [Hint: $\lim\limits_{t \to 0} \dfrac{\sin t}{t} = 1.$]

Problems

12. By approaching the origin along the positive x-axis and the positive y-axis, show that the following limit does not exist:
$$\lim_{(x,y) \to (0,0)} \frac{2x - y^2}{2x + y^2}.$$

13. By approaching the origin along the positive x-axis and the positive y-axis, show that the following limit does not exist:
$$\lim_{(x,y) \to (0,0)} \frac{x + y^2}{2x + y}.$$

14. Is the following function continuous at $(0, 0)$?
$$f(x, y) = \begin{cases} x^2 + y^2 & \text{if } (x, y) \neq (0, 0) \\ 2 & \text{if } (x, y) = (0, 0) \end{cases}$$

15. What value of c makes the following function continuous at $(0, 0)$?
$$f(x, y) = \begin{cases} x^2 + y^2 + 1 & \text{if } (x, y) \neq (0, 0) \\ c & \text{if } (x, y) = (0, 0) \end{cases}$$

16. (a) Use a computer to draw the graph and the contour diagram of the following function:
$$f(x, y) = \begin{cases} \dfrac{xy(x^2 - y^2)}{x^2 + y^2}, & (x, y) \neq (0, 0), \\ 0, & (x, y) = (0, 0). \end{cases}$$

   (b) Do your answers to part (a) suggest that f is continuous at $(0, 0)$? Explain your answer.

17. The function f, whose graph and contour diagram are in Figures 12.84 and 12.85, is given by
$$f(x, y) = \begin{cases} \dfrac{xy}{x^2 + y^2}, & (x, y) \neq (0, 0), \\ 0, & (x, y) = (0, 0). \end{cases}$$

   (a) Show that $f(0, y)$ and $f(x, 0)$ are each continuous functions of one variable.
   (b) Show that rays emanating from the origin are contained in contours of f.
   (c) Is f continuous at $(0, 0)$?

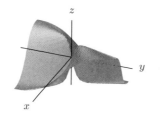

Figure 12.84: Graph of $z = xy/(x^2 + y^2)$

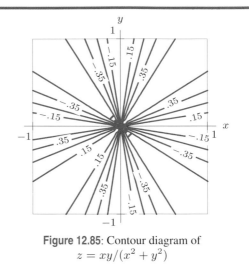

Figure 12.85: Contour diagram of $z = xy/(x^2 + y^2)$

18. Show that the function f does not have a limit at $(0, 0)$ by examining the limits of f as $(x, y) \to (0, 0)$ along the line $y = x$ and along the parabola $y = x^2$:
$$f(x, y) = \frac{x^2 y}{x^4 + y^2}, \qquad (x, y) \neq (0, 0).$$

19. Show that the function f does not have a limit at $(0, 0)$ by examining the limits of f as $(x, y) \to (0, 0)$ along the curve $y = kx^2$ for different values of k:
$$f(x, y) = \frac{x^2}{x^2 + y}, \qquad x^2 + y \neq 0.$$

20. Explain why the following function is not continuous along the line $y = 0$:
$$f(x, y) = \begin{cases} 1 - x, & y \geq 0, \\ -2, & y < 0, \end{cases}$$

In Problems 21–22, determine whether there is a value for c making the function continuous everywhere. If so, find it. If not, explain why not.

21. $f(x, y) = \begin{cases} c + y, & x \leq 3, \\ 5 - x, & x > 3. \end{cases}$

22. $f(x, y) = \begin{cases} c + y, & x \leq 3, \\ 5 - y, & x > 3. \end{cases}$

CHAPTER SUMMARY (see also Ready Reference at the end of the book)

- **3-Space**
  Cartesian coordinates, x-, y- and z-axes, xy-, xz- and yz-planes, distance formula.
- **Functions of Two Variables**
  Represented by: tables, graphs, formulas, cross-sections (one variable fixed), contours (function value fixed); cylinders (one variable missing).
- **Linear Functions**

  Recognizing linear functions from tables, graphs, contour diagrams, formulas. Converting from one representation to another.
- **Functions of Three Variables**
  Sketching level surfaces (function value fixed) in 3-space; graph of $z = f(x, y)$ is same as level surface $g(x, y, z) = f(x, y) - z = 0$.
- **Continuity**

REVIEW EXERCISES AND PROBLEMS FOR CHAPTER TWELVE

Exercises

Decide if the statements in Exercises 1–5 must be true, might be true, or could not be true. The function $z = f(x, y)$ is defined everywhere.

1. The level curves corresponding to $z = 1$ and $z = -1$ cross at the origin.

2. The level curve $z = 1$ consists of the circle $x^2 + y^2 = 2$ and the circle $x^2 + y^2 = 3$, but no other points.

3. The level curve $z = 1$ consists of two lines which intersect at the origin.

4. If $z = e^{-(x^2+y^2)}$, there is a level curve for every value of z.

5. If $z = e^{-(x^2+y^2)}$, there is a level curve through every point (x, y).

For each of the functions in Exercises 6–9, make a contour plot in the region $-2 < x < 2$ and $-2 < y < 2$. In each case, what is the equation and the shape of the contour lines?

6. $z = 3x - 5y + 1$

7. $z = \sin y$

8. $z = 2x^2 + y^2$

9. $z = e^{-2x^2-y^2}$

Find equations for the linear functions with the contour diagrams in Exercises 10–11.

10.

11.

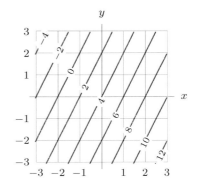

12. Match the functions with the level surfaces in Figure 12.86.

(a) $f(x, y, z) = x^2 + y^2 + z^2$
(b) $g(x, y, z) = x^2 + z^2$.

Figure 12.86

13. Describe the set of points whose x coordinate is 2 and whose y coordinate is 1.

14. Find the equation of the plane through the points $(0, 0, 2), (0, 3, 0), (5, 0, 0)$.

15. Find the center and radius of the sphere with equation $x^2 + 4x + y^2 - 6y + z^2 + 12z = 0$.

Problems

16. Sketch cross-sections of $f(r, h) = \pi r^2 h$, first keeping h fixed, then keeping r fixed.

17. Find a linear function whose graph is the plane that intersects the xy-plane along the line $y = 2x + 2$ and contains the point $(1, 2, 2)$.

18. Figure 12.87 shows the graph of $z = f(x, y)$.

    (a) Suppose y is fixed and positive. Does z increase or decrease as x increases? Graph z against x.

    (b) Suppose x is fixed and positive. Does z increase or decrease as y increases? Graph z against y.

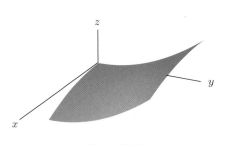

Figure 12.87

19. Match the following descriptions of a company's success with the graphs in Figure 12.88.

    (a) Our success is measured in dollars, plain and simple. More hard work won't hurt, but it also won't help.

    (b) No matter how much money or hard work we put into the company, we just could not make a go of it.

    (c) Although we are not always totally successful, it seems that the amount of money invested does not matter. As long as we put hard work into the company our success will increase.

    (d) The company's success is based on both hard work and investment.

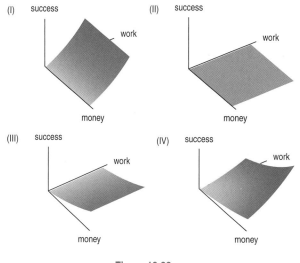

Figure 12.88

20. Figure 12.89 is the contour diagram of $f(x, y)$. Match the functions (a)–(f) with their contour diagrams (I)–(VI). All diagrams are drawn on the same region in the xy-plane.

  (a) $2f(x, y)$ **(b)** $f(x, y) + 2$ **(c)** $f(x - 1, y)$

  (d) $f(x, -y)$ **(e)** $f(2x, y)$ **(f)** $f(y, x)$

Figure 12.89

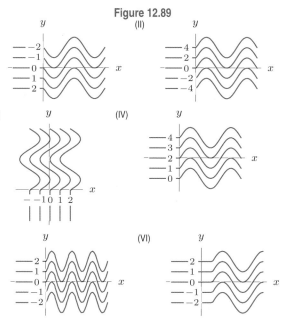

21. Figure 12.90 is the contour diagram of a function $f(x, y)$. Sketch a contour diagram of

  (a) $g(x, y) = (f(x, y))^2$ **(b)** $h(x, y) = \sin(f(x, y))$.

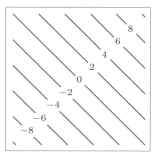

Figure 12.90

22. Values of $f(x, y) = \frac{1}{2}(x + y - 2)(x + y - 1) + y$ are in Table 12.13.

   (a) Find a pattern in the table. Make a conjecture and use it to complete Table 12.13 without computation. Check by using the formula for f.

   (b) Using the formula, check that the pattern holds for all $x \geq 1$ and $y \geq 1$.

Table 12.13

		1	2	3	4	5	6
	1	1	3	6	10	15	21
	2	2	5	9	14	20	
x	3	4	8	13	19		
	4	7	12	18			
	5	11	17				
	6	16					

(top row label: y)

23. You are in a room 30 feet long with a heater at one end. In the morning the room is $65°$F. You turn on the heater, which quickly warms up to $85°$F. Let $H(x, t)$ be the temperature x feet from the heater, t minutes after the heater is turned on. Figure 12.91 shows the contour diagram for H. How warm is it 10 feet from the heater 5 minutes after it was turned on? 10 minutes after it was turned on?

t (minutes)

Figure 12.91

24. Using the contour diagram in Figure 12.91, sketch the graphs of the one-variable functions $H(x, 5)$ and $H(x, 20)$. Interpret the two graphs in practical terms, and explain the difference between them.

25. Describe in words the level surfaces of the function $g(x, y, z) = \cos(x + y + z)$.

26. **(a)** Sketch the level curves of $z = \cos \sqrt{x^2 + y^2}$.

   (b) Sketch a cross-section through the surface $z = \cos \sqrt{x^2 + y^2}$ in the plane containing the x- and z-axes. Put units on your axes.

   (c) Sketch the cross-section through the surface $z = \cos \sqrt{x^2 + y^2}$ in the plane containing the z-axis and the line $y = x$ in the xy-plane.

27. **(a)** Sketch level curves of $g(x, y) = \sqrt{x^2 + y^2} - x$ for $g = 1, 2, 3$.

   (b) For what values of c can level curves of $g = c$ be drawn?

Represent the surfaces in Problems 28–31 as graphs of functions, $f(x, y)$, and as level surfaces of the form $g(x, y, z) = c$. (There are many possible answers.)

28. Paraboloid obtained by shifting $z = x^2 + y^2$ vertically 5 units

29. Plane with intercepts $x = 2$, $y = 3$, $z = 4$.

30. Upper half of unit sphere centered at the origin

31. Lower half of sphere of radius 2 centered at $(3, 0, 0)$.

Problems 32–35 concern a vibrating guitar string. Snapshots of the guitar string at millisecond intervals are shown in Figure 12.92.

Figure 12.92: A vibrating guitar string: $f(x, t) = \cos t \sin x$ for four t values.

The guitar string is stretched tight along the x-axis from $x = 0$ to $x = \pi$. Each point on the string has an x-value, $0 \leq x \leq \pi$. As the string vibrates, each point on the string moves back and forth on either side of the x-axis. Let $y = f(x, t)$ be the displacement at time t of the point on the string located x units from the left end. A possible formula is

$$y = f(x, t) = \cos t \sin x, \quad 0 \leq x \leq \pi, \quad t \text{ in milliseconds.}$$

32. Explain what the functions $f(x, 0)$ and $f(x, 1)$ represent in terms of the vibrating string.

33. Explain what the functions $f(0, t)$ and $f(1, t)$ represent in terms of the vibrating string.

34. **(a)** Sketch graphs of y versus x for fixed t values, $t = 0$, $\pi/4$, $\pi/2$, $3\pi/4$, π.

   (b) Use your graphs to explain why this function could represent a vibrating guitar string.

35. Describe the motion of the guitar strings whose displacements are given by the following:

   (a) $y = g(x, t) = \cos 2t \sin x$

   (b) $y = h(x, t) = \cos t \sin 2x$

For Problems 36–39, use a computer or calculator to sketch the graph of a function with the given shapes. Include the axes and the equation used to generate it in your sketch.

36. A bowl which opens upward and has its vertex at 5 on the z-axis.

37. A plane which has its x, y, and z intercepts all positive.

38. A parabolic cylinder opening upward from along the line $y = x$ in the xy-plane.

39. A cone of circular cross-section opening downward and with its vertex at the origin.

CAS Challenge Problems

40. Let $A = (0, 0, 0)$ and $B = (2, 0, 0)$.

(a) Find a point C in the xy-plane that is a distance 2 from both A and B.

(b) Find a point D in 3-space that is a distance 2 from each of A, B, and C.

(c) Describe the figure obtained by joining A, B, C, and D with straight lines.

41. Let $f(x, y) = 3 + x + 2y$.

(a) Find formulas for $f(x, f(x, y)), f(x, f(x, f(x, y)))$ by hand.

(b) Consider $f(x, f(x, f(x, f(x, f(x, f(x, y))))))$. Conjecture a formula for this function and check your answer with a computer algebra system.

42. A function $f(x, y, z)$ has the property that $f(1, 0, 1) = 20$, $f(1, 1, 1) = 16$, and $f(1, 1, 2) = 21$.

(a) Estimate $f(1, 1, 3)$ and $f(1, 2, 1)$, assuming f is a linear function of each variable with the other variables held fixed.

(b) Suppose in fact that $f(x, y, z) = ax^2 + byz + czx^3 + d2^{x-y}$, for constants a, b, c and d. Which of your estimates in part (a) do you expect to be exact?

(c) Suppose in addition that $f(0, 0, 1) = 6$. Find an exact formula for f by solving for a, b, c, and d.

(d) Use the formula in part (c) to evaluate $f(1, 1, 3)$ and $f(1, 2, 1)$ exactly. Do the values confirm your answer to part (b)?

CHECK YOUR UNDERSTANDING

Are the statements in Problems 1–60 true or false? Give reasons for your answer.

1. If $f(x, y)$ is a function of two variables defined for all x and y, then $f(10, y)$ is a function of one variable.

2. The volume V of a box of height h and square base of side length s is a function of h and s.

3. If $H = f(t, d)$ is the function giving the water temperature $H°C$ of a lake at time t hours after midnight and depth d meters, then t is a function of d and H.

4. A table for a function $f(x, y)$ cannot have any values of f appearing twice.

5. The function given by the formula $f(v, w) = e^v/w$ is an increasing function of v when w is a nonzero constant.

6. If $f(x)$ and $g(y)$ are both functions of a single variable, then the product $f(x) \cdot g(y)$ is a function of two variables.

7. Two isotherms representing distinct temperatures on a weather map cannot intersect.

8. On a weather map, there can be two isotherms which represent the same temperature but that do not intersect.

9. A function $f(x, y)$ can be an increasing function of x with y held fixed, and be a decreasing function of y with x held fixed.

10. A function $f(x, y)$ can have the property that $g(x) = f(x, 5)$ is increasing, whereas $h(x) = f(x, 10)$ is decreasing.

11. The plane $x + 2y - 3z = 1$ passes through the origin.

12. The point $(1, 2, 3)$ lies above the plane $z = 2$.

13. The graph of the equation $z = 2$ is a plane parallel to the xz-plane.

14. The points $(1, 0, 1)$ and $(0, -1, 1)$ are the same distance from the origin.

15. The plane $x + y + z = 3$ intersects the x-axis when $x = 3$.

16. The point $(2, -1, 3)$ lies on the graph of the sphere $(x - 2)^2 + (y + 1)^2 + (z - 3)^2 = 25$.

17. There is only one point in the yz-plane that is a distance 3 from the point $(3, 0, 0)$.

18. There is only one point in the yz-plane that is distance 5 from the point $(3, 0, 0)$.

19. The sphere $x^2 + y^2 + z^2 = 10$ intersects the plane $x = 10$.

20. If the point $(0, b, 0)$ has distance 4 from the plane $y = 0$, then b must be 4.

21. The cross-section of the function $f(x, y) = x + y^2$ with $y = 1$ is a line.

22. The function $g(x, y) = 1 - y^2$ has identical parabolas for all cross-sections with $x = c$.

23. The function $g(x, y) = 1 - y^2$ has lines for all cross-sections with $y = c$.

24. The graphs of $f(x, y) = \sin(xy)$ and $g(x, y) = \sin(xy) + 2$ never intersect.

25. The graphs of $f(x, y) = x^2 + y^2$ and $g(x, y) = 1 - x^2 - y^2$ intersect in a circle.

26. If all of the $x = c$ cross-sections of the graph of $f(x, y)$ are lines, then the graph of f is a plane.

27. The only point of intersection of the graphs of $f(x, y)$ and $-f(x, y)$ is the origin.

28. A line parallel to the z-axis can intersect the graph of $f(x, y)$ at most once.

29. A line parallel to the y-axis can intersect the graph of $f(x, y)$ at most once.

30. The point $(0, 0, 10)$ is the highest point on the graph of the function $f(x, y) = 10 - x^2 - y^2$.

31. The contours of the function $f(x, y) = y^2 + (x - 2)^2$ are either circles or a single point.

32. Two contours of $f(x, y)$ with different heights never intersect.

33. If the contours of $g(x, y)$ are concentric circles, then the graph of g is a cone.

34. If the contours for $f(x, y)$ get closer together in a certain direction, then f is increasing in that direction.

35. If all of the contours of $f(x, y)$ are parallel lines, then the graph of f is a plane.

36. The $h = 1$ contour of the function $h(x, y) = xy$ is a hyperbola.

37. If the $f = 10$ contour of the function $f(x, y)$ is identical to the $g = 10$ contour of the function $g(x, y)$, then $f(x, y) = g(x, y)$ for all (x, y).

38. A function $f(x, y)$ has a contour $f = c$ for every value of c.

39. The $f = 5$ contour of the function $f(x, y)$ is identical to the $g = 0$ contour of the function $g(x, y) = f(x, y) - 5$.

40. Contours of $f(x, y) = 3x + 2y$ are lines with slope 3.

41. If the contours of f are all parallel lines, then f is linear.

42. If f is linear, then the contours of f are parallel lines.

43. The function f satisfying $f(0, 0) = 1, f(0, 1) = 4, f(0, 3) = 5$ cannot be linear.

44. The graph of a linear function is always a plane.

45. The cross-section $x = c$ of a linear function $f(x, y)$ is always a line.

46. There is no linear function $f(x, y)$ with a graph parallel to the xy-plane.

47. There is no linear function $f(x, y)$ with a graph parallel to the xz-plane.

48. A linear function $f(x, y) = 2x + 3y - 5$, has exactly one point (a, b) satisfying $f(a, b) = 0$.

49. In a table of values of a linear function, the columns have the same slope as the rows.

50. There is exactly one linear function $f(x, y)$ whose $f = 0$ contour is $y = 2x + 1$.

51. The graph of the function $f(x, y) = x^2 + y^2$ is the same as the level surface $g(x, y, z) = x^2 + y^2 - z = 0$.

52. The graph of $f(x, y) = \sqrt{1 - x^2 - y^2}$ is the same as the level surface $g(x, y, z) = x^2 + y^2 + z^2 = 1$.

53. Any surface which is the graph of a two-variable function $f(x, y)$ can also be represented as the level surface of a three-variable function $g(x, y, z)$.

54. Any surface which is the level surface of a three-variable function $g(x, y, z)$ can also be represented as the graph of a two-variable function $f(x, y)$.

55. The level surfaces of the function $g(x, y, z) = x + 2y + z$ are parallel planes.

56. The level surfaces of $g(x, y, z) = x^2 + y + z^2$ are cylinders with axis along the y-axis.

57. A level surface of a function $g(x, y, z)$ cannot be a single point.

58. If $g(x, y, z) = ax + by + cz + d$, where a, b, c, d are nonzero constants, then the level surfaces of g are planes.

59. If the level surfaces of g are planes, then $g(x, y, z) = ax + by + cz + d$, where a, b, c, d are constants.

60. If the level surfaces $g(x, y, z) = k_1$ and $g(x, y, z) = k_2$ are the same surface, then $k_1 = k_2$.

PROJECTS FOR CHAPTER TWELVE

1. A Heater in a Room

Figure 12.93 shows the contours of the temperature along one wall of a heated room through one winter day, with time indicated as on a 24-hour clock. The room has a heater located at the left-most corner of the wall and one window in the wall. The heater is controlled by a thermostat about 2 feet from the window.

(a) Where is the window?

(b) When is the window open?

(c) When is the heat on?

(d) Draw graphs of the temperature along the wall of the room at 6 am, at 11 am, at 3 pm (15 hours) and at 5 pm (17 hours).

(e) Draw a graph of the temperature as a function of time at the heater, at the window and midway between them.

(f) The temperature at the window at 5 pm (17 hours) is less than at 11 am. Why do you think this might be?

(g) To what temperature do you think the thermostat is set? How do you know?

(h) Where is the thermostat?

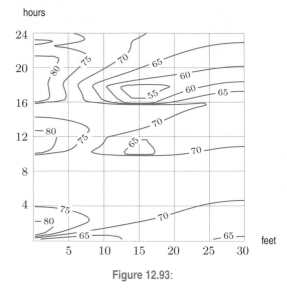

hours

Figure 12.93:

2. Light in a Wave-guide

Figure 12.94 shows the contours of light intensity as a function of location and time in a microscopic wave-guide.

time (nanoseconds)

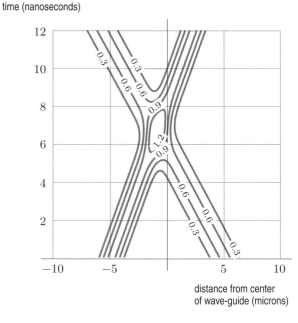

distance from center
of wave-guide (microns)

Figure 12.94

(a) Draw graphs showing intensity as a function of location at times 0, 2, 4, 6, 8, and 10 nanoseconds.
(b) If you could create an animation showing how the graph of intensity as a function of location varies as time passes, what would it look like?
(c) Draw a graph of intensity as a function of time at locations −5, 0, and 5 microns from center of wave-guide.
(d) Describe what the light beams are doing in this wave-guide.

Chapter Thirteen

A FUNDAMENTAL TOOL: VECTORS

In one-variable calculus we represented quantities such as velocity by numbers. However, to specify the velocity of a moving object in space, we need to say how fast it is moving and in what direction it is moving. In this chapter *vectors* are used to represent quantities that have direction as well as magnitude.

13.1 DISPLACEMENT VECTORS

Suppose you are a pilot planning a flight from Dallas to Pittsburgh. There are two things you must know: the distance to be traveled (so you have enough fuel to make it) and in what direction to go (so you don't miss Pittsburgh). Both these quantities together specify the displacement or *displacement vector* between the two cities.

> The **displacement vector** from one point to another is an arrow with its tail at the first point and its tip at the second. The **magnitude** (or length) of the displacement vector is the distance between the points, and is represented by the length of the arrow. The **direction** of the displacement vector is the direction of the arrow.

Figure 13.1 shows a map with the displacement vectors from Dallas to Pittsburgh, from Albuquerque to Oshkosh, and from Los Angeles to Buffalo, SD. These displacement vectors have the same length and the same direction. We say that the displacement vectors between the corresponding cities are the same, even though they do not coincide. In other words

> Displacement vectors which point in the same direction and have the same magnitude are considered to be the same, even if they do not coincide.

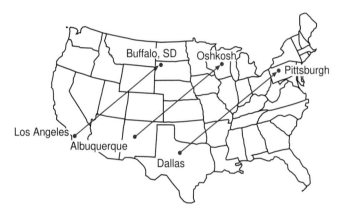

Figure 13.1: Displacement vectors between cities

Notation and Terminology

The displacement vector is our first example of a vector. Vectors have both magnitude and direction; in comparison, a quantity specified only by a number, but no direction, is called a *scalar*.[1] For instance, the time taken by the flight from Dallas to Pittsburgh is a scalar quantity. Displacement is a vector since it requires both distance and direction to specify it.

In this book, vectors are written with an arrow over them, $\vec{v}$, to distinguish them from scalars. Other books use a bold **v** to denote a vector. We use the notation $\overrightarrow{PQ}$ to denote the displacement vector from a point P to a point Q. The magnitude, or length, of a vector $\vec{v}$ is written $\|\vec{v}\|$.

Addition and Subtraction of Displacement Vectors

Suppose NASA commands a robot on Mars to move 75 meters in one direction and then 50 meters in another direction. (See Figure 13.2.) Where does the robot end up? Suppose the displacements are represented by the vectors $\vec{v}$ and $\vec{w}$, respectively. Then the sum $\vec{v} + \vec{w}$ gives the final position.

---

[1] So named by W. R. Hamilton because they are merely numbers on the *scale* from $-\infty$ to ∞.

The **sum**, $\vec{v} + \vec{w}$, of two vectors $\vec{v}$ and $\vec{w}$ is the combined displacement resulting from first applying $\vec{v}$ and then $\vec{w}$. (See Figure 13.3.) The sum $\vec{w} + \vec{v}$ gives the same displacement.

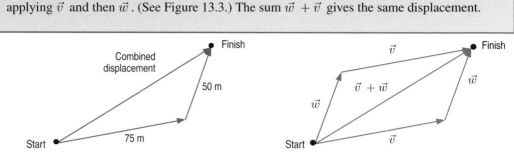

Figure 13.2: Sum of displacements of robots on Mars   Figure 13.3: The sum $\vec{v} + \vec{w} = \vec{w} + \vec{v}$

Suppose two different robots start from the same location. One moves along a displacement vector $\vec{v}$ and the second along a displacement vector $\vec{w}$. What is the displacement vector, $\vec{x}$, from the first robot to the second? (See Figure 13.4.) Since $\vec{v} + \vec{x} = \vec{w}$, we define $\vec{x}$ to be the difference $\vec{x} = \vec{w} - \vec{v}$. In other words, $\vec{w} - \vec{v}$ gets you from $\vec{v}$ to $\vec{w}$.

The **difference**, $\vec{w} - \vec{v}$, is the displacement vector which, when added to $\vec{v}$, gives $\vec{w}$. That is, $\vec{w} = \vec{v} + (\vec{w} - \vec{v})$. (See Figure 13.4.)

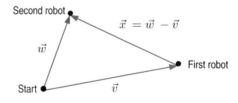

Figure 13.4: The difference $\vec{w} - \vec{v}$

If the robot ends up where it started, then its total displacement vector is the *zero vector*, $\vec{0}$. The zero vector has no direction.

The **zero vector**, $\vec{0}$, is a displacement vector with zero length.

Scalar Multiplication of Displacement Vectors

If $\vec{v}$ represents a displacement vector, the vector $2\vec{v}$ represents a displacement of twice the magnitude in the same direction as $\vec{v}$. Similarly, $-2\vec{v}$ represents a displacement of twice the magnitude in the opposite direction. (See Figure 13.5.)

Figure 13.5: Scalar multiples of the vector $\vec{v}$

> If λ is a scalar and $\vec{v}$ is a displacement vector, the **scalar multiple of $\vec{v}$ by** λ, written $\lambda\vec{v}$, is the displacement vector with the following properties:
> - The displacement vector $\lambda\vec{v}$ is parallel to $\vec{v}$, pointing in the same direction if $\lambda > 0$, and in the opposite direction if $\lambda < 0$.
> - The magnitude of $\lambda\vec{v}$ is $|\lambda|$ times the magnitude of $\vec{v}$, that is, $\|\lambda\vec{v}\| = |\lambda| \|\vec{v}\|$.

Note that $|\lambda|$ represents the absolute value of the scalar λ while $\|\lambda\vec{v}\|$ represents the magnitude of the vector $\lambda\vec{v}$.

Example 1   Explain why $\vec{w} - \vec{v} = \vec{w} + (-1)\vec{v}$.

Solution   The vector $(-1)\vec{v}$ has the same magnitude as $\vec{v}$, but points in the opposite direction. Figure 13.6 shows that the combined displacement $\vec{w} + (-1)\vec{v}$ is the same as the displacement $\vec{w} - \vec{v}$.

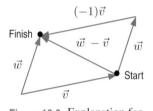

Figure 13.6: Explanation for why $\vec{w} - \vec{v} = \vec{w} + (-1)\vec{v}$

Parallel Vectors

Two vectors $\vec{v}$ and $\vec{w}$ are *parallel* if one is a scalar multiple of the other, that is, if $\vec{w} = \lambda\vec{v}$, for some scalar λ.

Components of Displacement Vectors: The Vectors $\vec{i}$, $\vec{j}$, and $\vec{k}$

Suppose that you live in a city with equally spaced streets running east-west and north-south and that you want to tell someone how to get from one place to another. You'd be likely to tell them how many blocks east-west and how many blocks north-south to go. For example, to get from P to Q in Figure 13.7, we go 4 blocks east and 1 block south. If $\vec{i}$ and $\vec{j}$ are as shown in Figure 13.7, then the displacement vector from P to Q is $4\vec{i} - \vec{j}$.

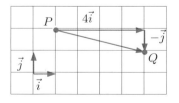

Figure 13.7: The displacement vector from P to Q is $4\vec{i} - \vec{j}$

We extend the same idea to 3-dimensions. First we choose a Cartesian system of coordinate axes. The three vectors of length 1 shown in Figure 13.8 are the vector $\vec{i}$, which points along the positive x-axis, the vector $\vec{j}$, along the positive y-axis, and the vector $\vec{k}$, along the positive z-axis.

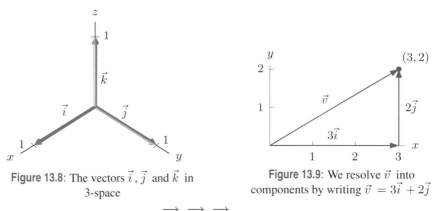

Figure 13.8: The vectors $\vec{i}$, $\vec{j}$ and $\vec{k}$ in 3-space

Figure 13.9: We resolve $\vec{v}$ into components by writing $\vec{v} = 3\vec{i} + 2\vec{j}$

Writing Displacement Vectors Using $\overrightarrow{i}$, $\overrightarrow{j}$, $\overrightarrow{k}$

Any displacement in 3-space or the plane can be expressed as a combination of displacements in the coordinate directions. For example, Figure 13.9 shows that the displacement vector $\vec{v}$ from the origin to the point $(3, 2)$ can be written as a sum of displacement vectors along the x and y-axes:

$$\vec{v} = 3\vec{i} + 2\vec{j}.$$

This is called *resolving $\vec{v}$ into components*. In general:

We **resolve** $\vec{v}$ into components by writing $\vec{v}$ in the form

$$\vec{v} = v_1\vec{i} + v_2\vec{j} + v_3\vec{k}.$$

We call $v_1\vec{i}$, $v_2\vec{j}$, and $v_3\vec{k}$ the **components** of $\vec{v}$.

An Alternative Notation for Vectors

Many people write a vector in 3-dimensions as a string of three numbers, that is, as

$$\vec{v} = (v_1, v_2, v_3) \quad \text{instead of} \quad \vec{v} = v_1\vec{i} + v_2\vec{j} + v_3\vec{k}.$$

Since the first notation can be confused with a point and the second cannot, we usually use the second form.

Example 2 Resolve the displacement vector, $\vec{v}$, from the point $P_1 = (2, 4, 10)$ to the point $P_2 = (3, 7, 6)$ into components.

Solution To get from P_1 to P_2, we move 1 unit in the positive x-direction, 3 units in the positive y-direction, and 4 units in the negative z-direction. Hence $\vec{v} = \vec{i} + 3\vec{j} - 4\vec{k}$.

Example 3 Decide whether the vector $\vec{v} = 2\vec{i} + 3\vec{j} + 5\vec{k}$ is parallel to each of the following vectors:
$$\vec{w} = 4\vec{i} + 6\vec{j} + 10\vec{k}, \quad \vec{a} = -\vec{i} - 1.5\vec{j} - 2.5\vec{k}, \quad \vec{b} = 4\vec{i} + 6\vec{j} + 9\vec{k}.$$

Solution Since $\vec{w} = 2\vec{v}$ and $\vec{a} = -0.5\vec{v}$, the vectors $\vec{v}$, $\vec{w}$, and $\vec{a}$ are parallel. However, $\vec{b}$ is not a multiple of $\vec{v}$ (since, for example, $4/2 \neq 9/5$), so $\vec{v}$ and $\vec{b}$ are not parallel.

In general, Figure 13.10 shows us how to express the displacement vector between two points in components:

Components of Displacement Vectors

The displacement vector from the point $P_1 = (x_1, y_1, z_1)$ to the point $P_2 = (x_2, y_2, z_2)$ is given in components by

$$\overrightarrow{P_1 P_2} = (x_2 - x_1)\vec{i} + (y_2 - y_1)\vec{j} + (z_2 - z_1)\vec{k}.$$

Position Vectors: Displacement of a Point from the Origin

A displacement vector whose tail is at the origin is called a *position vector*. Thus, any point (x_0, y_0, z_0) in space has associated with it the position vector $\vec{r}_0 = x_0\vec{i} + y_0\vec{j} + z_0\vec{k}$. (See Figure 13.11.) In general, a position vector gives the displacement of a point from the origin.

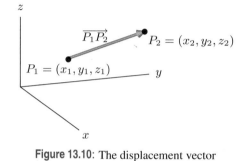

Figure 13.10: The displacement vector
$\overrightarrow{P_1 P_2} = (x_2 - x_1)\vec{i} + (y_2 - y_1)\vec{j} + (z_2 - z_1)\vec{k}$

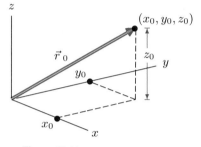

Figure 13.11: The position vector
$\vec{r}_0 = x_0\vec{i} + y_0\vec{j} + z_0\vec{k}$

The Components of the Zero Vector

The zero displacement vector has magnitude equal to zero and is written $\vec{0}$. So $\vec{0} = 0\vec{i} + 0\vec{j} + 0\vec{k}$.

The Magnitude of a Vector in Components

For a vector, $\vec{v} = v_1\vec{i} + v_2\vec{j}$, the Pythagorean theorem is used to find its magnitude, $\|\vec{v}\|$. (See Figure 13.12.)

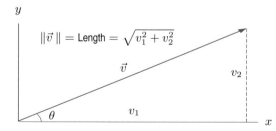

Figure 13.12: Magnitude, $\|\vec{v}\|$, of a 2-dimensional vector, $\vec{v}$

In 3-dimensions, for a vector $\vec{v} = v_1\vec{i} + v_2\vec{j} + v_3\vec{k}$, we have

---

Magnitude of $\vec{v}$ $= \|\vec{v}\| =$ Length of the arrow $= \sqrt{v_1^2 + v_2^2 + v_3^2}.$

For instance, if $\vec{v} = 3\vec{i} - 4\vec{j} + 5\vec{k}$, then $\|\vec{v}\| = \sqrt{3^2 + (-4)^2 + 5^2} = \sqrt{50}$.

Addition and Scalar Multiplication of Vectors in Components

Suppose the vectors $\vec{v}$ and $\vec{w}$ are given in components:

$$\vec{v} = v_1\vec{i} + v_2\vec{j} + v_3\vec{k} \quad \text{and} \quad \vec{w} = w_1\vec{i} + w_2\vec{j} + w_3\vec{k}.$$

Then

$$\vec{v} + \vec{w} = (v_1 + w_1)\vec{i} + (v_2 + w_2)\vec{j} + (v_3 + w_3)\vec{k},$$

and

$$\lambda\vec{v} = \lambda v_1\vec{i} + \lambda v_2\vec{j} + \lambda v_3\vec{k}.$$

Figures 13.13 and 13.14 illustrate these properties in two dimensions. Finally, $\vec{v} - \vec{w} = \vec{v} + (-1)\vec{w}$, so we can write $\vec{v} - \vec{w} = (v_1 - w_1)\vec{i} + (v_2 - w_2)\vec{j} + (v_3 - w_3)\vec{k}$.

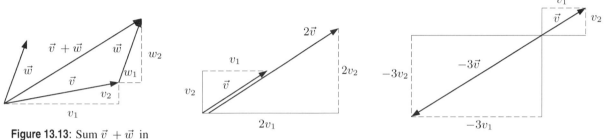

Figure 13.13: Sum $\vec{v} + \vec{w}$ in components

Figure 13.14: Scalar multiples of vectors showing $\vec{v}$, $2\vec{v}$, and $-3\vec{v}$

How to Resolve a Vector into Components

You may wonder how we find the components of a 2-dimensional vector, given its length and direction. Suppose the vector $\vec{v}$ has length v and makes an angle of θ with the x-axis, measured counterclockwise, as in Figure 13.15. If $\vec{v} = v_1\vec{i} + v_2\vec{j}$, Figure 13.15 shows that

$$v_1 = v\cos\theta \quad \text{and} \quad v_2 = v\sin\theta.$$

Thus, we resolve $\vec{v}$ into components by writing

$$\vec{v} = (v\cos\theta)\vec{i} + (v\sin\theta)\vec{j}.$$

Vectors in 3-space are resolved using direction cosines; see Problem 37 on page 681.

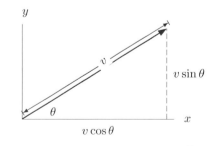

Figure 13.15: Resolving a vector: $\vec{v} = (v\cos\theta)\vec{i} + (v\sin\theta)\vec{j}$

Example 4   Resolve $\vec{v}$ into components if $\|\vec{v}\| = 2$ and $\theta = \pi/6$.

Solution   We have $\vec{v} = 2\cos(\pi/6)\vec{i} + 2\sin(\pi/6)\vec{j} = 2(\sqrt{3}/2)\vec{i} + 2(1/2)\vec{j} = \sqrt{3}\vec{i} + \vec{j}$.

Unit Vectors

A *unit vector* is a vector whose magnitude is 1. The vectors $\vec{i}$, $\vec{j}$, and $\vec{k}$ are unit vectors in the directions of the coordinate axes. It is often helpful to find a unit vector in the same direction as a given vector $\vec{v}$. Suppose that $\|\vec{v}\| = 10$; a unit vector in the same direction as $\vec{v}$ is $\vec{v}/10$. In general, a unit vector in the direction of any nonzero vector $\vec{v}$ is

$$\vec{u} = \frac{\vec{v}}{\|\vec{v}\|}.$$

Example 5   Find a unit vector, $\vec{u}$, in the direction of the vector $\vec{v} = \vec{i} + 3\vec{j}$.

Solution   If $\vec{v} = \vec{i} + 3\vec{j}$, then $\|\vec{v}\| = \sqrt{1^2 + 3^2} = \sqrt{10}$. Thus, a unit vector in the same direction is given by

$$\vec{u} = \frac{\vec{v}}{\sqrt{10}} = \frac{1}{\sqrt{10}}(\vec{i} + 3\vec{j}) = \frac{1}{\sqrt{10}}\vec{i} + \frac{3}{\sqrt{10}}\vec{j} \approx 0.32\vec{i} + 0.95\vec{j}.$$

Example 6   Find a unit vector at the point (x, y, z) that points radially outward away from the origin.

Solution   The vector from the origin to (x, y, z) is the position vector

$$\vec{r} = x\vec{i} + y\vec{j} + z\vec{k}.$$

Thus, if we put its tail at (x, y, z) it will point away from the origin. Its magnitude is

$$\|\vec{r}\| = \sqrt{x^2 + y^2 + z^2},$$

so a unit vector pointing in the same direction is

$$\frac{\vec{r}}{\|\vec{r}\|} = \frac{x\vec{i} + y\vec{j} + z\vec{k}}{\sqrt{x^2 + y^2 + z^2}} = \frac{x}{\sqrt{x^2 + y^2 + z^2}}\vec{i} + \frac{y}{\sqrt{x^2 + y^2 + z^2}}\vec{j} + \frac{z}{\sqrt{x^2 + y^2 + z^2}}\vec{k}.$$

Exercises and Problems for Section 13.1

Exercises

For Exercises 1–4, perform the indicated computation.

1. $(4\vec{i} + 2\vec{j}) - (3\vec{i} - \vec{j})$
2. $(\vec{i} + 2\vec{j}) + (-3)(2\vec{i} + \vec{j})$
3. $-4(\vec{i} - 2\vec{j}) - 0.5(\vec{i} - \vec{k})$
4. $2(0.45\vec{i} - 0.9\vec{j} - 0.01\vec{k}) - 0.5(1.2\vec{i} - 0.1\vec{k})$

Find the length of the vectors in Exercises 5–8.

5. $\vec{v} = \vec{i} - \vec{j} + 3\vec{k}$ 6. $\vec{v} = \vec{i} - \vec{j} + 2\vec{k}$

7. $\vec{v} = 1.2\vec{i} - 3.6\vec{j} + 4.1\vec{k}$
8. $\vec{v} = 7.2\vec{i} - 1.5\vec{j} + 2.1\vec{k}$

For Exercises 9–14, perform the indicated operations on the following vectors:

$$\vec{a} = 2\vec{j} + \vec{k}, \quad \vec{b} = -3\vec{i} + 5\vec{j} + 4\vec{k}, \quad \vec{c} = \vec{i} + 6\vec{j},$$
$$\vec{x} = -2\vec{i} + 9\vec{j}, \quad \vec{y} = 4\vec{i} - 7\vec{j}, \quad \vec{z} = \vec{i} - 3\vec{j} - \vec{k}.$$

9. $5\vec{b}$ 10. $\vec{a} + \vec{z}$ 11. $2\vec{c} + \vec{x}$
12. $\|\vec{z}\|$ 13. $\|\vec{y}\|$ 14. $2\vec{a} + 7\vec{b} - 5\vec{z}$

15. **(a)** Draw the position vector for $\vec{v} = 5\vec{i} - 7\vec{j}$.
    (b) What is $\|\vec{v}\|$?
    (c) Find the angle between $\vec{v}$ and the positive x-axis.

16. Resolve vector $\vec{v}$ into components if $\|\vec{v}\| = 8$ and the direction of $\vec{v}$ is shown in Figure 13.16.

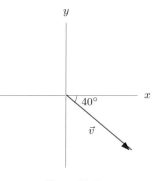

Figure 13.16

17. Resolve the vectors in Figure 13.17 into components.

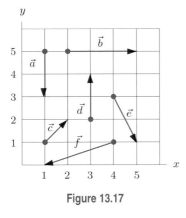

Figure 13.17

Resolve the vectors in Exercises 18–20 into components.

18. A vector starting at the point $Q = (4, 6)$ and ending at the point $P = (1, 2)$.

19.   **20.**

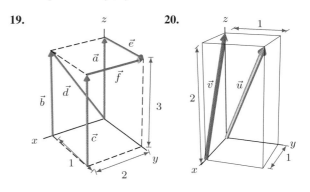

Problems

21. Which of the following vectors are parallel?

$$\vec{u} = 2\vec{i} + 4\vec{j} - 2\vec{k}, \quad \vec{p} = \vec{i} + \vec{j} + \vec{k},$$
$$\vec{v} = \vec{i} - \vec{j} + 3\vec{k}, \quad \vec{q} = 4\vec{i} - 4\vec{j} + 12\vec{k},$$
$$\vec{w} = -\vec{i} - 2\vec{j} + \vec{k}, \quad \vec{r} = \vec{i} - \vec{j} + \vec{k}.$$

22. Find a vector with length 2 that points in the same direction as $\vec{i} - \vec{j} + 2\vec{k}$.

23. **(a)** Find a unit vector from the point $P = (1, 2)$ and toward the point $Q = (4, 6)$.
    (b) Find a vector of length 10 pointing in the same direction.

24. Find all vectors $\vec{v}$ in 2 dimensions having $\|\vec{v}\| = 5$ and the $\vec{i}$-component of $\vec{v}$ is $3\vec{i}$.

25. In Figure 13.18, a ship is at point B, a submarine is 6 units below point C, a helicopter is 10 units above point A.

    (a) Find each of the following displacement vectors: from the submarine to the ship, from the helicopter to the ship, and from the submarine to the helicopter.
    (b) Which two objects are closest together?

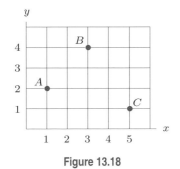

Figure 13.18

A cat on the ground at the point $(1, 4, 0)$ watches a squirrel at the top of a tree. The tree is one unit high with its base at $(2, 4, 0)$. Find the displacement vectors in Problems 26–29.

26. From the origin to the cat.

27. From the bottom of the tree to the squirrel.

28. From the bottom of the tree to the cat.

29. From the cat to the squirrel.

30. Figure 13.19 shows a rectangular box containing several vectors. Are the following statements true or false? Explain.

    (a) $\vec{c} = \vec{f}$ **(b)** $\vec{a} = \vec{d}$ **(c)** $\vec{a} = -\vec{b}$
    (d) $\vec{g} = \vec{f} + \vec{a}$ **(e)** $\vec{e} = \vec{a} - \vec{b}$ **(f)** $\vec{d} = \vec{g} - \vec{c}$

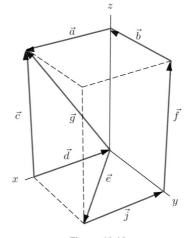

Figure 13.19

31. A truck is traveling due north at 30 km/hr approaching a crossroad. On a perpendicular road a police car is traveling west toward the intersection at 40 km/hr. Both vehicles will reach the crossroad in exactly one hour. Find the vector currently representing the displacement of the truck with respect to the police car.

32. Show that the medians of a triangle intersect at a point $\frac{1}{3}$ of the way along each median from the side it bisects.

33. Show that the lines joining the centroid (the intersection point of the medians) of a face of the tetrahedron and the opposite vertex meet at a point $\frac{1}{4}$ of the way from each centroid to its opposite vertex.

34. Figure 13.20 shows a molecule with four atoms at O, A, B and C. Verify that every atom in the molecule is 2 units away from every other atom.

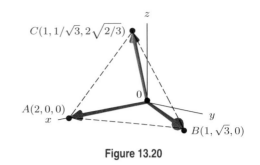

Figure 13.20

13.2 VECTORS IN GENERAL

Besides displacement, there are many quantities that have both magnitude and direction and are added and multiplied by scalars in the same way as displacements. Any such quantity is called a *vector* and is represented by an arrow in the same manner we represent displacements. The length of the arrow is the *magnitude* of the vector, and the direction of the arrow is the direction of the vector.

Velocity Versus Speed

The speed of a moving body tells us how fast it is moving, say 80 km/hr. The speed is just a number; it is therefore a scalar. The velocity, on the other hand, tells us both how fast the body is moving and the direction of motion; it is a vector. For instance, if a car is heading northeast at 80 km/hr, then its velocity is a vector of length 80 pointing northeast.

> The **velocity vector** of a moving object is a vector whose magnitude is the speed of the object and whose direction is the direction of its motion.

Example 1 A car is traveling north at a speed of 100 km/hr, while a plane above is flying horizontally south-west at a speed of 500 km/hr. Draw the velocity vectors of the car and the plane.

Solution Figure 13.21 shows the velocity vectors. The plane's velocity vector is five times as long as the car's, because its speed is five times as great.

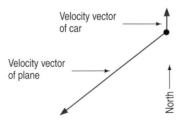

Figure 13.21: Velocity vector of the car is 100 km/hr north and of the plane is 500 km/hr south-west

The next example illustrates that the velocity vectors for two motions add to give the velocity vector for the combined motion, just as displacements do.

Example 2   A riverboat is moving with velocity $\vec{v}$ and a speed of 8 km/hr relative to the water. In addition, the river has a current $\vec{c}$ and a speed of 1 km/hr. (See Figure 13.22.) What is the physical significance of the vector $\vec{v} + \vec{c}$?

$\vec{c}$ = Velocity of current
$\|\vec{c}\| = 1$ km/hr

$\vec{v} + \vec{c}$

$\vec{v}$ = Velocity relative to water
$\|\vec{v}\| = 8$ km/hr

Figure 13.22: Boat's velocity relative to the river bed is the sum, $\vec{v} + \vec{c}$

Solution   The vector $\vec{v}$ shows how the boat is moving relative to the water, while $\vec{c}$ shows how the water is moving relative to the riverbed. During an hour, imagine that the boat first moves 8 km relative to the water, which remains still; this displacement is represented by $\vec{v}$. Then imagine the water moving 1 km while the boat remains stationary relative to the water; this displacement is represented by $\vec{c}$. The combined displacement is represented by $\vec{v} + \vec{c}$. Thus, the vector $\vec{v} + \vec{c}$ is the velocity of the boat relative to the riverbed.

Note that the effective speed of the boat is not necessarily 9 km/hr unless the boat is moving in the direction of the current. Although we add the velocity vectors, we do not necessarily add their lengths.

Scalar multiplication also makes sense for velocity vectors. For example, if $\vec{v}$ is a velocity vector, then $-2\vec{v}$ represents a velocity of twice the magnitude in the opposite direction.

Example 3   A ball is moving with velocity $\vec{v}$ when it hits a wall at a right angle and bounces straight back, with its speed reduced by 20%. Express its new velocity in terms of the old one.

Solution   The new velocity is $-0.8\vec{v}$, where the negative sign expresses the fact that the new velocity is in the direction opposite to the old.

We can represent velocity vectors in components in the same way we did on page 655.

Example 4   Represent the velocity vectors of the car and the plane in Example 1 using components. Take north to be the positive y-axis, east to be the positive x-axis, and upward to be the positive z-axis.

Solution   The car is traveling north at 100 km/hr, so the y-component of its velocity is $100\vec{j}$ and the x-component is $0\vec{i}$. Since it is traveling horizontally, the z-component is $0\vec{k}$. So we have

$$\text{Velocity of car} = 0\vec{i} + 100\vec{j} + 0\vec{k} = 100\vec{j}.$$

The plane's velocity vector also has $\vec{k}$ component equal to zero. Since it is traveling southwest, its $\vec{i}$ and $\vec{j}$ components have negative coefficients (north and east are positive). Since the plane is traveling at 500 km/hr, in one hour it is displaced $500/\sqrt{2} \approx 354$ km to the west and 354 km to the south. (See Figure 13.23.) Thus,

Distance traveled by the car in one hour — 100

Distance traveled by the plane in one hour — 500

North

$500/\sqrt{2} \approx 354$

$45°$

$500/\sqrt{2}$

Figure 13.23: Distance traveled by the plane and car in one hour

$$\text{Velocity of plane} = -(500\cos 45°)\vec{i} - (500\sin 45°)\vec{j} \approx -354\vec{i} - 354\vec{j}.$$

Of course, if the car were climbing a hill or if the plane were descending for a landing, then the $\vec{k}$ component would not be zero.

Acceleration

Another example of a vector quantity is acceleration. Acceleration, like velocity, is specified by both a magnitude and a direction — for example, the acceleration due to gravity is 9.81 m/sec^2 vertically downward.

Force

Force is another example of a vector quantity. Suppose you push on an open door. The result depends both on how hard you push and in what direction. Thus, to specify a force we must give its magnitude (or strength) and the direction in which it is acting. For example, the gravitational force exerted on an object by the earth is a vector pointing from the object toward the center of the earth; its magnitude is the strength of the gravitational force.

Example 5   The earth travels around the sun in an ellipse. The gravitational force on the earth and the velocity of the earth are governed by the following laws:

Newton's Law of Gravitation: The magnitude of the gravitational attraction, F, between two masses m_1 and m_2 at a distance r apart is given by $F = Gm_1m_2/r^2$, where G is a constant. The force vector lies along the line between the masses.

Kepler's Second Law: The line joining a planet to the sun sweeps out equal areas in equal times.

(a) Sketch vectors representing the gravitational force of the sun on the earth at two different positions in the earth's orbit.

(b) Sketch the velocity vector of the earth at two points in its orbit.

Solution

(a) Figure 13.24 shows the earth orbiting the sun. Note that the gravitational force vector always points toward the sun and is larger when the earth is closer to the sun because of the r^2 term in the denominator. (In fact, the real orbit looks much more like a circle than we have shown here.)

(b) The velocity vector points in the direction of motion of the earth. Thus, the velocity vector is tangent to the ellipse. See Figure 13.25. Furthermore, the velocity vector is longer at points of the orbit where the planet is moving quickly, because the magnitude of the velocity vector is the speed. Kepler's Second Law enables us to determine when the earth is moving quickly and when it is moving slowly. Over a fixed period of time, say one month, the line joining the earth to the sun sweeps out a sector having a certain area. Figure 13.25 shows two sectors swept out in two different one-month time-intervals. Kepler's law says that the areas of the two sectors are the same. Thus, the earth must move farther in a month when it is close to the sun than when it is far from the sun. Therefore, the earth moves faster when it is closer to the sun and slower when it is farther away.

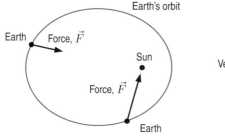

Figure 13.24: Gravitational force, $\vec{F}$, exerted by the sun on the earth: Greater magnitude closer to sun

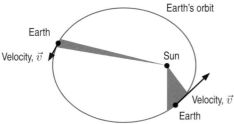

Figure 13.25: The velocity vector, $\vec{v}$, of the earth: Greater magnitude closer to the sun

Properties of Addition and Scalar Multiplication

In general, vectors add, subtract, and are multiplied by scalars in the same way as displacement vectors. Thus, for any vectors $\vec{u}$, $\vec{v}$, and $\vec{w}$ and any scalars α and β, we have the following properties:

Commutativity
1. $\vec{v} + \vec{w} = \vec{w} + \vec{v}$

Associativity
2. $(\vec{u} + \vec{v}) + \vec{w} = \vec{u} + (\vec{v} + \vec{w})$
3. $\alpha(\beta\vec{v}) = (\alpha\beta)\vec{v}$

Distributivity
4. $(\alpha + \beta)\vec{v} = \alpha\vec{v} + \beta\vec{v}$
5. $\alpha(\vec{v} + \vec{w}) = \alpha\vec{v} + \alpha\vec{w}$

Identity
6. $1 \cdot \vec{v} = \vec{v}$
7. $0 \cdot \vec{v} = \vec{0}$
8. $\vec{v} + \vec{0} = \vec{v}$
9. $\vec{w} + (-1) \cdot \vec{v} = \vec{w} - \vec{v}$

Problems 24–30 at the end of this section ask for a justification of these results in terms of displacement vectors.

Using Components

Example 6   A plane, heading due east at an airspeed of 600 km/hr, experiences a wind of 50 km/hr blowing toward the northeast. Find the plane's direction and ground speed.

Solution   We choose a coordinate system with the x-axis pointing east and the y-axis pointing north. See Figure 13.26.

The airspeed tells us the speed of the plane relative to still air. Thus, the plane is moving due east with velocity $\vec{v} = 600\vec{i}$ relative to still air. In addition, the air is moving with a velocity $\vec{w}$. Writing $\vec{w}$ in components, we have

$$\vec{w} = (50\cos45°)\vec{i} + (50\sin45°)\vec{j} = 35.4\vec{i} + 35.4\vec{j}.$$

The vector $\vec{v} + \vec{w}$ represents the displacement of the plane in one hour relative to the ground. Therefore, $\vec{v} + \vec{w}$ is the velocity of the plane relative to the ground. In components, we have

$$\vec{v} + \vec{w} = 600\vec{i} + \left(35.4\vec{i} + 35.4\vec{j}\right) = 635.4\vec{i} + 35.4\vec{j}.$$

The direction of the plane's motion relative to the ground is given by the angle θ in Figure 13.26, where

$$\tan\theta = \frac{35.4}{635.4}$$

so

$$\theta = \arctan\left(\frac{35.4}{635.4}\right) = 3.2°.$$

The ground speed is the speed of the plane relative to the ground, so

$$\text{Groundspeed} = \sqrt{635.4^2 + 35.4^2} = 636.4 \text{ km/hr}.$$

Thus, the speed of the plane relative to the ground has been increased slightly by the wind. (This is as we would expect, as the wind has a positive component in the direction in which the plane is traveling.) The angle θ shows how far the plane is blown off course by the wind.

Figure 13.26: Plane's velocity relative to the ground is the sum $\vec{v} + \vec{w}$

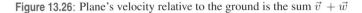

Vectors in n Dimensions

Using the alternative notation $\vec{v} = (v_1, v_2, v_3)$ for a vector in 3-space, we can define a vector in n dimensions as a string of n numbers. Thus, a vector in n dimensions can be written as

$$\vec{c} = (c_1, c_2, \ldots, c_n).$$

Addition and scalar multiplication are defined by the formulas

$$\vec{v} + \vec{w} = (v_1, v_2, \ldots, v_n) + (w_1, w_2, \ldots, w_n) = (v_1 + w_1, v_2 + w_2, \ldots, v_n + w_n)$$

and

$$\lambda\vec{v} = \lambda(v_1, v_2, \ldots, v_n) = (\lambda v_1, \lambda v_2, \ldots, \lambda v_n).$$

Why Do We Want Vectors in n Dimensions?

Vectors in two and three dimensions can be used to model displacement, velocities, or forces. But what about vectors in n dimensions? There is another interpretation of 3-dimensional vectors (or 3-vectors) which is useful: they can be thought of as listing 3 different quantities — for example, the displacements parallel to the x, y, and z axes. Similarly, the n-vector

$$\vec{c} = (c_1, c_2, \ldots, c_n)$$

can be thought of as a way of keeping n different quantities organized. For example, a *population* vector $\vec{N}$ shows the number of children and adults in a population:

$$\vec{N} = (\text{Number of children, Number of adults}),$$

or, if we are interested in a more detailed breakdown of ages, we might give the number in each ten-year age bracket in the population (up to age 110) in the form

$$\vec{N} = (N_1, N_2, N_3, N_4, \ldots, N_{10}, N_{11}),$$

where N_1 is the population aged 0–9, and N_2 is the population aged 10–19, and so on.
    A *consumption* vector ,

$$\vec{q} = (q_1, q_2, \ldots, q_n)$$

shows the quantities q_1, q_2, ..., q_n consumed of each of n different goods. A *price* vector

$$\vec{p} = (p_1, p_2, \ldots, p_n)$$

contains the prices of n different items.
    In 1907, Hermann Minkowski used vectors with four components when he introduced *space-time coordinates*, whereby each event is assigned a vector position $\vec{v}$ with four coordinates, three for its position in space and one for time:

$$\vec{v} = (x, y, z, t).$$

Example 7   Suppose the vector $\vec{I}$ represents the number of copies, in thousands, made by each of four copy centers in the month of December and $\vec{J}$ represents the number of copies made at the same four copy centers during the previous eleven months (the "year-to-date"). If $\vec{I} = (25, 211, 818, 642)$, and $\vec{J} = (331, 3227, 1377, 2570)$, compute $\vec{I} + \vec{J}$. What does this sum represent?

Solution   The sum is

$$\vec{I} + \vec{J} = (25 + 331, 211 + 3227, 818 + 1377, 642 + 2570) = (356, 3438, 2195, 3212).$$

Each term in $\vec{I} + \vec{J}$ represents the sum of the number of copies made in December plus those in the previous eleven months, that is, the total number of copies made during the entire year at that particular copy center.

Example 8   The price vector $\vec{p} = (p_1, p_2, p_3)$ represents the prices in dollars of three goods. Write a vector which gives the prices of the same goods in cents.

Solution   The prices in cents are $100p_1$, $100p_2$, and $100p_3$ respectively, so the new price vector is

$$(100p_1, 100p_2, 100p_3) = 100\vec{p}.$$

Exercises and Problems for Section 13.2

Exercises

In Exercises 1–5, say whether the given quantity is a vector or a scalar.

1. The population of the US.

2. The distance from Seattle to St. Louis.

3. The magnetic field at a point on the earth's surface.

4. The temperature at a point on the earth's surface.

5. The populations of each of the 50 states.

6. Which is traveling faster, a car whose velocity vector is $21\vec{i} + 35\vec{j}$, or a car whose velocity vector is $40\vec{i}$, assuming that the units are the same for both directions?

7. Give the components of the velocity vector of a boat which is moving at 40 km/hr in a direction $20°$ south of west. (Assume north is in the positive y-direction.)

8. Give the components of the velocity vector for wind blowing at 10 km/hr toward the southeast. (Assume north is the positive y-direction.)

9. A car is traveling at a speed of 50 km/hr. The positive y-axis is north and the positive x-axis is east. Resolve the car's velocity vector (in 2-space) into components if the car is traveling in each of the following directions:

   (a) East   (b) South
   (c) Southeast   (d) Northwest.

10. What angle does a force of $\vec{F} = 15\vec{i} + 18\vec{j}$ make with the x-axis?

Problems

11. Shortly after takeoff, a plane is climbing northwest through still air at an airspeed of 200 km/hr, and rising at a rate of 300 m/min. Resolve its velocity vector into components. The x-axis points east, the y-axis points north, and the z-axis points up.

12. A boat is heading due east at 25 km/hr (relative to the water). The current is moving toward the southwest at 10 km/hr.

    (a) Give the vector representing the actual movement of the boat.
    (b) How fast is the boat going, relative to the ground?
    (c) By what angle does the current push the boat off of its due east course?

13. An airplane is flying at an airspeed of 500 km/hr in a wind blowing at 60 km/hr toward the southeast. In what direction should the plane head to end up going due east? What is the airplane's speed relative to the ground?

14. A plane is heading due east and climbing at the rate of 80 km/hr. If its airspeed is 480 km/hr and there is a wind blowing 100 km/hr to the northeast, what is the ground speed of the plane?

15. An airplane is flying at an airspeed of 600 km/hr in a cross-wind that is blowing from the northeast at a speed of 50 km/hr. In what direction should the plane head to end up going due east?

16. A model rocket is shot into the air at an angle with the earth of about $60°$. The rocket is going fast initially but slows down as it reaches its highest point. It picks up speed again as it falls to earth.

    (a) Sketch a graph showing the path of the rocket. Draw several velocity vectors on your graph.
    (b) A second rocket has a parachute that deploys as it begins its descent. How do the velocity vectors from part (a) change for this rocket?

17. A car drives clockwise around the track in Figure 13.27, slowing down at the curves and speeding up along the straight portions. Sketch velocity vectors at the points P, Q, and R.

Figure 13.27

18. A racing car drives clockwise around the track shown in Figure 13.27 at a constant speed. At what point on the track does the car have the longest acceleration vector, and in roughly what direction is it pointing? (Recall that acceleration is the rate of change of velocity.)

19. There are five students in a class. Their scores on the midterm (out of 100) are given by the vector $\vec{v} = (73, 80, 91, 65, 84)$. Their scores on the final (out of 100) are given by $\vec{w} = (82, 79, 88, 70, 92)$. If the final counts twice as much as the midterm, find a vector giving the total scores (as a percentage) of the students.

20. The price vector of beans, rice, and tofu is $(0.30, 0.20, 0.50)$ in dollars per pound. Express it in dollars per ounce.

21. Two forces, represented by the vectors $\vec{F}_1 = 8\vec{i} - 6\vec{j}$ and $\vec{F}_2 = 3\vec{i} + 2\vec{j}$, are acting on an object. Give a vector representing the force that must be applied to the object if it is to remain stationary.

22. One force is pushing an object in a direction 50° south of east with a force of 25 newtons. A second force is simultaneously pushing the object in a direction 70° north of west with a force of 60 newtons. If the object is to remain stationary, give the direction and magnitude of the third force which must be applied to the object to counterbalance the first two.

23. An object is to be moved vertically upward by a crane. As the crane cannot get directly above the object, three ropes are attached to guide the object. One rope is pulled parallel to the ground with a force of 100 newtons in a direction 30° north of east. The second rope is pulled parallel to the ground with a force of 70 newtons in a direction 80° south of east. If the crane is attached to the third rope and can pull with a total force of 3000 newtons, find the force vector for the crane. What is the resulting (total) force on the object? (Assume vector $\vec{i}$ points east, vector $\vec{j}$ points north, and vector $\vec{k}$ points vertically up.)

Use the geometric definition of addition and scalar multiplication to explain each of the properties in Problems 24–31.

24. $\vec{w} + \vec{v} = \vec{v} + \vec{w}$ 25. $(\alpha + \beta)\vec{v} = \alpha\vec{v} + \beta\vec{v}$

26. $\alpha(\vec{v} + \vec{w}) = \alpha\vec{v} + \alpha\vec{w}$ 27. $\alpha(\beta\vec{v}) = (\alpha\beta)\vec{v}$

28. $\vec{v} + \vec{0} = \vec{v}$ 29. $1\vec{v} = \vec{v}$

30. $\vec{v} + (-1)\vec{w} = \vec{v} - \vec{w}$

31. $(\vec{u} + \vec{v}) + \vec{w} = \vec{u} + (\vec{v} + \vec{w})$

13.3   THE DOT PRODUCT

We have seen how to add vectors; can we multiply two vectors together? In the next two sections we will see two different ways of doing so: the *scalar product* (or *dot product*) which produces a scalar, and the *vector product* (or *cross product*), which produces a vector.

Definition of the Dot Product

The dot product links geometry and algebra. We already know how to calculate the length of a vector from its components; the dot product gives us a way of computing the angle between two vectors. For any two vectors $\vec{v} = v_1\vec{i} + v_2\vec{j} + v_3\vec{k}$ and $\vec{w} = w_1\vec{i} + w_2\vec{j} + w_3\vec{k}$, shown in Figure 13.28, we define a scalar as follows:

The following two definitions of the **dot product**, or **scalar product**, $\vec{v} \cdot \vec{w}$, are equivalent:
- **Geometric definition**

  $\vec{v} \cdot \vec{w} = \|\vec{v}\|\|\vec{w}\|\cos\theta$ where θ is the angle between $\vec{v}$ and $\vec{w}$ and $0 \le \theta \le \pi$.
- **Algebraic definition**

  $\vec{v} \cdot \vec{w} = v_1 w_1 + v_2 w_2 + v_3 w_3$.

Notice that the dot product of two vectors is a *number*.

Why don't we give just one definition of $\vec{v} \cdot \vec{w}$? The reason is that both definitions are equally important; the geometric definition gives us a picture of what the dot product means and the algebraic definition gives us a way of calculating it.

How do we know the two definitions are equivalent — that is, they really do define the same thing? First, we observe that the two definitions give the same result in a particular example. Then we show why they are equivalent in general.

Figure 13.28: The vectors $\vec{v}$ and $\vec{w}$

Figure 13.29: Calculating the dot product of the vectors $v = \vec{i}$ and $\vec{w} = 2\vec{i} + 2\vec{j}$ geometrically and algebraically gives the same result

Example 1   Suppose $\vec{v} = \vec{i}$ and $\vec{w} = 2\vec{i} + 2\vec{j}$. Compute $\vec{v} \cdot \vec{w}$ both geometrically and algebraically.

Solution   To use the geometric definition, see Figure 13.29. The angle between the vectors is $\pi/4$, or $45°$, and the lengths of the vectors are given by

$$\|\vec{v}\| = 1 \quad \text{and} \quad \|\vec{w}\| = 2\sqrt{2}.$$

Thus,

$$\vec{v} \cdot \vec{w} = \|\vec{v}\|\|\vec{w}\|\cos\theta = 1 \cdot 2\sqrt{2}\cos\left(\frac{\pi}{4}\right) = 2.$$

Using the algebraic definition, we get the same result:

$$\vec{v} \cdot \vec{w} = 1 \cdot 2 + 0 \cdot 2 = 2.$$

Why the Two Definitions of the Dot Product Give the Same Result

In the previous example, the two definitions give the same value for the dot product. To show that the geometric and algebraic definitions of the dot product always give the same result, we must show that, for any vectors $\vec{v} = v_1\vec{i} + v_2\vec{j} + v_3\vec{k}$ and $\vec{w} = w_1\vec{i} + w_2\vec{j} + w_3\vec{k}$ with an angle θ between them:

$$\|\vec{v}\|\|\vec{w}\|\cos\theta = v_1w_1 + v_2w_2 + v_3w_3.$$

One method follows; a method which does not use trigonometry is given in Problem 46 on page 672.

Using the Law of Cosines. Suppose that $0 < \theta < \pi$, so that the vectors $\vec{v}$ and $\vec{w}$ form a triangle. (See Figure 13.30.) By the Law of Cosines, we have

$$\|\vec{v} - \vec{w}\|^2 = \|\vec{v}\|^2 + \|\vec{w}\|^2 - 2\|\vec{v}\|\|\vec{w}\|\cos\theta.$$

This result is also true for $\theta = 0$ and $\theta = \pi$. We calculate the lengths using components:

$$\|\vec{v}\|^2 = v_1^2 + v_2^2 + v_3^2$$
$$\|\vec{w}\|^2 = w_1^2 + w_2^2 + w_3^2$$
$$\|\vec{v} - \vec{w}\|^2 = (v_1 - w_1)^2 + (v_2 - w_2)^2 + (v_3 - w_3)^2$$
$$= v_1^2 - 2v_1w_1 + w_1^2 + v_2^2 - 2v_2w_2 + w_2^2 + v_3^2 - 2v_3w_3 + w_3^2.$$

Figure 13.30: Triangle used in the justification of $\|\vec{v}\|\|\vec{w}\|\cos\theta = v_1w_1 + v_2w_2 + v_3w_3$

Substituting into the Law of Cosines and canceling, we see that

$$-2v_1w_1 - 2v_2w_2 - 2v_3w_3 = -2\|\vec{v}\|\|\vec{w}\|\cos\theta.$$

Therefore we have the result we wanted, namely that:

$$v_1w_1 + v_2w_2 + v_3w_3 = \|\vec{v}\|\|\vec{w}\|\cos\theta.$$

Properties of the Dot Product

The following properties of the dot product can be justified using the algebraic definition; see Problem 39 on page 672. For a geometric interpretation of Property 3, see Problem 43.

> **Properties of the Dot Product.** For any vectors $\vec{u}$, $\vec{v}$, and $\vec{w}$ and any scalar λ,
> 1. $\vec{v} \cdot \vec{w} = \vec{w} \cdot \vec{v}$
> 2. $\vec{v} \cdot (\lambda\vec{w}) = \lambda(\vec{v} \cdot \vec{w}) = (\lambda\vec{v}) \cdot \vec{w}$
> 3. $(\vec{v} + \vec{w}) \cdot \vec{u} = \vec{v} \cdot \vec{u} + \vec{w} \cdot \vec{u}$

Perpendicularity, Magnitude, and Dot Products

Two vectors are perpendicular if the angle between them is $\pi/2$ or $90°$. Since $\cos(\pi/2) = 0$, if $\vec{v}$ and $\vec{w}$ are perpendicular, then $\vec{v} \cdot \vec{w} = 0$. Conversely, provided that $\vec{v} \cdot \vec{w} = 0$, then $\cos\theta = 0$, so $\theta = \pi/2$ and the vectors are perpendicular. Thus, we have the following result:

> Two nonzero vectors $\vec{v}$ and $\vec{w}$ are **perpendicular**, or **orthogonal**, if and only if
>
> $$\vec{v} \cdot \vec{w} = 0.$$

For example: $\vec{i} \cdot \vec{j} = 0, \vec{j} \cdot \vec{k} = 0, \vec{i} \cdot \vec{k} = 0$.

If we take the dot product of a vector with itself, then $\theta = 0$ and $\cos\theta = 1$. For any vector $\vec{v}$:

> Magnitude and dot product are related as follows:
>
> $$\vec{v} \cdot \vec{v} = \|\vec{v}\|^2.$$

For example: $\vec{i} \cdot \vec{i} = 1, \vec{j} \cdot \vec{j} = 1, \vec{k} \cdot \vec{k} = 1$.

Using the Dot Product

Depending on the situation, one definition of the dot product may be more convenient to use than the other. In Example 2 which follows, the geometric definition is the only one which can be used because we are not given components. In Example 3, the algebraic definition is used.

Example 2   Suppose the vector $\vec{b}$ is fixed and has length 2; the vector $\vec{a}$ is free to rotate and has length 3. What are the maximum and minimum values of the dot product $\vec{a} \cdot \vec{b}$ as the vector $\vec{a}$ rotates through all possible positions? What positions of $\vec{a}$ and $\vec{b}$ lead to these values?

Solution   The geometric definition gives $\vec{a} \cdot \vec{b} = \|\vec{a}\|\|\vec{b}\|\cos\theta = 3 \cdot 2\cos\theta = 6\cos\theta$. Thus, the maximum value of $\vec{a} \cdot \vec{b}$ is 6, and it occurs when $\cos\theta = 1$ so $\theta = 0$, that is, when $\vec{a}$ and $\vec{b}$ point in the same direction. The minimum value of $\vec{a} \cdot \vec{b}$ is -6, and it occurs when $\cos\theta = -1$ so $\theta = \pi$, that is, when $\vec{a}$ and $\vec{b}$ point in opposite directions. (See Figure 13.31.)

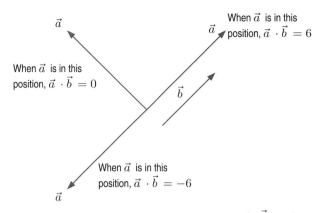

Figure 13.31: Maximum and minimum values of $\vec{a} \cdot \vec{b}$ obtained from a fixed vector $\vec{b}$ of length 2 and rotating vector $\vec{a}$ of length 3

Example 3 Which pairs from the following list of 3-dimensional vectors are perpendicular to one another?

$$\vec{u} = \vec{i} + \sqrt{3}\,\vec{k}, \quad \vec{v} = \vec{i} + \sqrt{3}\,\vec{j}, \quad \vec{w} = \sqrt{3}\,\vec{i} + \vec{j} - \vec{k}.$$

Solution The geometric definition tells us that two vectors are perpendicular if and only if their dot product is zero. Since the vectors are given in components, we calculate dot products using the algebraic definition:

$$\vec{v} \cdot \vec{u} = (\vec{i} + \sqrt{3}\,\vec{j} + 0\vec{k}) \cdot (\vec{i} + 0\vec{j} + \sqrt{3}\,\vec{k}) = 1 \cdot 1 + \sqrt{3} \cdot 0 + 0 \cdot \sqrt{3} = 1,$$

$$\vec{v} \cdot \vec{w} = (\vec{i} + \sqrt{3}\,\vec{j} + 0\vec{k}) \cdot (\sqrt{3}\,\vec{i} + \vec{j} - \vec{k}) = 1 \cdot \sqrt{3} + \sqrt{3} \cdot 1 + 0(-1) = 2\sqrt{3},$$

$$\vec{w} \cdot \vec{u} = (\sqrt{3}\,\vec{i} + \vec{j} - \vec{k}) \cdot (\vec{i} + 0\vec{j} + \sqrt{3}\,\vec{k}) = \sqrt{3} \cdot 1 + 1 \cdot 0 + (-1) \cdot \sqrt{3} = 0.$$

So the only two vectors which are perpendicular are $\vec{w}$ and $\vec{u}$.

Normal Vectors and the Equation of a Plane

In Section 12.4 we wrote the equation of a plane given its x-slope, y-slope and z-intercept. Now we write the equation of a plane using a vector and a point lying on the plane. A *normal vector* to a plane is a vector that is perpendicular to the plane, that is, it is perpendicular to every displacement vector between any two points in the plane. Let $\vec{n} = a\vec{i} + b\vec{j} + c\vec{k}$ be a normal vector to the plane, let $P_0 = (x_0, y_0, z_0)$ be a fixed point in the plane, and let $P = (x, y, z)$ be any other point in the plane. Then $\overrightarrow{P_0P} = (x - x_0)\vec{i} + (y - y_0)\vec{j} + (z - z_0)\vec{k}$ is a vector whose head and tail both lie in the plane. (See Figure 13.32.) Thus, the vectors $\vec{n}$ and $\overrightarrow{P_0P}$ are perpendicular, so $\vec{n} \cdot \overrightarrow{P_0P} = 0$. The algebraic definition of the dot product gives $\vec{n} \cdot \overrightarrow{P_0P} = a(x - x_0) + b(y - y_0) + c(z - z_0)$, so we obtain the following result:

Figure 13.32: Plane with normal $\vec{n}$ and containing a fixed point (x_0, y_0, z_0)

The **equation of the plane** with normal vector $\vec{n} = a\vec{i} + b\vec{j} + c\vec{k}$ and containing the point $P_0 = (x_0, y_0, z_0)$ is

$$a(x - x_0) + b(y - y_0) + c(z - z_0) = 0.$$

Letting $d = ax_0 + by_0 + cz_0$ (a constant), we can write the equation of the plane in the form

$$ax + by + cz = d.$$

Example 4   Find the equation of the plane perpendicular to $-\vec{i} + 3\vec{j} + 2\vec{k}$ and passing through the point $(1, 0, 4)$.

Solution   The equation of the plane is

$$-(x - 1) + 3(y - 0) + 2(z - 4) = 0,$$

which simplifies to

$$-x + 3y + 2z = 7.$$

Example 5   Find a normal vector to the plane with equation (a) $x - y + 2z = 5$ (b) $z = 0.5x + 1.2y$.

Solution   (a) Since the coefficients of $\vec{i}$, $\vec{j}$, and $\vec{k}$ in a normal vector are the coefficients of x, y, z in the equation of the plane, a normal vector is $\vec{n} = \vec{i} - \vec{j} + 2\vec{k}$.

(b) Before we can find a normal vector, we rewrite the equation of the plane in the form

$$0.5x + 1.2y - z = 0.$$

Thus, a normal vector is $\vec{n} = 0.5\vec{i} + 1.2\vec{j} - \vec{k}$.

The Dot Product in n Dimensions

The algebraic definition of the dot product can be extended to vectors in higher dimensions.

If $\vec{u} = (u_1, \ldots, u_n)$ and $\vec{v} = (v_1, \ldots, v_n)$ then the dot product of $\vec{u}$ and $\vec{v}$ is the **scalar**

$$\vec{u} \cdot \vec{v} = u_1 v_1 + \ldots + u_n v_n.$$

Example 6   A video store sells videos, tapes, CDs, and computer games. We define the quantity vector $\vec{q} = (q_1, q_2, q_3, q_4)$, where q_1, q_2, q_3, q_4 denote the quantities sold of each of the items, and the price vector $\vec{p} = (p_1, p_2, p_3, p_4)$, where p_1, p_2, p_3, p_4 denote the price per unit of each item. What does the dot product $\vec{p} \cdot \vec{q}$ represent?

Solution   The dot product is $\vec{p} \cdot \vec{q} = p_1 q_1 + p_2 q_2 + p_3 q_3 + p_4 q_4$. The quantity $p_1 q_1$ represents the revenue received by the store for the videos, $p_2 q_2$ represents the revenue for the tapes, and so on. The dot product represents the total revenue received by the store for the sale of these four items.

Resolving a Vector into Components: Projections

In Section 13.1, we resolved a vector into components parallel to the axes. Now we see how to resolve a vector, $\vec{v}$, into components, called $\vec{v}_{\text{parallel}}$ and $\vec{v}_{\text{perp}}$, which are parallel and perpendicular, respectively, to a given nonzero vector, $\vec{u}$. (See Figure 13.33.)

Figure 13.33: Resolving $\vec{v}$ into components parallel and perpendicular to $\vec{u}$
(a) $0 < \theta < \pi/2$ (b) $\pi/2 < \theta < \pi$

The projection of $\vec{v}$ on $\vec{u}$, written $\vec{v}_{\text{parallel}}$, measures (in some sense) how much the vector $\vec{v}$ is aligned with the vector $\vec{u}$. The length of $\vec{v}_{\text{parallel}}$ is the length of the shadow cast by $\vec{v}$ on a line in the direction of $\vec{u}$.

To compute $\vec{v}_{\text{parallel}}$, we assume $\vec{u}$ is a unit vector. (If not, create one by dividing by its length.) Then Figure 13.33(a) shows that, if $0 \le \theta \le \pi/2$:

$$\|\vec{v}_{\text{parallel}}\| = \|\vec{v}\| \cos\theta = \vec{v} \cdot \vec{u} \qquad (\text{since } \|\vec{u}\| = 1).$$

Now $\vec{v}_{\text{parallel}}$ is a scalar multiple of $\vec{u}$, and since $\vec{u}$ is a unit vector,

$$\vec{v}_{\text{parallel}} = (\|\vec{v}\| \cos\theta)\vec{u} = (\vec{v} \cdot \vec{u})\vec{u}.$$

A similar argument shows that if $\pi/2 < \theta \le \pi$, as in Figure 13.33(b), this formula for $\vec{v}_{\text{parallel}}$ still holds. The vector $\vec{v}_{\text{perp}}$ is specified by

$$\vec{v}_{\text{perp}} = \vec{v} - \vec{v}_{\text{parallel}}.$$

Thus, we have the following results:

Projection of $\vec{v}$ on the Line in the Direction of the Unit Vector $\vec{u}$

If $\vec{v}_{\text{parallel}}$ and $\vec{v}_{\text{perp}}$ are components of $\vec{v}$ which are parallel and perpendicular, respectively, to $\vec{u}$, then

$$\begin{array}{l}\text{Projection of} \\ \vec{v} \text{ on to } \vec{u}\end{array} = \vec{v}_{\text{parallel}} = (\vec{v} \cdot \vec{u})\vec{u} \qquad \text{provided } \|\vec{u}\| = 1$$

and $\qquad \vec{v} = \vec{v}_{\text{parallel}} + \vec{v}_{\text{perp}} \qquad$ so $\qquad \vec{v}_{\text{perp}} = \vec{v} - \vec{v}_{\text{parallel}}.$

Example 7   Figure 13.34 shows the force the wind exerts on the sail of a sailboat. Find the component of the force in the direction in which the sailboat is traveling.

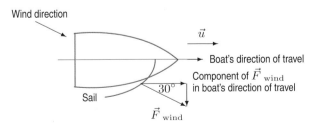

Figure 13.34: Wind moving a sailboat

Solution   Let $\vec{u}$ be a unit vector in the direction of travel. The force of the wind on the sail makes an angle of $30°$ with $\vec{u}$. Thus, the component of this force in the direction of $\vec{u}$ is

$$\vec{F}_{\text{parallel}} = (\vec{F} \cdot \vec{u})\vec{u} = \|\vec{F}\|(\cos 30°)\vec{u} = 0.87\|\vec{F}\|\vec{u}.$$

Thus, the boat is being pushed forward with about 87% of the total force due to the wind. (In fact, the interaction of wind and sail is much more complex than this model suggests.)

A Physical Interpretation of the Dot Product: Work

In physics, the word "work" has a slightly different meaning from its everyday meaning. In physics, when a force of magnitude F acts on an object through a distance d, we say the *work*, W, done by the force is

$$W = Fd,$$

provided the force and the displacement are in the same direction. For example, if a 1 kg body falls 10 meters under the force of gravity, which is 9.8 newtons, then the work done by gravity is

$$W = (9.8 \text{ newtons}) \cdot (10 \text{ meters}) = 98 \text{ joules}.$$

What if the force and the displacement are not in the same direction? Suppose a force $\vec{F}$ acts on an object as it moves along a displacement vector $\vec{d}$. Let θ be the angle between $\vec{F}$ and $\vec{d}$. First, we assume $0 \leq \theta \leq \pi/2$. Figure 13.35 shows how we can resolve $\vec{F}$ into components that are parallel and perpendicular to $\vec{d}$:

$$\vec{F} = \vec{F}_{\text{parallel}} + \vec{F}_{\text{perp}},$$

Then the work done by $\vec{F}$ is defined to be

$$W = \|\vec{F}_{\text{parallel}}\| \, \|\vec{d}\|.$$

We see from Figure 13.35 that $\vec{F}_{\text{parallel}}$ has magnitude $\|\vec{F}\| \cos\theta$. So the work is given by the dot product:

$$W = (\|\vec{F}\| \cos\theta)\|\vec{d}\| = \|\vec{F}\|\|\vec{d}\| \cos\theta = \vec{F} \cdot \vec{d}.$$

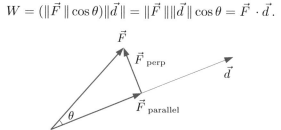

Figure 13.35: Resolving the force $\vec{F}$ into two forces, one parallel to $\vec{d}$, one perpendicular to $\vec{d}$

The formula $W = \vec{F} \cdot \vec{d}$ holds when $\pi/2 < \theta \leq \pi$ also. In that case, the work done by the force is negative and the object is moving against the force. Thus, we have the following definition:

The **work**, W, done by a force $\vec{F}$ acting on an object through a displacement $\vec{d}$ is given by

$$W = \vec{F} \cdot \vec{d}.$$

Notice that if the vectors $\vec{F}$ and $\vec{d}$ are parallel and in the same direction, with magnitudes F and d, then $\cos\theta = \cos 0 = 1$, so $W = \|\vec{F}\|\|\vec{d}\| = Fd$, which is the original definition. When the vectors are perpendicular, $\cos\theta = \cos\frac{\pi}{2} = 0$, so $W = 0$ and no work is done in the technical definition of the word. For example, if you carry a heavy box across the room at the same horizontal height, no work is done by gravity because the force of gravity is vertical but the motion is horizontal.

Exercises and Problems for Section 13.3

Exercises

For Exercises 1–9, perform the following operations on the given 3-dimensional vectors.

$$\vec{a} = 2\vec{j} + \vec{k} \qquad \vec{b} = -3\vec{i} + 5\vec{j} + 4\vec{k} \qquad \vec{c} = \vec{i} + 6\vec{j}$$

$$\vec{y} = 4\vec{i} - 7\vec{j} \qquad \vec{z} = \vec{i} - 3\vec{j} - \vec{k}$$

1. $\vec{a} \cdot \vec{y}$

2. $\vec{c} \cdot \vec{y}$

3. $\vec{a} \cdot \vec{b}$

4. $\vec{a} \cdot \vec{z}$

5. $\vec{a} \cdot (\vec{c} + \vec{y})$

6. $\vec{c} \cdot \vec{a} + \vec{a} \cdot \vec{y}$

7. $(\vec{a} \cdot \vec{b})\vec{a}$

8. $(\vec{a} \cdot \vec{y})(\vec{c} \cdot \vec{z})$

9. $((\vec{c} \cdot \vec{c})\vec{a}) \cdot \vec{a}$

In Exercises 10–13, find a normal vector to the plane.

10. $2x + y - z = 5$

11. $z = 3x + 4y - 7$

12. $z - 5(x - 2) = 3(5 - y)$

13. $2(x - z) = 3(x + y)$

Problems

14. Let θ be the angle between $\vec{v}$ and $\vec{w}$, with $0 < \theta < \pi/2$. What is the effect on $\vec{v} \cdot \vec{w}$ of increasing each of the following quantities? Does $\vec{v} \cdot \vec{w}$ increase or decrease?

(a) $\|\vec{v}\|$
(b) θ

15. (a) Give a vector that is parallel to, but not equal to, $\vec{v} = 4\vec{i} + 3\vec{j}$.
(b) Give a vector that is perpendicular to $\vec{v}$.

16. Compute the angle between the vectors $\vec{i} + \vec{j} + \vec{k}$ and $\vec{i} - \vec{j} - \vec{k}$.

17. Which pairs of the vectors $\sqrt{3}\vec{i} + \vec{j}, 3\vec{i} + \sqrt{3}\vec{j}, \vec{i} - \sqrt{3}\vec{j}$ are parallel and which are perpendicular?

18. For what values of t are $\vec{u} = t\vec{i} - \vec{j} + \vec{k}$ and $\vec{v} = t\vec{i} + t\vec{j} - 2\vec{k}$ perpendicular? Are there values of t for which $\vec{u}$ and $\vec{v}$ are parallel?

19. List any vectors which are parallel to each other and any vectors which are perpendicular to each other:

$$\vec{v}_1 = \vec{i} - 2\vec{j}, \qquad \vec{v}_2 = 2\vec{i} + 4\vec{j}, \qquad \vec{v}_3 = 3\vec{i} + 1.5\vec{j},$$
$$\vec{v}_4 = -1.2\vec{i} + 2.4\vec{j}, \vec{v}_5 = -5\vec{i} - 2.5\vec{j}, \vec{v}_6 = 12\vec{i} - 12\vec{j},$$
$$\vec{v}_7 = 4\vec{i} + 2\vec{j}, \qquad \vec{v}_8 = 3\vec{i} - 6\vec{j}, \qquad \vec{v}_9 = 0.70\vec{i} - 0.35\vec{j}$$

20. Which pairs (if any) of vectors from the following list

(a) Are perpendicular?
(b) Are parallel?
(c) Have an angles less than $\pi/2$ between them?
(d) Have an angle of more than $\pi/2$ between them?

$$\vec{a} = \vec{i} - 3\vec{j} - \vec{k} \qquad \vec{b} = \vec{i} + \vec{j} + 2\vec{k}$$
$$\vec{c} = -2\vec{i} - \vec{j} + \vec{k} \qquad \vec{d} = -\vec{i} - \vec{j} + \vec{k}$$

21. Write $\vec{a} = 3\vec{i} + 2\vec{j} - 6\vec{k}$ as the sum of two vectors, one parallel, and one perpendicular, to $\vec{d} = 2\vec{i} - 4\vec{j} + \vec{k}$.

In Problems 22–26, find an equation of a plane that satisfies the given conditions.

22. Perpendicular to the vector $-\vec{i} + 2\vec{j} + \vec{k}$ and passing through the point $(1, 0, 2)$.

23. Perpendicular to the vector $5\vec{i} + \vec{j} - 2\vec{k}$ and passing through the point $(0, 1, -1)$.

24. Perpendicular to the vector $2\vec{i} - 3\vec{j} + 7\vec{k}$ and passing through the point $(1, -1, 2)$.

25. Parallel to the plane $2x + 4y - 3z = 1$ and through the point $(1, 0, -1)$.

26. Through $(-2, 3, 2)$ and parallel to $3x + y + z = 4$.

27. A plane has equation $z = 5x - 2y + 7$.

(a) Find a value of λ making the vector $\lambda\vec{i} + \vec{j} + 0.5\vec{k}$ normal to the plane.
(b) Find a value of a so that the point $(a + 1, a, a - 1)$ lies on the plane.

28. Find the points where the plane $z = 5x - 4y + 3$ intersects each of the coordinate axes. Find the lengths of the sides and the angles of the triangle formed by these points.

29. Find the angle between the planes $5(x - 1) + 3(y + 2) + 2z = 0$ and $x + 3(y - 1) + 2(z + 4) = 0$.

30. Find angle BAC if $A = (2, 2, 2)$, $B = (4, 2, 1)$, and $C = (2, 3, 1)$.

31. Let $A = (0, 4)$, $B = (-1, -3)$, and $C = (-5, 1)$. Draw triangle ABC and find each of its interior angles.

32. Let S be the triangle with vertices $A = (2, 2, 2)$, $B = (4, 2, 1)$, and $C = (2, 3, 1)$.

(a) Find the length of the shortest side of S.
(b) Find the cosine of the angle BAC at vertex A.

33. A 100-meter dash is run on a track in the direction of the vector $\vec{v} = 2\vec{i} + 6\vec{j}$. The wind velocity $\vec{w}$ is $5\vec{i} + \vec{j}$ km/hr. The rules say that a legal wind speed measured in the direction of the dash must not exceed 5 km/hr. Will the race results be disqualified due to an illegal wind? Justify your answer.

34. A basketball gymnasium is 25 meters high, 80 meters wide and 200 meters long. For a half time stunt, the cheerleaders want to run two strings, one from each of the two corners above one basket to the diagonally opposite corners of the gym floor. What is the cosine of the angle made by the strings as they cross?

35. A course has four exams, weighted 10%, 15%, 25%, 50%, respectively. The class average on each of these exams is 75%, 91%, 84%, 87%, respectively. What do the vectors $\vec{a} = (0.75, 0.91, 0.84, 0.87)$ and $\vec{w} = (0.1, 0.15, 0.25, 0.5)$ represent, in terms of the course? Calculate the dot product $\vec{w} \cdot \vec{a}$. What does it represent, in terms of the course?

36. A street vendor sells six items, with prices p_1 dollars per unit, p_2 dollars per unit, and so on. The vendor's price vector is $\vec{p} = (p_1, p_2, p_3, p_4, p_5, p_6) = (1.00, 3.50, 4.00, 2.75, 5.00, 3.00)$. The vendor sells q_1 units of the first item, q_2 units of the second item, and so on. The vendor's quantity vector is $\vec{q} = (q_1, q_2, q_3, q_4, q_5, q_6) = (43, 57, 12, 78, 20, 35)$. Find $\vec{p} \cdot \vec{q}$, give its units, and explain its significance to the vendor.

37. A consumption vector of three goods is defined by $\vec{x} = (x_1, x_2, x_3)$, where x_1, x_2 and x_3 are the quantities consumed of the three goods. A budget constraint is represented by the equation $\vec{p} \cdot \vec{x} = k$, where $\vec{p}$ is the price vector of the three goods and k is a constant. Show that the difference between two consumption vectors corresponding to points satisfying the same budget constraint is perpendicular to the price vector $\vec{p}$.

38. Consider a point P and plane through the point P_0 with normal vector $\vec{n}$.

   (a) What do all the points satisfying the following condition have in common:

   $$\vec{n} \cdot \overrightarrow{P_0P} > 0?$$

   What about all the points satisfying $\vec{n} \cdot \overrightarrow{P_0P} < 0$?
   (b) Express the conditions $\vec{n} \cdot \overrightarrow{P_0P} > 0$ and $\vec{n} \cdot \overrightarrow{P_0P} < 0$ in terms of coordinates.
   (c) Use parts (a) and (b) to decide which of the points

   $$P = (-1, -1, 1), \quad Q = (-1, -1, -1)$$
   $$R = (1, 1, 1)$$

   are on the same side of the plane $2x - 3y + 4z = 4$.

39. Show why each of the properties of the dot product in the box on page 666 follows from the algebraic definition of the dot product:

$$\vec{v} \cdot \vec{w} = v_1 w_1 + v_2 w_2 + v_3 w_3$$

40. What does Property 2 of the dot product in the box on page 666 say geometrically?

41. Show that the vectors $(\vec{b} \cdot \vec{c})\vec{a} - (\vec{a} \cdot \vec{c})\vec{b}$ and $\vec{c}$ are perpendicular.

42. Show that if $\vec{u}$ and $\vec{v}$ are two vectors such that

$$\vec{u} \cdot \vec{w} = \vec{v} \cdot \vec{w}$$

for every vector $\vec{w}$, then

$$\vec{u} = \vec{v}.$$

43. Figure 13.36 shows that, given three vectors $\vec{u}$, $\vec{v}$, and $\vec{w}$, the sum of the components of $\vec{v}$ and $\vec{w}$ in the direction of $\vec{u}$ is the component of $\vec{v} + \vec{w}$ in the direction of $\vec{u}$. (Although the figure is drawn in two dimensions, this result is also true in three dimensions.) Use this figure to explain why the geometric definition of the dot product satisfies $(\vec{v} + \vec{w}) \cdot \vec{u} = \vec{v} \cdot \vec{u} + \vec{w} \cdot \vec{u}$.

Figure 13.36: Component of $\vec{v} + \vec{w}$ in the direction of $\vec{u}$ is the sum of the components of $\vec{v}$ and $\vec{w}$ in that direction

44. (a) Using the geometric definition of the dot product, show that

   $$\vec{u} \cdot (-\vec{v}) = -(\vec{u} \cdot \vec{v}).$$

   [Hint: What happens to the angle when you multiply $\vec{v}$ by -1?]
   (b) Using the geometric definition of the dot product, show that for any negative scalar λ

   $$\vec{u} \cdot (\lambda \vec{v}) = \lambda(\vec{u} \cdot \vec{v})$$
   $$(\lambda \vec{u}) \cdot \vec{v} = \lambda(\vec{u} \cdot \vec{v}).$$

45. The Law of Cosines for a triangle with side lengths a, b, and c, and with angle C opposite side c, says

$$c^2 = a^2 + b^2 - 2ab \cos C.$$

On page 665, we used the Law of Cosines to show that the two definitions of the dot product are equivalent. In this problem, use the geometric definition of the dot product and its properties in the box on page 666 to prove the Law of Cosines. [Hint: Let $\vec{u}$ and $\vec{v}$ be the displacement vectors from C to the other two vertices, and express c^2 in terms of $\vec{u}$ and $\vec{v}$.]

46. Use the following steps and the results of Problems 43–44 to show (without trigonometry) that the geometric and algebraic definitions of the dot product are equivalent.
      Let $\vec{u} = u_1\vec{i} + u_2\vec{j} + u_3\vec{k}$ and $\vec{v} = v_1\vec{i} + v_2\vec{j} + v_3\vec{k}$ be any vectors. Write $(\vec{u} \cdot \vec{v})_{\text{geom}}$ for the result of the dot product computed geometrically. Substitute $\vec{u} = u_1\vec{i} + u_2\vec{j} + u_3\vec{k}$ and use Problems 43–44 to expand $(\vec{u} \cdot \vec{v})_{\text{geom}}$. Substitute for $\vec{v}$ and expand. Then calculate the dot products $\vec{i} \cdot \vec{i}$, $\vec{i} \cdot \vec{j}$, etc. geometrically.

47. For any vectors $\vec{v}$ and $\vec{w}$, consider the following function of t:

$$q(t) = (\vec{v} + t\vec{w}) \cdot (\vec{v} + t\vec{w}).$$

(a) Explain why $q(t) \geq 0$ for all real t.

(b) Expand $q(t)$ as a quadratic polynomial in t using the properties on page 666.

(c) Using the discriminant of the quadratic, show that,

$$|\vec{v} \cdot \vec{w}| \leq \|\vec{v}\|\|\vec{w}\|.$$

13.4 THE CROSS PRODUCT

In the previous section we combined two vectors to get a number, the dot product. In this section we see another way of combining two vectors, this time to get a vector, the *cross product*. Any two vectors in 3-space form a parallelogram. We define the cross product using this parallelogram.

The Area of a Parallelogram

Consider the parallelogram formed by the vectors $\vec{v}$ and $\vec{w}$ with an angle of θ between them. Then Figure 13.37 shows

$$\text{Area of parallelogram} = \text{Base} \cdot \text{Height} = \|\vec{v}\|\|\vec{w}\|\sin\theta.$$

How would we compute the area of the parallelogram if we were given $\vec{v}$ and $\vec{w}$ in components, $\vec{v} = v_1\vec{i} + v_2\vec{j} + v_3\vec{k}$ and $\vec{w} = w_1\vec{i} + w_2\vec{j} + w_3\vec{k}$? Project 1 on page 682 shows that if $\vec{v}$ and $\vec{w}$ are in the xy-plane, so $v_3 = w_3 = 0$, then

$$\text{Area of parallelogram} = |v_1 w_2 - v_2 w_1|.$$

What if $\vec{v}$ and $\vec{w}$ do not lie in the xy-plane? The cross product will enable us to compute the area of the parallelogram formed by any two vectors.

Figure 13.37: Parallelogram formed by $\vec{v}$ and $\vec{w}$ has
Area $= \|\vec{v}\|\|\vec{w}\|\sin\theta$

Definition of the Cross Product

We define the cross product of the vectors $\vec{v}$ and $\vec{w}$, written $\vec{v} \times \vec{w}$, to be a vector perpendicular to both $\vec{v}$ and $\vec{w}$. The magnitude of this vector is the area of the parallelogram formed by the two vectors. The direction of $\vec{v} \times \vec{w}$ is given by the normal vector, $\vec{n}$, to the plane defined by $\vec{v}$ and $\vec{w}$. If we require that $\vec{n}$ be a unit vector, there are two choices for $\vec{n}$, pointing out of the plane in opposite directions. We pick one by the following rule (see Figure 13.38):

The right-hand rule: Place $\vec{v}$ and $\vec{w}$ so that their tails coincide and curl the fingers of your right hand through the smaller of the two angles from $\vec{v}$ to $\vec{w}$; your thumb points in the direction of the normal vector, $\vec{n}$.

Like the dot product, there are two equivalent definitions of the cross product:

The following two definitions of the **cross product** or **vector product** $\vec{v} \times \vec{w}$ are equivalent:
- **Geometric definition**

  If $\vec{v}$ and $\vec{w}$ are not parallel, then

  $$\vec{v} \times \vec{w} = \left(\begin{array}{c}\text{Area of parallelogram} \\ \text{with edges } \vec{v} \text{ and } \vec{w}\end{array}\right) \vec{n} = (\|\vec{v}\|\|\vec{w}\| \sin \theta)\vec{n}\,,$$

  where $0 \leq \theta \leq \pi$ is the angle between $\vec{v}$ and $\vec{w}$ and $\vec{n}$ is the unit vector perpendicular to $\vec{v}$ and $\vec{w}$ pointing in the direction given by the right-hand rule. If $\vec{v}$ and $\vec{w}$ are parallel, then $\vec{v} \times \vec{w} = \vec{0}$.

- **Algebraic definition**

  $$\vec{v} \times \vec{w} = (v_2 w_3 - v_3 w_2)\vec{i} + (v_3 w_1 - v_1 w_3)\vec{j} + (v_1 w_2 - v_2 w_1)\vec{k}$$

  where $\vec{v} = v_1\vec{i} + v_2\vec{j} + v_3\vec{k}$ and $\vec{w} = w_1\vec{i} + w_2\vec{j} + w_3\vec{k}$.

Notice that the magnitude of the $\vec{k}$ component is the area of a 2-dimensional parallelogram and the other components have a similar form. Problems 33 and 31 at the end of this section show that the geometric and algebraic definitions of the cross product give the same result.

Figure 13.38: Area of parallelogram $= \|\vec{v} \times \vec{w}\|$

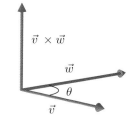

Figure 13.39: The cross product $\vec{v} \times \vec{w}$

Unlike the dot product, the cross product is only defined for three-dimensional vectors. The geometric definition shows us that the cross product is *rotation invariant*. Imagine the two vectors $\vec{v}$ and $\vec{w}$ as two metal rods welded together. Attach a third rod whose direction and length correspond to $\vec{v} \times \vec{w}$. (See Figure 13.39.) Then, no matter how we turn this set of rods, the third will still be the cross product of the first two.

The algebraic definition is more easily remembered by writing it as a 3×3 determinant. (See Appendix D.)

$$\vec{v} \times \vec{w} = \begin{vmatrix} \vec{i} & \vec{j} & \vec{k} \\ v_1 & v_2 & v_3 \\ w_1 & w_2 & w_3 \end{vmatrix} = (v_2 w_3 - v_3 w_2)\vec{i} + (v_3 w_1 - v_1 w_3)\vec{j} + (v_1 w_2 - v_2 w_1)\vec{k}\,.$$

Example 1   Find $\vec{i} \times \vec{j}$ and $\vec{j} \times \vec{i}$.

Solution   The vectors $\vec{i}$ and $\vec{j}$ both have magnitude 1 and the angle between them is $\pi/2$. By the right-hand rule, the vector $\vec{i} \times \vec{j}$ is in the direction of $\vec{k}$, so $\vec{n} = \vec{k}$ and we have

$$\vec{i} \times \vec{j} = \left(\|\vec{i}\|\|\vec{j}\| \sin \frac{\pi}{2}\right) \vec{k} = \vec{k}\,.$$

Similarly, the right-hand rule says that the direction of $\vec{j} \times \vec{i}$ is $-\vec{k}$, so

$$\vec{j} \times \vec{i} = (\|\vec{j}\|\|\vec{i}\| \sin \frac{\pi}{2})(-\vec{k}) = -\vec{k}\,.$$

Similar calculations show that $\vec{j} \times \vec{k} = \vec{i}$ and $\vec{k} \times \vec{i} = \vec{j}$.

Example 2   For any vector $\vec{v}$, find $\vec{v} \times \vec{v}$.

Solution   Since $\vec{v}$ is parallel to itself, $\vec{v} \times \vec{v} = \vec{0}$.

Example 3   Find the cross product of $\vec{v} = 2\vec{i} + \vec{j} - 2\vec{k}$ and $\vec{w} = 3\vec{i} + \vec{k}$ and check that the cross product is perpendicular to both $\vec{v}$ and $\vec{w}$.

Solution   Writing $\vec{v} \times \vec{w}$ as a determinant and expanding it into three two-by-two determinants, we have

$$\vec{v} \times \vec{w} = \begin{vmatrix} \vec{i} & \vec{j} & \vec{k} \\ 2 & 1 & -2 \\ 3 & 0 & 1 \end{vmatrix} = \vec{i} \begin{vmatrix} 1 & -2 \\ 0 & 1 \end{vmatrix} - \vec{j} \begin{vmatrix} 2 & -2 \\ 3 & 1 \end{vmatrix} + \vec{k} \begin{vmatrix} 2 & 1 \\ 3 & 0 \end{vmatrix}$$

$$= \vec{i}\,(1(1) - 0(-2)) - \vec{j}\,(2(1) - 3(-2)) + \vec{k}\,(2(0) - 3(1))$$
$$= \vec{i} - 8\vec{j} - 3\vec{k}.$$

To check that $\vec{v} \times \vec{w}$ is perpendicular to $\vec{v}$, we compute the dot product:

$$\vec{v} \cdot (\vec{v} \times \vec{w}) = (2\vec{i} + \vec{j} - 2\vec{k}) \cdot (\vec{i} - 8\vec{j} - 3\vec{k}) = 2 - 8 + 6 = 0.$$

Similarly,

$$\vec{w} \cdot (\vec{v} \times \vec{w}) = (3\vec{i} + 0\vec{j} + \vec{k}) \cdot (\vec{i} - 8\vec{j} - 3\vec{k}) = 3 + 0 - 3 = 0.$$

Thus, $\vec{v} \times \vec{w}$ is perpendicular to both $\vec{v}$ and $\vec{w}$.

Properties of the Cross Product

The right-hand rule tells us that $\vec{v} \times \vec{w}$ and $\vec{w} \times \vec{v}$ point in opposite directions. The magnitudes of $\vec{v} \times \vec{w}$ and $\vec{w} \times \vec{v}$ are the same, so $\vec{w} \times \vec{v} = -(\vec{v} \times \vec{w})$. (See Figure 13.40.)

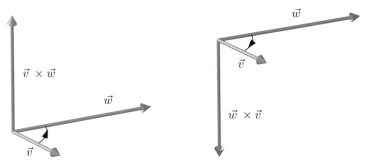

Figure 13.40: Diagram showing $\vec{v} \times \vec{w} = -(\vec{w} \times \vec{v})$

This explains the first of the following properties. The other two are derived in Problems 25, 28, and 33 at the end of this section.

Properties of the Cross Product

For vectors $\vec{u}$, $\vec{v}$, $\vec{w}$ and scalar λ

1. $\vec{w} \times \vec{v} = -(\vec{v} \times \vec{w})$.
2. $(\lambda \vec{v}) \times \vec{w} = \lambda(\vec{v} \times \vec{w}) = \vec{v} \times (\lambda \vec{w})$
3. $\vec{u} \times (\vec{v} + \vec{w}) = \vec{u} \times \vec{v} + \vec{u} \times \vec{w}$.

The Equivalence of the Two Definitions of the Cross Product

Problem 33 on page 678 uses geometric arguments to show that the cross product distributes over addition. Problem 31 then shows how the formula in the algebraic definition of the cross product can be derived from the geometric definition.

The Equation of a Plane Through Three Points

The equation of a plane is determined by a point $P_0 = (x_0, y_0, z_0)$ on the plane, and a normal vector, $\vec{n} = a\vec{i} + b\vec{j} + c\vec{k}$:

$$a(x - x_0) + b(y - y_0) + c(z - z_0) = 0.$$

However, a plane can also be determined by three points on it (provided they do not lie on a line). In that case we can find an equation of the plane by first determining two vectors in the plane and then finding a normal vector using the cross product, as in the following example.

Example 4   Find an equation of the plane containing the points $P = (1, 3, 0)$, $Q = (3, 4, -3)$, and $R = (3, 6, 2)$.

Solution   Since the points P and Q are in the plane, the displacement vector between them, $\overrightarrow{PQ}$, is in the plane, where

$$\overrightarrow{PQ} = (3 - 1)\vec{i} + (4 - 3)\vec{j} + (-3 - 0)\vec{k} = 2\vec{i} + \vec{j} - 3\vec{k}.$$

The displacement vector $\overrightarrow{PR}$ is also in the plane, where

$$\overrightarrow{PR} = (3 - 1)\vec{i} + (6 - 3)\vec{j} + (2 - 0)\vec{k} = 2\vec{i} + 3\vec{j} + 2\vec{k}.$$

Thus, a normal vector, $\vec{n}$, to the plane is given by

$$\vec{n} = \overrightarrow{PQ} \times \overrightarrow{PR} = \begin{vmatrix} \vec{i} & \vec{j} & \vec{k} \\ 2 & 1 & -3 \\ 2 & 3 & 2 \end{vmatrix} = 11\vec{i} - 10\vec{j} + 4\vec{k}.$$

Since the point $(1, 3, 0)$ is on the plane, the equation of the plane is

$$11(x - 1) - 10(y - 3) + 4(z - 0) = 0,$$

which simplifies to

$$11x - 10y + 4z = -19.$$

You should check that P, Q, and R satisfy the equation of the plane.

Areas and Volumes Using the Cross Product and Determinants

We can use the cross product to calculate the area of the parallelogram with sides $\vec{v}$ and $\vec{w}$. We say that $\vec{v} \times \vec{w}$ is the *area vector* of the parallelogram. The geometric definition of the cross product tells us that $\vec{v} \times \vec{w}$ is normal to the parallelogram and gives us the following result:

Area of a parallelogram with edges $\vec{v} = v_1\vec{i} + v_2\vec{j} + v_3\vec{k}$ and $\vec{w} = w_1\vec{i} + w_2\vec{j} + w_3\vec{k}$ is given by

$$\text{Area} = \|\vec{v} \times \vec{w}\|, \qquad \text{where} \quad \vec{v} \times \vec{w} = \begin{vmatrix} \vec{i} & \vec{j} & \vec{k} \\ v_1 & v_2 & v_3 \\ w_1 & w_2 & w_3 \end{vmatrix}.$$

Example 5   Find the area of the parallelogram with edges $\vec{v} = 2\vec{i} + \vec{j} - 3\vec{k}$ and $\vec{w} = \vec{i} + 3\vec{j} + 2\vec{k}$.

Solution   We calculate the cross product:

$$\vec{v} \times \vec{w} = \begin{vmatrix} \vec{i} & \vec{j} & \vec{k} \\ 2 & 1 & -3 \\ 1 & 3 & 2 \end{vmatrix} = (2+9)\vec{i} - (4+3)\vec{j} + (6-1)\vec{k} = 11\vec{i} - 7\vec{j} + 5\vec{k}.$$

The area of the parallelogram with edges $\vec{v}$ and $\vec{w}$ is the magnitude of the vector $\vec{v} \times \vec{w}$:

$$\text{Area} = \|\vec{v} \times \vec{w}\| = \sqrt{11^2 + (-7)^2 + 5^2} = \sqrt{195}.$$

Volume of a Parallelepiped

Consider the parallelepiped with sides formed by $\vec{a}$, $\vec{b}$, and $\vec{c}$. (See Figure 13.41.) Since the base is formed by the vectors $\vec{b}$ and $\vec{c}$, we have

$$\text{Area of base of parallelepiped} = \|\vec{b} \times \vec{c}\|.$$

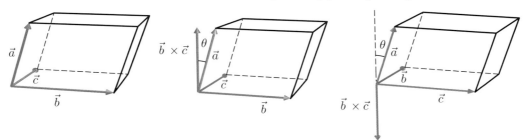

Figure 13.41: Volume of a   Figure 13.42: The vectors $\vec{a}$, $\vec{b}$, $\vec{c}$ are   Figure 13.43: The vectors $\vec{a}$, $\vec{b}$, $\vec{c}$ are
Parallelepiped   called a right-handed set   called a left-handed set

The vectors $\vec{a}$, $\vec{b}$, and $\vec{c}$ can be arranged either as in Figure 13.42 or as in Figure 13.43. In either case,

$$\text{Height of parallelepiped} = \|\vec{a}\| \cos\theta,$$

where θ is the angle shown in the figures. In Figure 13.42 the angle θ is less than $\pi/2$, so the product, $(\vec{b} \times \vec{c}) \cdot \vec{a}$, called the *triple product*, is positive. Thus, in this case

$$\text{Volume of parallelepiped} = \text{Base} \cdot \text{Height} = \|\vec{b} \times \vec{c}\| \cdot \|\vec{a}\| \cos\theta = (\vec{b} \times \vec{c}) \cdot \vec{a}.$$

In Figure 13.43, the angle, $\pi - \theta$, between $\vec{a}$ and $\vec{b} \times \vec{c}$ is more than $\pi/2$, so the product $(\vec{b} \times \vec{c}) \cdot \vec{a}$ is negative. Thus, in this case we have

$$\text{Volume} = \text{Base} \cdot \text{Height} = \|\vec{b} \times \vec{c}\| \cdot \|\vec{a}\| \cos\theta = -\|\vec{b} \times \vec{c}\| \cdot \|\vec{a}\| \cos(\pi - \theta)$$
$$= -(\vec{b} \times \vec{c}) \cdot \vec{a} = \left| (\vec{b} \times \vec{c}) \cdot \vec{a} \right|.$$

Therefore, in both cases the volume is given by $\left| (\vec{b} \times \vec{c}) \cdot \vec{a} \right|$. Using determinants, we can write

Volume of a parallelepiped with edges $\vec{a}$, $\vec{b}$, $\vec{c}$ is given by

$$\text{Volume} = \left| (\vec{b} \times \vec{c}) \cdot \vec{a} \right| = \text{Absolute value of the determinant} \begin{vmatrix} a_1 & a_2 & a_3 \\ b_1 & b_2 & b_3 \\ c_1 & c_2 & c_3 \end{vmatrix}.$$

Exercises and Problems for Section 13.4

Exercises

In Exercises 1–6, use the algebraic definition to find $\vec{v} \times \vec{w}$.

1. $\vec{v} = \vec{k}, \vec{w} = \vec{j}$

2. $\vec{v} = -\vec{i}, \vec{w} = \vec{j} + \vec{k}$

3. $\vec{v} = \vec{i} + \vec{k}, \vec{w} = \vec{i} + \vec{j}$

4. $\vec{v} = \vec{i} + \vec{j} + \vec{k}, \vec{w} = \vec{i} + \vec{j} + -\vec{k}$

5. $\vec{v} = 2\vec{i} - 3\vec{j} + \vec{k}, \vec{w} = \vec{i} + 2\vec{j} - \vec{k}$

6. $\vec{v} = 2\vec{i} - \vec{j} - \vec{k}, \vec{w} = -6\vec{i} + 3\vec{j} + 3\vec{k}$

In Exercises 7–8, use the properties on page 675 to find:

7. $\left((\vec{i} + \vec{j}) \times \vec{i} \right) \times \vec{j}$

8. $(\vec{i} + \vec{j}) \times (\vec{i} \times \vec{j})$

Use the geometric definition in Exercises 9–10 to find:

9. $2\vec{i} \times (\vec{i} + \vec{j})$

10. $(\vec{i} + \vec{j}) \times (\vec{i} - \vec{j})$

11. If $\vec{v} = 3\vec{i} - 2\vec{j} + 4\vec{k}$ and $\vec{w} = \vec{i} + 2\vec{j} - \vec{k}$, find $\vec{v} \times \vec{w}$ and $\vec{w} \times \vec{v}$. What is the relation between the two answers?

12. For $\vec{a} = 3\vec{i} + \vec{j} - \vec{k}$ and $\vec{b} = \vec{i} - 4\vec{j} + 2\vec{k}$, find $\vec{a} \times \vec{b}$ and check that it is perpendicular to both $\vec{a}$ and $\vec{b}$.

Find an equation for the plane through the points in Exercises 13–14.

13. $(1, 0, 0), (0, 1, 0), (0, 0, 1)$.

14. $(3, 4, 2), (-2, 1, 0), (0, 2, 1)$.

Problems

15. Suppose $\vec{v} \cdot \vec{w} = 5$ and $\|\vec{v} \times \vec{w}\| = 3$, and the angle between $\vec{v}$ and $\vec{w}$ is θ. Find

    (a) $\tan \theta$ **(b)** θ.

16. If $\vec{v} \times \vec{w} = 2\vec{i} - 3\vec{j} + 5\vec{k}$, and $\vec{v} \cdot \vec{w} = 3$, find $\tan \theta$ where θ is the angle between $\vec{v}$ and $\vec{w}$.

17. If $\vec{v}$ and $\vec{w}$ are both parallel to the xy-plane, what can you conclude about $\vec{v} \times \vec{w}$? Explain.

18. Let $A = (-1, 3, 0)$, $B = (3, 2, 4)$, and $C = (1, -1, 5)$.

    (a) Find an equation for the plane that passes through these three points.

    (b) Find the area of the triangle determined by these three points.

19. Let $P = (0, 1, 0)$, $\quad Q = (-1, 1, 2)$, $\quad R = (2, 1, -1)$. Find

    (a) The area of the triangle PQR.

    (b) The equation for a plane that contains P, Q, and R.

20. Find a vector parallel to the line of intersection of the two planes $4x - 3y + 2z = 12$ and $x + 5y - z = 25$.

21. Find a vector parallel to the intersection of the planes $2x - 3y + 5z = 2$ and $4x + y - 3z = 7$.

22. Find the equation of the plane through the origin which is perpendicular to the line of intersection of the planes in Problem 21.

23. Find the equation of the plane through the point $(4, 5, 6)$ which is perpendicular to the line of intersection of the planes in Problem 21.

24. Find the equation of a plane through the origin and perpendicular to $x - y + z = 5$ and $2x + y - 2z = 7$.

25. Use the algebraic definition to check that

$$\vec{a} \times (\vec{b} + \vec{c}) = (\vec{a} \times \vec{b}) + (\vec{a} \times \vec{c}).$$

26. In this problem, we arrive at the algebraic definition for the cross product by a different route. Let $\vec{a} = a_1\vec{i} + a_2\vec{j} + a_3\vec{k}$ and $\vec{b} = b_1\vec{i} + b_2\vec{j} + b_3\vec{k}$. We seek a vector $\vec{v} = x\vec{i} + y\vec{j} + z\vec{k}$ which is perpendicular to both $\vec{a}$ and $\vec{b}$. Use this requirement to construct two equations for x, y, and z. Eliminate x and solve for y in terms of z. Then eliminate y and solve for x in terms of z. Since z can be any value whatsoever (the direction of $\vec{v}$ is unaffected), select the value for z which eliminates the denominator in the equation you obtained. How does the resulting expression for $\vec{v}$ compare to the formula we derived on page 674?

27. Suppose $\vec{a}$ and $\vec{b}$ are vectors in the xy-plane, such that $\vec{a} = a_1\vec{i} + a_2\vec{j}$ and $\vec{b} = b_1\vec{i} + b_2\vec{j}$ with $0 < a_2 < a_1$ and $0 < b_1 < b_2$.

    (a) Sketch $\vec{a}$ and $\vec{b}$ and the vector $\vec{c} = -a_2\vec{i} + a_1\vec{j}$. Shade the parallelogram formed by $\vec{a}$ and $\vec{b}$.

    (b) What is the relation between $\vec{a}$ and $\vec{c}$? [Hint: Find $\vec{c} \cdot \vec{a}$ and $\vec{c} \cdot \vec{c}$.]

    (c) Find $\vec{c} \cdot \vec{b}$.

    (d) Explain why $\vec{c} \cdot \vec{b}$ gives the area of the parallelogram formed by $\vec{a}$ and $\vec{b}$.

    (e) Check that in this case $\vec{a} \times \vec{b} = (a_1b_2 - a_2b_1)\vec{k}$.

28. If $\vec{v}$ and $\vec{w}$ are nonzero vectors, use the geometric definition of the cross product to explain why

$$(\lambda\vec{v}) \times \vec{w} = \lambda(\vec{v} \times \vec{w}) = \vec{v} \times (\lambda\vec{w}).$$

Consider the cases $\lambda > 0$, and $\lambda = 0$, and $\lambda < 0$ separately.

29. Use a parallelepiped to show that $\vec{a} \cdot (\vec{b} \times \vec{c}) = (\vec{a} \times \vec{b}) \cdot \vec{c}$ for any vectors $\vec{a}$, $\vec{b}$, and $\vec{c}$.

30. Show that $\|\vec{a} \times \vec{b}\|^2 = \|\vec{a}\|^2\|\vec{b}\|^2 - (\vec{a} \cdot \vec{b})^2$.

31. Use the fact that $\vec{i} \times \vec{i} = \vec{0}, \vec{i} \times \vec{j} = \vec{k}, \vec{i} \times \vec{k} = -\vec{j}$, and so on, together with the properties on page 675 to derive the algebraic definition for the cross product.

32. For vectors $\vec{a}$ and $\vec{b}$, let $\vec{c} = \vec{a} \times (\vec{b} \times \vec{a})$.

  (a) Show that $\vec{c}$ lies in the plane containing $\vec{a}$ and $\vec{b}$.
  (b) Use Problems 29 and 30 to show that $\vec{a} \cdot \vec{c} = 0$ and $\vec{b} \cdot \vec{c} = \|\vec{a}\|^2\|\vec{b}\|^2 - (\vec{a} \cdot \vec{b})^2$.
  (c) Show that
$$\vec{a} \times (\vec{b} \times \vec{a}) = \|\vec{a}\|^2\vec{b} - (\vec{a} \cdot \vec{b})\vec{a}.$$

33. Use the result of Problem 29 to show that the cross product distributes over addition. First, use distributivity for the dot product to show that for any vector $\vec{d}$,
$$[(\vec{a} + \vec{b}) \times \vec{c}] \cdot \vec{d} = [(\vec{a} \times \vec{c}) + (\vec{b} \times \vec{c})] \cdot \vec{d}.$$

Next, show that for any vector $\vec{d}$,
$$[((\vec{a} + \vec{b}) \times \vec{c}) - (\vec{a} \times \vec{c}) - (\vec{b} \times \vec{c})] \cdot \vec{d} = 0.$$

Finally, explain why you can conclude that
$$(\vec{a} + \vec{b}) \times \vec{c} = (\vec{a} \times \vec{c}) + (\vec{b} \times \vec{c}).$$

34. Figure 13.44 shows the tetrahedron determined by three vectors $\vec{a}, \vec{b}, \vec{c}$. The *area vector* of a face is a vector perpendicular to the face, pointing outward, whose magnitude is the area of the face. Show that the sum of the four outward pointing area vectors of the faces equals the zero vector.

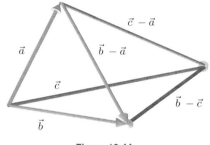

Figure 13.44

CHAPTER SUMMARY (see also Ready Reference at the end of the book)

- **Vectors**
  Geometric definition of vector addition, subtraction and scalar multiplication, resolving into $\vec{i}, \vec{j}$, and $\vec{k}$ components, magnitude of a vector, algebraic properties of addition and scalar multiplication.

- **Dot Product**
  Geometric and algebraic definition, algebraic properties, using dot products to find angles and determine

  perpendicularity, the equation of a plane with given normal vector passing through a given point, projection of a vector in a direction given by a unit vector.

- **Cross Product**
  Geometric and algebraic definition, algebraic properties, cross product and volume, finding the equation of a plane through three points.

REVIEW EXERCISES AND PROBLEMS FOR CHAPTER THIRTEEN

Exercises

In Exercises 1–10, use $\vec{v} = 2\vec{i} + 3\vec{j} - \vec{k}$ and $\vec{w} = \vec{i} - \vec{j} + 2\vec{k}$ to calculate the given quantities.

1. $\vec{v} + 2\vec{w}$

2. $3\vec{v} - \vec{w} - \vec{v}$

3. $\vec{v} \cdot \vec{w}$

4. $\vec{v} \times \vec{w}$

5. $\vec{v} \times \vec{v}$

6. $\|\vec{v} + \vec{w}\|$

7. $(\vec{v} \cdot \vec{w})\vec{v}$

8. $(\vec{v} \times \vec{w}) \times \vec{w}$

9. $(\vec{v} \times \vec{w}) \cdot \vec{w}$

10. $(\vec{v} \times \vec{w}) \times (\vec{v} \times \vec{w})$

11. Let $\vec{v} = 3\vec{i} + 2\vec{j} - 2\vec{k}$ and $\vec{w} = 4\vec{i} - 3\vec{j} + \vec{k}$. Find each of the following:

  (a) $\vec{v} \cdot \vec{w}$

  (b) $\vec{v} \times \vec{w}$
  (c) A vector of length 5 parallel to vector $\vec{v}$
  (d) The angle between vectors $\vec{v}$ and $\vec{w}$
  (e) The component of vector $\vec{v}$ in the direction of vector $\vec{w}$
  (f) A vector perpendicular to vector $\vec{v}$
  (g) A vector perpendicular to both vectors $\vec{v}$ and $\vec{w}$

In Exercises 12–15, find a vector with the given property.

12. Length 10, parallel to $2\vec{i} + 3\vec{j} - \vec{k}$.

13. Unit vector perpendicular to $\vec{i} + \vec{j}$ and $\vec{i} - \vec{j} - \vec{k}$

14. Unit vector in the xy-plane perpendicular to $3\vec{i} - 2\vec{j}$.

15. The vector obtained from $4\vec{i} + 3\vec{j}$ by rotating it $90°$ counterclockwise.

Problems

16. For what values of t are the following pairs of vectors parallel?

   (a) $2\vec{i} + (t^2 + \frac{2}{3}t + 1)\vec{j} + t\vec{k}, \qquad 6\vec{i} + 8\vec{j} + 3\vec{k}$
   (b) $t\vec{i} + \vec{j} + (t-1)\vec{k}, \qquad 2\vec{i} - 4\vec{j} + \vec{k}$
   (c) $2t\vec{i} + t\vec{j} + t\vec{k}, \qquad 6\vec{i} + 3\vec{j} + 3\vec{k}.$

17. Suppose $\vec{v} \cdot \vec{w} = 8$ and $\vec{v} \times \vec{w} = 12\vec{i} - 3\vec{j} + 4\vec{k}$ and that the angle between $\vec{v}$ and $\vec{w}$ is θ. Find

   (a) $\tan\theta$ **(b)** θ

18. Find a vector normal to $4(x - 1) + 6(z + 3) = 12$.

19. Find the equation of the plane through the origin which is parallel to $z = 4x - 3y + 8$.

20. Consider the plane $5x - y + 7z = 21$.

   (a) Find a point on the x-axis on this plane.
   (b) Find two other points on the plane.
   (c) Find a vector perpendicular to the plane.
   (d) Find a vector parallel to the plane.

21. Find the area of the triangle with vertices $P = (-2, 2, 0)$, $Q = (1, 3, -1)$, and $R = (-4, 2, 1)$.

22. A plane is drawn through the points $A = (2, 1, 0)$, $B = (0, 1, 3)$ and $C = (1, 0, 1)$. Find

   (a) Two vectors lying in the plane.
   (b) A vector perpendicular to the plane.
   (c) The equation of the plane.

23. Given the points $P = (1, 2, 3)$, $Q = (3, 5, 7)$, and $R = (2, 5, 3)$, find:

   (a) A unit vector perpendicular to a plane containing P, Q, R.
   (b) The angle between PQ and PR.
   (c) The area of the triangle PQR.
   (d) The distance from R to the line through P and Q.

24. Find the distance from the point $P = (2, -1, 3)$ to the plane $2x + 4y - z = -1$.

25. Find an equation of the plane passing through the three points $(1, 1, 1)$, $(1, 4, 5)$, $(-3, -2, 0)$. Find the distance from the origin to the plane.

26. An airplane heads northeast at an airspeed of 700 km/hr, but there is a wind blowing from the west at 60 km/hr. In what direction does the plane end up flying? What is its speed relative to the ground?

27. A man wishes to row the shortest possible distance from north to south across a river which is flowing at 4 km/hr from the east. He can row at 5 km/hr.

   (a) In which direction should he steer?
   (b) If there is a wind of 10 km/hr from the southwest, in which direction should he steer to try and go directly across the river? What happens?

28. A particle moving with speed v hits a barrier at an angle of $60°$ and bounces off at an angle of $60°$ in the opposite direction with speed reduced by 20 percent. See Figure 13.45. Find the velocity vector of the object after impact.

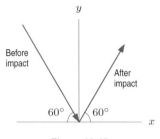

Figure 13.45

29. An airport is at the point $(200, 10, 0)$ and an approaching plane is at the point $(550, 60, 4)$. Assume that the xy-plane is horizontal, with the x-axis pointing eastward and the y-axis pointing northward. Also assume that the z-axis is upward and that all distances are measured in kilometers. The plane flies due west at a constant altitude at a speed of 500 km/hr for half an hour. It then descends at 200 km/hr, heading straight for the airport.

   (a) Find the velocity vector of the plane while it is flying at constant altitude.
   (b) Find the coordinates of the point at which the plane starts to descend.
   (c) Find a vector representing the velocity of the plane when it is descending.

30. The earth is at the origin, the moon is at the point $(384, 0)$, and a spaceship is at $(280, 90)$, where distance is in thousands of kilometers.

   (a) What is the displacement vector of the moon relative to the earth? Of the spaceship relative to the earth? Of the spaceship relative to the moon?
   (b) How far is the spaceship from the earth? From the moon?
   (c) The gravitational force on the spaceship from the earth is 461 newtons, and force from the moon is 26 newtons. What is the resulting force?

31. A large ship is being towed by two tugs. The larger tug exerts a force which is 25% greater than the smaller tug and at an angle of 30 degrees north of east. Which direction must the smaller tug pull to ensure that the ship travels due east?

32. The current in a river is pushing a boat in direction $25°$ north of east with a speed of 12 km/hr. The wind is pushing the same boat in a direction $80°$ south of east with a speed of 7 km/hr. Find the velocity vector of the boat's engine (relative to the water) if the boat actually moves due east at a speed of 40 km/hr relative to the ground.

33. Three people are trying to hold a ferocious lion still for the veterinarian. The lion, in the center, is wearing a collar with three ropes attached to it and each person has hold of a rope. Charlie is pulling in the direction $62°$ west of north with a force of 175 newtons and Sam is pulling in the direction $43°$ east of north with a force of 200 newtons. What is the direction and magnitude of the force which must be exerted by Alice on the third rope to counterbalance Sam and Charlie?

34. An object is moving counterclockwise at a constant speed around the circle $x^2 + y^2 = 1$, where x and y are measured in meters. It completes one revolution every minute.

    (a) What is its speed?
    (b) What is its velocity vector 30 seconds after it passes the point $(1, 0)$? Does your answer change if the object is moving clockwise? Explain.

35. An object is attached by a string to a fixed point and rotates 30 times per minute in a horizontal plane. Show that

the speed of the object is constant but the velocity is not. What does this imply about the acceleration?

36. Find the vector $\vec{v}$ with all of the following properties:

    - Magnitude 10
    - Angle of $45°$ with positive x-axis
    - Angle of $75°$ with positive y-axis
    - Positive $\vec{k}$ -component.

37. (a) A vector $\vec{v}$ of magnitude v makes an angle α with the positive x-axis, angle β with the positive y-axis, and angle γ with the positive z-axis. Show that

    $$\vec{v} = v \cos \alpha \vec{i} + v \cos \beta \vec{j} + v \cos \gamma \vec{k}.$$

    (b) $\cos \alpha$, $\cos \beta$, and $\cos \gamma$ are called *direction cosines*. Show that

    $$\cos^2 \alpha + \cos^2 \beta + \cos^2 \gamma = 1.$$

38. Two lines in space are skew if they are not parallel and do not intersect. Determine the minimum distance between two such lines.

CAS Challenge Problems

39. Let $\vec{a} = x\vec{i} + y\vec{j} + z\vec{k}$, $\vec{b} = u\vec{i} + v\vec{j} + w\vec{k}$, and $\vec{c} = m\vec{a} + n\vec{b}$. Compute $(\vec{a} \times \vec{b}) \cdot \vec{c}$ and $(\vec{a} \times \vec{b}) \times (\vec{a} \times \vec{c})$, and explain the geometric meaning of your answers.

40. Let $\vec{a} = x\vec{i} + y\vec{j} + z\vec{k}$, $\vec{b} = u\vec{i} + v\vec{j} + w\vec{k}$ and $\vec{c} = r\vec{i} + s\vec{j} + t\vec{k}$. Show that the parallelepiped with edges $\vec{a}$, $\vec{b}$, $\vec{c}$ has the same volume as the parallelepiped with edges $\vec{a}$, $\vec{b}$, $2\vec{a} - \vec{b} + \vec{c}$. Explain this result geometrically.

41. Let $\vec{a} = \vec{i} + 2\vec{j} + 3\vec{k}$, $\vec{b} = 2\vec{i} + \vec{j} + 2\vec{k}$ and let θ be the angle between $\vec{a}$ and $\vec{b}$.

    (a) For $\vec{c} = x\vec{i} + y\vec{j} + z\vec{k}$, write the following conditions as equations in x, y, z and solve them:

    $$\vec{a} \cdot \vec{c} = 0, \quad \vec{b} \cdot \vec{c} = 0, \quad \|\vec{c}\|^2 = \|\vec{a}\|^2 \|\vec{b}\|^2 \sin^2 \theta.$$

    [Hint: Use the dot product to find $\sin^2 \theta$.]

    (b) Compute the cross product $\vec{a} \times \vec{b}$ and compare with your answer in part (a). What do you notice? Explain.

42. Let $A = (0, 0, 0)$, $B = (2, 0, 0)$, $C = (1, \sqrt{3}, 0)$ and $D = (1, 1/\sqrt{3}, 2\sqrt{2/3})$.

    (a) Show that A, B, C, D are all the same distance from each other.
    (b) Find the point $P = (x, y, z)$ which is equidistant from A, B, C and D by setting up and solving three equations in x, y, and z.
    (c) Use the dot product to find the angle APB. (In chemistry, this angle is often approximated by $109.5°$. A methane molecule can be represented by four hydrogen atoms at points A, B, C and D, and a carbon atom at P.)

43. Let $P = (x, y, z)$, $Q = (u, v, w)$ and $R = (r, s, t)$ be points on the plane $ax + by + cz = d$.

    (a) What is the relation between $\overrightarrow{PQ} \times \overrightarrow{PR}$ and the normal vector to the plane, $a\vec{i} + b\vec{j} + c\vec{k}$?
    (b) Express $\overrightarrow{PQ} \times \overrightarrow{PR}$ in terms of $x, y, z, u, v, w, r, s, t$.
    (c) Use the equation for the plane to eliminate z, w, and t from the expression you obtained in part (b), and simplify. Does your answer agree with what you said in part (a)?

CHECK YOUR UNDERSTANDING

Are the statements in Problems 1–30 true or false? Give reasons for your answer.

1. There is exactly one unit vector parallel to a given nonzero vector $\vec{v}$.

2. The vector $\dfrac{1}{\sqrt{3}}\vec{i} + \dfrac{-1}{\sqrt{3}}\vec{j} + \dfrac{2}{\sqrt{3}}\vec{k}$ is a unit vector.

3. The length of the vector $2\vec{v}$ is twice the length of the vector $\vec{v}$.

4. If $\vec{v}$ and $\vec{w}$ are any two vectors, then $\|\vec{v} + \vec{w}\| = \|\vec{v}\| + \|\vec{w}\|$.

5. If $\vec{v}$ and $\vec{w}$ are any two vectors, then $\|\vec{v} - \vec{w}\| = \|\vec{v}\| - \|\vec{w}\|$.

6. The vectors $2\vec{i} - \vec{j} + \vec{k}$ and $\vec{i} - 2\vec{j} + \vec{k}$ are parallel.

7. The vector $\vec{u} + \vec{v}$ is always larger in magnitude than both $\vec{u}$ and $\vec{v}$.

8. For any scalar c and vector $\vec{v}$ we have $\|c\vec{v}\| = c\|\vec{v}\|$.

9. The displacement vector from $(1, 1, 1)$ to $(1, 2, 3)$ is $-\vec{j} - 2\vec{k}$.

10. The displacement vector from (a, b) to (c, d) is the same as the displacement vector from (c, d) to (a, b).

11. The quantity $\vec{u} \cdot \vec{v}$ is a vector.

12. The plane $x + 2y - 3z = 5$ has normal vector $\vec{i} + 2\vec{j} - 3\vec{k}$.

13. If $\vec{u} \cdot \vec{v} < 0$ then the angle between $\vec{u}$ and $\vec{v}$ is greater than $\pi/2$.

14. An equation of the plane with normal vector $\vec{i} + \vec{j} + \vec{k}$ containing the point $(1, 2, 3)$ is $z = x + y$.

15. The triangle in 3-space with vertices $(1, 1, 0), (0, 1, 0)$ and $(0, 1, 1)$ has a right angle.

16. The dot product $\vec{v} \cdot \vec{v}$ is never negative.

17. If $\vec{u} \cdot \vec{v} = 0$ then either $\vec{u} = 0$ or $\vec{v} = 0$.

18. If $\vec{u}, \vec{v}$ and $\vec{w}$ are all nonzero, and $\vec{u} \cdot \vec{v} = \vec{u} \cdot \vec{w}$, then $\vec{v} = \vec{w}$.

19. For any vectors $\vec{u}$ and $\vec{v}$: $(\vec{u} + \vec{v}) \cdot (\vec{u} - \vec{v}) = \|\vec{u}\|^2 - \|\vec{v}\|^2$.

20. If $\|\vec{u}\| = 1$, then the vector $\vec{v} - (\vec{v} \cdot \vec{u})\vec{u}$ is perpendicular to $\vec{u}$.

21. $\vec{u} \times \vec{v}$ is a vector.

22. $\vec{u} \times \vec{v}$ has direction parallel to both $\vec{u}$ and $\vec{v}$.

23. $\|\vec{u} \times \vec{v}\| = \|\vec{u}\|\|\vec{v}\|$.

24. $(\vec{i} \times \vec{j}) \cdot \vec{k} = \vec{i} \cdot (\vec{j} \times \vec{k})$.

25. If $\vec{v}$ is a nonzero vector and $\vec{v} \times \vec{u} = \vec{v} \times \vec{w}$, then $\vec{u} = \vec{w}$.

26. The value of $\vec{v} \cdot (\vec{v} \times \vec{w})$ is always 0.

27. The value of $\vec{v} \times \vec{w}$ is never the same as $\vec{v} \cdot \vec{w}$.

28. The area of the triangle with two sides given by $\vec{i} + \vec{j}$ and $\vec{j} + 2\vec{k}$ is $3/2$.

29. Given a nonzero vector $\vec{v}$ in 3-space, there is a nonzero vector $\vec{w}$ such that $\vec{v} \times \vec{w} = \vec{0}$.

30. It is never true that $\vec{v} \times \vec{w} = \vec{w} \times \vec{v}$.

PROJECTS FOR CHAPTER THIRTEEN

1. Cross Product of Vectors in the Plane

Let $\vec{a} = a_1\vec{i} + a_2\vec{j}$ and $\vec{b} = b_1\vec{i} + b_2\vec{j}$ be two nonparallel vectors in 2-space, as in Figure 13.46.

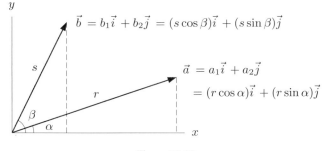

Figure 13.46

(a) Use the identity $\sin(\beta - \alpha) = (\sin\beta\cos\alpha - \cos\beta\sin\alpha)$ to derive the formula for the area of the parallelogram formed by $\vec{a}$ and $\vec{b}$:

$$\text{Area of parallelogram} = |a_1 b_2 - a_2 b_1|.$$

(b) Show that $a_1 b_2 - a_2 b_1$ is positive when the rotation from $\vec{a}$ to $\vec{b}$ is counterclockwise, and negative when it is clockwise.

(c) Use parts (a) and (b) to show that the geometric and algebraic definitions of $\vec{a} \times \vec{b}$ give the same result.

2. The Dot Product in Genetics[2]

We define[3] the angle between two n-dimensional vectors, $\vec{v}$ and $\vec{w}$, using the dot product:

$$\cos\theta = \frac{\vec{v}\cdot\vec{w}}{\|\vec{v}\|\|\vec{w}\|} = \frac{v_1w_1 + v_2w_2 + \cdots + v_nw_n}{\|\vec{v}\|\|\vec{w}\|}, \qquad \text{provided } \|\vec{v}\|, \|\vec{w}\| \neq 0.$$

We use this idea of angle to measure how close two populations are to one another genetically. The table shows the relative frequencies of four alleles (variants of a gene) in four populations.

Allele	Eskimo	Bantu	English	Korean
A_1	0.29	0.10	0.20	0.22
A_2	0.00	0.08	0.06	0.00
B	0.03	0.12	0.06	0.20
O	0.67	0.69	0.66	0.57

Let $\vec{a}_1$ be the 4-vector showing the relative frequencies of the alleles in the Eskimo population. Let $\vec{a}_2, \vec{a}_3, \vec{a}_4$ be the corresponding vectors for the Bantu, English, and Korean populations, respectively. The genetic distance between two populations is defined as the angle between the corresponding vectors.

(a) Using this definition, is the English population closer genetically to the Bantus or to the Koreans? Explain.

(b) Is the English population closer to a half Eskimo, half Bantu population than to the Bantu population alone?

(c) Among all possible populations that are a mix of Eskimo and Bantu, find the mix that is closest to the English population.

[2]Adapted from Cavalli-Sforza and Edwards, "Models and Estimation Procedures," Am J. Hum. Genet., Vol. 19 (1967), pp. 223-57.
[3]The result of Problem 47 on page 673 shows that the quantity on the right-hand side of this equation is between -1 and 1, so this definition makes sense.

Chapter Fourteen

DIFFERENTIATING FUNCTIONS OF SEVERAL VARIABLES

For a function of one variable, $y = f(x)$, the derivative $dy/dx = f'(x)$ gives the rate of change of y with respect to x. For a function of two variables, $z = f(x, y)$, there is no such thing as *the* rate of change, since x and y can each vary while the other is held fixed or both can vary at once. However, we can consider the rate of change with respect to each one of the independent variables. This chapter introduces these *partial derivatives* and several ways they can be used to get a complete picture of the way the function varies.

14.1   THE PARTIAL DERIVATIVE

The derivative of a one-variable function measures its rate of change. In this section we see how a two variable function has two rates of change: one as x changes (with y held constant) and one as y changes (with x held constant).

Rate of Change of Temperature in a Metal Rod: a One-Variable Problem

Imagine an unevenly heated metal rod lying along the x-axis, with its left end at the origin and x measured in meters. (See Figure 14.1.) Let $u(x)$ be the temperature (in °C) of the rod at the point x. Table 14.1 gives values of $u(x)$. We see that the temperature increases as we move along the rod, reaching its maximum at $x = 4$, after which it starts to decrease.

Figure 14.1: Unevenly heated metal rod

Table 14.1   *Temperature $u(x)$ of the rod*

x (m)	0	1	2	3	4	5
$u(x)$ (°C)	125	128	135	160	175	160

Example 1   Estimate the derivative $u'(2)$ using Table 14.1 and explain what the answer means in terms of temperature.

Solution   The derivative $u'(2)$ is defined as a limit of difference quotients:

$$u'(2) = \lim_{h \to 0} \frac{u(2 + h) - u(2)}{h}.$$

Choosing $h = 1$ so that we can use the data in Table 14.1, we get

$$u'(2) \approx \frac{u(2 + 1) - u(2)}{1} = \frac{160 - 135}{1} = 25.$$

This means that the temperature increases at a rate of approximately 25°C per meter as we go from left to right, past $x = 2$.

Rate of Change of Temperature in a Metal Plate

Imagine an unevenly heated thin rectangular metal plate lying in the xy-plane with its lower left corner at the origin and x and y measured in meters. The temperature (in °C) at the point (x, y) is $T(x, y)$. See Figure 14.2 and Table 14.2. How does T vary near the point $(2, 1)$? We consider the horizontal line $y = 1$ containing the point $(2, 1)$. The temperature along this line is the cross-section, $T(x, 1)$, of the function $T(x, y)$ with $y = 1$. Suppose we write $u(x) = T(x, 1)$.

Figure 14.2: Unevenly heated metal plate

Table 14.2   *Temperature (°C) of a metal plate*

y (m)	3	85	90	110	135	155	180
	2	100	110	120	145	190	170
	1	125	128	135	160	175	160
	0	120	135	155	160	160	150
		0	1	2	3	4	5

x (m)

What is the meaning of the derivative $u'(2)$? It is the rate of change of temperature T *in the x-direction* at the point $(2, 1)$, keeping y fixed. Denote this rate of change by $T_x(2, 1)$, so that

$$T_x(2, 1) = u'(2) = \lim_{h \to 0} \frac{u(2 + h) - u(2)}{h} = \lim_{h \to 0} \frac{T(2 + h, 1) - T(2, 1)}{h}.$$

We call $T_x(2, 1)$ the *partial derivative of T with respect to x at the point* $(2, 1)$. Taking $h = 1$, we can read values of T from the row with $y = 1$ in Table 13.2, giving

$$T_x(2, 1) \approx \frac{T(3, 1) - T(2, 1)}{1} = \frac{160 - 135}{1} = 25°\text{C/m}.$$

The fact that $T_x(2, 1)$ is positive means that the temperature of the plate is increasing as we move past the point $(2, 1)$ in the direction of increasing x (that is, horizontally from left to right in Figure 14.2).

Example 2   Estimate the rate of change of T in the y-direction at the point $(2, 1)$.

Solution   The temperature along the line $x = 2$ is the cross-section of T with $x = 2$, that is, the function $v(y) = T(2, y)$. If we denote the rate of change of T in the y-direction at $(2, 1)$ by $T_y(2, 1)$, then

$$T_y(2, 1) = v'(1) = \lim_{h \to 0} \frac{v(1 + h) - v(1)}{h} = \lim_{h \to 0} \frac{T(2, 1 + h) - T(2, 1)}{h}.$$

We call $T_y(2, 1)$ the *partial derivative of T with respect to y at the point* $(2, 1)$. Taking $h = 1$ so that we can use the column with $x = 2$ in Table 13.2, we get

$$T_y(2, 1) \approx \frac{T(2, 1 + 1) - T(2, 1)}{1} = \frac{120 - 135}{1} = -15°\text{C/m}.$$

The fact that $T_y(2, 1)$ is negative means that the temperature decreases as y increases.

Definition of the Partial Derivative

We study the influence of x and y separately on the value of the function $f(x, y)$ by holding one fixed and letting the other vary. This leads to the following definitions.

Partial Derivatives of f With Respect to x and y

For all points at which the limits exist, we define the **partial derivatives at the point** $(\mathbf{a}, \mathbf{b})$ by

$$f_x(a, b) = \begin{array}{c} \text{Rate of change of } f \text{ with respect to } x \\ \text{at the point } (a, b) \end{array} = \lim_{h \to 0} \frac{f(a + h, b) - f(a, b)}{h},$$

$$f_y(a, b) = \begin{array}{c} \text{Rate of change of } f \text{ with respect to } y \\ \text{at the point } (a, b) \end{array} = \lim_{h \to 0} \frac{f(a, b + h) - f(a, b)}{h}.$$

If we let a and b vary, we have the **partial derivative functions** $f_x(x, y)$ and $f_y(x, y)$.

Just as with ordinary derivatives, there is an alternative notation:

Alternative Notation for Partial Derivatives

If $z = f(x, y)$, we can write

$$f_x(x, y) = \frac{\partial z}{\partial x} \quad \text{and} \quad f_y(x, y) = \frac{\partial z}{\partial y},$$

$$f_x(a, b) = \frac{\partial z}{\partial x}\bigg|_{(a,b)} \quad \text{and} \quad f_y(a, b) = \frac{\partial z}{\partial y}\bigg|_{(a,b)}.$$

We use the symbol ∂ to distinguish partial derivatives from ordinary derivatives. In cases where the independent variables have names different from x and y, we adjust the notation accordingly. For example, the partial derivatives of $f(u, v)$ are denoted by f_u and f_v.

Visualizing Partial Derivatives on a Graph

The ordinary derivative of a one-variable function is the slope of its graph. How do we visualize the partial derivative $f_x(a, b)$? The graph of the one-variable function $f(x, b)$ is the curve where the vertical plane $y = b$ cuts the graph of $f(x, y)$. (See Figure 14.3.) Thus, $f_x(a, b)$ is the slope of the tangent line to this curve at $x = a$.

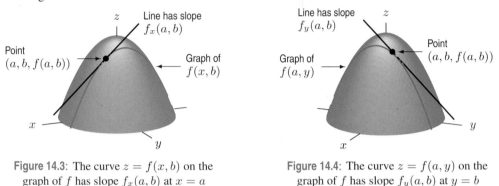

Figure 14.3: The curve $z = f(x, b)$ on the graph of f has slope $f_x(a, b)$ at $x = a$

Figure 14.4: The curve $z = f(a, y)$ on the graph of f has slope $f_y(a, b)$ at $y = b$

Similarly, the graph of the function $f(a, y)$ is the curve where the vertical plane $x = a$ cuts the graph of f, and the partial derivative $f_y(a, b)$ is the slope of this curve at $y = b$. (See Figure 14.4.)

Example 3   At each point labeled on the graph of the surface $z = f(x, y)$ in Figure 14.5, say whether each partial derivative is positive or negative.

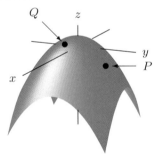

Figure 14.5: Decide the signs of f_x and f_y at P and Q

Solution   The positive x-axis points out of the page. Imagine heading off in this direction from the point marked P; we descend steeply. So the partial derivative with respect to x is negative at P, with quite a large absolute value. The same is true for the partial derivative with respect to y at P, since there is also a steep descent in the positive y-direction.

At the point marked Q, heading in the positive x-direction results in a gentle descent, whereas heading in the positive y-direction results in a gentle ascent. Thus, the partial derivative f_x at Q is negative but small (that is, near zero), and the partial derivative f_y is positive but small.

Estimating Partial Derivatives From a Contour Diagram

The graph of a function $f(x, y)$ often makes clear the sign of the partial derivatives. However, numerical estimates of these derivatives are more easily made from a contour diagram than a surface graph. If we move parallel to one of the axes on a contour diagram, the partial derivative is the rate of change of the value of the function on the contours. For example, if the values on the contours are increasing as we move in the positive direction, then the partial derivative must be positive.

Example 4   Figure 14.6 shows the contour diagram for the temperature $H(x, t)$ (in °C) in a room as a function of distance x (in meters) from a heater and time t (in minutes) after the heater has been turned on. What are the signs of $H_x(10, 20)$ and $H_t(10, 20)$? Estimate these partial derivatives and explain the answers in practical terms.

t (minutes)

x (meters)

Figure 14.6: Temperature in a heated room: Heater at $x = 0$ is turned on at $t = 0$

Solution   The point $(10, 20)$ is nearly on the $H = 25$ contour. As x increases, we move toward the $H = 20$ contour, so H is decreasing and $H_x(10, 20)$ is negative. This makes sense because the $H = 30$ contour is to the left: As we move further from the heater, the temperature drops. On the other hand, as t increases, we move toward the $H = 30$ contour, so H is increasing; as t decreases H decreases. Thus, $H_t(10, 20)$ is positive. This says that as time passes, the room warms up.

To estimate the partial derivatives, use a difference quotient. Looking at the contour diagram, we see there is a point on the $H = 20$ contour about 14 units to the right of the point $(10, 20)$. Hence, H decreases by 5 when x increases by 14, so we find

$$\text{Rate of change of } H \text{ with respect to } x = H_x(10, 20) \approx \frac{-5}{14} \approx -0.36°\text{C/meter.}$$

This means that near the point 10 m from the heater, after 20 minutes the temperature drops about 0.36, or one third, of a degree, for each meter we move away from the heater.

To estimate $H_t(10, 20)$, we notice that the $H = 30$ contour is about 32 units directly above the point $(10, 20)$. So H increases by 5 when t increases by 32. Hence,

$$\text{Rate of change of } H \text{ with respect to } t = H_t(10, 20) \approx \frac{5}{32} = 0.16°\text{C/minute.}$$

This means that after 20 minutes the temperature is going up about 0.16, or 1/6, of a degree each minute at the point 10 m from the heater.

Using Units to Interpret Partial Derivatives

The meaning of a partial derivative can often be explained using units.

Example 5   Suppose that your weight w in pounds is a function $f(c, n)$ of the number c of calories you consume daily and the number n of minutes you exercise daily. Using the units for w, c and n, interpret in everyday terms the statements

$$\frac{\partial w}{\partial c}(2000, 15) = 0.02 \quad \text{and} \quad \frac{\partial w}{\partial n}(2000, 15) = -0.025.$$

Solution   The units of $\partial w / \partial c$ are pounds per calorie. The statement

$$\frac{\partial w}{\partial c}(2000, 15) = 0.02$$

means that if you are presently consuming 2000 calories daily and exercising 15 minutes daily, you will weigh 0.02 pounds more for each extra calorie you consume daily, or about 2 pounds for each extra 100 calories per day. The units of $\partial w/\partial n$ are pounds per minute. The statement

$$\frac{\partial w}{\partial n}(2000, 15) = -0.025$$

means that for the same calorie consumption and number of minutes of exercise, you will weigh 0.025 pounds less for each extra minute you exercise daily, or about 1 pound less for each extra 40 minutes per day. So if you eat an extra 100 calories each day and exercise about 80 minutes more each day, your weight should remain roughly steady.

Exercises and Problems for Section 14.1

Exercises

1. Using difference quotients, estimate $f_x(3, 2)$ and $f_y(3, 2)$ for the function given by

$$f(x, y) = \frac{x^2}{y + 1}.$$

   [Recall: A difference quotient is an expression of the form $(f(a + h, b) - f(a, b))/h$.]

2. Use difference quotients with $\Delta x = 0.1$ and $\Delta y = 0.1$ to estimate $f_x(1, 3)$ and $f_y(1, 3)$ where

$$f(x, y) = e^{-x} \sin y.$$

   Then give better estimates by using $\Delta x = 0.01$ and $\Delta y = 0.01$.

3. The quantity, Q, of beef purchased at a store, in kilograms per week, is a function of the price of beef, b, and the price of chicken, c, both in dollars per kilogram.

   (a) Do you expect $\partial Q/\partial b$ to be positive or negative? Explain.
   (b) Do you expect $\partial Q/\partial c$ to be positive or negative? Explain.
   (c) Interpret the statement $\partial Q/\partial b = -213$ in terms of quantity of beef purchased.

4. The sales of a product, $S = f(p, a)$, is a function of the price, p, of the product (in dollars per unit) and the amount, a, spent on advertising (in thousands of dollars).

   (a) Do you expect f_p to be positive or negative? Why?
   (b) Explain the meaning of the statement $f_a(8, 12) = 150$ in terms of sales.

5. A drug is injected into a patient's blood vessel. The function $c = f(x, t)$ represents the concentration of the drug at a distance x mm in the direction of the blood flow measured from the point of injection and at time t seconds since the injection. What are the units of the following partial derivatives? What are their practical interpretations? What do you expect their signs to be?

   (a) $\partial c/\partial x$ (b) $\partial c/\partial t$

6. The monthly mortgage payment in dollars, P, for a house is a function of three variables

$$P = f(A, r, N),$$

where A is the amount borrowed in dollars, r is the interest rate, and N is the number of years before the mortgage is paid off.

   (a) $f(92000, 14, 30) = 1090.08$. What does this tell you, in financial terms?
   (b) $\left.\dfrac{\partial P}{\partial r}\right|_{(92000, 14, 30)} = 72.82$. What is the financial significance of the number 72.82?
   (c) Would you expect $\partial P/\partial A$ to be positive or negative? Why?
   (d) Would you expect $\partial P/\partial N$ to be positive or negative? Why?

7. Suppose that x is the average price of a new car and that y is the average price of a gallon of gasoline. Then q_1, the number of new cars bought in a year, depends on both x and y, so $q_1 = f(x, y)$. Similarly, if q_2 is the quantity of gas bought in a year, then $q_2 = g(x, y)$.

   (a) What do you expect the signs of $\partial q_1/\partial x$ and $\partial q_2/\partial y$ to be? Explain.
   (b) What do you expect the signs of $\partial q_1/\partial y$ and $\partial q_2/\partial x$ to be? Explain.

8. Figure 14.7 is a contour diagram of $f(x, y)$. In each of the following cases, list the marked points in the diagram (there may be none or more than one) at which

   (a) $f_x < 0$ (b) $f_y > 0$
   (c) $f_{xx} > 0$ (d) $f_{yy} < 0$

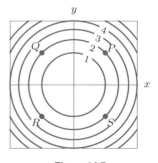

Figure 14.7

9. Figure 14.8 is a contour diagram for $z = f(x, y)$. Is f_x positive or negative? Is f_y positive or negative? Estimate $f(2, 1)$, $f_x(2, 1)$, and $f_y(2, 1)$.

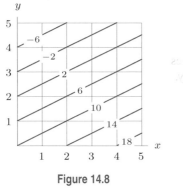

Figure 14.8

For Exercises 10–12, refer to Table 12.2 on page 611 giving the temperature adjusted for wind-chill, C, in °F, as a function $f(w, T)$ of the wind speed, w, in mph, and the temperature, T, in °F. The temperature adjusted for wind-chill tells you how cold it feels, as a result of the combination of wind and temperature.

10. Estimate $f_w(10, 25)$. What does your answer mean in practical terms?

11. Estimate $f_T(5, 20)$. What does your answer mean in practical terms?

12. From Table 12.2 you can see that when the temperature is 20°F, the temperature adjusted for wind-chill drops by an average of about 0.8°F with every 1 mph increase in wind speed from 5 mph to 10 mph. Which partial derivative is this telling you about?

Problems

13. Figure 14.9 gives a contour diagram for the number n of foxes per square kilometer in southwestern England. Estimate $\partial n / \partial x$ and $\partial n / \partial y$ at the points A, B, and C, where x is kilometers east and y is kilometers north.

Figure 14.9

14. The cardiac output, represented by c, is the volume of blood flowing through a person's heart, per unit time. The systemic vascular resistance (SVR), represented by s, is the resistance to blood flowing through veins and arteries. Let p be a person's blood pressure. Then p is a function of c and s, so $p = f(c, s)$.

(a) What does $\partial p / \partial c$ represent?

Suppose now that $p = kcs$, where k is a constant.

(b) Sketch the level curves of p. What do they represent? Label your axes.
(c) For a person with a weak heart, it is desirable to have the heart pumping against less resistance, while maintaining the same blood pressure. Such a person may be given the drug nitroglycerine to decrease

the SVR and the drug Dopamine to increase the cardiac output. Represent this on a graph showing level curves. Put a point A on the graph representing the person's state before drugs are given and a point B for after.
(d) Right after a heart attack, a patient's cardiac output drops, thereby causing the blood pressure to drop. A common mistake made by medical residents is to get the patient's blood pressure back to normal by using drugs to increase the SVR, rather than by increasing the cardiac output. On a graph of the level curves of p, put a point D representing the patient before the heart attack, a point E representing the patient right after the heart attack, and a third point F representing the patient after the resident has given the drugs to increase the SVR.

15. Figure 14.10 shows a contour diagram for the monthly payment P as a function of the interest rate, $r\%$, and the amount, L, of a 5-year loan. Estimate $\partial P / \partial r$ and $\partial P / \partial L$ at the following points. In each case, give the units and the everyday meaning of your answer.

(a) $r = 8$, $L = 4000$ (b) $r = 8$, $L = 6000$
(c) $r = 13$, $L = 7000$

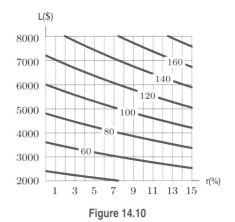

Figure 14.10

16. An experiment to measure the toxicity of formaldehyde yielded the data in Table 14.3. The values show the percent, $P = f(t, c)$, of rats surviving an exposure to formaldehyde at a concentration of c (in parts per million, ppm) after t months. Estimate $f_t(18, 6)$ and $f_c(18, 6)$. Interpret your answers in terms of formaldehyde toxicity.

Table 14.3

		\multicolumn{6}{c}{Time t (months)}
		14	16	18	20	22	24
	0	100	100	100	99	97	95
Conc. c	2	100	99	98	97	95	92
(ppm)	6	96	95	93	90	86	80
	15	96	93	82	70	58	36

17. The surface $z = f(x, y)$ is shown in Figure 14.11. The points A and B are in the xy-plane.

(a) What is the sign of

  (i) $f_x(A)$?   (ii) $f_y(A)$?

(b) The point P in the xy-plane moves along a straight line from A to B. How does the sign of $f_x(P)$ change? How does the sign of $f_y(P)$ change?

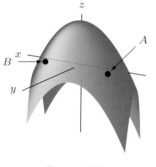

Figure 14.11

18. Figure 14.12 shows the saddle-shaped surface $z = f(x, y)$.

(a) What is the sign of $f_x(0, 5)$?
(b) What is the sign of $f_y(0, 5)$?

Figure 14.12

19. Figure 14.13 shows contours of $f(x, y)$ with values of f on the contours omitted. If $f_x(P) > 0$, find the sign of

(a) $f_y(P)$ **(b)** $f_y(Q)$ **(c)** $f_x(Q)$

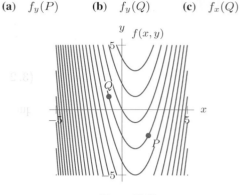

Figure 14.13

20. In each case, give a possible contour diagram for the function $f(x, y)$ if

(a) $f_x > 0$ and $f_y > 0$ **(b)** $f_x > 0$ and $f_y < 0$
(c) $f_x < 0$ and $f_y > 0$ **(d)** $f_x < 0$ and $f_y < 0$

An airport can be cleared of fog by heating the air. The amount of heat required depends on the air temperature and the wetness of the fog. Problems 21–23 involve Figure 14.14 which shows the heat $H(T, w)$ required (in calories per cubic meter of fog) as a function of the temperature T (in degrees Celsius) and the water content w (in grams per cubic meter of fog). Note that Figure 14.14 is not a contour diagram, but shows cross-sections of H with w fixed at 0.1, 0.2, 0.3, and 0.4.

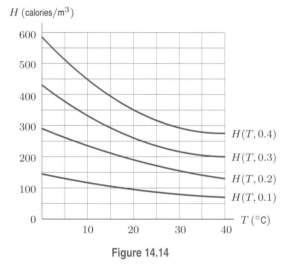

Figure 14.14

21. Use Figure 14.14 to estimate $H_T(10, 0.1)$. Interpret the partial derivative in practical terms.

22. Make a table of values for $H(T, w)$ from Figure 14.14, and use it to estimate $H_T(T, w)$ for $T = 10, 20$, and 30 and $w = 0.1, 0.2$, and 0.3.

23. Repeat Problem 22 for $H_w(T, w)$ at $T = 10, 20$, and 30 and $w = 0.1, 0.2$, and 0.3. What is the practical meaning of these partial derivatives?

14.2 COMPUTING PARTIAL DERIVATIVES ALGEBRAICALLY

Since the partial derivative $f_x(x, y)$ is the ordinary derivative of the function $f(x, y)$ with y held constant and $f_y(x, y)$ is the ordinary derivative of $f(x, y)$ with x held constant, we can use all the differentiation formulas from one-variable calculus to find partial derivatives.

Example 1   Let $f(x, y) = \dfrac{x^2}{y + 1}$. Find $f_x(3, 2)$ algebraically.

Solution   We use the fact that $f_x(3, 2)$ equals the derivative of $f(x, 2)$ at $x = 3$. Since

$$f(x, 2) = \frac{x^2}{2 + 1} = \frac{x^2}{3},$$

differentiating with respect to x, we have

$$f_x(x, 2) = \frac{\partial}{\partial x}\left(\frac{x^2}{3}\right) = \frac{2x}{3}, \qquad \text{and so} \qquad f_x(3, 2) = 2.$$

Example 2   Compute the partial derivatives with respect to x and with respect to y for the following functions.
   (a) $f(x, y) = y^2 e^{3x}$ (b) $z = (3xy + 2x)^5$ (c) $g(x, y) = e^{x+3y} \sin(xy)$

Solution   (a) This is the product of a function of x (namely e^{3x}) and a function of y (namely y^2). When we differentiate with respect to x, we think of the function of y as a constant, and vice versa. Thus,

$$f_x(x, y) = y^2 \frac{\partial}{\partial x}\left(e^{3x}\right) = 3y^2 e^{3x},$$

$$f_y(x, y) = e^{3x} \frac{\partial}{\partial y}(y^2) = 2ye^{3x}.$$

(b) Here we use the chain rule:

$$\frac{\partial z}{\partial x} = 5(3xy + 2x)^4 \frac{\partial}{\partial x}(3xy + 2x) = 5(3xy + 2x)^4(3y + 2),$$

$$\frac{\partial z}{\partial y} = 5(3xy + 2x)^4 \frac{\partial}{\partial y}(3xy + 2x) = 5(3xy + 2x)^4 3x = 15x(3xy + 2x)^4.$$

(c) Since each function in the product is a function of both x and y, we need to use the product rule for each partial derivative:

$$g_x(x, y) = \left(\frac{\partial}{\partial x}(e^{x+3y})\right)\sin(xy) + e^{x+3y}\frac{\partial}{\partial x}(\sin(xy)) = e^{x+3y}\sin(xy) + e^{x+3y}y\cos(xy),$$

$$g_y(x, y) = \left(\frac{\partial}{\partial y}(e^{x+3y})\right)\sin(xy) + e^{x+3y}\frac{\partial}{\partial y}(\sin(xy)) = 3e^{x+3y}\sin(xy) + e^{x+3y}x\cos(xy).$$

For functions of three or more variables, we find partial derivatives by the same method: Differentiate with respect to one variable, regarding the other variables as constants. For a function $f(x, y, z)$, the partial derivative $f_x(a, b, c)$ gives the rate of change of f with respect to x along the line $y = b$, $z = c$.

Example 3   Find all the partial derivatives of $f(x, y, z) = \dfrac{x^2 y^3}{z}$.

Solution   To find $f_x(x, y, z)$, for example, we consider y and z as fixed, giving

$$f_x(x, y, z) = \frac{2xy^3}{z}, \quad \text{and} \quad f_y(x, y, z) = \frac{3x^2 y^2}{z}, \quad \text{and} \quad f_z(x, y, z) = -\frac{x^2 y^3}{z^2}.$$

Interpretation of Partial Derivatives

Example 4   A vibrating guitar string, originally at rest along the x-axis, is shown in Figure 14.15. Let x be the distance in meters from the left end of the string. At time t seconds the point x has been displaced $y = f(x, t)$ meters vertically from its rest position, where

$$y = f(x, t) = 0.003 \sin(\pi x) \sin(2765t).$$

Evaluate $f_x(0.3, 1)$ and $f_t(0.3, 1)$ and explain what each means in practical terms.

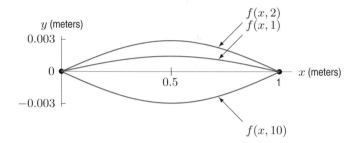

Figure 14.15: The position of a vibrating guitar string at several different times: Graph of $f(x, t)$ for $t = 1, 2, 10$.

Solution   Differentiating $f(x, t) = 0.003 \sin(\pi x) \sin(2765t)$ with respect to x, we have

$$f_x(x, t) = 0.003\pi \cos(\pi x) \sin(2765t).$$

In particular, substituting $x = 0.3$ and $t = 1$ gives

$$f_x(0.3, 1) = 0.003\pi \cos(\pi(0.3)) \sin(2765) \approx 0.002.$$

To see what $f_x(0.3, 1)$ means, think about the function $f(x, 1)$. The graph of $f(x, 1)$ in Figure 14.16 is a snapshot of the string at the time $t = 1$. Thus, the derivative $f_x(0.3, 1)$ is the slope of the string at the point $x = 0.3$ at the instant when $t = 1$.

Similarly, taking the derivative of $f(x, t) = 0.003 \sin(\pi x) \sin(2765t)$ with respect to t, we get

$$f_t(x, t) = (0.003)(2765) \sin(\pi x) \cos(2765t) = 8.3 \sin(\pi x) \cos(2765t).$$

Since $f(x, t)$ is in meters and t is in seconds, the derivative $f_t(0.3, 1)$ is in m/sec. Thus, substituting $x = 0.3$ and $t = 1$,

$$f_t(0.3, 1) = 8.3 \sin(\pi(0.3)) \cos(2765(1)) \approx 6 \text{ m/sec.}$$

Figure 14.16: Graph of $f(x, 1)$: Snapshot of the shape of the string at $t = 1$ sec

To see what $f_t(0.3, 1)$ means, think about the function $f(0.3, t)$. The graph of $f(0.3, t)$ is a position versus time graph that tracks the up and down movement of the point on the string where $x = 0.3$.

(See Figure 14.17.) The derivative $f_t(0.3, 1) = 6$ m/sec is the velocity of that point on the string at time $t = 1$. The fact that $f_t(0.3, 1)$ is positive indicates that the point is moving upward when $t = 1$.

Figure 14.17: Graph of $f(0.3, t)$: Position versus time graph of the point $x = 0.3$ m from the end of the guitar string

Exercises and Problems for Section 14.2

Exercises

Find the partial derivatives in Exercises 1–35. Assume the variables are restricted to a domain on which the function is defined.

1. f_x and f_y if $f(x, y) = 5x^2 y^3 + 8xy^2 - 3x^2$

2. $f_x(1, 2)$ and $f_y(1, 2)$ if $f(x, y) = x^3 + 3x^2 y - 2y^2$

3. $\dfrac{\partial}{\partial y}(3x^5 y^7 - 32x^4 y^3 + 5xy)$

4. $\dfrac{\partial z}{\partial x}$ and $\dfrac{\partial z}{\partial y}$ if $z = (x^2 + x - y)^7$

5. z_x if $z = x^2 y + 2x^5 y$ **6.** V_r if $V = \frac{1}{3}\pi r^2 h$

7. z_y if $z = \dfrac{3x^2 y^7 - y^2}{15xy - 8}$ **8.** $\dfrac{\partial}{\partial T}\left(\dfrac{2\pi r}{T}\right)$

9. $\dfrac{\partial}{\partial x}(a\sqrt{x})$ **10.** $\dfrac{\partial}{\partial x}(xe^{\sqrt{xy}})$

11. F_m if $F = mg$ **12.** $z_x(2, 3)$ if $z = (\cos x) + y$

13. $\dfrac{\partial A}{\partial h}$ if $A = \frac{1}{2}(a + b)h$ **14.** $\dfrac{\partial}{\partial m}\left(\frac{1}{2}mv^2\right)$

15. $\dfrac{\partial}{\partial B}\left(\dfrac{1}{u_0}B^2\right)$ **16.** $\dfrac{\partial}{\partial r}\left(\dfrac{2\pi r}{v}\right)$

17. F_v if $F = \dfrac{mv^2}{r}$ **18.** $\dfrac{\partial}{\partial v_0}(v_0 + at)$

19. $\dfrac{\partial F}{\partial m_2}$ if $F = \dfrac{Gm_1 m_2}{r^2}$

20. a_v if $a = \dfrac{v^2}{r}$

21. $\dfrac{\partial}{\partial t}\left(v_0 t + \frac{1}{2}at^2\right)$

22. $\dfrac{\partial f_0}{\partial L}$ if $f_0 = \dfrac{1}{2\pi\sqrt{LC}}$

23. $\dfrac{\partial T}{\partial l}$ if $T = 2\pi\sqrt{\dfrac{l}{g}}$

24. $\dfrac{\partial}{\partial M}\left(\dfrac{2\pi r^{3/2}}{\sqrt{GM}}\right)$

25. f_a if $f(a, b) = e^a \sin(a + b)$

26. z_x if $z = \sin(5x^3 y - 3xy^2)$

27. g_x if $g(x, y) = \ln(ye^{xy})$

28. $\dfrac{\partial V}{\partial r}$ and $\dfrac{\partial V}{\partial h}$ if $V = \frac{4}{3}\pi r^2 h$

29. u_E if $u = \dfrac{1}{2}\epsilon_0 E^2 + \dfrac{1}{2\mu_0}B^2$

30. $\dfrac{\partial}{\partial x}\left(\dfrac{1}{\sqrt{2\pi}\sigma}e^{-(x-\mu)^2/(2\sigma^2)}\right)$

31. $\dfrac{\partial Q}{\partial K}$ if $Q = c(a_1 K^{b_1} + a_2 L^{b_2})^\gamma$

32. z_x and z_y for $z = x^7 + 2^y + x^y$

33. $\dfrac{\partial m}{\partial v}$ if $m = \dfrac{m_0}{\sqrt{1 - v^2/c^2}}$

34. $\left.\dfrac{\partial z}{\partial y}\right|_{(1, 0.5)}$ if $z = e^{x+2y}\sin y$

35. $\left.\dfrac{\partial f}{\partial x}\right|_{(\pi/3, 1)}$ if $f(x, y) = x\ln(y\cos x)$

Problems

36. (a) Let $f(u, v) = u(u^2 + v^2)^{3/2}$. Use a difference quotient to approximate $f_u(1, 3)$ with $h = 0.001$.

(b) Now evaluate $f_u(1, 3)$ exactly. Was the approximation in part (a) reasonable?

37. (a) Let $f(x, y) = x^2 + y^2$. Estimate $f_x(2, 1)$ and $f_y(2, 1)$ using the contour diagram for f in Figure 14.18.

(b) Estimate $f_x(2, 1)$ and $f_y(2, 1)$ from a table of values for f with $x = 1.9, 2, 2.1$ and $y = 0.9, 1, 1.1$.

(c) Compare your estimates in parts (a) and (b) with the exact values of $f_x(2, 1)$ and $f_y(2, 1)$ found algebraically.

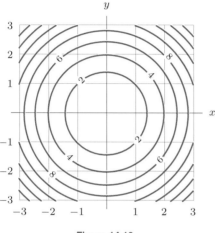

Figure 14.18

38. The energy, E, of a body of mass m moving with speed v is given by the formula

$$E = mc^2 \left(\frac{1}{\sqrt{1 - v^2/c^2}} - 1 \right).$$

The speed, v, is nonnegative and less than the speed of light, c, which is a constant.

(a) Find $\partial E/\partial m$. What would you expect the sign of $\partial E/\partial m$ to be? Explain.

(b) Find $\partial E/\partial v$. Explain what you would expect the sign of $\partial E/\partial v$ to be and why.

39. A one-meter long bar is heated unevenly, with temperature in °C at a distance x meters from one end at time t given by

$$H(x, t) = 100e^{-0.1t} \sin(\pi x) \qquad 0 \le x \le 1.$$

(a) Sketch a graph of H against x for $t = 0$ and $t = 1$.

(b) Calculate $H_x(0.2, t)$ and $H_x(0.8, t)$. What is the practical interpretation (in terms of temperature) of these two partial derivatives? Explain why each one has the sign it does.

(c) Calculate $H_t(x, t)$. What is its sign? What is its interpretation in terms of temperature?

40. Money in a bank account earns interest at a continuous rate, r. The amount of money, $\$B$, in the account depends on the amount deposited, $\$P$, and the time, t, it has been in the bank according to the formula

$$B = Pe^{rt}.$$

Find $\partial B/\partial t$ and $\partial B/\partial P$ and interpret each in financial terms.

41. The gravitational force, F newtons, exerted on a mass of m kg at a distance of r meters from the center of the earth is given by

$$F = \frac{GMm}{r^2}$$

where the mass of the earth $M = 6 \cdot 10^{24}$ kilograms, and $G = 6.67 \cdot 10^{-11}$. Find the gravitational force on a person with mass 70 kg at the surface of the earth ($r = 6.4 \cdot 10^6$). Calculate $\partial F/\partial m$ and $\partial F/\partial r$ for these values of m and r. Interpret these partial derivatives in terms of gravitational force.

42. Let $h(x, t) = 5 + \cos(0.5x - t)$ describe a wave. The value of $h(x, t)$ gives the depth of the water in cm at a distance x meters from a fixed point and at time t seconds. Evaluate $h_x(2, 5)$ and $h_t(2, 5)$ and interpret each in terms of the wave.

43. The Dubois formula relates a person's surface area, s, in m², to weight, w, in kg, and height, h, in cm, by

$$s = f(w, h) = 0.01w^{0.25}h^{0.75}.$$

Find $f(65, 160)$, $f_w(65, 160)$, and $f_h(65, 160)$. Interpret your answers in terms of surface area, height, and weight.

44. Show that the Cobb-Douglas function

$$Q = bK^\alpha L^{1-\alpha} \quad \text{where} \quad 0 < \alpha < 1$$

satisfies the equation

$$K\frac{\partial Q}{\partial K} + L\frac{\partial Q}{\partial L} = Q.$$

45. Is there a function f which has the following partial derivatives? If so what is it? Are there any others?

$$f_x(x, y) = 4x^3y^2 - 3y^4,$$
$$f_y(x, y) = 2x^4y - 12xy^3.$$

14.3 LOCAL LINEARITY AND THE DIFFERENTIAL

In Sections 14.1 and 14.2 we studied a function of two variables by allowing one variable at a time to change. We now let both variables change at once to develop a linear approximation for functions of two variables.

Zooming In to See Local Linearity

For a function of one variable, local linearity means that as we zoom in on the graph, it looks like a straight line. As we zoom in on the graph of a two-variable function, the graph usually looks like a plane, which is the graph of a linear function of two variables. (See Figure 14.19.)

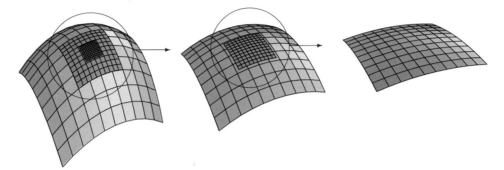

Figure 14.19: Zooming in on the graph of a function of two variables until the graph looks like a plane

Similarly, Figure 14.20 shows three successive views of the contours near a point. As we zoom in, the contours look more like equally spaced parallel lines, which are the contours of a linear function. (As we zoom in, we have to add more contours.)

Figure 14.20: Zooming in on a contour diagram until the lines look parallel and equally spaced

This effect can also be seen numerically by zooming in with tables of values. Table 14.4 shows three tables of values for $f(x, y) = x^2 + y^3$ near $x = 2$, $y = 1$, each one a closer view than the previous one. Notice how each table looks more like the table of a linear function.

Table 14.4   *Zooming in on values of $f(x, y) = x^2 + y^3$ near $(2, 1)$ until the table looks linear*

		y		
		0	1	2
x	1	1	2	9
	2	4	5	12
	3	9	10	17

		y		
		0.9	1.0	1.1
x	1.9	4.34	4.61	4.94
	2.0	4.73	5.00	5.33
	2.1	5.14	5.41	5.74

		y		
		0.99	1.00	1.01
x	1.99	4.93	4.96	4.99
	2.00	4.97	5.00	5.03
	2.01	5.01	5.04	5.07

Zooming in Algebraically: Differentiability

Seeing a plane when we zoom in at a point tells us (provided the plane is not vertical) that $f(x, y)$ is closely approximated near that point by a linear function, $L(x, y)$:

$$f(x, y) \approx L(x, y).$$

The graph of the function $z = L(x, y)$ is the tangent plane at that point. Provided the approximation is sufficiently good, we say that $f(x, y)$ is *differentiable* at the point. Section 14.8 on page 734 defines precisely what is meant by the approximation being sufficiently good. The functions we encounter are differentiable at most points in their domain. We usually only consider the question of differentiability at points that are in the interior of the domain of f, not on the boundary[1].

The Tangent Plane

The plane that we see when we zoom in on a surface is called the *tangent plane* to the surface at the point. Figure 14.21 shows the graph of a function with a tangent plane.

What is the equation of the tangent plane? At the point (a, b), the x-slope of the graph of f is the partial derivative $f_x(a, b)$ and the y-slope is $f_y(a, b)$. Thus, using the equation for a plane on page 629 of Chapter 12, we have the following result:

Tangent Plane to the Surface $z = f(x, y)$ at the Point (a, b)

Assuming f is differentiable at (a, b), the equation of the tangent plane is

$$z = f(a, b) + f_x(a, b)(x - a) + f_y(a, b)(y - b).$$

Here we are thinking of a and b as fixed, so $f(a, b)$, and $f_x(a, b)$, and $f_y(a, b)$ are constants. Thus, the right side of the equation is a linear function of x and y.

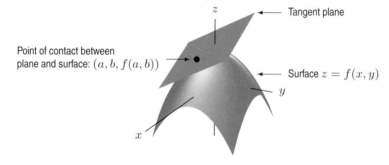

Point of contact between
plane and surface: $(a, b, f(a, b))$

Tangent plane

Surface $z = f(x, y)$

Figure 14.21: The tangent plane to the surface $z = f(x, y)$ at the point (a, b)

Example 1   Find the equation for the tangent plane to the surface $z = x^2 + y^2$ at the point $(3, 4)$.

Solution   We have $f_x(x, y) = 2x$, so $f_x(3, 4) = 6$, and $f_y(x, y) = 2y$, so $f_y(3, 4) = 8$. Also, $f(3, 4) = 3^2 + 4^2 = 25$. Thus, the equation for the tangent plane at $(3, 4)$ is

$$z = 25 + 6(x - 3) + 8(y - 4) = -25 + 6x + 8y.$$

Local Linearization

Since the tangent plane lies close to the surface near the point at which they meet, z-values on the tangent plane are close to values of $f(x, y)$ for points near (a, b). Thus, replacing z by $f(x, y)$ in the equation of the tangent plane, we get the following approximation:

---

[1] Interior points and boundary points are defined precisely on page 761

Tangent Plane Approximation to $f(x, y)$ for (x, y) Near the Point (a, b)

Provided f is differentiable at (a, b), we can approximate $f(x, y)$:

$$f(x, y) \approx f(a, b) + f_x(a, b)(x - a) + f_y(a, b)(y - b).$$

We are thinking of a and b as fixed, so the expression on the right side is linear in x and y. The right side of this approximation is called the **local linearization** of f near $x = a$, $y = b$.

Figure 14.22 shows the tangent plane approximation graphically.

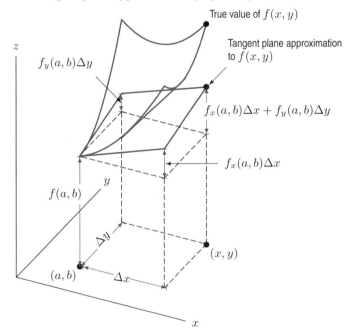

Figure 14.22: Local linearization: Approximating $f(x, y)$ by the z-value from the tangent plane

Example 2   Find the local linearization of $f(x, y) = x^2 + y^2$ at the point $(3, 4)$. Estimate $f(2.9, 4.2)$ and $f(2, 2)$ using the linearization and compare your answers to the true values.

Solution   Let $z = f(x, y) = x^2 + y^2$. In Example 1 on page 698, we found the equation of the tangent plane at $(3, 4)$ to be

$$z = 25 + 6(x - 3) + 8(y - 4).$$

Therefore, for (x, y) near $(3, 4)$, we have the local linearization

$$f(x, y) \approx 25 + 6(x - 3) + 8(y - 4).$$

Substituting $x = 2.9, y = 4.2$ gives

$$f(2.9, 4.2) \approx 25 + 6(-0.1) + 8(0.2) = 26.$$

This compares favorably with the true value $f(2.9, 4.2) = (2.9)^2 + (4.2)^2 = 26.05$.

However, the local linearization does not give a good approximation at points far away from $(3, 4)$. For example, if $x = 2$, $y = 2$, the local linearization gives

$$f(2, 2) \approx 25 + 6(-1) + 8(-2) = 3,$$

whereas the true value of the function is $f(2, 2) = 2^2 + 2^2 = 8$.

Example 3   Designing safe boilers depends on knowing how steam behaves under changes in temperature and pressure. Steam tables, such as Table 14.5, are published giving values of the function $V = f(T, P)$ where V is the volume (in ft^3) of one pound of steam at a temperature T (in °F) and pressure P (in lb/in^2).

(a) Give a linear function approximating $V = f(T, P)$ for T near 500°F and P near 24 lb/in^2.

(b) Estimate the volume of a pound of steam at a temperature of 505°F and a pressure of 24.3 lb/in^2.

Table 14.5 *Volume (in cubic feet) of one pound of steam at various temperatures and pressures*

	Pressure P (lb/in^2)			
	20	22	24	26
480	27.85	25.31	23.19	21.39
Temperature 500	28.46	25.86	23.69	21.86
T 520	29.06	26.41	24.20	22.33
(°F) 540	29.66	26.95	24.70	22.79

Solution   (a) We want the local linearization around the point $T = 500$, $P = 24$, which is

$$f(T, P) \approx f(500, 24) + f_T(500, 24)(T - 500) + f_P(500, 24)(P - 24).$$

We read the value $f(500, 24) = 23.69$ from the table.

Next we approximate $f_T(500, 24)$ by a difference quotient. From the $P = 24$ column, we compute the average rate of change between $T = 500$ and $T = 520$:

$$f_T(500, 24) \approx \frac{f(520, 24) - f(500, 24)}{520 - 500} = \frac{24.20 - 23.69}{20} = 0.0255.$$

Note that $f_T(500, 24)$ is positive, because steam expands when heated.

Next we approximate $f_P(500, 24)$ by looking at the $T = 500$ row and computing the average rate of change between $P = 24$ and $P = 26$:

$$f_P(500, 24) \approx \frac{f(500, 26) - f(500, 24)}{26 - 24} = \frac{21.86 - 23.69}{2} = -0.915.$$

Note that $f_P(500, 24)$ is negative, because increasing the pressure on steam decreases its volume. Using these approximations for the partial derivatives, we obtain the local linearization:

$$V = f(T, P) \approx 23.69 + 0.0255(T - 500) - 0.915(P - 24) \text{ ft}^3 \quad \begin{array}{l} \text{for } T \text{ near } 500 \text{ °F} \\ \text{and } P \text{ near } 24 \text{ lb/in}^2. \end{array}$$

(b) We are interested in the volume at $T = 505$°F and $P = 24.3$ lb/in^2. Since these values are close to $T = 500$°F and $P = 24$ lb/in^2, we use the linear relation obtained in part (a).

$$V \approx 23.69 + 0.0255(505 - 500) - 0.915(24.3 - 24) = 23.54 \text{ ft}^3.$$

Local Linearity With Three or More Variables

Local linear approximations for functions of three or more variables follow the same pattern as for functions of two variables. The local linearization of $f(x, y, z)$ at (a, b, c) is given by

$$f(x, y, z) \approx f(a, b, c) + f_x(a, b, c)(x - a) + f_y(a, b, c)(y - b) + f_z(a, b, c)(z - c).$$

The Differential

We are often interested in the change in the value of the function as we move from the point (a, b) to a nearby point (x, y). Then we use the notation

$$\Delta f = f(x, y) - f(a, b) \quad \text{and} \quad \Delta x = x - a \quad \text{and} \quad \Delta y = y - b$$

to rewrite the tangent plane approximation

$$f(x, y) \approx f(a, b) + f_x(a, b)(x - a) + f_y(a, b)(y - b)$$

in the form

$$\Delta f \approx f_x(a, b)\Delta x + f_y(a, b)\Delta y.$$

For fixed a and b, the right side of this is a linear function of Δx and Δy that can be used to estimate Δf. We call this linear function the *differential*. To define the differential in general, we introduce new variables dx and dy to represent changes in x and y.

The Differential of a Function $z = f(x, y)$

The **differential**, df (or dz), at a point (a, b) is the linear function of dx and dy given by the formula

$$df = f_x(a, b)\, dx + f_y(a, b)\, dy.$$

The differential at a general point is often written $df = f_x\, dx + f_y\, dy.$

Note that the differential, df, is a function of four variables a, b, and dx, dy.

Example 4   Compute the differentials of the following functions.

(a) $f(x, y) = x^2 e^{5y}$ (b) $z = x \sin(xy)$ (c) $f(x, y) = x \cos(2x)$

Solution   (a) Since $f_x(x, y) = 2xe^{5y}$ and $f_y(x, y) = 5x^2 e^{5y}$, we have

$$df = 2xe^{5y}\, dx + 5x^2 e^{5y}\, dy.$$

(b) Since $\partial z/\partial x = \sin(xy) + xy \cos(xy)$ and $\partial z/\partial y = x^2 \cos(xy)$, we have

$$dz = (\sin(xy) + xy \cos(xy))\, dx + x^2 \cos(xy)\, dy.$$

(c) Since $f_x(x, y) = \cos(2x) - 2x \sin(2x)$ and $f_y(x, y) = 0$, we have

$$df = (\cos(2x) - 2x \sin(2x))\, dx + 0\, dy = (\cos(2x) - 2x \sin(2x))\, dx.$$

Example 5   The density ρ (in g/cm^3) of carbon dioxide gas CO_2 depends upon its temperature T (in °C) and pressure P (in atmospheres). The ideal gas model for CO_2 gives what is called the state equation

$$\rho = \frac{0.5363P}{T + 273.15}.$$

Compute the differential $d\rho$. Explain the signs of the coefficients of dT and dP.

Solution   The differential for $\rho = f(T, P)$ is

$$d\rho = f_T(T, P)\, dT + f_P(T, P)dP = \frac{-0.5363P}{(T + 273.15)^2}\, dT + \frac{0.5363}{T + 273.15}\, dP.$$

The coefficient of dT is negative because increasing the temperature expands the gas (if the pressure is kept constant) and therefore decreases its density. The coefficient of dP is positive because increasing the pressure compresses the gas (if the temperature is kept constant) and therefore increases its density.

Where Does the Notation for the Differential Come From?

We write the differential as a linear function of the new variables dx and dy. You may wonder why we chose these names for our variables. The reason is historical: The people who invented calculus thought of dx and dy as "infinitesimal" changes in x and y. The equation

$$df = f_x dx + f_y dy$$

was regarded as an infinitesimal version of the local linear approximation

$$\Delta f \approx f_x \Delta x + f_y \Delta y.$$

In spite of the problems with defining exactly what "infinitesimal" means, some mathematicians, scientists, and engineers think of the differential in terms of infinitesimals.

Figure 14.23 illustrates a way of thinking about differentials that combines the definition with this informal point of view. It shows the graph of f along with a view of the graph around the point $(a, b, f(a, b))$ under a microscope. Since f is locally linear at the point, the magnified view looks like the tangent plane. Under the microscope, we use a magnified coordinate system with its origin at the point $(a, b, f(a, b))$ and with coordinates dx, dy, and dz along the three axes. The graph of the differential df is the tangent plane, which has equation $dz = f_x(a, b)\, dx + f_y(a, b)\, dy$ in the magnified coordinates.

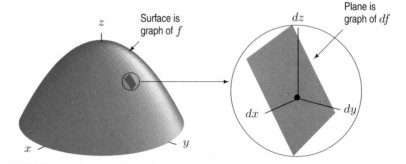

Figure 14.23: The graph of f, with a view through a microscope showing the tangent plane in the magnified coordinate system

Exercises and Problems for Section 14.3

Exercises

For the functions in Exercises 1–4, find the equation of the tangent plane at the given point.

1. $z = \frac{1}{2}(x^2 + 4y^2)$ at the point $(2, 1, 4)$.

2. $z = ye^{x/y}$ at the point $(1, 1, e)$.

3. $z = e^y + x + x^2 + 6$ at the point $(1, 0, 9)$.

4. $z = \ln(x^2 + 1) + y^2$ at the point $(0, 3, 9)$

Find the differentials of the functions in Exercises 5–8.

5. $g(u, v) = u^2 + uv$ **6.** $f(x, y) = \sin(xy)$

7. $z = e^{-x}\cos y$

8. $h(x, t) = e^{-3t}\sin(x + 5t)$

Find differentials of the functions in Exercises 9–12 at the given point.

9. $f(x, y) = xe^{-y}$ at $(1, 0)$

10. $F(m, r) = Gm/r^2$ at $(100, 10)$

11. $g(x, t) = x^2 \sin(2t)$ at $(2, \pi/4)$

12. $P(L, K) = 1.01L^{0.25}K^{0.75}$ at $(100, 1)$

Problems

13. Find the local linearization of the function $f(x, y) = x^2 y$ at the point $(3, 1)$.

14. A student was asked to find the equation of the tangent plane to the surface $z = x^3 - y^2$ at the point $(x, y) = (2, 3)$. The student's answer was

$$z = 3x^2(x - 2) - 2y(y - 3) - 1.$$

(a) At a glance, how do you know this is wrong?
(b) What mistake did the student make?
(c) Answer the question correctly.

15. (a) Check the local linearity of $f(x, y) = e^{-x} \sin y$ near $x = 1$, $y = 2$ by making a table of values of f for $x = 0.9$, 1.0, 1.1 and $y = 1.9$, 2.0, 2.1. Express values of f with 4 digits after the decimal point. Then make a table of values for $x = 0.99$, 1.00, 1.01 and $y = 1.99$, 2.00. 2.01, again showing 4 digits after the decimal point. Do both tables look nearly linear? Does the second table look more linear than the first?

    (b) Give the local linearization of $f(x, y) = e^{-x} \sin y$ at $(1, 2)$, first using your tables, and second using the fact that $f_x(x, y) = -e^{-x} \sin y$ and $f_y(x, y) = e^{-x} \cos y$.

16. Find the differential of $f(x, y) = \sqrt{x^2 + y^3}$ at the point $(1, 2)$. Use it to estimate $f(1.04, 1.98)$.

17. Give the local linearization for the monthly car-loan payment function at each of the points investigated in Problem 15 on page 691.

18. An unevenly heated plate has temperature $T(x, y)$ in $°C$ at the point (x, y). If $T(2, 1) = 135$, and $T_x(2, 1) = 16$, and $T_y(2, 1) = -15$, estimate the temperature at the point $(2.04, 0.97)$.

19. In Example 3 on page 700 we found a linear approximation for $V = f(T, P)$ near $(500, 24)$. Now find a linear approximation near $(480, 20)$.

20. In Example 3 on page 700 we found a linear approximation for $V = f(T, p)$ near $(500, 24)$.

    (a) Test the accuracy of this approximation by comparing its predicted value with the four neighboring values in the table. What do you notice? Which predicted values are accurate? Which are not? Explain your answer.

    (b) Suggest a linear approximation for $f(T, p)$ near $(500, 24)$ that does not have the property you noticed in part (a). [Hint: Estimate the partial derivatives in a different way.]

21. A company uses x hours of unskilled labor and y hours of skilled labor to produce $F(x, y) = 60x^{2/3}y^{1/3}$ units of output. It currently employs 400 hours of unskilled labor and 50 hours of skilled labor. The company is planning to hire an additional 5 hours of skilled labor.

    (a) Use a linear approximation to decide by about how much the company can reduce its use of unskilled labor and keep its output at current level.

    (b) Calculate the exact value of the reduction.

22. The gas equation for one mole of oxygen relates its pressure, P (in atmospheres), its temperature, T (in K), and its volume, V (in cubic decimeters, dm^3):

$$T = 16.574\frac{1}{V} - 0.52754\frac{1}{V^2} - 0.3879P + 12.187VP.$$

    (a) Find the temperature T and differential dT if the volume is 25 dm^3 and the pressure is 1 atmosphere.

    (b) Use your answer to part (a) to estimate how much the volume would have to change if the pressure increased by 0.1 atmosphere and the temperature remained constant.

23. The coefficient, β, of thermal expansion of a liquid relates the change in the volume V (in m^3) of a fixed quantity of a liquid to an increase in its temperature T (in $°C$):

$$dV = \beta V \, dT.$$

    (a) Let ρ be the density (in kg/m^3) of water as a function of temperature. Write an expression for $d\rho$ in terms of ρ and dT.

    (b) The graph in Figure 14.24 shows density of water as a function of temperature. Use it to estimate β when $T = 20°C$ and when $T = 80°C$.

Figure 14.24

24. One mole of ammonia gas is contained in a vessel which is capable of changing its volume (a compartment sealed by a piston, for example). The total energy U (in joules) of the ammonia, is a function of the volume V (in m^3) of the container, and the temperature T (in K) of the gas. The differential dU is given by

$$dU = 840 \, dV + 27.32 \, dT.$$

    (a) How does the energy change if the volume is held constant and the temperature is increased slightly?

    (b) How does the energy change if the temperature is held constant and the volume is increased slightly?

    (c) Find the approximate change in energy if the gas is compressed by 100 cm^3 and heated by 2 K.

25. A fluid moves through a tube of radius $r = 0.005 \pm 0.00025$ m, and length 1 m under a pressure $p = 10^5 \pm 1000$ pascals, at a rate $v = 0.625 \cdot 10^{-9}$ m^3 per unit time. Find the maximum error in the viscosity η given by

$$\eta = \frac{\pi}{8}\frac{pr^4}{v}.$$

26. The period, T, of oscillation in seconds of a pendulum clock is given by $T = 2\pi\sqrt{l/g}$, where g is the acceleration due to gravity. The length of the pendulum, l, depends on the temperature, t, according to the formula $l = l_0(1 + \alpha t)$ where l_0 is the length of the pendulum at temperature t_0 and α is a constant which characterizes the clock. The clock is set to the correct period at the temperature t_0. How many seconds a day does the clock lose or gain when the temperature is $t_0 + \Delta t$? Show that this loss or gain is independent of l_0.

27. (a) Write a formula for the number π using only the perimeter L and the area A of a circle.
(b) Suppose that L and A are determined experimentally. Show that if the relative, or percent, errors in the measured values of L and A are λ and μ, respectively, then the resulting relative, or percent, error in π is $2\lambda - \mu$.

14.4 GRADIENTS AND DIRECTIONAL DERIVATIVES IN THE PLANE

The Rate of Change in an Arbitrary Direction: The Directional Derivative

The partial derivatives of a function f tell us the rate of change of f in the directions parallel to the coordinate axes. In this section we see how to compute the rate of change of f in an arbitrary direction.

Example 1   Figure 14.25 shows the temperature, in °C, at the point (x, y). Estimate the average rate of change of temperature as we walk from point A to point B.

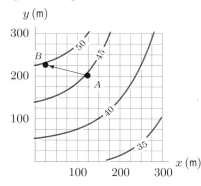

Figure 14.25: Estimating rate of change on a temperature map

Solution   At the point A we are on the $H = 45°$C contour. At B we are on the $H = 50°$C contour. The displacement vector from A to B has x component approximately $-100\vec{i}$ and y component approximately $25\vec{j}$, so its length is $\sqrt{(-100)^2 + 25^2} \approx 103$. Thus the temperature rises by 5°C as we move 103 meters, so the average rate of change of the temperature in that direction is about $5/103 \approx 0.05°$C/m.

Suppose we want to compute the rate of change of a function $f(x, y)$ at the point $P = (a, b)$ in the direction of the unit vector $\vec{u} = u_1\vec{i} + u_2\vec{j}$. For $h > 0$, consider the point $Q = (a + hu_1, b + hu_2)$ whose displacement from P is $h\vec{u}$. (See Figure 14.26.) Since $\|\vec{u}\| = 1$, the distance from P to Q is h. Thus,

$$\begin{array}{l}\text{Average rate of change}\\ \text{in } f \text{ from } P \text{ to } Q\end{array} = \frac{\text{Change in } f}{\text{Distance from } P \text{ to } Q} = \frac{f(a + hu_1, b + hu_2) - f(a, b)}{h}.$$

Figure 14.26: Displacement of $h\vec{u}$ from the point (a, b)

Taking the limit as $h \to 0$ gives the instantaneous rate of change and the following definition:

Directional Derivative of f at (a, b) in the Direction of a Unit Vector $\vec{u}$

If $\vec{u} = u_1 \vec{i} + u_2 \vec{j}$ is a unit vector, we define the directional derivative, $f_{\vec{u}}$, by

$$f_{\vec{u}}(a,b) = \begin{array}{c} \text{Rate of change} \\ \text{of } f \text{ in direction} \\ \text{of } \vec{u} \text{ at } (a,b) \end{array} = \lim_{h \to 0} \frac{f(a + hu_1, b + hu_2) - f(a, b)}{h},$$

provided the limit exists.

Notice that if $\vec{u} = \vec{i}$, so $u_1 = 1, u_2 = 0$, then the directional derivative is f_x, since

$$f_{\vec{i}}(a,b) = \lim_{h \to 0} \frac{f(a + h, b) - f(a, b)}{h} = f_x(a, b).$$

Similarly, if $\vec{u} = \vec{j}$ then the directional derivative $f_{\vec{j}} = f_y$.

What If We Do Not Have a Unit Vector?

We defined $f_{\vec{u}}$ for $\vec{u}$ a unit vector. If $\vec{v}$ is not a unit vector, $\vec{v} \neq \vec{0}$, we construct a unit vector $\vec{u} = \vec{v}/\|\vec{v}\|$ in the same direction as $\vec{v}$ and define the rate of change of f in the direction of $\vec{v}$ as $f_{\vec{u}}$.

Example 2   For each of the functions f, g, and h in Figure 14.27, decide whether the directional derivative at the indicated point is positive, negative, or zero, in the direction of the vector $\vec{v} = \vec{i} + 2\vec{j}$, and in the direction of the vector $\vec{w} = 2\vec{i} + \vec{j}$.

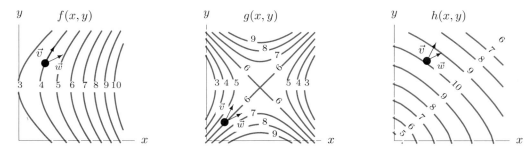

Figure 14.27: Contour diagrams of three functions with direction vectors $\vec{v} = \vec{i} + 2\vec{j}$ and $\vec{w} = 2\vec{i} + \vec{j}$ marked on each

Solution   On the contour diagram for f, the vector $\vec{v} = \vec{i} + 2\vec{j}$ appears to be tangent to the contour. Thus, in this direction, the value of the function is not changing, so the directional derivative in the direction of $\vec{v}$ is zero. The vector $\vec{w} = 2\vec{i} + \vec{j}$ points from the contour marked 4 toward the contour marked 5. Thus, the values of the function are increasing and the directional derivative in the direction of $\vec{w}$ is positive.

On the contour diagram for g, the vector $\vec{v} = \vec{i} + 2\vec{j}$ points from the contour marked 6 toward the contour marked 5, so the function is decreasing in that direction. Thus, the rate of change is negative. On the other hand, the vector $\vec{w} = 2\vec{i} + \vec{j}$ points from the contour marked 6 toward the contour marked 7, and hence the directional derivative in the direction of $\vec{w}$ is positive.

Finally, on the contour diagram for h, both vectors point from the $h = 10$ contour to the $h = 9$ contour, so both directional derivatives are negative.

Example 3   Calculate the directional derivative of $f(x, y) = x^2 + y^2$ at $(1, 0)$ in the direction of the vector $\vec{i} + \vec{j}$.

Solution   First we have to find the unit vector in the same direction as the vector $\vec{i} + \vec{j}$. Since this vector has magnitude $\sqrt{2}$, the unit vector is

$$\vec{u} = \frac{1}{\sqrt{2}}(\vec{i} + \vec{j}) = \frac{1}{\sqrt{2}}\vec{i} + \frac{1}{\sqrt{2}}\vec{j}.$$

Thus,

$$f_{\vec{u}}(1, 0) = \lim_{h \to 0} \frac{f(1 + h/\sqrt{2}, h/\sqrt{2}) - f(1, 0)}{h} = \lim_{h \to 0} \frac{(1 + h/\sqrt{2})^2 + (h/\sqrt{2})^2 - 1}{h}$$

$$= \lim_{h \to 0} \frac{\sqrt{2}h + h^2}{h} = \lim_{h \to 0} (\sqrt{2} + h) = \sqrt{2}.$$

Computing Directional Derivatives From Partial Derivatives

If f is differentiable, we will now see how to use local linearity to find a formula for the directional derivative which does not involve a limit. If $\vec{u}$ is a unit vector, the definition of $f_{\vec{u}}$ says

$$f_{\vec{u}}(a, b) = \lim_{h \to 0} \frac{f(a + hu_1, b + hu_2) - f(a, b)}{h} = \lim_{h \to 0} \frac{\Delta f}{h},$$

where $\Delta f = f(a + hu_1, b + hu_2) - f(a, b)$ is the change in f. We write Δx for the change in x, so $\Delta x = (a + hu_1) - a = hu_1$; similarly $\Delta y = hu_2$. Using local linearity, we have

$$\Delta f \approx f_x(a, b)\Delta x + f_y(a, b)\Delta y = f_x(a, b)hu_1 + f_y(a, b)hu_2.$$

Thus, dividing by h gives

$$\frac{\Delta f}{h} \approx \frac{f_x(a, b)hu_1 + f_y(a, b)hu_2}{h} = f_x(a, b)u_1 + f_y(a, b)u_2.$$

This approximation becomes exact as $h \to 0$, so we have the following formula:

$$f_{\vec{u}}(a, b) = f_x(a, b)u_1 + f_y(a, b)u_2.$$

Example 4   Use the preceding formula to compute the directional derivative in Example 3. Check that we get the same answer as before.

Solution   We calculate $f_{\vec{u}}(1, 0)$, where $f(x, y) = x^2 + y^2$ and $\vec{u} = \frac{1}{\sqrt{2}}\vec{i} + \frac{1}{\sqrt{2}}\vec{j}$.

The partial derivatives are $f_x(x, y) = 2x$ and $f_y(x, y) = 2y$. So, as before

$$f_{\vec{u}}(1, 0) = f_x(1, 0)u_1 + f_y(1, 0)u_2 = (2)\left(\frac{1}{\sqrt{2}}\right) + (0)\left(\frac{1}{\sqrt{2}}\right) = \sqrt{2}.$$

The Gradient Vector

Notice that the expression for $f_{\vec{u}}(a, b)$ can be written as a dot product of $\vec{u}$ and a new vector:

$$f_{\vec{u}}(a, b) = f_x(a, b)u_1 + f_y(a, b)u_2 = (f_x(a, b)\vec{i} + f_y(a, b)\vec{j}) \cdot (u_1\vec{i} + u_2\vec{j}).$$

The new vector, $f_x(a, b)\vec{i} + f_y(a, b)\vec{j}$, turns out to be important. Thus, we make the following definition:

The Gradient Vector of a differentiable function f at the point (a, b) is

$$\text{grad } f(a, b) = f_x(a, b)\vec{i} + f_y(a, b)\vec{j}$$

The formula for the directional derivative can be written in terms of the gradient as follows:

The Directional Derivative and the Gradient

If f is differentiable at (a, b) and $\vec{u} = u_1\vec{i} + u_2\vec{j}$ is a unit vector, then

$$f_{\vec{u}}(a, b) = f_x(a, b)u_1 + f_y(a, b)u_2 = \text{grad } f(a, b) \cdot \vec{u}.$$

Example 5 Find the gradient vector of $f(x, y) = x + e^y$ at the point $(1, 1)$.

Solution Using the definition we have

$$\text{grad } f = f_x\vec{i} + f_y\vec{j} = \vec{i} + e^y\vec{j},$$

so at the point $(1, 1)$

$$\text{grad } f(1, 1) = \vec{i} + e\vec{j}.$$

Alternative Notation for the Gradient

You can think of $\dfrac{\partial f}{\partial x}\vec{i} + \dfrac{\partial f}{\partial y}\vec{j}$ as the result of applying the vector operator (pronounced "del")

$$\nabla = \frac{\partial}{\partial x}\vec{i} + \frac{\partial}{\partial y}\vec{j}$$

to the function f. Thus, we get the alternative notation

$$\text{grad } f = \nabla f.$$

If $z = f(x, y)$, we can write $\text{grad } z$ or ∇z for $\text{grad } f$ or for ∇f.

What Does the Gradient Tell Us?

The fact that $f_{\vec{u}} = \text{grad } f \cdot \vec{u}$ enables us to see what the gradient vector represents. Suppose θ is the angle between the vectors $\text{grad} f$ and $\vec{u}$. At the point (a, b), we have

$$f_{\vec{u}} = \text{grad } f \cdot \vec{u} = \| \text{grad } f\| \underbrace{\|\vec{u}\|}_{1} \cos\theta = \| \text{grad } f\| \cos\theta.$$

Imagine that $\text{grad } f$ is fixed and that $\vec{u}$ can rotate. (See Figure 14.28.) The maximum value of $f_{\vec{u}}$ occurs when $\cos\theta = 1$, so $\theta = 0$ and $\vec{u}$ is pointing in the direction of $\text{grad } f$. Then

$$\text{Maximum } f_{\vec{u}} = \| \text{grad } f\| \cos 0 = \| \text{grad } f\|.$$

The minimum value of $f_{\vec{u}}$ occurs when $\cos\theta = -1$, so $\theta = \pi$ and $\vec{u}$ is pointing in the direction opposite to $\text{grad } f$. Then

$$\text{Minimum } f_{\vec{u}} = \| \text{grad } f\| \cos\pi = -\| \text{grad } f\|.$$

When $\theta = \pi/2$ or $3\pi/2$, so $\cos\theta = 0$, the directional derivative is zero.

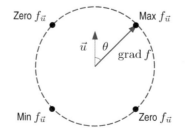

Figure 14.28: Values of the directional derivative at different angles to the gradient

Properties of The Gradient Vector

We have seen that the gradient vector points in the direction of the greatest rate of change at a point and the magnitude of the gradient vector is that rate of change.

Figure 14.29 shows that the gradient vector at a point is perpendicular to the contour through that point. If the contours represent equally spaced f-values and f is differentiable, local linearity tells us that the contours of f around a point appear straight, parallel, and equally spaced. The greatest rate of change is obtained by moving in the direction that takes us to the next contour in the shortest possible distance; that is, perpendicular to the contour. Thus, we have the following:

Geometric Properties of the Gradient Vector in the Plane

If f is a differentiable function at the point (a, b) and grad $f(a, b) \neq \vec{0}$, then:
- The direction of grad $f(a, b)$ is

    - Perpendicular to the contour of f through (a, b)
    - In the direction of increasing f
- The magnitude of the gradient vector, $\| \text{grad } f \|$, is

    - The maximum rate of change of f at that point
    - Large when the contours are close together and small when they are far apart.

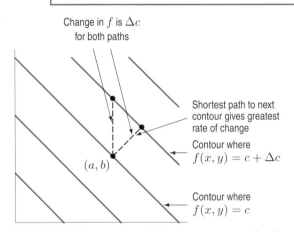

Change in f is Δc
for both paths

Shortest path to next
contour gives greatest
rate of change

Contour where
$f(x, y) = c + \Delta c$

(a, b)

Contour where
$f(x, y) = c$

Figure 14.29: Close-up view of the contours around (a, b), showing the gradient is perpendicular to the contours

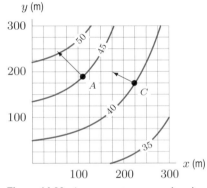

Figure 14.30: A temperature map showing directions and relative magnitudes of two gradient vectors

Examples of Directional Derivatives and Gradient Vectors

Example 6   Explain why the gradient vectors at points A and C in Figure 14.30 have the direction and the relative magnitudes they do.

Solution   The gradient vector points in the direction of greatest increase of the function. This means that in Figure 14.30, the gradient points directly toward warmer temperatures. The magnitude of the gradient vector measures the rate of change. The gradient vector at A is longer than the gradient vector at C because the contours are closer together at A, so the rate of change is larger.

Example 2 on page 705 shows how the contour diagram can tell us the sign of the directional derivative. In the next example we compute the directional derivative in three directions, two close to that of the gradient vector and one that is not.

Example 7   Use the gradient to find the directional derivative of $f(x, y) = x + e^y$ at the point $(1, 1)$ in the direction of the vectors $\vec{i} - \vec{j}, \vec{i} + 2\vec{j}, \vec{i} + 3\vec{j}$.

Solution   In Example 5 we found

$$\operatorname{grad} f(1, 1) = \vec{i} + e\vec{j}.$$

A unit vector in the direction of $\vec{i} - \vec{j}$ is $\vec{s} = (\vec{i} - \vec{j})/\sqrt{2}$, so

$$f_{\vec{s}}(1, 1) = \operatorname{grad} f(1, 1) \cdot \vec{s} = (\vec{i} + e\vec{j}) \cdot \left(\frac{\vec{i} - \vec{j}}{\sqrt{2}}\right) = \frac{1 - e}{\sqrt{2}} \approx -1.215.$$

A unit vector in the direction of $\vec{i} + 2\vec{j}$ is $\vec{v} = (\vec{i} + 2\vec{j})/\sqrt{5}$, so

$$f_{\vec{v}}(1, 1) = \operatorname{grad} f(1, 1) \cdot \vec{v} = (\vec{i} + e\vec{j}) \cdot \left(\frac{\vec{i} + 2\vec{j}}{\sqrt{5}}\right) = \frac{1 + 2e}{\sqrt{5}} \approx 2.879.$$

A unit vector in the direction of $\vec{i} + 3\vec{j}$ is $\vec{w} = (\vec{i} + 3\vec{j})/\sqrt{10}$, so

$$f_{\vec{w}}(1, 1) = \operatorname{grad} f(1, 1) \cdot \vec{w} = (\vec{i} + e\vec{j}) \cdot \left(\frac{\vec{i} + 3\vec{j}}{\sqrt{10}}\right) = \frac{1 + 3e}{\sqrt{10}} \approx 2.895.$$

Now look back at the answers and compare with the value of $\|\operatorname{grad} f\| = \sqrt{1 + e^2} \approx 2.896$. One answer is not close to this value; the other two, $f_{\vec{v}} = 2.879$ and $f_{\vec{w}} = 2.895$, are close but slightly smaller than $\|\operatorname{grad} f\|$. Since $\|\operatorname{grad} f\|$ is the maximum rate of change of f at the point, we would expect for *any* unit vector $\vec{u}$:

$$f_{\vec{u}}(1, 1) \le \|\operatorname{grad} f\|.$$

with equality when $\vec{u}$ is in the direction of $\operatorname{grad} f$. Since $e \approx 2.718$, the vectors $\vec{i} + 2\vec{j}$ and $\vec{i} + 3\vec{j}$ both point roughly, but not exactly, in the direction of the gradient vector $\operatorname{grad} f(1, 1) = \vec{i} + e\vec{j}$. Thus, the values of $f_{\vec{v}}$ and $f_{\vec{w}}$ are both close to the value of $\|\operatorname{grad} f\|$. The direction of the vector $\vec{i} - \vec{j}$ is not close to the direction of $\operatorname{grad} f$ and the value of $f_{\vec{s}}$ is not close to the value of $\|\operatorname{grad} f\|$.

Exercises and Problems for Section 14.4

Exercises

In Exercises 1–6, use the contour diagram of f in Figure 14.31 to decide if the specified directional derivative is positive, negative, or approximately zero.

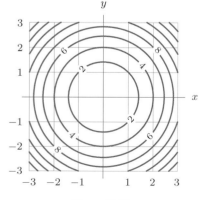

Figure 14.31

1. At point $(-2, 2)$, in direction $\vec{i}$.

2. At point $(0, -2)$, in direction $\vec{j}$.

3. At point $(-1, 1)$, in direction $\vec{i} + \vec{j}$.

4. At point $(-1, 1)$, in direction $-\vec{i} + \vec{j}$.

5. At point $(0, -2)$, in direction $\vec{i} + 2\vec{j}$.

6. At point $(0, -2)$, in direction $\vec{i} - 2\vec{j}$.

In Exercises 7–14, use the contour diagram of f in Figure 14.31 to find the approximate direction of the gradient vector at the given point.

7. $(2, 0)$ **8.** $(0, 2)$ **9.** $(-2, 0)$ **10.** $(0, -2)$

11. $(2, 2)$ **12.** $(2, -2)$ **13.** $(-2, 2)$ **14.** $(-2, -2)$

In Exercises 15–28 find the gradient of the given function. Assume the variables are restricted to a domain on which the function is defined.

15. $f(x, y) = \frac{3}{2}x^5 - \frac{4}{7}y^6$ **16.** $f(m, n) = m^2 + n^2$

17. $Q = 50K + 100L$ **18.** $z = xe^y$

19. $z = (x + y)e^y$ **20.** $f(K, L) = K^{0.3}L^{0.7}$

21. $f(x, y) = \sqrt{x^2 + y^2}$ **22.** $f(r, h) = \pi r^2 h$

23. $f(r, \theta) = r \sin \theta$

24. $z = \sin(x^2 + y^2)$

25. $z = \sin(x/y)$

26. $z = \tan^{-1}(x/y)$

27. $f(\alpha, \beta) = \dfrac{2\alpha + 3\beta}{2\alpha - 3\beta}$

28. $z = x\dfrac{e^y}{x + y}$

In Exercises 29–34, find the gradient at the given point.

29. $f(x, y) = x^2 y + 7xy^3$, at $(1, 2)$

30. $f(m, n) = 5m^2 + 3n^4$, at $(5, 2)$

31. $f(r, h) = 2\pi rh + \pi r^2$, at $(2, 3)$

32. $f(x, y) = \sin(x^2) + \cos y$, at $(\frac{\sqrt{\pi}}{2}, 0)$

33. $f(x, y) = 1/(x^2 + y^2)$, at $(-1, 3)$

34. $f(x, y) = \sqrt{\tan x + y}$, at $(0, 1)$

In Exercises 35–38, find the directional derivative $f_{\vec{u}}(1, 2)$ for the function f with $\vec{u} = (3\vec{i} - 4\vec{j})/5$.

35. $f(x, y) = 3x - 4y$

36. $f(x, y) = x^2 - y^2$

37. $f(x, y) = xy + y^3$

38. $f(x, y) = \sin(2x - y)$

Problems

In Problems 39–44, use the contour diagram for $f(x, y)$ in Figure 14.32 to estimate the directional derivative of $f(x, y)$ in the direction $\vec{v}$ at the point given.

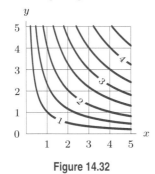

Figure 14.32

39. $\vec{v} = \vec{i}$ at $(1, 1)$

40. $\vec{v} = \vec{j}$ at $(1, 1)$

41. $\vec{v} = \vec{i} + \vec{j}$ at $(1, 1)$

42. $\vec{v} = \vec{i} + \vec{j}$ at $(4, 1)$

43. $\vec{v} = -2\vec{i} + \vec{j}$ at $(3, 3)$

44. $\vec{v} = -2\vec{i} + \vec{j}$ at $(4, 1)$

45. Figure 14.32 shows the level curves of $f(x, y)$. At the points $(1, 1)$ and $(1, 4)$, draw a vector representing grad f. Explain how you know the direction and length of each vector.

46. Figure 14.33 shows grad $f(x, y)$. In each of the following cases, list the marked points (if any) at which

(a) $f_x > 0$

(b) $f_y < 0$

(c) $f_{xx} > 0$

(d) $f_{yy} < 0$

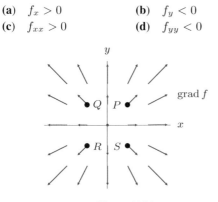

Figure 14.33

47. Figure 14.33 shows grad $f(x, y)$; the scales along the x and y-axes are equal. In each of the following cases, list the marked points at which

(a) grad $f \cdot (\vec{i} + \vec{j})$ is maximum.

(b) The rate of change of $f(x, y)$ is maximum in the $\vec{i} - \vec{j}$ direction.

(c) The rate of change of $f(x, y)$ is zero in the $\vec{i} + \vec{j}$ direction.

(d) The rate of change of $f(x, y)$ is negative in the $\vec{i} + \vec{j}$ direction.

48. The surface $z = g(x, y)$ is in Figure 14.34. What is the sign of each of the following directional derivatives?

(a) $g_{\vec{u}}(2, 5)$ where $\vec{u} = (\vec{i} - \vec{j})/\sqrt{2}$.

(b) $g_{\vec{u}}(2, 5)$ where $\vec{u} = (\vec{i} + \vec{j})/\sqrt{2}$.

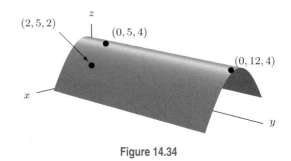

Figure 14.34

49. The table gives values of a differentiable function $f(x, y)$. At the point $(1.2, 0)$, into which quadrant does the gradient vector of f point? Justify your answer.

		y		
		-1	0	1
	1.0	0.7	0.1	-0.5
x	1.2	4.8	4.2	3.6
	1.4	8.9	8.3	7.7

50. Figure 14.35 represents the level curves $f(x, y) = c$; the values of f on each curve are marked. In each of the following parts, decide whether the given quantity is positive, negative or zero. Explain your answer.

(a) The value of $\nabla f \cdot \vec{i}$ at P.
(b) The value of $\nabla f \cdot \vec{j}$ at P.
(c) $\partial f / \partial x$ at Q.
(d) $\partial f / \partial y$ at Q.

Figure 14.35

51. In Figure 14.35, which is larger: $\|\nabla f\|$ at P or $\|\nabla f\|$ at Q? Explain how you know.

52. (a) Let $f(x, y) = (x+y)/(1+x^2)$. Find the directional derivative of f at $P = (1, -2)$ in the direction of:

 (i) $\vec{v} = 3\vec{i} - 2\vec{j}$ (ii) $\vec{v} = -\vec{i} + 4\vec{j}$

(b) What is the direction of greatest increase of f at P?

53. (a) Let $f(x, y) = x^2 + \ln y$. Find the average rate of change of f as you go from $(3, 1)$ to $(1, 2)$.

(b) Find the instantaneous rate of change of f as you leave the point $(3, 1)$ heading toward $(1, 2)$.

54. (a) What is the rate of change of $f(x, y) = 3xy + y^2$ at the point $(2, 3)$ in the direction $\vec{v} = 3\vec{i} - \vec{j}$?

(b) What is the direction of maximum rate of change of f at $(2, 3)$?

(c) What is the maximum rate of change?

55. Let $f(x, y) = x^2 y^3$. At the point $(-1, 2)$, find a vector

(a) In the direction of maximum rate of change.
(b) In the direction of minimum rate of change.
(c) In a direction in which the rate of change is zero.

56. Let P be a fixed point in the plane and let $f(x, y)$ be the distance from P to (x, y). Answer the following questions using geometric interpretations, not formulas.

(a) What are the level curves of f?
(b) What direction does grad $f(x, y)$ point?
(c) What is the magnitude $\| \operatorname{grad} f(x, y) \|$?

57. The directional derivative of $z = f(x, y)$ at $(2, 1)$ in the direction toward the point $(1, 3)$ is $-2/\sqrt{5}$, and the directional derivative in the direction toward the point $(5, 5)$ is 1. Compute $\partial z / \partial x$ and $\partial z / \partial y$ at $(2, 1)$.

58. Consider the function $f(x, y)$. If you start at the point $(4, 5)$ and move toward the point $(5, 6)$, the directional derivative is 2. Starting at the point $(4, 5)$ and moving toward the point $(6, 6)$ gives a directional derivative of 3. Find ∇f at the point $(4, 5)$.

59. Find the directional derivative of $z = x^2 y$ at $(1, 2)$ in the direction making an angle of $5\pi/4$ with the x-axis. In which direction is the directional derivative the largest?

In Problems 60–63, check that the point $(2, 3)$ lies on the curve. Then, viewing the curve as a contour of $f(x, y)$, use grad $f(2, 3)$ to find a vector normal to the curve at $(2, 3)$ and an equation for the tangent line to the curve at $(2, 3)$.

60. $x^2 + y^2 = 13$ **61.** $xy = 6$

62. $y = x^2 - 1$ **63.** $(y - x)^2 + 2 = xy - 3$

64. An ant is crawling across a heated plate with velocity $\vec{v}$ cm/sec, and the temperature of the plate at position (x, y) is $H(x, y)$ degrees, where x and y are in centimeters. Which of the following (if any) is correct? The rate of change in deg/sec of the temperature felt by the ant is:

(a) $\| \operatorname{grad} H \| \|\vec{v}\|$, because it is the product of the ant's speed and the rate of change of H with respect to distance.

(b) $\operatorname{grad} H \cdot \vec{v}$, because it is the product of the ant's speed and the directional derivative of H in the direction of $\vec{v}$.

(c) $H_{\vec{u}}$, where $\vec{u} = \vec{v} / \|\vec{v}\|$, because it is the rate of change of H in the direction of $\vec{v}$.

65. The depth, in feet, of a lake at a point x miles east and y miles north of a buoy is given by

$$h(x, y) = 150 - 30x^2 - 20y^2.$$

(a) A rowboat is 1 mile east and 2 miles south of the buoy. At what rate is the depth changing with respect to distance in the direction of the buoy?

(b) The boat starts moving toward the buoy at a rate of 3 mph. At what rate is the depth of the lake beneath the boat changing with respect to time?

66. The temperature at any point in the plane is given by the function

$$T(x, y) = \frac{100}{x^2 + y^2 + 1}.$$

(a) What shape are the level curves of T?
(b) Where on the plane is it hottest? What is the temperature at that point?
(c) Find the direction of the greatest increase in temperature at the point $(3, 2)$. What is the magnitude of that greatest increase?
(d) Find the direction of the greatest decrease in temperature at the point $(3, 2)$.
(e) Find a direction at the point $(3, 2)$ in which the temperature does not increase or decrease.

67. At a certain point on a heated plate, the greatest rate of temperature increase, $5°$ C per meter, is toward the northeast. If an object at this point moves directly north, at what rate is the temperature increasing?

68. You are climbing a mountain by the steepest route at a slope of $20°$ when you come upon a trail branching off at a $30°$ angle from yours. What is the angle of ascent of the branch trail?

69. Figure 14.37 is a graph of the directional derivative, $f_{\vec{u}}$, at the point (a, b) versus θ, the angle in Figure 14.36.

    (a) Which points on the graph in Figure 14.37 correspond to the greatest rate of increase of f? The greatest rate of decrease?

    (b) Mark points on the circle in Figure 14.36 corresponding to the points P, Q, R, S.

    (c) What is the amplitude of the function graphed in Figure 14.37? What is its formula?

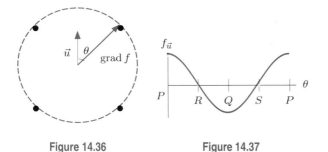

Figure 14.36   Figure 14.37

70. In this problem we see another way of obtaining the formula $f_{\vec{u}}(a, b) = \operatorname{grad} f(a, b) \cdot \vec{u}$. Imagine zooming in on a function $f(x, y)$ at a point (a, b). By local linearity, the contours around (a, b) look like the contours of a linear function. See Figure 14.38. Suppose you want to find the directional derivative $f_{\vec{u}}(a, b)$ in the direction of a unit vector $\vec{u}$. If you move from P to Q, a small distance

h in the direction of $\vec{u}$, then the directional derivative is approximated by the difference quotient

$$\frac{\text{Change in } f \text{ between } P \text{ and } Q}{h}.$$

    (a) Use the gradient to show that

$$\text{Change in } f \approx \| \operatorname{grad} f \|(h \cos \theta).$$

    (b) Use part (a) to obtain $f_{\vec{u}}(a, b) = \operatorname{grad} f(a, b) \cdot \vec{u}$.

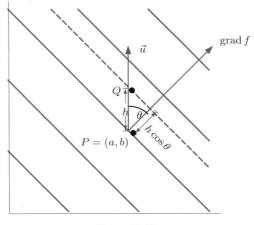

Figure 14.38

14.5 GRADIENTS AND DIRECTIONAL DERIVATIVES IN SPACE

Directional Derivatives of Functions of Three Variables

We calculate directional derivatives of a function of three variables in the same way as for a function of two variables. If the function f is differentiable at the point (a, b, c), then the rate of change of $f(x, y, z)$ at the point (a, b, c) in the direction of a unit vector $\vec{u} = u_1\vec{i} + u_2\vec{j} + u_3\vec{k}$ is

$$f_{\vec{u}}(a, b, c) = f_x(a, b, c)u_1 + f_y(a, b, c)u_2 + f_z(a, b, c)u_3.$$

This can be justified using local linearity in the same way as for functions of two variables.

Example 1   Find the directional derivative of $f(x, y, z) = xy + z$ at the point $(-1, 0, 1)$ in the direction of the vector $\vec{v} = 2\vec{i} + \vec{k}$.

Solution   The magnitude of $\vec{v}$ is $\|\vec{v}\| = \sqrt{2^2 + 1} = \sqrt{5}$, so a unit vector in the same direction as $\vec{v}$ is

$$\vec{u} = \frac{\vec{v}}{\|\vec{v}\|} = \frac{2}{\sqrt{5}}\vec{i} + 0\vec{j} + \frac{1}{\sqrt{5}}\vec{k}.$$

The partial derivatives of f are $f_x(x, y, z) = y$ and $f_y(x, y, z) = x$ and $f_z(x, y, z) = 1$. Thus,

$$f_{\vec{u}}(-1, 0, 1) = f_x(-1, 0, 1)u_1 + f_y(-1, 0, 1)u_2 + f_z(-1, 0, 1)u_3$$

$$= (0)\left(\frac{2}{\sqrt{5}}\right) + (-1)(0) + (1)\left(\frac{1}{\sqrt{5}}\right) = \frac{1}{\sqrt{5}}.$$

The Gradient Vector of a Function of Three Variables

The gradient of a function of three variables is defined in the same way as for two variables:

$$\text{grad } f(a, b, c) = f_x(a, b, c)\vec{i} + f_y(a, b, c)\vec{j} + f_z(a, b, c)\vec{k}.$$

Directional derivatives are related to gradients in the same way as for functions of two variables:

$$f_{\vec{u}}(a, b, c) = f_x(a, b, c)u_1 + f_y(a, b, c)u_2 + f_z(a, b, c)u_3 = \text{grad } f(a, b, c) \cdot \vec{u}.$$

Since $\text{grad } f(a, b, c) \cdot \vec{u} = \| \text{grad } f(a, b, c)\| \cos\theta$, where θ is the angle between $\text{grad } f(a, b, c)$ and $\vec{u}$, the value of $f_{\vec{u}}(a, b, c)$ is largest when $\theta = 0$, that is, when $\vec{u}$ is in the same direction as $\text{grad } f(a, b, c)$. In addition, $f_{\vec{u}}(a, b, c) = 0$ when $\theta = \pi/2$, so $\text{grad } f(a, b, c)$ is perpendicular to the level surface of f. The properties of gradients in space are similar to those in the plane:

Properties of the Gradient Vector in Space

If f is differentiable at (a, b, c) and $\vec{u}$ is a unit vector, then

$$f_{\vec{u}}(a, b, c) = \text{grad } f(a, b, c) \cdot \vec{u}.$$

If, in addition, $\text{grad } f(a, b, c) \neq \vec{0}$, then
- $\text{grad } f(a, b, c)$ is in the direction of the greatest rate of increase of f
- $\text{grad } f(a, b, c)$ is perpendicular to the level surface of f at (a, b, c)
- $\| \text{grad } f(a, b, c)\|$ is the maximum rate of change of f at (a, b, c).

Example 2   Let $f(x, y, z) = x^2 + y^2$ and $g(x, y, z) = -x^2 - y^2 - z^2$. What can we say about the direction of the following vectors?

(a) $\text{grad } f(0, 1, 1)$ (b) $\text{grad } f(1, 0, 1)$ (c) $\text{grad } g(0, 1, 1)$ (d) $\text{grad } g(1, 0, 1)$.

Solution   The cylinder $x^2 + y^2 = 1$ in Figure 14.39 is a level surface of f and contains both the points $(0, 1, 1)$ and $(1, 0, 1)$. Since the value of f does not change at all in the z-direction, all the gradient vectors are horizontal. They are perpendicular to the cylinder and point outward because the value of f increases as we move out.

Similarly, the points $(0, 1, 1)$ and $(1, 0, 1)$ also lie on the same level surface of g, namely $g(x, y, z) = -x^2 - y^2 - z^2 = -2$, which is the sphere $x^2 + y^2 + z^2 = 2$. Part of this level surface is shown in Figure 14.40. This time the gradient vectors point inward, since the negative signs mean that the function increases (from large negative values to small negative values) as we move inward.

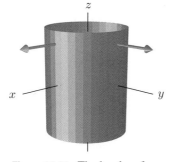

Figure 14.39: The level surface $f(x, y, z) = x^2 + y^2 = 1$ with two gradient vectors

Figure 14.40: The level surface $g(x, y, z) = -x^2 - y^2 - z^2 = -2$ with two gradient vectors

Example 3   Consider the functions $f(x,y) = 4-x^2-2y^2$ and $g(x,y) = 4-x^2$. Calculate a vector perpendicular to each of the following:

(a)   The level curve of f at the point $(1,1)$ (b)   The surface $z = f(x,y)$ at the point $(1,1,1)$
(c)   The level curve of g at the point $(1,1)$ (d)   The surface $z = g(x,y)$ at the point $(1,1,3)$

Solution   (a) The vector we want is a 2-vector in the plane. Since grad $f = -2x\vec{i} - 4y\vec{j}$, we have

$$\text{grad } f(1,1) = -2\vec{i} - 4\vec{j}.$$

Any nonzero multiple of this vector is perpendicular to the level curve at the point $(1,1)$.

(b) In this case we want a 3-vector in space. To find it we rewrite $z = 4 - x^2 - 2y^2$ as the level surface of the function F, where

$$F(x,y,z) = 4 - x^2 - 2y^2 - z = 0.$$

Then

$$\text{grad } F = -2x\vec{i} - 4y\vec{j} - \vec{k},$$

so

$$\text{grad } F(1,1,1) = -2\vec{i} - 4\vec{j} - \vec{k},$$

and grad $F(1,1,1)$ is perpendicular to the surface $z = 4-x^2-2y^2$ at the point $(1,1,1)$. Notice that $-2\vec{i} - 4\vec{j} - \vec{k}$ is not the only possible answer: any multiple of this vector will do.

(c) We are looking for a 2-vector. Since grad $g = -2x\vec{i} + 0\vec{j}$, we have

$$\text{grad } g(1,1) = -2\vec{i}.$$

Any multiple of this vector is perpendicular to the level curve also.

(d) We are looking for a 3-vector. We rewrite $z = 4 - x^2$ as the level surface of the function G, where

$$G(x,y,z) = 4 - x^2 - z = 0.$$

Then

$$\text{grad } G = -2x\vec{i} - \vec{k}$$

So

$$\text{grad } G(1,1,3) = -2\vec{i} - \vec{k},$$

and any multiple of grad $G(1,1,3)$ is perpendicular to the surface $z = 4 - x^2$ at this point.

Example 4   (a) A hiker on the surface $f(x,y) = 4 - x^2 - 2y^2$ at the point $(1,-1,1)$ starts to climb along the path of steepest ascent. What is the relation between the vector grad $f(1,-1)$ and a vector tangent to the path at the point $(1,-1,1)$ and pointing uphill?

(b) Consider the surface $g(x,y) = 4 - x^2$. What is the relation between grad $g(-1,-1)$ and a vector tangent to the path of steepest ascent at $(-1,-1,3)$?

(c) At the point $(1,-1,1)$ on the surface $f(x,y) = 4 - x^2 - 2y^2$, calculate a vector perpendicular to the surface and a vector, $\vec{T}$, tangent to the curve of steepest ascent.

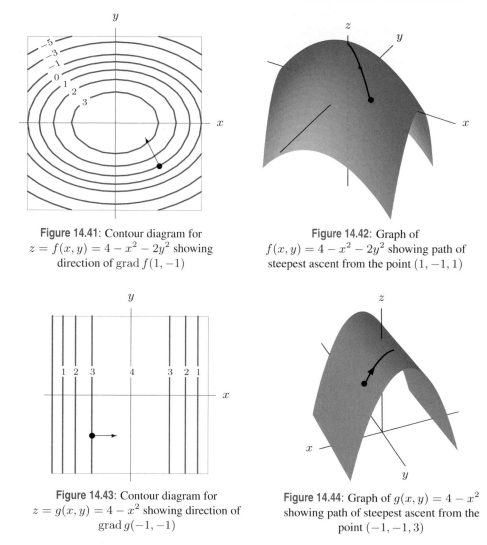

Figure 14.41: Contour diagram for $z = f(x, y) = 4 - x^2 - 2y^2$ showing direction of grad $f(1, -1)$

Figure 14.42: Graph of $f(x, y) = 4 - x^2 - 2y^2$ showing path of steepest ascent from the point $(1, -1, 1)$

Figure 14.43: Contour diagram for $z = g(x, y) = 4 - x^2$ showing direction of grad $g(-1, -1)$

Figure 14.44: Graph of $g(x, y) = 4 - x^2$ showing path of steepest ascent from the point $(-1, -1, 3)$

Solution

(a) The hiker at the point $(1, -1, 1)$ lies directly above the point $(1, -1)$ in the xy-plane. The vector grad $f(1, -1)$ lies in 2-space, pointing like a compass in the direction in which f increases most rapidly. Therefore, grad $f(1, -1)$ lies directly under a vector tangent to the hiker's path at $(1, -1, 1)$ and pointing uphill. (See Figures 14.41 and 14.42.)

(b) The point $(-1, -1, 3)$ lies above the point $(-1, -1)$. The vector grad $g(-1, -1)$ points in the direction in which g increases most rapidly and lies directly under the path of steepest ascent. (See Figures 14.43 and 14.44.)

(c) The surface is represented by $F(x, y, z) = 4 - x^2 - 2y^2 - z = 0$. Since grad $F = -2x\vec{i} - 4y\vec{j} - \vec{k}$, the normal, $\vec{N}$, to the surface is given by

$$\vec{N} = \text{grad } F(1, -1, 1) = -2(1)\vec{i} - 4(-1)\vec{j} - \vec{k} = -2\vec{i} + 4\vec{j} - \vec{k}.$$

We take the $\vec{i}$ and $\vec{j}$ components of $\vec{T}$ to be the vector grad $f(1, -1) = -2\vec{i} + 4\vec{j}$. Thus we have that, for some $a > 0$,

$$\vec{T} = -2\vec{i} + 4\vec{j} + a\vec{k}$$

We want $\vec{N} \cdot \vec{T} = 0$, so

$$\vec{N} \cdot \vec{T} = (-2\vec{i} + 4\vec{j} - \vec{k}) \cdot (-2\vec{i} + 4\vec{j} + a\vec{k}) = 4 + 16 - a = 0$$

So $a = 20$ and hence

$$\vec{T} = -2\vec{i} + 4\vec{j} + 20\vec{k}.$$

Example 5   Find the equation of the tangent plane to the sphere $x^2 + y^2 + z^2 = 14$ at the point $(1, 2, 3)$.

Solution   We write the sphere as a level surface as follows:

$$f(x, y, z) = x^2 + y^2 + z^2 = 14.$$

We have

$$\operatorname{grad} f = 2x\vec{i} + 2y\vec{j} + 2z\vec{k},$$

so the vector

$$\operatorname{grad} f(1, 2, 3) = 2\vec{i} + 4\vec{j} + 6\vec{k}$$

is perpendicular to the sphere at the point $(1, 2, 3)$. Since the vector $\operatorname{grad} f(1, 2, 3)$ is normal to the tangent plane, the equation of the plane is

$$2x + 4y + 6z = 2(1) + 4(2) + 6(3) = 28 \quad \text{or} \quad x + 2y + 3z = 14.$$

The equation of a tangent plane to a level surface can always be found by the method of Example 5. If the level surface $f(x, y, z) = k$ cannot be solved for z in terms of x and y, this is the only method.

Tangent Plane to a Level Surface

If $f(x, y, z)$ is differentiable at (a, b, c), then an equation for the tangent plane to the level surface of f at the point (a, b, c) is

$$f_x(a, b, c)(x - a) + f_y(a, b, c)(y - b) + f_z(a, b, c)(z - c) = 0.$$

Caution: Units and the Geometric Interpretation of the Gradient

When we interpreted the gradient of a function geometrically (page 708), we tacitly assumed that the scales along the x and y axes were the same. If they are not, the gradient vector may not look perpendicular to the contours. Consider the function $f(x, y) = x^2 + y$ with gradient vector given by $\operatorname{grad} f = 2x\vec{i} + \vec{j}$. Figure 14.45 shows the gradient vector at $(1, 1)$ using the same scales in the x and y directions. As expected, the gradient vector is perpendicular to the contour line. Figure 14.46 shows contours of the same function with unequal scales on the two axes. Notice that the gradient vector no longer appears perpendicular to the contour lines. Thus we see that the geometric interpretation of the gradient vector requires that the same scale be used on the two axes.

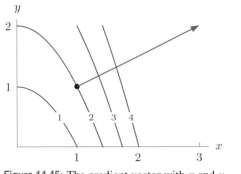

Figure 14.45: The gradient vector with x and y scales equal

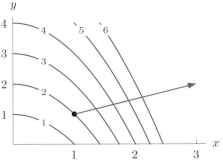

Figure 14.46: The gradient vector with x and y scales unequal

Exercises and Problems for Section 14.5

Exercises

In Exercises 1–10, find the gradient of the given function.

1. $f(x, y, z) = x^2 + y^3 - z^4$

2. $f(x, y, z) = \cos(x + y) + \sin(y + z)$

3. $f(x, y, z) = 1/(x^2 + y^2 + z^2)$

4. $f(x, y, z) = \sqrt{x^2 + y^2 + z^2}$

5. $f(x, y, z) = xe^y + \ln(xz)$

6. $f(x, y, z) = xe^y \sin z$

7. $f(x_1, x_2, x_3) = x_1^2 x_2^3 x_3^4$

8. $f(p, q, r) = e^p + \ln q + e^{r^2}$

9. $f(x, y, z) = xyz$ at $(1, 2, 3)$

10. $f(x, y, z) = e^{xy} \cos z$ at $(1, 0, 0)$

In Exercises 11–16, find the directional derivative using $f(x, y, z) = xy + z^2$.

11. At $(1, 2, 3)$ in the direction of $\vec{i} + \vec{j} + \vec{k}$.

12. At $(1, 1, 1)$ in the direction of $\vec{i} + 2\vec{j} + 3\vec{k}$.

13. As you leave the point $(1, 1, 0)$ heading in the direction of the point $(0, 1, 1)$.

14. As you arrive at $(0, 1, 1)$ from the direction of $(1, 1, 0)$.

15. At the point $(2, 3, 4)$ in the direction of a vector making an angle of $3\pi/4$ with grad $f(2, 3, 4)$.

16. At the point $(2, 3, 4)$ in the direction of the maximum rate of change of f.

In Exercises 17–22, check that the point $(-1, 1, 2)$ lies on the given surface. Then, viewing the surface as a level surface for a function $f(x, y, z)$, find a vector normal to the surface and an equation for the tangent plane to the surface at $(-1, 1, 2)$.

17. $x^2 - y^2 + z^2 = 4$

18. $x^2 - xyz = 3$

19. $z = x^2 + y^2$

20. $y^2 = z^2 - 3$

21. $\cos(x + y) = e^{xz+2}$

22. $y = 4/(2x + 3z)$

Problems

23. Find the directional derivative of $f(x, y, z) = 3x^2y^2 + 2yz$ at the point $(-1, 0, 4)$ in the following directions
(a) $\vec{i} - \vec{k}$ (b) $-\vec{i} + 3\vec{j} + 3\vec{k}$.

24. Let $f(x, y, z) = e^x \sin y + z$.

(a) Find a vector at the point $(0, \pi/2, 2)$ pointing in the direction in which f

(i) Increases fastest   (ii) Decreases fastest

(b) Is there a direction at the point $(0, \pi/2, 2)$ in which the function does not change initially? If so, find a vector pointing in that direction.

25. You are standing above the point $(1, 3)$ on the surface $z = 20 - (2x^2 + y^2)$.

(a) In which direction should you walk to descend fastest? (Give your answer as a 2-vector.)

(b) If you start to move in this direction, what is the slope of your path?

26. (a) Find the tangent plane to the surface $x^2 + y^2 + 3z^2 = 4$ at the point $(0.6, 0.8, 1)$.

(b) Is there a point on the surface $x^2 + y^2 + 3z^2 = 4$ at which the tangent plane is parallel to the plane $8x + 6y + 30z = 1$? If so, find it. If not, explain why not.

27. The surface S is represented by the equation $F = 0$ where $F(x, y, z) = x^2 - (y/z^2)$.

(a) Find the unit vectors $\vec{u}_1$ and $\vec{u}_2$ pointing in the direction of maximum increase of F at the points

$(0, 0, 1)$ and $(1, 1, 1)$ respectively.

(b) Find the tangent plane to S at the points $(0, 0, 1)$ and $(1, 1, 1)$.

(c) Find all points on S where a normal vector is parallel to the xy-plane.

28. Find the equation of the tangent plane at $(2, 3, 1)$ to the surface $x^2 + y^2 - xyz = 7$. Do this in two ways:

(a) Viewing the surface as the level set of a function of three variables, $F(x, y, z)$.

(b) Viewing the surface as the graph of a function of two variables $z = f(x, y)$.

29. Consider the function $f(x, y) = (e^x - x) \cos y$. Suppose S is the surface $z = f(x, y)$.

(a) Find a vector which is perpendicular to the level curve of f through the point $(2, 3)$ in the direction in which f decreases most rapidly.

(b) Suppose $\vec{v} = 5\vec{i} + 4\vec{j} + a\vec{k}$ is a vector in 3-space which is tangent to the surface S at the point P lying on the surface above $(2, 3)$. What is a?

30. A differentiable function $f(x, y)$ has the property that $f(1, 3) = 7$ and grad $f(1, 3) = 2\vec{i} - 5\vec{j}$.

(a) Find the equation of the tangent line to the level curve of f through the point $(1, 3)$.

(b) Find the equation of the tangent plane to the surface $z = f(x, y)$ at the point $(1, 3, 7)$.

31. Let $g_x(2,1,7) = 3$, $g_y(2,1,7) = 10$, $g_z(2,1,7) = -5$. Find the equation of the tangent plane to $g(x,y,z) = 0$ at the point $(2,1,7)$.

32. The temperature of a gas at the point (x,y,z) is given by $G(x,y,z) = x^2 - 5xy + y^2 z$.

   (a) What is the rate of change in the temperature at the point $(1,2,3)$ in the direction $\vec{v} = 2\vec{i} + \vec{j} - 4\vec{k}$?
   (b) What is the direction of maximum rate of change of temperature at the point $(1,2,3)$?
   (c) What is the maximum rate of change at the point $(1,2,3)$?

33. The temperature at the point (x,y,z) in 3-space is given, in degrees Celsius, by $T(x,y,z) = e^{-(x^2+y^2+z^2)}$.

   (a) Describe in words the shape of surfaces on which the temperature is constant.
   (b) Find grad T.
   (c) You travel from the point $(1,0,0)$ to the point $(2,1,0)$ at a speed of 3 units per second. Find the instantaneous rate of change of the temperature as you leave the point $(1,0,0)$. Give units.

34. At an altitude of h feet above sea level, the air pressure, P, in inches of mercury (in Hg), is given by

$$P = 30e^{-3.23\times 10^{-5}h}.$$

An unpressurized seaplane takes off at an angle of $30°$ to the horizontal and a speed of 100 mph. What is the rate of change of pressure in the plane with respect to time at take-off, in inches of mercury per second?

35. A spaceship is plunging into the atmosphere of a planet. The pressure of the atmosphere at a position (x,y,z), with respect to the center of the planet, where the coordinates are in miles, is given by

$$P = 5e^{-0.1\sqrt{x^2+y^2+z^2}} \text{ atmospheres.}$$

The velocity vector, in miles/sec, of the spaceship at $(0,0,1)$ is $\vec{v} = \vec{i} - 2.5\vec{k}$. At $(0,0,1)$. What is the rate of change with respect to time of the pressure experienced by the spacecraft?

36. The earth has mass M and is located at the origin in 3-space, while the moon has mass m. Newton's Law of Gravitation states that if the moon is located at the point (x,y,z) then the attractive force exerted by the earth on the moon is given by the vector

$$\vec{F} = -GMm\frac{\vec{r}}{\|\vec{r}\|^3},$$

where $\vec{r} = x\vec{i} + y\vec{j} + z\vec{k}$. Show that $\vec{F} = \operatorname{grad}\varphi$, where φ is the function given by

$$\varphi(x,y,z) = \frac{GMm}{\|\vec{r}\|}.$$

37. Two surfaces are said to be *tangential* at a point P if they have the same tangent plane at P. Show that the surfaces $z = \sqrt{2x^2 + 2y^2 - 25}$ and $z = \frac{1}{5}(x^2 + y^2)$ are tangential at the point $(4,3,5)$.

38. Two surfaces are said to be *orthogonal* to each other at a point P if the normals to their tangent planes are perpendicular at P. Show that the surfaces $z = \frac{1}{2}(x^2 + y^2 - 1)$ and $z = \frac{1}{2}(1 - x^2 - y^2)$ are orthogonal at all points of intersection.

39. Let $\vec{r}$ be the position vector of the point (x,y,z). If $\vec{\mu} = \mu_1\vec{i} + \mu_2\vec{j} + \mu_3\vec{k}$ is a constant vector, show that

$$\operatorname{grad}(\vec{\mu} \cdot \vec{r}) = \vec{\mu}.$$

40. Let $\vec{r}$ be the position vector of the point (x,y,z). Show that, if a is a constant,

$$\operatorname{grad}(\|\vec{r}\|^a) = a\|\vec{r}\|^{a-2}\vec{r}, \qquad \vec{r} \neq \vec{0}.$$

41. Let f and g be functions on 3-space. Suppose f is differentiable and that

$$\operatorname{grad} f(x,y,z) = (x\vec{i} + y\vec{j} + z\vec{k})g(x,y,z).$$

Explain why f must be constant on any sphere centered at the origin.

14.6   THE CHAIN RULE

Composition of Functions of Many Variables and Rates of Change

The chain rule enables us to differentiate *composite functions*. If we have a function of two variables $z = f(x,y)$ and we substitute $x = g(t), y = h(t)$ into $z = f(x,y)$, then we have a composite function in which z is a function of t:

$$z = f(g(t), h(t)).$$

If, on the other hand, we substitute $x = g(u,v), y = h(u,v)$, then we have a different composite function in which z is a function of u and v:

$$z = f(g(u,v), h(u,v)).$$

The next example shows how to calculate the rate of change of a composite function.

Example 1 Corn production, C, depends on annual rainfall, R, and average temperature, T, so $C = f(R, T)$. Global warming predicts that both rainfall and temperature depend on time. Suppose that according to a particular model of global warming, rainfall is decreasing at 0.2 cm per year and temperature is increasing at $0.1°C$ per year. Use the fact that at current levels of production, $f_R = 3.3$ and $f_T = -5$ to estimate the current rate of change, dC/dt.

Solution By local linearity, we know that changes ΔR and ΔT generate a change, ΔC, in C given approximately by

$$\Delta C \approx f_R \Delta R + f_T \Delta T = 3.3 \Delta R - 5 \Delta T.$$

We want to know how ΔC depends on the time increment, Δt. A change Δt causes changes ΔR and ΔT, which in turn cause a change ΔC. The model of global warming tells us that

$$\frac{dR}{dt} = -0.2 \quad \text{and} \quad \frac{dT}{dt} = 0.1.$$

Thus, a time increment, Δt, generates changes of ΔR and ΔT given by

$$\Delta R \approx -0.2 \Delta t \quad \text{and} \quad \Delta T \approx 0.1 \Delta t.$$

Substituting for ΔR and ΔT in the expression for ΔC gives us

$$\Delta C \approx 3.3(-0.2\Delta t) - 5(0.1\Delta t) = -1.16\Delta t.$$

Thus,

$$\frac{\Delta C}{\Delta t} \approx -1.16 \quad \text{and, therefore,} \quad \frac{dC}{dt} \approx -1.16.$$

The relationship between ΔC and Δt, which gives the value of dC/dt, is an example of the *chain rule*. The argument in Example 1 leads to more general versions of the chain rule.

The Chain Rule for $z = f(x, y)$, $x = g(t)$, $y = h(t)$

Since $z = f(g(t), h(t))$ is a function of t, we can consider the derivative dz/dt. The chain rule gives dz/dt in terms of the derivatives of f, g, and h. Since dz/dt represents the rate of change of z with t, we look at the change Δz generated by a small change, Δt.

We substitute the local linearizations

$$\Delta x \approx \frac{dx}{dt} \Delta t \quad \text{and} \quad \Delta y \approx \frac{dy}{dt} \Delta t$$

into the local linearization

$$\Delta z \approx \frac{\partial z}{\partial x} \Delta x + \frac{\partial z}{\partial y} \Delta y,$$

yielding

$$\Delta z \approx \frac{\partial z}{\partial x} \frac{dx}{dt} \Delta t + \frac{\partial z}{\partial y} \frac{dy}{dt} \Delta t$$

$$= \left(\frac{\partial z}{\partial x} \frac{dx}{dt} + \frac{\partial z}{\partial y} \frac{dy}{dt} \right) \Delta t.$$

Thus,

$$\frac{\Delta z}{\Delta t} \approx \frac{\partial z}{\partial x} \frac{dx}{dt} + \frac{\partial z}{\partial y} \frac{dy}{dt}.$$

Taking the limit as $\Delta t \to 0$, we get the following result.

If f, g, and h are differentiable and if $z = f(x, y)$, and $x = g(t)$, and $y = h(t)$, then

$$\frac{dz}{dt} = \frac{\partial z}{\partial x} \frac{dx}{dt} + \frac{\partial z}{\partial y} \frac{dy}{dt}.$$

Visualizing the Chain Rule with a Diagram

The diagram in Figure 14.47 provides a way of remembering the chain rule. It shows the chain of dependence: z depends on x and y, which in turn depend on t. Each line in the diagram is labeled with a derivative relating the variables at its ends.

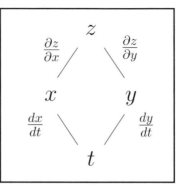

Figure 14.47: Diagram for $z = f(x, y)$, $x = g(t)$, $y = h(t)$. Lines represent dependence of z on x and y, and of x and y on t

The diagram keeps track of how a change in t propagates through the chain of composed functions. There are two paths from t to z, one through x and one through y. For each path, we multiply together the derivatives along the path. Then, to calculate dz/dt, we add the contributions from the two paths.

Example 2   Suppose that $z = f(x, y) = x \sin y$, where $x = t^2$ and $y = 2t + 1$. Let $z = g(t)$. Compute $g'(t)$ directly and using the chain rule.

Solution   Since $z = g(t) = f(t^2, 2t + 1) = t^2 \sin(2t + 1)$, it is possible to compute $g'(t)$ directly by one-variable methods:

$$g'(t) = t^2 \frac{d}{dt}(\sin(2t + 1)) + \left(\frac{d}{dt}(t^2)\right) \sin(2t + 1) = 2t^2 \cos(2t + 1) + 2t \sin(2t + 1).$$

The chain rule provides an alternative route to the same answer. We have

$$\frac{dz}{dt} = \frac{\partial z}{\partial x}\frac{dx}{dt} + \frac{\partial z}{\partial y}\frac{dy}{dt} = (\sin y)(2t) + (x \cos y)(2) = 2t \sin(2t + 1) + 2t^2 \cos(2t + 1).$$

Example 3   The capacity, C, of a communication channel, such as a telephone line, to carry information depends on the ratio of the signal strength, S, to the noise, N. For some positive constant k,

$$C = k \ln\left(1 + \frac{S}{N}\right).$$

Suppose that the signal and noise are given as a function of time, t in seconds, by

$$S(t) = 4 + \cos(4\pi t) \qquad N(t) = 2 + \sin(2\pi t).$$

What is dC/dt one second after transmission started? Is the capacity increasing or decreasing at that instant?

Solution   By the chain rule

$$\frac{dC}{dt} = \frac{\partial C}{\partial S}\frac{dS}{dt} + \frac{\partial C}{\partial N}\frac{dN}{dt}$$

$$= \frac{k}{1 + S/N} \cdot \frac{1}{N}(-4\pi \sin 4\pi t) + \frac{k}{1 + S/N}\left(-\frac{S}{N^2}\right)(2\pi \cos 2\pi t).$$

When $t = 1$, the first term is zero, $S(1) = 5$, and $N(1) = 2$, so

$$\frac{dC}{dt} = \frac{k}{1 + S(1)/N(1)} \left(-\frac{S(1)}{(N(1))^2} \right) \cdot 2\pi = \frac{k}{1 + \frac{5}{2}} \left(-\frac{5}{4} \right) \cdot 2\pi.$$

Since dC/dt is negative, the capacity is decreasing at time $t = 1$ second.

How to Formulate a General Chain Rule

A diagram can be used to write the chain rule for general compositions.

To find the rate of change of one variable with respect to another in a chain of composed differentiable functions:
- Draw a diagram expressing the relationship between the variables, and label each link in the diagram with the derivative relating the variables at its ends.
- For each path between the two variables, multiply together the derivatives from each step along the path.
- Add the contributions from each path.

The diagram keeps track of all the ways in which a change in one variable can cause a change in another; the diagram generates all the terms we would get from the appropriate substitutions into the local linearizations.

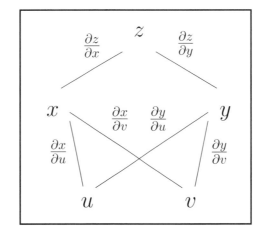

Figure 14.48: Diagram for $z = f(x, y)$, $x = g(u, v)$, $y = h(u, v)$. Lines represent dependence of z on x and y, and of x and y on u and v

For example, we can use Figure 14.48 to find formulas for $\partial z/\partial u$ and $\partial z/\partial v$. Adding the contributions for the two paths from z to u, we get the following results:

If f, g, h are differentiable and if $z = f(x, y)$, with $x = g(u, v)$ and $y = h(u, v)$, then

$$\frac{\partial z}{\partial u} = \frac{\partial z}{\partial x} \frac{\partial x}{\partial u} + \frac{\partial z}{\partial y} \frac{\partial y}{\partial u},$$

$$\frac{\partial z}{\partial v} = \frac{\partial z}{\partial x} \frac{\partial x}{\partial v} + \frac{\partial z}{\partial y} \frac{\partial y}{\partial v}.$$

Example 4   Let $w = x^2 e^y$, $x = 4u$, and $y = 3u^2 - 2v$. Compute $\partial w/\partial u$ and $\partial w/\partial v$ using the chain rule.

Solution   Using the previous result, we have

$$\frac{\partial w}{\partial u} = \frac{\partial w}{\partial x}\frac{\partial x}{\partial u} + \frac{\partial w}{\partial y}\frac{\partial y}{\partial u} = 2xe^y(4) + x^2e^y(6u) = (8x + 6x^2u)e^y$$

$$= (32u + 96u^3)e^{3u^2-2v}.$$

Similarly,

$$\frac{\partial w}{\partial v} = \frac{\partial w}{\partial x}\frac{\partial x}{\partial v} + \frac{\partial w}{\partial y}\frac{\partial y}{\partial v} = 2xe^y(0) + x^2e^y(-2) = -2x^2e^y$$

$$= -32u^2 e^{3u^2-2v}.$$

Example 5   A quantity z can be expressed either as a function of x and y, so that $z = f(x, y)$, or as a function of u and v, so that $z = g(u, v)$. The two coordinate systems are related by

$$x = u + v, \quad y = u - v.$$

(a) Use the chain rule to express $\partial z/\partial u$ and $\partial z/\partial v$ in terms of $\partial z/\partial x$ and $\partial z/\partial y$.
(b) Solve the equations in part (a) for $\partial z/\partial x$ and $\partial z/\partial y$.
(c) Show that the expressions we get in part (b) are the same as we get by expressing u and v in terms of x and y and using the chain rule.

Solution   (a) We have $\partial x/\partial u = 1$ and $\partial x/\partial v = 1$, and also $\partial y/\partial u = 1$ and $\partial y/\partial v = -1$. Thus,

$$\frac{\partial z}{\partial u} = \frac{\partial z}{\partial x}(1) + \frac{\partial z}{\partial y}(1) = \frac{\partial z}{\partial x} + \frac{\partial z}{\partial y}$$

and

$$\frac{\partial z}{\partial v} = \frac{\partial z}{\partial x}(1) + \frac{\partial z}{\partial y}(-1) = \frac{\partial z}{\partial x} - \frac{\partial z}{\partial y}.$$

(b) Adding together the equations for $\partial z/\partial u$ and $\partial z/\partial v$ we get

$$\frac{\partial z}{\partial u} + \frac{\partial z}{\partial v} = 2\frac{\partial z}{\partial x}, \quad \text{so} \quad \frac{\partial z}{\partial x} = \frac{1}{2}\frac{\partial z}{\partial u} + \frac{1}{2}\frac{\partial z}{\partial v}.$$

Similarly, subtracting the equations for $\partial z/\partial u$ and $\partial z/\partial v$ yields

$$\frac{\partial z}{\partial y} = \frac{1}{2}\frac{\partial z}{\partial u} - \frac{1}{2}\frac{\partial z}{\partial v}.$$

(c) Alternatively, we can solve the equations

$$x = u + v, \quad y = u - v$$

for u and v, which yields

$$u = \frac{1}{2}x + \frac{1}{2}y, \quad v = \frac{1}{2}x - \frac{1}{2}y.$$

Now we can think of z as a function of u and v, and u and v as functions of x and y, and apply the chain rule again. This gives us

$$\frac{\partial z}{\partial x} = \frac{\partial z}{\partial u}\frac{\partial u}{\partial x} + \frac{\partial z}{\partial v}\frac{\partial v}{\partial x} = \frac{1}{2}\frac{\partial z}{\partial u} + \frac{1}{2}\frac{\partial z}{\partial v}$$

and

$$\frac{\partial z}{\partial y} = \frac{\partial z}{\partial u}\frac{\partial u}{\partial y} + \frac{\partial z}{\partial v}\frac{\partial v}{\partial y} = \frac{1}{2}\frac{\partial z}{\partial u} - \frac{1}{2}\frac{\partial z}{\partial v}.$$

These are the same expressions we got in part (b).

An Application to Physical Chemistry

A chemist investigating the properties of a gas such as carbon dioxide may want to know how the internal energy U of a given quantity of the gas depends on its temperature, T, pressure, P, and volume, V. The three quantities T, P, and V are not independent, however. For instance, according to the ideal gas law, they satisfy the equation

$$PV = kT$$

where k is a constant which depends only upon the quantity of the gas. The internal energy can then be thought of as a function of any two of the three quantities T, P, and V:

$$U = U_1(T, P) = U_2(T, V) = U_3(P, V).$$

The chemist writes, for example, $\left(\frac{\partial U}{\partial T}\right)_P$ to indicate the partial derivative of U with respect to T *holding P constant*, signifying that for this computation U is viewed as a function of T and P. Thus, we interpret $\left(\frac{\partial U}{\partial T}\right)_P$ as

$$\left(\frac{\partial U}{\partial T}\right)_P = \frac{\partial U_1(T, P)}{\partial T}.$$

If U is to be viewed as a function of T and V, the chemist writes $\left(\frac{\partial U}{\partial T}\right)_V$ for the partial derivative of U with respect to T holding V constant: thus, $\left(\frac{\partial U}{\partial T}\right)_V = \frac{\partial U_2(T,V)}{\partial T}$.

Each of the functions U_1, U_2, U_3 gives rise to one of the following formulas for the differential dU:

$$dU = \left(\frac{\partial U}{\partial T}\right)_P dT + \left(\frac{\partial U}{\partial P}\right)_T dP \qquad \text{corresponds to } U_1$$

$$dU = \left(\frac{\partial U}{\partial T}\right)_V dT + \left(\frac{\partial U}{\partial V}\right)_T dV \qquad \text{corresponds to } U_2,$$

$$dU = \left(\frac{\partial U}{\partial P}\right)_V dP + \left(\frac{\partial U}{\partial V}\right)_P dV \qquad \text{corresponds to } U_3.$$

All the six partial derivatives appearing in formulas for dU have physical meaning, but they are not all equally easy to measure experimentally. A relationship among the partial derivatives, usually derived from the chain rule, may make it possible to evaluate one of the partials in terms of others that are more easily measured.

Example 6 Suppose a gas satisfies the equation $PV = 2T$ and $P = 3$ when $V = 4$. If $\left(\frac{\partial U}{\partial P}\right)_V = 7$ and $\left(\frac{\partial U}{\partial V}\right)_P = 8$, find the values of $\left(\frac{\partial U}{\partial P}\right)_T$ and $\left(\frac{\partial U}{\partial T}\right)_P$.

Solution   Since we know the values of $\left(\frac{\partial U}{\partial P}\right)_V$ and $\left(\frac{\partial U}{\partial V}\right)_P$, we think of U as a function of P and V and use the function U_3 to write

$$dU = \left(\frac{\partial U}{\partial P}\right)_V dP + \left(\frac{\partial U}{\partial V}\right)_P dV$$
$$dU = 7dP + 8dV.$$

To calculate $\left(\frac{\partial U}{\partial P}\right)_T$ and $\left(\frac{\partial U}{\partial T}\right)_P$, we think of U as a function of T and P. Thus we want to substitute for dV in terms of dT and dP. Since $PV = 2T$, we have

$$P\,dV + V\,dP = 2dT,$$
$$3dV + 4dP = 2dT.$$

Solving gives $dV = (2dT - 4dP)/3$ so

$$dU = 7dP + 8\left(\frac{2dT - 4dP}{3}\right)$$
$$dU = -\frac{11}{3}dP + \frac{16}{3}dT.$$

Comparing with the formula for dU obtained from U_1:

$$dU = \left(\frac{\partial U}{\partial T}\right)_P dT + \left(\frac{\partial U}{\partial P}\right)_T dP,$$

we have

$$\left(\frac{\partial U}{\partial T}\right)_P = \frac{16}{3} \quad \text{and} \quad \left(\frac{\partial U}{\partial P}\right)_T = -\frac{11}{3}.$$

In Example 6, we could have substituted for dP instead of dV, leading to values of $\left(\frac{\partial U}{\partial T}\right)_V$ and $\left(\frac{\partial U}{\partial V}\right)_T$. See Problem 31.

In general, if for some particular P, V, and T, we can measure two of the six quantities $\left(\frac{\partial U}{\partial P}\right)_V$, $\left(\frac{\partial U}{\partial V}\right)_P$, $\left(\frac{\partial U}{\partial P}\right)_T$, $\left(\frac{\partial U}{\partial T}\right)_P$, $\left(\frac{\partial U}{\partial V}\right)_T$, $\left(\frac{\partial U}{\partial T}\right)_V$, then we can compute the other four using the relationship between dP, dV, and dT given by the gas law.

General formulas for each partial derivative in terms of others can be obtained in the same way. See the following example and Problem 31.

Example 7 Express $\left(\frac{\partial U}{\partial T}\right)_P$ in terms of $\left(\frac{\partial U}{\partial T}\right)_V$ and $\left(\frac{\partial U}{\partial V}\right)_T$ and $\left(\frac{\partial V}{\partial T}\right)_P$

Solution Since we are interested in the derivatives $\left(\frac{\partial U}{\partial T}\right)_V$ and $\left(\frac{\partial U}{\partial V}\right)_T$, we think of U as a function of T and V and use the formula

$$dU = \left(\frac{\partial U}{\partial T}\right)_V dT + \left(\frac{\partial U}{\partial V}\right)_T dV \qquad \text{corresponding to } U_2.$$

We want to find a formula for $\left(\frac{\partial U}{\partial T}\right)_P$, which means thinking of U as a function of T and P. Thus, we want to substitute for dV. Since V is a function of T and P, we have

$$dV = \left(\frac{\partial V}{\partial T}\right)_P dT + \left(\frac{\partial V}{\partial P}\right)_T dP.$$

Substituting for dV into the formula for dU corresponding to U_2 gives

$$dU = \left(\frac{\partial U}{\partial T}\right)_V dT + \left(\frac{\partial U}{\partial V}\right)_T \left(\left(\frac{\partial V}{\partial T}\right)_P dT + \left(\frac{\partial V}{\partial P}\right)_T dP\right).$$

Collecting the terms containing dT and the terms containing dP gives

$$dU = \left(\left(\frac{\partial U}{\partial T}\right)_V + \left(\frac{\partial U}{\partial V}\right)_T \left(\frac{\partial V}{\partial T}\right)_P\right) dT + \left(\frac{\partial U}{\partial V}\right)_T \left(\frac{\partial V}{\partial P}\right)_T dP.$$

But we also have the formula

$$dU = \left(\frac{\partial U}{\partial T}\right)_P dT + \left(\frac{\partial U}{\partial P}\right)_T dP \qquad \text{corresponding to } U_1.$$

We now have two formulas for dU in terms of dT and dP. The coefficients of dT must be identical, so we conclude

$$\left(\frac{\partial U}{\partial T}\right)_P = \left(\frac{\partial U}{\partial T}\right)_V + \left(\frac{\partial U}{\partial V}\right)_T \left(\frac{\partial V}{\partial T}\right)_P.$$

Example 7 expresses $\left(\frac{\partial U}{\partial T}\right)_P$ in terms of three other partial derivatives. Two of them, namely $\left(\frac{\partial U}{\partial T}\right)_V$, the constant-volume heat capacity, and $\left(\frac{\partial V}{\partial T}\right)_P$, the expansion coefficient, can be easily measured experimentally. The third, the internal pressure, $\left(\frac{\partial U}{\partial V}\right)_T$, cannot be measured directly but can be related to $\left(\frac{\partial P}{\partial T}\right)_V$, which is measurable. Thus, $\left(\frac{\partial U}{\partial T}\right)_P$ can be determined indirectly using this identity.

Exercises and Problems for Section 14.6

Exercises

For Exercises 1–6, find dz/dt using the chain rule. Assume the variables are restricted to domains on which the functions are defined.

1. $z = xy^2$, $x = e^{-t}$, $y = \sin t$

2. $z = \ln(x^2 + y^2)$, $x = 1/t$, $y = \sqrt{t}$

3. $z = xe^y$, $x = 2t$, $y = 1 - t^2$

4. $z = (x + y)e^y$, $x = 2t$, $y = 1 - t^2$

5. $z = x \sin y + y \sin x$, $x = t^2$, $y = \ln t$

6. $z = \sin(x/y)$, $x = 2t$, $y = 1 - t^2$

For Exercises 7–14, find $\partial z/\partial u$ and $\partial z/\partial v$. The variables are restricted to domains on which the functions are defined.

7. $z = xe^y$, $x = \ln u$, $y = v$

8. $z = (x + y)e^y$, $x = \ln u$, $y = v$

9. $z = xe^y$, $x = u^2 + v^2$, $y = u^2 - v^2$

10. $z = (x + y)e^y$, $x = u^2 + v^2$, $y = u^2 - v^2$

11. $z = \sin(x/y)$, $x = \ln u$, $y = v$

12. $z = \tan^{-1}(x/y)$, $x = u^2 + v^2$, $y = u^2 - v^2$

13. $z = xe^{-y} + ye^{-x}$, $x = u \sin v$, $y = v \cos u$

14. $z = \cos(x^2 + y^2)$, $x = u \cos v$, $y = u \sin v$

Problems

15. Suppose $w = f(x, y, z)$ and that x, y, z are functions of u and v. Use a tree diagram to write down the chain rule formula for $\partial w/\partial u$ and $\partial w/\partial v$.

16. Suppose $w = f(x, y, z)$ and that x, y, z are all functions of t. Use a tree diagram to write down the chain rule for dw/dt.

17. Corn production, C, is a function of rainfall, R, and temperature, T. Figures 14.49 and 14.50 show how rainfall and temperature are predicted to vary with time because of global warming. Suppose we know that $\Delta C \approx 3.3\Delta R - 5\Delta T$. Use this to estimate the change in corn production between the year 2020 and the year 2021. Hence, estimate dC/dt when $t = 2020$.

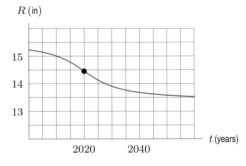

Figure 14.49: Rainfall as a function of time

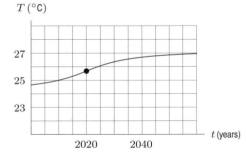

Figure 14.50: Temperature as a function of time

18. Let $z = g(u, v, w)$ and $u = u(s, t)$, $v = v(s, t)$, $w = w(s, t)$. How many terms are there in the expression for $\partial z/\partial t$?

19. The voltage, V, (in volts) across a circuit is given by Ohm's law: $V = IR$, where I is the current (in amps) flowing through the circuit and R is the resistance (in ohms). If we place two circuits, with resistance R_1 and R_2, in parallel, then their combined resistance, R, is given by

$$\frac{1}{R} = \frac{1}{R_1} + \frac{1}{R_2}.$$

Suppose the current is 2 amps and increasing at 10^{-2} amp/sec and R_1 is 3 ohms and increasing at 0.5 ohm/sec, while R_2 is 5 ohms and decreasing at 0.1 ohm/sec. Calculate the rate at which the voltage is changing.

20. Air pressure decreases at a rate of 2 pascals per kilometer in the eastward direction. In addition, the air pressure is dropping at a constant rate with respect to time everywhere. A ship sailing eastward at 10 km/hour past an island takes barometer readings and records a pressure drop of 50 pascals in 2 hours. Estimate the time rate of change of air pressure on the island. (A pascal is a unit of air pressure.)

21. Given $z = f(x, y)$, $x = x(u, v)$, $y = y(u, v)$ and $x(1, 2) = 5$, $y(1, 2) = 3$, calculate $z_u(1, 2)$ in terms of some of the numbers a, b, c, d, e, k, p, q, where

$$\begin{array}{llll} f_x(1, 2) = a & f_y(1, 2) = c & x_u(1, 2) = e & y_u(1, 2) = p \\ f_x(5, 3) = b & f_y(5, 3) = d & x_v(1, 2) = k & y_v(1, 2) = q \end{array}$$

22. Let $F(u, v)$ be a function of two variables. Find $f'(x)$ if

(a) $f(x) = F(x, 3)$ (b) $f(x) = F(3, x)$

(c) $f(x) = F(x, x)$ (d) $f(x) = F(5x, x^2)$

23. Let $F(u, v, w)$ be a function of three variables. Find $G_x(x, y)$ if

(a) $G(x, y) = F(x, y, 3)$ (b) $G(x, y) = F(3, y, x)$

(c) $G(x, y) = F(x, y, x)$ (d) $G(x, y) = F(x, y, xy)$

24. The equation $f(x, y) = f(a, b)$ defines a level curve through a point (a, b) where grad $f(a, b) \neq \vec{0}$. Use implicit differentiation and the chain rule to show that the slope of the line tangent to this curve at the point (a, b) is $-f_x(a, b)/f_y(a, b)$ if $f_y(a, b) \neq 0$.

25. A function $f(x, y)$ is *homogeneous of degree p* if $f(tx, ty) = t^p f(x, y)$ for all t. Show that any differentiable, homogeneous function of degree p satisfies Euler's Theorem:

$$x \, f_x(x, y) + y \, f_y(x, y) = p \, f(x, y).$$

[Hint: Define $g(t) = f(tx, ty)$ and compute $g'(1)$.]

26. Let $F(x, y, z)$ be a function and define a function $z = f(x, y)$ implicitly by letting $F(x, y, f(x, y)) = 0$. Use the chain rule to show that

$$\frac{\partial z}{\partial x} = -\frac{\partial F/\partial x}{\partial F/\partial z} \quad \text{and} \quad \frac{\partial z}{\partial y} = -\frac{\partial F/\partial y}{\partial F/\partial z}.$$

For Problems 27–28, suppose the quantity z can be expressed either as a function of Cartesian coordinates (x, y) or as a function of polar coordinates (r, θ), so that $z = f(x, y) = g(r, \theta)$. [Recall that $x = r\cos\theta, y = r\sin\theta$ and $r = \sqrt{x^2 + y^2}, \theta = \arctan(y/x)$]

27. (a) Use the chain rule to find $\partial z/\partial r$ and $\partial z/\partial \theta$ in terms of $\partial z/\partial x$ and $\partial z/\partial y$.

(b) Solve the equations you have just written down for $\partial z/\partial x$ and $\partial z/\partial y$ in terms of $\partial z/\partial r$ and $\partial z/\partial \theta$.

(c) Show that the expressions you get in part (b) are the same as you would get by using the chain rule to find $\partial z/\partial x$ and $\partial z/\partial y$ in terms of $\partial z/\partial r$ and $\partial z/\partial \theta$.

28. Show that

$$\left(\frac{\partial z}{\partial x}\right)^2 + \left(\frac{\partial z}{\partial y}\right)^2 = \left(\frac{\partial z}{\partial r}\right)^2 + \frac{1}{r^2}\left(\frac{\partial z}{\partial \theta}\right)^2.$$

Problems 29–35 are continuations of the physical chemistry example on page 724.

29. Write $\left(\frac{\partial U}{\partial P}\right)_V$ as a partial derivative of one of the functions $U_1, U_2,$ or U_3.

30. Write $\left(\frac{\partial U}{\partial P}\right)_T$ as a partial derivative of one of the functions U_1, U_2, U_3.

31. For the gas in Example 6, find $\left(\frac{\partial U}{\partial T}\right)_V$ and $\left(\frac{\partial U}{\partial V}\right)_T$. [Hint: Use the same method as the example, but substitute for dP instead of dV.]

32. Show that $\left(\frac{\partial T}{\partial V}\right)_P = 1 / \left(\frac{\partial V}{\partial T}\right)_P$.

33. Express $\left(\frac{\partial U}{\partial P}\right)_T$ in terms of $\left(\frac{\partial U}{\partial V}\right)_T$ and $\left(\frac{\partial V}{\partial P}\right)_T$.

34. Use Example 7 and Problem 32 to show that

$$\left(\frac{\partial U}{\partial V}\right)_P = \left(\frac{\partial U}{\partial V}\right)_T + \frac{\left(\frac{\partial U}{\partial T}\right)_V}{\left(\frac{\partial V}{\partial T}\right)_P}.$$

35. In Example 6, we calculated values of $(\partial U/\partial T)_P$ and $(\partial U/\partial P)_T$ using the relationship $PV = 2T$ for a specific gas. In this problem, you will derive general relationships for these two partial derivatives.

(a) Think of V as a function of P and T and write an expression for dV.

(b) Substitute for dV into the following formula for dU (thinking of U as a function of P and V):

$$dU = \left(\frac{\partial U}{\partial P}\right)_V dP + \left(\frac{\partial U}{\partial V}\right)_P dV.$$

(c) Thinking of U as a function of P and T, write an expression for dU.

(d) By comparing coefficients of dP and dT in your answers to parts (b) and (c), show that

$$\left(\frac{\partial U}{\partial T}\right)_P = \left(\frac{\partial U}{\partial V}\right)_P \cdot \left(\frac{\partial V}{\partial T}\right)_P$$

$$\left(\frac{\partial U}{\partial P}\right)_T = \left(\frac{\partial U}{\partial P}\right)_V + \left(\frac{\partial U}{\partial V}\right)_P \cdot \left(\frac{\partial V}{\partial P}\right)_T.$$

Problems 36–38 concern differentiating an integral in one variable, y, which also involves another variable x, either in the integrand, or in the limits, or both:

$$\int_0^5 (x^2 y + 4)\, dy \quad \text{or} \quad \int_0^x (y+4)\, dy \quad \text{or} \quad \int_0^x (xy+4)\, dy.$$

To differentiate the first integral with respect to x, it can be shown that in most cases we can differentiate with respect to x inside the integral:

$$\frac{d}{dx}\left(\int_0^5 (x^2 y + 4)\, dy\right) = \int_0^5 2xy\, dy.$$

Differentiating the second integral with respect to x uses the Fundamental Theorem of Calculus:

$$\frac{d}{dx}\int_0^x (y+4)\, dy = x + 4.$$

Differentiating the third integral involves the chain rule, as shown in Problem 38. Assume that the function F is continuously differentiable and b is constant throughout.

36. Let $f(x) = \int_0^b F(x, y)\, dy$. Find $f'(x)$.

37. Let $f(x) = \int_0^x F(b, y)\, dy$. Find $f'(x)$.

38. Let $f(x) = \int_0^x F(x, y)\, dy$. Use Problem 36 and Problem 37 to find $f'(x)$ by the following steps:

(a) Let $G(u, w) = \int_0^w F(u, y)\, dy$. Find $G_u(u, w)$ and $G_w(u, w)$.

(b) Use part (a) and the chain rule applied to $G(x, x) = f(x)$ to show:

$$f'(x) = \int_0^x F_u(x, y)\, dy + F(x, x).$$

14.7 SECOND-ORDER PARTIAL DERIVATIVES

Since the partial derivatives of a function are themselves functions, we can differentiate them, giving *second-order partial derivatives*. A function $z = f(x, y)$ has two first-order partial derivatives, f_x and f_y, and four second-order partial derivatives.

The Second-Order Partial Derivatives of $z = f(x, y)$

$$\frac{\partial^2 z}{\partial x^2} = f_{xx} = (f_x)_x, \qquad \frac{\partial^2 z}{\partial x \partial y} = f_{yx} = (f_y)_x,$$

$$\frac{\partial^2 z}{\partial y \partial x} = f_{xy} = (f_x)_y, \qquad \frac{\partial^2 z}{\partial y^2} = f_{yy} = (f_y)_y.$$

It is usual to omit the parentheses, writing f_{xy} instead of $(f_x)_y$ and $\dfrac{\partial^2 z}{\partial y \, \partial x}$ instead of $\dfrac{\partial}{\partial y}\left(\dfrac{\partial z}{\partial x}\right)$.

Example 1 Compute the four second-order partial derivatives of $f(x, y) = xy^2 + 3x^2 e^y$.

Solution From $f_x(x, y) = y^2 + 6xe^y$ we get

$$f_{xx}(x, y) = \frac{\partial}{\partial x}(y^2 + 6xe^y) = 6e^y \quad \text{and} \quad f_{xy}(x, y) = \frac{\partial}{\partial y}(y^2 + 6xe^y) = 2y + 6xe^y.$$

From $f_y(x, y) = 2xy + 3x^2 e^y$ we get

$$f_{yx}(x, y) = \frac{\partial}{\partial x}(2xy + 3x^2 e^y) = 2y + 6xe^y \quad \text{and} \quad f_{yy}(x, y) = \frac{\partial}{\partial y}(2xy + 3x^2 e^y) = 2x + 3x^2 e^y.$$

Observe that $f_{xy} = f_{yx}$ in this example.

Example 2 Use the values of the function $f(x, y)$ in Table 14.6 to estimate $f_{xy}(1, 2)$ and $f_{yx}(1, 2)$.

Table 14.6 *Values of $f(x, y)$*

$y \backslash x$	0.9	1.0	1.1
1.8	4.72	5.83	7.06
2.0	6.48	8.00	9.60
2.2	8.62	10.65	12.88

Solution Since $f_{xy} = (f_x)_y$, we first estimate f_x

$$f_x(1, 2) \approx \frac{f(1.1, 2) - f(1, 2)}{0.1} = \frac{9.60 - 8.00}{0.1} = 16.0,$$

$$f_x(1, 2.2) \approx \frac{f(1.1, 2.2) - f(1, 2.2)}{0.1} = \frac{12.88 - 10.65}{0.1} = 22.3.$$

Thus,

$$f_{xy}(1, 2) \approx \frac{f_x(1, 2.2) - f_x(1, 2)}{0.2} = \frac{22.3 - 16.0}{0.2} = 31.5.$$

Similarly,

$$f_{yx}(1,2) \approx \frac{f_y(1.1,2) - f_y(1,2)}{0.1} \approx \frac{1}{0.1}\left(\frac{f(1.1,2.2) - f(1.1,2)}{0.2} - \frac{f(1,2.2) - f(1,2)}{0.2}\right)$$

$$= \frac{1}{0.1}\left(\frac{12.88 - 9.60}{0.2} - \frac{10.65 - 8.00}{0.2}\right) = 31.5.$$

Observe that in this example also, $f_{xy} = f_{yx}$.

The Mixed Partial Derivatives Are Equal

It is not an accident that the estimates for $f_{xy}(1,2)$ and $f_{yx}(1,2)$ are equal in Example 2, because the same values of the function are used to calculate each one. The fact that $f_{xy} = f_{yx}$ in Examples 1 and 2 corroborates the following general result; Problem 36 suggests why you might expect it to be true.[2]

Theorem 14.1: Equality of Mixed Partial Derivatives

If f_{xy} and f_{yx} are continuous at (a,b), an interior point of their domain, then

$$f_{xy}(a,b) = f_{yx}(a,b).$$

For most functions f we encounter and most points (a,b) in their domains, not only are f_{xy} and f_{yx} continuous at (a,b), but all their higher order partial derivatives (such as f_{xxy} or f_{xyyy}) exist and are continuous at (a,b). In that case we say f is *smooth* at (a,b). We say f is smooth on a region R if it is smooth at every point of R.

What Do the Second-Order Partial Derivatives Tell Us?

Example 3   Let us return to the guitar string of Example 4, page 694. The string is 1 meter long and at time t seconds, the point x meters from one end is displaced $f(x,t)$ meters from its rest position, where

$$f(x,t) = 0.003 \sin(\pi x) \sin(2765t).$$

Compute the four second-order partial derivatives of f at the point $(x,t) = (0.3, 1)$ and describe the meaning of their signs in practical terms.

Solution   First we compute $f_x(x,t) = 0.003\pi \cos(\pi x) \sin(2765t)$, from which we get

$$f_{xx}(x,t) = \frac{\partial}{\partial x}(f_x(x,t)) = -0.003\pi^2 \sin(\pi x) \sin(2765t), \qquad \text{so} \qquad f_{xx}(0.3,1) \approx -0.01;$$

and

$$f_{xt}(x,t) = \frac{\partial}{\partial t}(f_x(x,t)) = (0.003)(2765)\pi \cos(\pi x) \cos(2765t), \qquad \text{so} \qquad f_{xt}(0.3,1) \approx 14.$$

On page 694 we saw that $f_x(x,t)$ gives the slope of the string at any point and time. Therefore, $f_{xx}(x,t)$ measures the concavity of the string. The fact that $f_{xx}(0.3,1) < 0$ means the string is concave down at the point $x = 0.3$ when $t = 1$. (See Figure 14.51.)

On the other hand, $f_{xt}(x,t)$ is the rate of change of the slope of the string with respect to time. Thus $f_{xt}(0.3,1) > 0$ means that at time $t = 1$ the slope at the point $x = 0.3$ is increasing. (See Figure 14.52.)

---

[2]For a proof, see M. Spivak, *Calculus on Manifolds*, p. 26 (New York: Benjamin, 1965).

The slope at B is less
than the slope at A

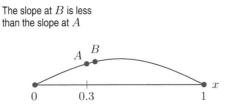

The slope at B is greater
than the slope at A

Figure 14.51: Interpretation of $f_{xx}(0.3, 1) < 0$:
The concavity of the string at $t = 1$

Figure 14.52: Interpretation of
$f_{xt}(0.3, 1) > 0$: The slope of one point on
the string at two different times

Now we compute $f_t(x, t) = (0.003)(2765) \sin(\pi x) \cos(2765t)$, from which we get

$$f_{tx}(x, t) = \frac{\partial}{\partial x}(f_t(x, t)) = (0.003)(2765)\pi \cos(\pi x) \cos(2765t), \quad \text{so} \quad f_{tx}(0.3, 1) \approx 14$$

and

$$f_{tt}(x, t) = \frac{\partial}{\partial t}(f_t(x, t)) = -(0.003)(2765)^2 \sin(\pi x) \sin(2765t), \quad \text{so} \quad f_{tt}(0.3, 1) \approx -7200.$$

On page 694 we saw that $f_t(x, t)$ gives the velocity of the string at any point and time. Therefore, $f_{tx}(x, t)$ and $f_{tt}(x, t)$ will both be rates of change of velocity. That $f_{tx}(0.3, 1) > 0$ means that at time $t = 1$ the velocities of points just to the right of $x = 0.3$ are greater than the velocity at $x = 0.3$. (See Figure 14.53.) That $f_{tt}(0.3, 1) < 0$ means that the velocity of the point $x = 0.3$ is decreasing at time $t = 1$. Thus, $f_{tt}(0.3, 1) = -7200$ m/sec^2 is the acceleration of this point. (See Figure 14.54.)

The velocity at B is greater
than the velocity at A

The velocity at B is less
than the velocity at A

Figure 14.53: Interpretation of
$f_{tx}(0.3, 1) > 0$: The velocity of different
points on the string at $t = 1$

Figure 14.54: Interpretation of
$f_{tt}(0.3, 1) < 0$: Negative acceleration. The
velocity of one point on the string at two
different times

Taylor Approximations

We use second derivatives to construct quadratic Taylor approximations. In Section 14.3, we saw how to approximate $f(x, y)$ by a linear function (its local linearization). We now see how to improve this approximation of $f(x, y)$ using a quadratic function.

Linear and Quadratic Approximations Near (0,0)

For a function of one variable, local linearity tells us that the best *linear* approximation is the degree 1 Taylor polynomial

$$f(x) \approx f(a) + f'(a)(x - a) \quad \text{for } x \text{ near } a.$$

A better approximation to $f(x)$ is given by the degree 2 Taylor polynomial:

$$f(x) \approx f(a) + f'(a)(x - a) + \frac{f''(a)}{2}(x - a)^2 \quad \text{for } x \text{ near } a.$$

For a function of two variables the local linearization for (x, y) near (a, b) is

$$f(x, y) \approx L(x, y) = f(a, b) + f_x(a, b)(x - a) + f_y(a, b)(y - b).$$

In the case $(a, b) = (0, 0)$, we have:

Taylor Polynomial of Degree 1 Approximating $f(x,y)$ for (x,y) near (0,0)
If f has continuous first-order partial derivatives, then

$$f(x,y) \approx L(x,y) = f(0,0) + f_x(0,0)x + f_y(0,0)y.$$

We get a better approximation to f by using a quadratic polynomial. We choose a quadratic polynomial $Q(x,y)$, with the same partial derivatives as the original function f. You can check that the following Taylor polynomial of degree 2 has this property.

Taylor Polynomial of Degree 2 Approximating $f(x,y)$ for (x,y) near (0,0) If f has continuous second-order partial derivatives, then

$$f(x,y) \approx Q(x,y)$$
$$= f(0,0) + f_x(0,0)x + f_y(0,0)y + \frac{f_{xx}(0,0)}{2}x^2 + f_{xy}(0,0)xy + \frac{f_{yy}(0,0)}{2}y^2.$$

Example 4   Let $f(x,y) = \cos(2x + y) + 3\sin(x + y)$
(a) Compute the linear and quadratic Taylor polynomials, L and Q, approximating f near $(0,0)$.
(b) Explain why the contour plots of L and Q for $-1 \le x \le 1$, $-1 \le y \le 1$ look the way they do.

Solution   (a) We have $f(0,0) = 1$. The derivatives we need are as follows:

$$
\begin{aligned}
f_x(x,y) &= -2\sin(2x+y) + 3\cos(x+y) &\text{so} \quad & f_x(0,0) = 3, \\
f_y(x,y) &= -\sin(2x+y) + 3\cos(x+y) &\text{so} \quad & f_y(0,0) = 3, \\
f_{xx}(x,y) &= -4\cos(2x+y) - 3\sin(x+y) &\text{so} \quad & f_{xx}(0,0) = -4, \\
f_{xy}(x,y) &= -2\cos(2x+y) - 3\sin(x+y) &\text{so} \quad & f_{xy}(0,0) = -2, \\
f_{yy}(x,y) &= -\cos(2x+y) - 3\sin(x+y) &\text{so} \quad & f_{yy}(0,0) = -1.
\end{aligned}
$$

Thus, the linear approximation, $L(x,y)$, to $f(x,y)$ at $(0,0)$ is given by

$$f(x,y) \approx L(x,y) = f(0,0) + f_x(0,0)x + f_y(0,0)y = 1 + 3x + 3y.$$

The quadratic approximation, $Q(x,y)$, to $f(x,y)$ near $(0,0)$ is given by

$$f(x,y) \approx Q(x,y)$$
$$= f(0,0) + f_x(0,0)x + f_y(0,0)y + \frac{f_{xx}(0,0)}{2}x^2 + f_{xy}(0,0)xy + \frac{f_{yy}(0,0)}{2}y^2$$
$$= 1 + 3x + 3y - 2x^2 - 2xy - \frac{1}{2}y^2.$$

Notice that the linear terms in $Q(x,y)$ are the same as the linear terms in $L(x,y)$. The quadratic terms in $Q(x,y)$ can be thought of as "correction terms" to the linear approximation.
(b) The contour plots of $f(x,y)$, $L(x,y)$, and $Q(x,y)$ are in Figures 14.55–14.57.

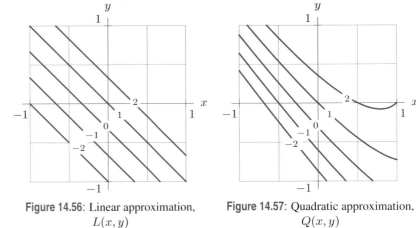

Figure 14.55: Original function, $f(x, y)$

Figure 14.56: Linear approximation, $L(x, y)$

Figure 14.57: Quadratic approximation, $Q(x, y)$

Notice that the contour plot of Q is more similar to the contour plot of f than is the contour plot of L. Since L is linear, the contour plot of L consists of parallel, equally spaced lines.

An alternative, and much quicker, way to find the Taylor polynomial in the previous example is to use the single-variable approximations. For example, since

$$\cos u = 1 - \frac{u^2}{2!} + \frac{u^4}{4!} + \cdots \quad \text{and} \quad \sin v = v - \frac{v^3}{3!} + \cdots,$$

we can substitute $u = 2x + y$ and $v = x + y$ and expand. We discard terms beyond the second (since we want the quadratic polynomial) getting

$$\cos(2x + y) = 1 - \frac{(2x + y)^2}{2!} + \frac{(2x + y)^4}{4!} + \cdots \approx 1 - \frac{1}{2}(4x^2 + 4xy + y^2) = 1 - 2x^2 - 2xy - \frac{1}{2}y^2$$

and

$$\sin(x + y) = (x + y) - \frac{(x + y)^3}{3!} + \cdots \approx x + y.$$

Combining these results, we get

$$\cos(2x + y) + 3\sin(x + y) \approx 1 - 2x^2 - 2xy - \frac{1}{2}y^2 + 3(x + y) = 1 + 3x + 3y - 2x^2 - 2xy - \frac{1}{2}y^2.$$

Linear and Quadratic Approximations Near (a, b)

The local linearization for a function $f(x, y)$ at a point (a, b) is

> **Taylor Polynomial of Degree 1 Approximating $f(x, y)$ for (x, y) near (a, b)**
> If f has continuous first-order partial derivatives, then
>
> $$f(x, y) \approx L(x, y) = f(a, b) + f_x(a, b)(x - a) + f_y(a, b)(y - b).$$

This suggests that a quadratic polynomial approximation $Q(x, y)$ for $f(x, y)$ near a point (a, b) should be written in terms of $(x - a)$ and $(y - b)$ instead of x and y. If we require that $Q(a, b) = f(a, b)$ and that the first- and second-order partial derivatives of Q and f at (a, b) be equal, then we get the following polynomial:

> **Taylor Polynomial of Degree 2 Approximating $f(x, y)$ for (x, y) near (a, b)**
> If f has continuous second-order partial derivatives, then
>
> $$f(x, y) \approx Q(x, y)$$
> $$= f(a, b) + f_x(a, b)(x - a) + f_y(a, b)(y - b)$$
> $$+ \frac{f_{xx}(a, b)}{2}(x - a)^2 + f_{xy}(a, b)(x - a)(y - b) + \frac{f_{yy}(a, b)}{2}(y - b)^2.$$

These coefficients are derived in exactly the same way as for $(a, b) = (0, 0)$.

Example 5   Find the Taylor polynomial of degree 2 at the point $(1, 2)$ for the function $f(x, y) = \dfrac{1}{xy}$.

Solution   Table 14.7 contains the partial derivatives and their values at the point $(1, 2)$.

Table 14.7   *Partial derivatives of $f(x, y) = 1/(xy)$*

Derivative	Formula	Value at $(1, 2)$	Derivative	Formula	Value at $(1, 2)$
$f(x, y)$	$1/(xy)$	$1/2$	$f_{xx}(x, y)$	$2/(x^3 y)$	1
$f_x(x, y)$	$-1/(x^2 y)$	$-1/2$	$f_{xy}(x, y)$	$1/(x^2 y^2)$	$1/4$
$f_y(x, y)$	$-1/(xy^2)$	$-1/4$	$f_{yy}(x, y)$	$2/(xy^3)$	$1/4$

So, the quadratic Taylor polynomial for f near $(1, 2)$ is

$$\frac{1}{xy} \approx Q(x, y)$$

$$= \frac{1}{2} - \frac{1}{2}(x - 1) - \frac{1}{4}(y - 2) + \frac{1}{2}(1)(x - 1)^2 + \frac{1}{4}(x - 1)(y - 2) + \left(\frac{1}{2}\right)\left(\frac{1}{4}\right)(y - 2)^2$$

$$= \frac{1}{2} - \frac{x - 1}{2} - \frac{y - 2}{4} + \frac{(x - 1)^2}{2} + \frac{(x - 1)(y - 2)}{4} + \frac{(y - 2)^2}{8}.$$

Exercises and Problems for Section 14.7

Exercises

In Exercises 1–10, calculate all four second-order partial derivatives and check that $f_{xy} = f_{yx}$. Assume the variables are restricted to a domain on which the function is defined.

1. $f(x, y) = 3x^2 y + 5xy^3$

2. $f(x, y) = (x + y)^2$

3. $f(x, y) = (x + y)^3$

4. $f(x, y) = xe^y$

5. $f(x, y) = e^{2xy}$

6. $f(x, y) = (x + y)e^y$

7. $f(x, y) = \sqrt{x^2 + y^2}$

8. $f(x, y) = \sin(x/y)$

9. $f(x, y) = \sin(x^2 + y^2)$

10. $f(x, y) = 3\sin 2x \cos 5y$

Find the quadratic Taylor polynomials about $(0, 0)$ for the functions in Exercises 11–18.

11. $(x - y + 1)^2$ **12.** $(y - 1)(x + 1)^2$

13. $e^{-2x^2 - y^2}$ **14.** $1/(1 + 2x - y)$

15. $e^x \cos y$ **16.** $\cos(x + 3y)$

17. $\ln(1 + x^2 - y)$ **18.** $\sin 2x + \cos y$

In Exercises 19–28, use the level curves of the function $z = f(x,y)$ to decide the sign (positive, negative, or zero) of each of the following partial derivatives at the point P. Assume the x- and y-axes are in the usual positions.

(a) $f_x(P)$ (b) $f_y(P)$ (c) $f_{xx}(P)$
(d) $f_{yy}(P)$ (e) $f_{xy}(P)$

27.

28.

In Exercises 29–32, find the linear, $L(x,y)$, and quadratic, $Q(x,y)$, Taylor polynomials valid near $(1,0)$. Compare the values of the approximations $L(0.9, 0.2)$ and $Q(0.9, 0.2)$ with the exact value of the function $f(0.9, 0.2)$.

29. $f(x,y) = x^2 y$ **30.** $f(x,y) = \sqrt{x + 2y}$

31. $f(x,y) = xe^{-y}$

32. $f(x,y) = \sin(x-1)\cos y$

Problems

33. Figure 14.58 shows a graph of $z = f(x,y)$. Is $f_{xx}(0,0)$ positive or negative? Is $f_{yy}(0,0)$ positive or negative? Give reasons for your answers.

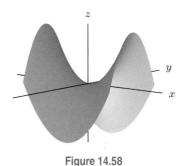

Figure 14.58

34. If $z = f(x) + yg(x)$, what can you say about z_{yy}? Explain your answer.

35. If $z_{xy} = 4y$, what can you say about the value of
(a) z_{yx}?   (b) z_{xyx}?   (c) z_{xyy}?

36. Give an explanation of why you might expect $f_{xy}(a,b) = f_{yx}(a,b)$ using the following steps.

(a) Write the definition of $f_x(a,b)$.
(b) Write a definition of $f_{xy}(a,b)$ as $(f_x)_y$.
(c) Substitute for f_x in the definition of f_{xy}.
(d) Write an expression for f_{yx} similar to the one for f_{xy} you obtained in part (c).
(e) Compare your answers to parts (c) and (d). What do you have to assume to conclude that f_{xy} and f_{yx} are equal?

37. A contour diagram for the smooth function $z = f(x, y)$ is in Figure 14.59.

    (a) Is z an increasing or decreasing function of x? Of y?
    (b) Is f_x positive or negative? How about f_y?
    (c) Is f_{xx} positive or negative? How about f_{yy}?
    (d) Sketch the direction of grad f at points P and Q.
    (e) Is grad f longer at P or at Q? How do you know?

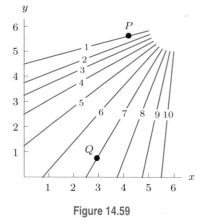

Figure 14.59

38. Figure 14.60 shows the level curves of a function $f(x, y)$ around a maximum or minimum, M. One of the points P and Q has coordinates (x_1, y_1) and the other has co-ordinates (x_2, y_2). Suppose $b > 0$ and $c > 0$. Consider the two linear approximations to f given by

$$f(x, y) \approx a + b(x - x_1) + c(y - y_1)$$
$$f(x, y) \approx k + m(x - x_2) + n(y - y_2).$$

    (a) What is the relationship between the values of a and k?
    (b) What are the coordinates of P?
    (c) Is M a maximum or a minimum?
    (d) What can you say about the sign of the constants m and n?

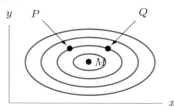

Figure 14.60

39. Consider the function $f(x, y) = (\sin x)(\sin y)$.

    (a) Find the Taylor polynomials of degree 2 for f about the points $(0, 0)$ and $(\pi/2, \pi/2)$.
    (b) Use the Taylor polynomials to sketch the contours of f close to each of the points $(0, 0)$ and $(\pi/2, \pi/2)$.

40. Let $f(x, y) = \sqrt{x + 2y + 1}$.

    (a) Compute the local linearization of f at $(0, 0)$.
    (b) Compute the quadratic Taylor polynomial for f at $(0, 0)$.
    (c) Compare the values of the linear and quadratic approximations in part (a) and part (b) with the true values for $f(x, y)$ at the points $(0.1, 0.1)$, $(-0.1, 0.1)$, $(0.1, -0.1)$, $(-0.1, -0.1)$. Which approximation gives the closest values?

41. Using a computer and your answer to Problem 40, draw the six contour diagrams of $f(x, y) = \sqrt{x + 2y + 1}$ and its linear and quadratic approximations, $L(x, y)$ and $Q(x, y)$, in the two windows $[-0.6, 0.6] \times [-0.6, 0.6]$ and $[-2, 2] \times [-2, 2]$. Explain the shape of the contours, their spacing, and the relationship between the contours of f, L, and Q.

42. Give contour diagrams of two different functions $f(x, y)$ and $g(x, y)$ that have the same quadratic approximations near $(0, 0)$.

43. Suppose that $f(x, y)$ has continuous partial derivatives f_x and f_y. Using the Fundamental Theorem of Calculus to evaluate the integrals, show that

$$f(a, b) = f(0, 0) + \int_{t=0}^{a} f_x(t, 0)dt + \int_{t=0}^{b} f_y(a, t)dt.$$

44. Suppose that $f(x, y)$ has continuous partial derivatives and that $f(0, 0) = 0$ and $|f_x(x, y)| \leq A$ and $|f_y(x, y)| \leq B$ for all points (x, y) in the plane. Use Problem 43 to show that $|f(x, y)| \leq A|x| + B|y|$.

This inequality shows how bounds on the partial derivatives of f limit the growth of f.

14.8 DIFFERENTIABILITY

In Section 14.3 we gave an informal introduction to the concept of differentiability. We called a function $f(x, y)$ *differentiable* at a point (a, b) if it is well-approximated by a linear function near (a, b). This section focuses on the precise meaning of the phrase "well-approximated." By looking at examples, we shall see that local linearity requires the existence of partial derivatives, but they do not tell the whole story. In particular, existence of partial derivatives at a point is not sufficient to guarantee local linearity at that point.

We begin by discussing the relation between continuity and differentiability. As an illustration, take a sheet of paper, crumple it into a ball and smooth it out again. Wherever there is a crease it would be difficult to approximate the surface by a plane—these are points of nondifferentiability of the function giving the height of the paper above the floor. Yet the sheet of paper models a graph which is continuous—there are no breaks. As in the case of one-variable calculus, continuity does not imply differentiability. But differentiability does *require* continuity: there cannot be linear approximations to a surface at points where there are abrupt changes in height.

Differentiability For Functions Of Two Variables

For a function of two variables, as for a function of one variable, we define differentiability at a point in terms of the error and the distance from the point. If the point is (a, b) and a nearby point is $(a + h, b + k)$, the distance between them is $\sqrt{h^2 + k^2}$. (See Figure 14.61.)

A function $f(x, y)$ is **differentiable at the point** (a, b) if there is a linear function $L(x, y) = f(a, b) + m(x - a) + n(y - b)$ such that if the *error* $E(x, y)$ is defined by

$$f(x, y) = L(x, y) + E(x, y),$$

and if $h = x - a, k = y - b$, then the *relative error* $E(a + h, b + k)/\sqrt{h^2 + k^2}$ satisfies

$$\lim_{\substack{h \to 0 \\ k \to 0}} \frac{E(a + h, b + k)}{\sqrt{h^2 + k^2}} = 0.$$

The function f is **differentiable on a region** R if it is differentiable at each point of R. The function $L(x, y)$ is called the *local linearization* of $f(x, y)$ near (a, b).

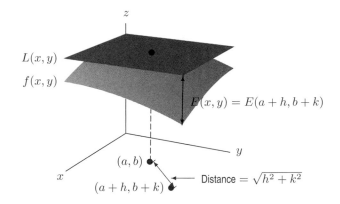

Figure 14.61: Graph of function $z = f(x, y)$ and its local linearization $z = L(x, y)$ near the point (a, b)

Partial Derivatives and Differentiability

In the next example, we show that this definition of differentiability is consistent with our previous notion — that is, that $m = f_x$ and $n = f_y$ and that the graph of $L(x, y)$ is the tangent plane.

Example 1   Show that if f is a differentiable function with local linearization $L(x, y) = f(a, b) + m(x - a) + n(y - b)$, then $m = f_x(a, b)$ and $n = f_y(a, b)$.

Solution

Since f is differentiable, we know that the relative error in $L(x, y)$ tends to 0 as we get close to (a, b). Suppose $h > 0$ and $k = 0$. Then we know that

$$
\begin{aligned}
0 = \lim_{h \to 0} \frac{E(a + h, b)}{\sqrt{h^2 + k^2}} &= \lim_{h \to 0} \frac{E(a + h, b)}{h} = \lim_{h \to 0} \frac{f(a + h, b) - L(a + h, b)}{h} \\
&= \lim_{h \to 0} \frac{f(a + h, b) - f(a, b) - mh}{h} \\
&= \lim_{h \to 0} \left(\frac{f(a + h, b) - f(a, b)}{h} \right) - m = f_x(a, b) - m.
\end{aligned}
$$

A similar result holds if $h < 0$, so we have $m = f_x(a, b)$. The result $n = f_y(a, b)$ is found in a similar manner.

The previous example shows that if a function is differentiable at a point, it has partial derivatives there. Therefore, if any of the partial derivatives fail to exist, then the function cannot be differentiable. This is what happens in the following example of a cone.

Example 2   Consider the function $f(x, y) = \sqrt{x^2 + y^2}$. Is f differentiable at the origin?

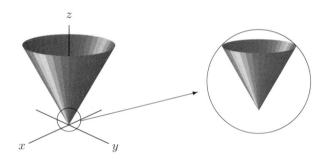

Figure 14.62: The function $f(x, y) = \sqrt{x^2 + y^2}$ is not locally linear at $(0, 0)$: Zooming in around $(0, 0)$ does not make the graph look like a plane

Solution

If we zoom in on the graph of the function $f(x, y) = \sqrt{x^2 + y^2}$ at the origin, as shown in Figure 14.62, the sharp point remains; the graph never flattens out to look like a plane. Near its vertex, the graph does not look like it is well approximated (in any reasonable sense) by any plane.

Judging from the graph of f, we would not expect f to be differentiable at $(0, 0)$. Let us check this by trying to compute the partial derivatives of f at $(0, 0)$:

$$
f_x(0, 0) = \lim_{h \to 0} \frac{f(h, 0) - f(0, 0)}{h} = \lim_{h \to 0} \frac{\sqrt{h^2 + 0} - 0}{h} = \lim_{h \to 0} \frac{|h|}{h}.
$$

Since $|h|/h = \pm 1$, depending on whether h approaches 0 from the left or right, this limit does not exist and so neither does the partial derivative $f_x(0, 0)$. Thus, f cannot be differentiable at the origin. If it were, both of the partial derivatives, $f_x(0, 0)$ and $f_y(0, 0)$, would exist.

Alternatively, we could show directly that there is no linear approximation near $(0, 0)$ that satisfies the small relative error criterion for differentiability. Any plane passing through the point $(0, 0, 0)$ has the form $L(x, y) = mx + ny$ for some constants m and n. If $E(x, y) = f(x, y) - L(x, y)$, then

$$
E(x, y) = \sqrt{x^2 + y^2} - mx - ny.
$$

Then for f to be differentiable at the origin, we would need to show that

$$\lim_{\substack{h \to 0 \\ k \to 0}} \frac{\sqrt{h^2 + k^2} - mh - nk}{\sqrt{h^2 + k^2}} = 0.$$

Taking $k = 0$ gives

$$\lim_{h \to 0} \frac{|h| - mh}{|h|} = 1 - m \lim_{h \to 0} \frac{h}{|h|}.$$

This limit exists only if $m = 0$ for the same reason as before. But then the value of the limit is 1 and not 0 as required. Thus, we again conclude f is not differentiable.

In Example 2 the partial derivatives f_x and f_y did not exist at the origin and this was sufficient to establish nondifferentiability there. We might expect that if both partial derivatives do exist, then f *is* differentiable. But the next example shows that this not necessarily true: the existence of both partial derivatives at a point is *not* sufficient to guarantee differentiability.

Example 3 Consider the function $f(x, y) = x^{1/3}y^{1/3}$. Show that the partial derivatives $f_x(0,0)$ and $f_y(0,0)$ exist, but that f is not differentiable at $(0,0)$.

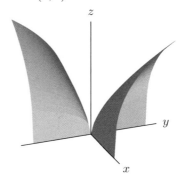

Figure 14.63: Graph of $z = x^{1/3}y^{1/3}$ for $z \geq 0$

Solution See Figure 14.63 for the part of the graph of $z = x^{1/3}y^{1/3}$ when $z \geq 0$. We have $f(0,0) = 0$ and we compute the partial derivatives using the definition:

$$f_x(0,0) = \lim_{h \to 0} \frac{f(h, 0) - f(0,0)}{h} = \lim_{h \to 0} \frac{0 - 0}{h} = 0,$$

and similarly

$$f_y(0,0) = 0.$$

So, if there did exist a linear approximation near the origin, it would have to be $L(x, y) = 0$. But we can show that this choice of $L(x, y)$ does not result in the small relative error that is required for differentiability. In fact, since $E(x, y) = f(x, y) - L(x, y) = f(x, y)$, we need to look at the limit

$$\lim_{\substack{h \to 0 \\ k \to 0}} \frac{h^{1/3}k^{1/3}}{\sqrt{h^2 + k^2}}.$$

If this limit exists, we get the same value no matter how h and k approach 0. Suppose we take $k = h > 0$. Then the limit becomes

$$\lim_{h \to 0} \frac{h^{1/3}h^{1/3}}{\sqrt{h^2 + h^2}} = \lim_{h \to 0} \frac{h^{2/3}}{h\sqrt{2}} = \lim_{h \to 0} \frac{1}{h^{1/3}\sqrt{2}}.$$

But this limit does not exist, since small values for h will make the fraction arbitrarily large. So the only possible candidate for a linear approximation at the origin does not have a sufficiently small relative error. Thus, this function is *not* differentiable at the origin, even though the partial derivatives $f_x(0,0)$ and $f_y(0,0)$ exist. Figure 14.63 confirms that near the origin the graph of $z = f(x, y)$ is not well approximated by any plane.

In summary,

- If a function is differentiable at a point, then both partial derivatives exist there.
- Having both partial derivatives at a point does not guarantee that a function is differentiable there.

Continuity and Differentiability

We know that differentiable functions of one variable are continuous. Similarly, it can be shown that if a function of two variables is differentiable at a point, then the function is continuous there.

In Example 3 the function f was continuous at the point where it was not differentiable. Example 4 shows that even if the partial derivatives of a function exist at a point, the function is not necessarily continuous at that point if it is not differentiable there.

Example 4   Suppose that f is the function of two variables defined by

$$f(x,y) = \begin{cases} \dfrac{xy}{x^2+y^2}, & (x,y) \neq (0,0), \\ 0, & (x,y) = (0,0). \end{cases}$$

Problem 17 on page 642 showed that $f(x,y)$ is not continuous at the origin. Show that the partial derivatives $f_x(0,0)$ and $f_y(0,0)$ exist. Could f be differentiable at $(0,0)$?

Solution   From the definition of the partial derivative we see that

$$f_x(0,0) = \lim_{h \to 0} \frac{f(h,0) - f(0,0)}{h} = \lim_{h \to 0} \left(\frac{1}{h} \cdot \frac{0}{h^2 + 0^2} \right) = \lim_{h \to 0} \frac{0}{h} = 0,$$

and similarly

$$f_y(0,0) = 0.$$

So, the partial derivatives $f_x(0,0)$ and $f_y(0,0)$ exist. However, f cannot be differentiable at the origin since it is not continuous there.

In summary,

- If a function is differentiable at a point, then it is continuous there.
- Having both partial derivatives at a point does not guarantee that a function is continuous there.

How Do We Know If a Function Is Differentiable?

Can we use partial derivatives to tell us if a function is differentiable? As we see from Examples 3 and 4, it is not enough that the partial derivatives exist. However, the following theorem gives conditions that *do* guarantee differentiability[3]:

Theorem 14.2: Continuity of Partial Derivatives Implies Differentiability

If the partial derivatives, f_x and f_y, of a function f exist and are continuous on a small disk centered at the point (a,b), then f is differentiable at (a,b).

We will not prove this theorem, although it provides a criterion for differentiability which is often simpler to use than the definition. It turns out that the requirement of continuous partial derivatives is more stringent than that of differentiability, so there exist differentiable functions which do

---

[3]For a proof, see M. Spivak, *Calculus on Manifolds*, p. 31, (New York: Benjamin, 1965).

not have continuous partial derivatives. However, most functions we encounter will have continuous partial derivatives. The class of functions with continuous partial derivatives is given the name C^1.

Example 5   Show that the function $f(x,y) = \ln(x^2 + y^2)$ is differentiable everywhere in its domain.

Solution   The domain of f is all of 2-space except for the origin. We shall show that f has continuous partial derivatives everywhere in its domain (that is, the function f is in C^1). The partial derivatives are

$$f_x = \frac{2x}{x^2 + y^2} \quad \text{and} \quad f_y = \frac{2y}{x^2 + y^2}.$$

Since each of f_x and f_y is the quotient of continuous functions, the partial derivatives are continuous everywhere except the origin (where the denominators are zero). Thus, f is differentiable everywhere in its domain.

Most functions built up from elementary functions have continuous partial derivatives, except perhaps at a few obvious points. Thus, in practice, we can often identify functions as being C^1 without explicitly computing the partial derivatives.

Exercises and Problems for Section 14.8

Exercises

In Exercises 1–10, list the points in the xy-plane, if any, at which the function $z = f(x,y)$ is not differentiable.

1. $z = -\sqrt{x^2 + y^2}$

2. $z = \sqrt{(x+1)^2 + y^2}$

3. $z = e^{-(x^2+y^2)}$

4. $z = |x| + |y|$

5. $z = x^{1/3} + y^2$

6. $z = |x+2| - |y-3|$

7. $z = (\sin x)(\cos|y|)$

8. $z = |x-3|^2 + y^3$

9. $z = 4 + \sqrt{(x-1)^2 + (y-2)^2}$

10. $z = 1 + \left((x-1)^2 + (y-2)^2\right)^2$

Problems

For the functions f in Problems 11–14 answer the following questions. Justify your answers.

(a) Use a computer to draw a contour diagram for f.

(b) Is f differentiable at all points $(x,y) \neq (0,0)$?

(c) Do the partial derivatives f_x and f_y exist and are they continuous at all points $(x,y) \neq (0,0)$?

(d) Is f differentiable at $(0,0)$?

(e) Do the partial derivatives f_x and f_y exist and are they continuous at $(0,0)$?

11. $f(x,y) = \begin{cases} \dfrac{x}{y} + \dfrac{y}{x}, & x \neq 0 \text{ and } y \neq 0, \\ 0, & x = 0 \text{ or } y = 0. \end{cases}$

12. $f(x,y) = \begin{cases} \dfrac{2xy}{(x^2+y^2)^2}, & (x,y) \neq (0,0), \\ 0, & (x,y) = (0,0). \end{cases}$

13. $f(x,y) = \begin{cases} \dfrac{xy}{\sqrt{x^2+y^2}}, & (x,y) \neq (0,0), \\ 0, & (x,y) = (0,0). \end{cases}$

14. $f(x,y) = \begin{cases} \dfrac{x^2 y}{x^4+y^2}, & (x,y) \neq (0,0), \\ 0, & (x,y) = (0,0). \end{cases}$

15. Consider the function

$$f(x,y) = \begin{cases} \dfrac{xy^2}{x^2+y^2}, & (x,y) \neq (0,0), \\ 0, & (x,y) = (0,0). \end{cases}$$

(a) Use a computer to draw the contour diagram for f.

(b) Is f differentiable for $(x,y) \neq (0,0)$?

(c) Show that $f_x(0,0)$ and $f_y(0,0)$ exist.

(d) Is f differentiable at $(0,0)$?

(e) Suppose $x(t) = at$ and $y(t) = bt$, where a and b are constants, not both zero. If $g(t) = f(x(t), y(t))$, show that

$$g'(0) = \frac{ab^2}{a^2+b^2}.$$

(f) Show that

$$f_x(0,0)x'(0) + f_y(0,0)y'(0) = 0.$$

Does the chain rule hold for the composite function $g(t)$ at $t = 0$? Explain.

(g) Show that the directional derivative $f_{\vec{u}}(0,0)$ exists for each unit vector $\vec{u}$. Does this imply that f is differentiable at $(0,0)$?

16. Consider the function

$$f(x,y) = \begin{cases} \dfrac{xy^2}{x^2+y^4}, & (x,y) \neq (0,0), \\ 0, & (x,y) = (0,0). \end{cases}$$

(a) Use a computer to draw the contour diagram for f.

(b) Show that the directional derivative $f_{\vec{u}}(0,0)$ exists for each unit vector $\vec{u}$.

(c) Is f continuous at $(0,0)$? Is f differentiable at $(0,0)$? Explain.

17. Consider the function $f(x,y) = \sqrt{|xy|}$.

(a) Use a computer to draw the contour diagram for f. Does the contour diagram look like that of a plane when we zoom in on the origin?

(b) Use a computer to draw the graph of f. Does the graph look like a plane when we zoom in on the origin?

(c) Is f differentiable for $(x,y) \neq (0,0)$?

(d) Show that $f_x(0,0)$ and $f_y(0,0)$ exist.

(e) Is f differentiable at $(0,0)$? [Hint: Consider the directional derivative $f_{\vec{u}}(0,0)$ for $\vec{u} = (\vec{i}+\vec{j})/\sqrt{2}$.]

18. Suppose a function f is differentiable at the point (a,b). Show that f is continuous at (a,b).

19. Suppose $f(x,y)$ is a function such that $f_x(0,0) = 0$ and $f_y(0,0) = 0$, and $f_{\vec{u}}(0,0) = 3$ for $\vec{u} = (\vec{i}+\vec{j})/\sqrt{2}$.

(a) Is f differentiable at $(0,0)$? Explain.

(b) Give an example of a function f defined on 2-space which satisfies these conditions. [Hint: The function f does not have to be defined by a single formula valid over all of 2-space.]

20. Consider the following function:

$$f(x,y) = \begin{cases} \dfrac{xy(x^2-y^2)}{x^2+y^2}, & (x,y) \neq (0,0), \\ 0, & (x,y) = (0,0). \end{cases}$$

The graph of f is shown in Figure 14.64, and the contour diagram of f is shown in Figure 14.65.

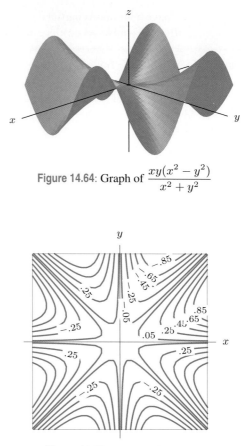

Figure 14.64: Graph of $\dfrac{xy(x^2-y^2)}{x^2+y^2}$

Figure 14.65: Contour diagram of
$$\dfrac{xy(x^2-y^2)}{x^2+y^2}$$

(a) Find $f_x(x,y)$ and $f_y(x,y)$ for $(x,y) \neq (0,0)$.

(b) Show that $f_x(0,0) = 0$ and $f_y(0,0) = 0$.

(c) Are the functions f_x and f_y continuous at $(0,0)$?

(d) Is f differentiable at $(0,0)$?

CHAPTER SUMMARY (see also Ready Reference at the end of the book)

- **Partial Derivatives**
  Definition as a difference quotient, visualizing on a graph, estimating from a contour diagram, computing from a formula, interpreting units, alternative notation.

- **Local Linearity**
  Zooming on a surface, contour diagram or table to see local linearity, the idea of tangent plane, formula for a tangent plane in terms of partials, the differential.

- **Directional Derivatives**
  Definition as a difference quotient, interpretation as a rate of change, computation using partial derivatives.

- **Gradient Vector**
  Definition in terms of partial derivatives, geometric

properties of gradient's length and direction, relation to directional derivative, relation to contours and level surfaces.

- **Chain Rule**
  Local linearity and differentials for composition of functions, tree diagrams, chain rule in general, application to physical chemistry.

- **Second and Higher Order Partial Derivatives**
  Interpretations, mixed partials are equal.

- **Taylor Approximations**
  Linear and quadratic polynomial approximations to functions near a point.

REVIEW EXERCISES AND PROBLEMS FOR CHAPTER FOURTEEN

Exercises

For Exercises 1–16, find the partial derivatives. Assume the variables are restricted to a domain on which the function is defined.

1. f_x and f_y if $f(x, y) = x^2 y + x^3 - 7xy^6$

2. $\dfrac{\partial w}{\partial h}$ if $w = 320\pi g h^2 (20 - h)$

3. F_L if $F(L, K) = 3\sqrt{LK}$

4. $\dfrac{\partial B}{\partial t}$ and $\dfrac{\partial B}{\partial r}$ when $B = P(1 + r)^t$.

5. f_x and f_y if $f(x, y) = \dfrac{x^2 y}{x^2 + y^2}$

6. $\dfrac{\partial F}{\partial r}$ and $\dfrac{\partial F}{\partial r}$ if $F = \dfrac{G\mu m y}{(r^2 + y^2)^{3/2}}$

7. f_p and f_q if $f(p, q) = e^{p/q}$

8. f_x if $f(x, y) = e^{xy}(\ln y)$

9. f_N if $f(N, V) = cN^\alpha V^\beta$

10. f_x and f_y if $f(x, y) = \sqrt{(x - a)^2 + (y - b)^2}$

11. $\dfrac{\partial}{\partial x}(e^x \cos(xy) + ay^2)$, $\dfrac{\partial}{\partial y}(e^x \cos(xy) + ay^2)$,

$\dfrac{\partial}{\partial a}(e^x \cos(xy) + ay^2)$

12. $\dfrac{\partial}{\partial \theta}(\sin(\pi \theta \phi) + \ln(\theta^2 + \phi))$

13. f_{xx} and f_{xy} if $f(x, y) = 1/\sqrt{x^2 + y^2}$

14. u_{xx} and u_{yy} if $u = e^x \sin y$

15. V_{rr} and V_{rh} if $V = \pi r^2 h$

16. f_{xxy} and f_{yxx} if $f(x, y) = \sin(x - 2y)$

In Exercises 17–20, find the gradient for the given function.

17. $f(x, y, z) = x^3 + z^3 - xyz$

18. $f(x, y) = \sin(y^2 - xy)$

19. $f(x, y) = \ln(x^2 + y^2)$

20. $f(\rho, \phi, \theta) = \rho \sin \phi \cos \theta$

In Exercises 21–26, find the directional derivative for the function.

21. $f(x, y) = x^3 - y^3$ at $(2, -1)$ in the direction of $\vec{i} - \vec{j}$

22. $f(x, y) = xe^y$ at $(3, 0)$ in the direction of $4\vec{i} - 3\vec{j}$

23. $f(x, y, z) = x^2 + y^2 - z^2$ at $(2, 3, 4)$ in the direction of $2\vec{i} - 2\vec{j} + \vec{k}$

24. $f(x, y, z) = 3x^2 y^2 + 2yz$ at $(-1, 0, 4)$ in the direction of $\vec{i} - \vec{k}$

25. $f(x, y, z) = 3x^2 y^2 + 2yz$ at $(-1, 0, 4)$ in the direction of $-\vec{i} + 3\vec{j} + 3\vec{k}$

26. $f(x, y, z) = e^{x+z} \cos y$ at $(1, 0, -1)$ in the direction of $\vec{i} + \vec{j} + \vec{k}$

In Exercises 27–29, find a vector normal to the curve or surface at the point.

27. $x^2 - y^2 = 3$ at $(2, 1)$

28. $xy + xz + yz = 11$ at $(1, 2, 3)$

29. $z^2 - 2xyz = x^2 + y^2$ at $(1, 2, -1)$

In Exercises 30–32, find the quadratic Taylor polynomial for the function.

30. $f(x, y) = (x + 1)^3 (y + 2)$ about $(0, 0)$

31. $f(x, y) = \cos x \cos 3y$ about $(0, 0)$

32. $f(x, y) = \sqrt{2x - y}$ about $(3, 5)$

Problems

33. (a) Let $f(w, z) = e^{w \ln z}$. Use difference quotients with $h = 0.01$ to approximate $f_w(2, 2)$ and $f_z(2, 2)$.
(b) Now evaluate $f_w(2, 2)$ and $f_z(2, 2)$ exactly.

For Problems 34–38 use Figure 14.66, showing level curves of $f(x, y)$, to estimate the directional derivatives.

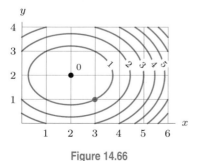

Figure 14.66

34. $f_{\vec{i}}(3, 1)$

35. $f_{\vec{j}}(3, 1)$

36. $f_{\vec{u}}(3, 1)$ where $\vec{u} = (\vec{i} - \vec{j})/\sqrt{2}$

37. $f_{\vec{u}}(3, 1)$ where $\vec{u} = (-\vec{i} + \vec{j})/\sqrt{2}$

38. $f_{\vec{u}}(3, 1)$ with $\vec{u} = (-2\vec{i} + \vec{j})/\sqrt{5}$

39. If $f(x, y) = x^2 y$ and $\vec{v} = 4\vec{i} - 3\vec{j}$, find the directional derivative at the point $(2, 6)$ in the direction of $\vec{v}$.

40. Find the directional derivative of $z = x^2 - y^2$ at the point $(3, -1)$ in the direction making an angle $\theta = \pi/4$ with the x-axis. In which direction is the directional derivative the largest?

The functions $u(x, t)$ in Problems 41–44 satisfy one of the equations (I)–(V), known as *partial differential equations,* for some positive real value of c. Match the function to an equation and give the value of c.

(I) $u_{xx} = c^2 u_{tt}$

(II) $u_{xx} = c^2 (u_t)^2$

(III) $u_{xx} = -c^2 u_{tt}$

(IV) $u_{xx} = -c^2 (u_t)^2$

(V) $(u_{xx})^2 = c^2 u_t$

41. $\sin(x - 2t)$

42. $e^{3t} \sin x$

43. e^{4x-9t}

44. $e^{3x} \sin 2t$

45. Find the point(s) on $x^2 + y^2 + z^2 = 8$ where the tangent plane is parallel to the plane $x - y + 3z = 0$.

46. The vector ∇f at point P and four unit vectors $\vec{u}_1, \vec{u}_2, \vec{u}_3, \vec{u}_4$ are shown in Figure 14.67. Arrange the following quantities in ascending order

$$f_{\vec{u}_1}, \quad f_{\vec{u}_2}, \quad f_{\vec{u}_3}, \quad f_{\vec{u}_4}, \quad \text{the number } 0.$$

The directional derivatives are all evaluated at the point P and the function $f(x, y)$ is differentiable at P.

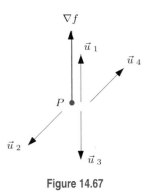

Figure 14.67

47. A differentiable function $f(x, y)$ has $f_x(4, 1) = 2$ and $f_y(4, 1) = -1$. Find the equation of the tangent line to the contour of f through the point $(4, 1)$.

48. The temperature at (x, y) is $T(x, y) = 100 - x^2 - y^2$. In which direction should a heat-seeking bug move from the point (x, y) to increase its temperature fastest?

49. A car is driving northwest at v mph across a sloping plain whose height, in feet above sea level, at a point N miles north and E miles east of a city is given by

$$h(N, E) = 2500 + 100N + 50E.$$

(a) At what rate is the height above sea level changing with respect to distance in the direction the car is driving?

(b) Express the rate of change of the height of the car with respect to time in terms of v.

50. The quantity Q (in pounds) of beef that a certain community buys during a week is a function $Q = f(b, c)$ of the prices of beef, b, and chicken, c, during the week. Do you expect $\partial Q/\partial b$ to be positive or negative? What about $\partial Q/\partial c$?

51. The cost of producing one unit of a product is given by

$$c = a + bx + ky,$$

where x is the amount of labor used (in man hours) and y is the amount of raw material used (by weight) and a and b and k are constants. What does $\partial c/\partial x = b$ mean? What is the practical interpretation of b?

52. People commuting to a city can choose to go either by bus or by train. The number of people who choose either method depends in part upon the price of each. Let $f(P_1, P_2)$ be the number of people who take the bus when P_1 is the price of a bus ride and P_2 is the price of a train ride. What can you say about the signs of $\partial f/\partial P_1$ and $\partial f/\partial P_2$? Explain your answers.

53. Figure 14.68 shows a contour diagram for the temperature T (in °C) along a wall in a heated room as a function of distance x along the wall and time t in minutes. Estimate $\partial T/\partial x$ and $\partial T/\partial t$ at the given points. Give units and interpret your answers.

(a) $x = 15, t = 20$ **(b)** $x = 5, t = 12$

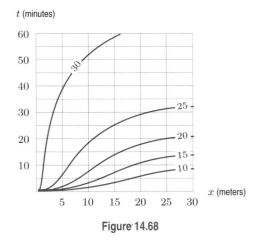

Figure 14.68

54. The acceleration g due to gravity, at a distance r from the center of a planet of mass m, is given by

$$g = \frac{Gm}{r^2},$$

where G is the universal gravitational constant.

(a) Find $\partial g/\partial m$ and $\partial g/\partial r$.

(b) Interpret each of the partial derivatives you found in part (a) as the slope of a graph in the plane and sketch the graph.

55. In analyzing a factory and deciding whether or not to hire more workers, it is useful to know under what circumstances productivity increases. Suppose $P = f(x_1, x_2, x_3)$ is the total quantity produced as a function of x_1, the number of workers, and any other variables x_2, x_3. We define the average productivity of a worker as P/x_1. Show that the average productivity increases as x_1 increases when marginal production, $\partial P/\partial x_1$, is greater than the average productivity, P/x_1.

56. For the Cobb-Douglas production function $P = 40L^{0.25}K^{0.75}$, find the differential dP when $L = 2$ and $K = 16$.

57. The period, T, of a pendulum is $T = 2\pi\sqrt{l/g}$. If the approximate length is $l = 2$ meters, find the approximate error in T if the true length is $l = 1.99$ and we take $g = 9.8$ as an approximation for $g = 9.81$ m/s^2.

58. Figure 14.69 shows the monthly payment, m, on a 5-year car loan if you borrow P dollars at r percent interest. Find a formula for a linear function which approximates m. What is the practical significance of the constants in your formula?

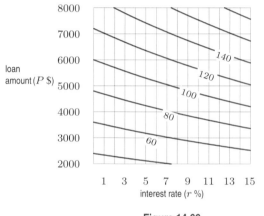

Figure 14.69

In Problems 59–66, the function f is differentiable and $f_x(2, 1) = -3$, $f_y(2, 1) = 4$, and $f(2, 1) = 7$.

59. **(a)** Give an equation for the tangent plane to the graph of f at $x = 2, y = 1$.
   (b) Give an equation for the tangent line to the contour for f at $x = 2, y = 1$.

60. **(a)** Find a vector perpendicular to the tangent plane to the graph of f at $x = 2, y = 1$.
   (b) Find a vector perpendicular to the tangent line to the contour for f at $x = 2, y = 1$.

61. Near $x = 2$ and $y = 1$, how far apart are the contours $f(x, y) = 7$ and $f(x, y) = 7.3$?

62. Give an approximate table of values of f for $x = 1.8, 2.0, 2.2$ and $y = 0.9, 1.0, 1, 1$.

63. Give an approximate contour diagram for f for $1 \le x \le 3$, $0 \le y \le 2$, using contour values $\ldots 5, 6, 7, 8, 9 \ldots$.

64. The function f gives temperature in °C and x and y are in centimeters. A bug leaves $(2, 1)$ at 3 cm/min so that it cools off as fast as possible. In which direction does the bug head? At what rate does it cool off, in °C/min?

65. Find $f_r(2, 1)$ and $f_\theta(2, 1)$, where r and θ are polar coordinates, $x = r\cos\theta$ and $y = r\sin\theta$. If $\vec{u}$ is the unit vector in the direction $2\vec{i} + \vec{j}$, show that $f_{\vec{u}}(2, 1) = f_r(2, 1)$ and explain why this should be the case.

66. Find approximately the largest value of f on or inside the circle of radius 0.1 about the point $(2, 1)$. At what point does f achieve this value?

67. Values of the function $f(x, y)$ near the point $x = 2$, $y = 3$ are given in Table 14.8. Estimate the following.

   (a) $\left.\dfrac{\partial f}{\partial x}\right|_{(2,3)}$ and $\left.\dfrac{\partial f}{\partial y}\right|_{(2,3)}$.
   (b) The rate of change of f at $(2, 3)$ in the direction of the vector $\vec{i} + 3\vec{j}$.
   (c) The maximum possible rate of change of f as you move away from the point $(2, 3)$. In which direction should you move to obtain this rate of change?
   (d) Write an equation for the level curve through the point $(2, 3)$.
   (e) Find a vector tangent to the level curve of f through the point $(2, 3)$.
   (f) Find the differential of f at the point $(2, 3)$. If $dx = 0.03$ and $dy = 0.04$, find df. What does df represent in this case?

Table 14.8

		x	
		2.00	2.01
y	3.00	7.56	7.42
	3.02	7.61	7.47

68. Find the quadratic Taylor polynomial about $(0, 0)$ for $f(x, y) = \cos(x + 2y)\sin(x - y)$.

69. Suppose $f(x, y) = e^{(x-1)^2 + (y-3)^2}$.

   (a) Find the first-order Taylor polynomial about $(0, 0)$.
   (b) Find the second-order (quadratic) Taylor polynomial about the point $(1, 3)$.
   (c) Find a 2-vector perpendicular to the level curve through $(0, 0)$.
   (d) Find a 3-vector perpendicular to the surface $z = f(x, y)$ at the point $(0, 0)$.

70. Figure 14.70 shows a contour diagram for a vibrating string function, $f(x, t)$.

    (a) Is $f_t(\pi/2, \pi/2)$ positive or negative? How about $f_t(\pi/2, \pi)$? What does the sign of $f_t(\pi/2, b)$ tell you about the motion of the point on the string at $x = \pi/2$ when $t = b$?

    (b) Find all t for which f_t is positive, for $0 \le t \le 5\pi/2$.

    (c) Find all x and t such that f_x is positive.

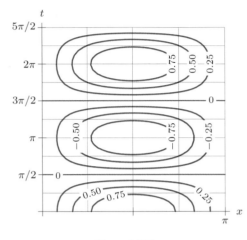

Figure 14.70

71. Each diagram (I) – (IV) in Figure 14.71 represents the level curves of a function $f(x, y)$. For each function f,

consider the point above P on the surface $z = f(x, y)$ and choose from the lists which follow:

    (a) A vector which could be the normal to the surface at that point;

    (b) An equation which could be the equation of the tangent plane to the surface at that point.

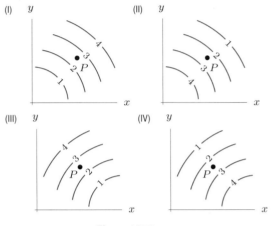

Figure 14.71

Vectors	Equations
(E) $2\vec{i} + 2\vec{j} - 2\vec{k}$	(J) $x + y + z = 4$
(F) $2\vec{i} + 2\vec{j} + 2\vec{k}$	(K) $2x - 2y - 2z = 2$
(G) $2\vec{i} - 2\vec{j} + 2\vec{k}$	(L) $-3x - 3y + 3z = 6$
(H) $-2\vec{i} + 2\vec{j} + 2\vec{k}$	(M) $-\dfrac{x}{2} + \dfrac{y}{2} - \dfrac{z}{2} = -7$

CAS Challenge Problems

72. (a) Find the quadratic Taylor polynomial about $(0, 0)$ of
$$f(x, y) = \frac{e^x (1 + \sin(3y))^2}{5 + e^{2x}}.$$

    (b) Find the quadratic Taylor polynomial about 0 of the one-variable functions $g(x) = e^x/(5 + e^{2x})$ and $h(y) = (1 + \sin(3y))^2$. Multiply these polynomials together and compare with your answer to part (a).

    (c) Show that if $f(x, y) = g(x)h(y)$, then the quadratic Taylor polynomial of f about $(0, 0)$ is the product of the quadratic Taylor polynomials of $g(x)$ and $h(y)$ about 0. [If you use a computer algebra system, make sure that f, g and h do not have any previously assigned formula.]

73. Let
$$f(x, y) = A_0 + A_1 x + A_2 y + A_3 x^2 + A_4 xy + A_5 y^2,$$
$$g(t) = 1 + B_1 t + B_2 t^2,$$
$$h(t) = 2 + C_1 t + C_2 t^2.$$

    (a) Find $L(x, y)$, the linear approximation to $f(x, y)$ at the point $(1, 2)$. Also find $m(t)$ and $n(t)$, the linear

approximations to $g(t)$ and $h(t)$ at $t = 0$.

    (b) Calculate $(d/dt)f(g(t), h(t))|_{t=0}$ and $(d/dt)L(m(t), n(t))|_{t=0}$. Describe what you notice and explain it in terms of the chain rule.

74. Let $f(x, y) = A_0 + A_1 x + A_2 y + A_3 x^2 + A_4 xy + A_5 y^2$.

    (a) Find the quadratic Taylor approximation for $f(x, y)$ near the point $(1, 2)$, and expand the result in powers of x and y.

    (b) Explain what you notice in part (a) and formulate a generalization to points other than $(1, 2)$.

    (c) Repeat part (a) for the linear approximation. How does it differ from the quadratic?

75. Suppose that $w = f(x, y, z)$, that x and y are functions of u and v, and that z, u, and v are functions of t. Use a computer algebra system to find the derivative

$$\frac{d}{dt} f(x(u(t), v(t)), y(u(t), v(t)), z(t))$$

and explain the answer using a tree diagram.

CHECK YOUR UNDERSTANDING

Are the statements in Problems 1–40 true or false? Give reasons for your answer.

1. If $f(x, y)$ is a function of two variables and $f_x(10, 20)$ is defined, then $f_x(10, 20)$ is a scalar.

2. If $f(x, y) = x^2 + y^2$, then $f_y(1, 1) < 0$.

3. If the graph of $f(x, y)$ is a hemisphere centered at the origin, then $f_x(0, 0) = f_y(0, 0) = 0$.

4. If $P = f(T, V)$ is a function expressing the pressure P (in grams/cm^3) of gas in a piston in terms of the temperature T (in degrees $^\circ$C) and volume V (in cm^3), then $\partial P / \partial V$ has units of grams.

5. If $f_x(a, b) > 0$, then the values of f decrease as we move in the negative x-direction near (a, b).

6. If $g(r, s) = r^2 + s$, then for fixed s, the partial derivative g_r increases as r increases.

7. If $g(u, v) = (u + v)^u$, then $2.3 \leq g_u(1, 1) \leq 2.4$.

8. Let $P = f(m, d)$ be the purchase price (in dollars) of a used car that has m miles on its engine and originally cost d dollars when new. Then $\partial P / \partial m$ and $\partial P / \partial d$ have the same sign.

9. If $f(x, y)$ is a function with the property that $f_x(x, y)$ and $f_y(x, y)$ are both constant, then f is linear.

10. If the point (a, b) is on the contour $f(x, y) = k$, then the slope of the line tangent to this contour at (a, b) is $f_y(a, b) / f_x(a, b)$.

11. There is a function $f(x, y)$ with $f_x(x, y) = y$ and $f_y(x, y) = x$.

12. The function $z(u, v) = u \cos v$ satisfies the equation

$$\cos v \frac{\partial z}{\partial u} - \frac{\sin v}{u} \frac{\partial z}{\partial v} = 1.$$

13. If $f(x, y)$ is a function of two variables and $g(x)$ is a function of a single variable, then

$$\frac{\partial}{\partial y} (g(x) f(x, y)) = g(x) f_y(x, y).$$

14. The function $k(r, s) = rse^s$ is increasing in the s-direction at the point $(r, s) = (-1, 2)$.

15. There is a function $f(x, y)$ with $f_x(x, y) = y^2$ and $f_y(x, y) = x^2$.

16. If f is a symmetric two-variable function, that is $f(x, y) = f(y, x)$, then $f_x(x, y) = f_y(x, y)$.

17. If f is a symmetric two-variable function, that is $f(x, y) = f(y, x)$, then $f_x(x, y) = f_y(y, x)$.

18. If $f(x, y)$ has $f_y(x, y) = 0$ then f must be a constant.

19. If $f(x, y) = ye^{g(x)}$ then $f_x = f$.

20. If $f(x, y)$ has $f_x(a, b) = f_y(a, b) = 0$ at the point (a, b), then f is constant everywhere.

21. An equation for the tangent plane to the surface $z = x^2 + y^3$ at $(1, 1)$ is $z = 2 + 2x(x - 1) + 3y^2(y - 1)$.

22. There is a function $f(x, y)$ which has a tangent plane with equation $z = 0$ at a point (a, b).

23. The tangent plane approximation of $f(x, y) = ye^{x^2}$ at the point $(0, 1)$ is $f(x, y) \approx y$.

24. If f is a function with differential $df = 2y\,dx + \sin(xy)\,dy$, then f changes by about -0.4 between the points $(1, 2)$ and $(0.9, 2.0002)$.

25. The local linearization of $f(x, y) = x^2 + y^2$ at $(1,1)$ gives an overestimate of the value of $f(x, y)$ at the point $(1.04, 0.95)$.

26. If two functions f and g have the same differential at the point $(1, 1)$, then $f = g$.

27. If two functions f and g have the same tangent plane at a point $(1, 1)$, then $f = g$.

28. If $f(x, y)$ is a constant function, then $df = 0$.

29. If $f(x, y)$ is a linear function, then df is a linear function of dx and dy.

30. If you zoom close enough near a point (a, b) on the contour diagram of any differentiable function, the contours will be *precisely* parallel and *exactly* equally spaced.

31. The gradient vector grad $f(a, b)$ is a vector in 3-space.

32. The gradient vector grad $f(a, b)$ is tangent to the contour of f at (a, b).

33. If you know the gradient vector of f at (a, b) then you can find the directional derivative $f_{\vec{u}}(a, b)$ for any unit vector $\vec{u}$.

34. If you know the directional derivative $f_{\vec{u}}(a, b)$ for all unit vectors $\vec{u}$ then you can find the gradient vector of f at (a, b).

35. The directional derivative $f_{\vec{u}}(a, b)$ is parallel to $\vec{u}$.

36. The gradient grad $f(3, 4)$ is perpendicular to the vector $3\vec{i} + 4\vec{j}$.

37. If grad $f(1, 2) = \vec{i}$, then f decreases in the $-\vec{i}$ direction at $(1, 2)$.

38. If grad $f(1, 2) = \vec{i}$, then $f(10, 2) > f(1, 2)$.

39. At the point $(3, 0)$, the function $g(x, y) = x^2 + y^2$ has the same maximal rate of increase as that of the function $h(x, y) = 2xy$.

40. If $f(x, y) = e^{x+y}$, then the directional derivative in any direction $\vec{u}$ (with $\|\vec{u}\| = 1$) at the point $(0, 0)$ is always less than or equal to $\sqrt{2}$.

Assume that $f(x, y)$ is a differentiable function. Are the statements in Problems 41–47 true or false? Explain your answer.

41. $f_{\vec{u}}(x_0, y_0)$ is a scalar.

42. $f_{\vec{u}}(a,b) = \|\nabla f(a,b)\|$

43. If $\vec{u}$ is tangent to the level curve of f at some point, then grad $f \cdot \vec{u} = 0$ there.

44. Suppose that f is differentiable at (a,b). Then there is always a direction in which the rate of change of f at (a,b) is 0.

45. There is a function with a point in its domain where

$\|$ grad $f\| = 0$ and where there is a nonzero directional derivative.

46. There is a function with $\|$ grad $f\| = 4$ and $f_{\vec{i}} = 5$ at some point.

47. There is a function with $\|$ grad $f\| = 4$ and $f_{\vec{j}} = -3$ at some point.

PROJECTS FOR CHAPTER FOURTEEN

1. Heat Equation

The function $T(x,y,z,t)$ is a solution to the *heat equation*

$$T_t = K(T_{xx} + T_{yy} + T_{zz}),$$

and gives the temperature at the point (x,y,z) in 3-space and time t. The constant K is the *thermal conductivity* of the medium through which the heat is flowing.

(a) Show that the function

$$T(x,y,z,t) = \frac{1}{(4\pi Kt)^{3/2}} e^{-(x^2+y^2+z^2)/4Kt}$$

is a solution to the heat equation for all (x,y,z) in 3-space and $t > 0$.

(b) For each fixed time t, what are the level surfaces of the function $T(x,y,z,t)$ in 3-space?

(c) Regard t as fixed and compute grad $T(x,y,z,t)$. What does grad $T(x,y,z,t)$ tell us about the direction and magnitude of the heat flow?

2. Matching Birthdays

Consider a class of m students and a year with n days. Let $q(m,n)$ denote the probability, expressed as a number between 0 and 1, that at least two students have the same birthday. Surprisingly, $q(23,365) \approx 0.5073$. (This means that there is slightly better than an even chance that at least two students in a class of 23 have the same birthday.) A general formula for q is complicated, but it can be shown that

$$\frac{\partial q}{\partial m} \approx +\frac{m}{n}(1-q) \qquad \text{and} \qquad \frac{\partial q}{\partial n} \approx -\frac{m^2}{2n^2}(1-q).$$

(These approximations hold when n is a good deal larger than m, and m is a good deal larger than 1.)

(a) Explain why the $+$ and $-$ signs in the approximations for $\partial q/\partial m$ and $\partial q/\partial n$ are to be expected.

(b) Suppose there are 21 students in a class. What is the approximate probability that at least two students in the class have the same birthday? (Assume that a year always has 365 days.)

(c) Suppose there is a class of 24 students and you know that no one was born in the first week of the year. (This has the effect of making $n = 358$.) What is the approximate value of q for this class?

(d) If you want to bet that a certain class of 23 students has at least two matching birthdays, would you prefer to have two more students added to the class or to be told that no one in the class was born in December?

(e) (Optional) Find the actual formula for q. [Hint: It's easier to find $1-q$. There are $n \cdot n \cdot n \cdots n = n^m$ different choices for the students' birthdays. How many such choices have no matching birthdays?]

Chapter Fifteen

OPTIMIZATION: LOCAL AND GLOBAL EXTREMA

In one-variable calculus we saw how to find the maximum and minimum values of a function of one variable. In practice, there are often several variables in an optimization problem. For example, you may have $10,000 to invest in new equipment and advertising for your business. What combination of equipment and advertising will yield the greatest profit? Or, what combination of drugs will lower a patient's temperature the most? In this chapter we consider optimization problems, where the variables are completely free to vary (unconstrained optimization) and where there is a constraint on the variables (for example, a budget constraint).

15.1 LOCAL EXTREMA

Functions of several variables, like functions of one variable, can have *local* and *global* extrema. (That is, local and global maxima and minima.) A function has a local extremum at a point where it takes on the largest or smallest value in a small region around the point. Global extrema are the largest or smallest values anywhere on the domain under consideration. (See Figures 15.1 and 15.2.)

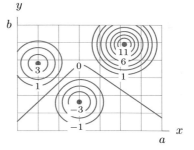

Figure 15.1: Local and global extrema for a function of two variables on $0 \leq x \leq a$, $0 \leq y \leq b$

Figure 15.2: Contour map of the function in Figure 15.1

More precisely, considering only points at which f is defined, we say:

- f has a **local maximum** at the point P_0 if $f(P_0) \geq f(P)$ for all points P near P_0.
- f has a **local minimum** at the point P_0 if $f(P_0) \leq f(P)$ for all points P near P_0.

How Do We Detect a Local Maximum or Minimum?

Recall that if the gradient vector of a function is defined and nonzero, then it points in a direction in which the function increases. Suppose that a function f has a local maximum at a point P_0 which is not on the boundary of the domain. If the vector grad $f(P_0)$ were defined and nonzero, then we could increase f by moving in the direction of grad $f(P_0)$. Since f has a local maximum at P_0, there is no direction in which f is increasing. Thus, if grad $f(P_0)$ is defined, we must have

$$\text{grad } f(P_0) = \vec{0}.$$

Similarly, suppose f has a local minimum at the point P_0. If grad $f(P_0)$ were defined and nonzero, then we could decrease f by moving in the direction opposite to grad $f(P_0)$, and so we must again have grad $f(P_0) = \vec{0}$ Therefore, we make the following definition:

Points where the gradient is either $\vec{0}$ or undefined are called **critical points** of the function.

If a function has a local maximum or minimum at a point P_0, not on the boundary of its domain, then P_0 is a critical point. For a function of two variables, we can also see that the gradient vector must be zero or undefined at a local maximum by looking at its contour diagram and a plot of its gradient vectors. (See Figures 15.3 and 15.4.) Around the maximum the vectors are all pointing inward, perpendicularly to the contours. At the maximum the gradient vector must be zero or undefined. A similar argument shows that the gradient must be zero at a local minimum.

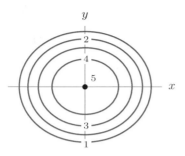

Figure 15.3: Contour diagram around a local maximum of a function

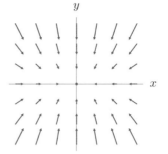

Figure 15.4: Gradients pointing toward the local maximum of the function in Figure 15.3

Finding and Analyzing Critical Points

To find critical points we set $\operatorname{grad} f = f_x\vec{i} + f_y\vec{j} + f_z\vec{k} = \vec{0}$, which means setting all the partial derivatives of f equal to zero. We must also look for the points where one or more of the partial derivatives is undefined.

Example 1   Find and analyze the critical points of $f(x, y) = x^2 - 2x + y^2 - 4y + 5$.

Solution   To find the critical points, we set both partial derivatives equal to zero:

$$f_x(x, y) = 2x - 2 = 0$$
$$f_y(x, y) = 2y - 4 = 0.$$

Solving these equations gives $x = 1$, $y = 2$. Hence, f has only one critical point, namely $(1, 2)$. To see the behavior of f near $(1, 2)$, look at the values of the function in Table 15.1.

Table 15.1   *Values of $f(x, y)$ near the point $(1, 2)$*

		0.8	0.9	1.0	1.1	1.2
	1.8	0.08	0.05	0.04	0.05	0.08
	1.9	0.05	0.02	0.01	0.02	0.05
y	2.0	0.04	0.01	0.00	0.01	0.04
	2.1	0.05	0.02	0.01	0.02	0.05
	2.2	0.08	0.05	0.04	0.05	0.08

The table suggests that the function has a local minimum value of 0 at $(1, 2)$. We can confirm this by completing the square:

$$f(x, y) = x^2 - 2x + y^2 - 4y + 5 = (x - 1)^2 + (y - 2)^2.$$

Figure 15.5 shows that the graph of f is a paraboloid with vertex at the point $(1, 2, 0)$. It is the same shape as the graph of $z = x^2 + y^2$ (see Figure 12.12 on page 612), except that the vertex has been shifted to $(1, 2)$. So the point $(1, 2)$ is a local minimum of f (as well as a global minimum).

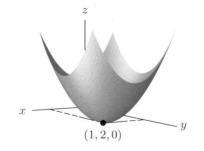

Figure 15.5: The graph of $f(x, y) = x^2 - 2x + y^2 - 4y + 5$ with a local minimum at the point $(1, 2)$

Example 2   Find and analyze any critical points of $f(x, y) = -\sqrt{x^2 + y^2}$.

Solution   We look for points where grad $f = \vec{0}$ or is undefined. The partial derivatives are given by

$$f_x(x, y) = -\frac{x}{\sqrt{x^2 + y^2}},$$

$$f_y(x, y) = -\frac{y}{\sqrt{x^2 + y^2}}.$$

These partial derivatives are never simultaneously zero; but they are undefined at $x = 0$, $y = 0$. Thus, $(0, 0)$ is a critical point and a possible extreme point. The graph of f (see Figure 15.6) is a cone, with vertex at $(0, 0)$. So f has a local and global maximum at $(0, 0)$.

Local maximum
Global maximum

Figure 15.6: Graph of $f(x, y) = -\sqrt{x^2 + y^2}$

Example 3   Find and analyze any critical points of $g(x, y) = x^2 - y^2$.

Solution   To find the critical points, we look for points where both partial derivatives are zero:

$$g_x(x, y) = 2x = 0$$

$$g_y(x, y) = -2y = 0.$$

Solving gives $x = 0$, $y = 0$, so the origin is the only critical point.

Figure 15.7 shows that near the origin g takes on both positive and negative values. Since $g(0, 0) = 0$, the origin is a critical point which is neither a local maximum nor a local minimum. The graph of g looks like a saddle.

The previous examples show that critical points can occur at local maxima or minima, or at points which are neither: The functions g and h in Figures 15.7 and 15.8 both have critical points at the origin. Figure 15.9 shows level curves of g. They are hyperbolas showing both positive and negative values of g near $(0, 0)$. Contrast this with the level curves of h near the local minimum in Figure 15.10.

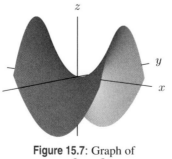

Figure 15.7: Graph of
$g(x, y) = x^2 - y^2$, showing
saddle shape at the origin

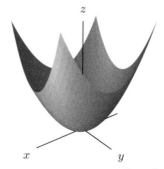

Figure 15.8: Graph of $h(x, y) = x^2 + y^2$, showing
minimum at the origin

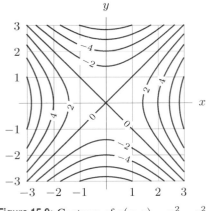

Figure 15.9: Contours of $g(x, y) = x^2 - y^2$, showing a saddle shape at the origin

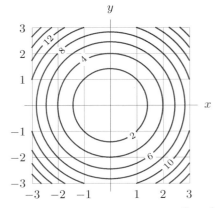

Figure 15.10: Contours of $h(x, y) = x^2 + y^2$, showing a local minimum at the origin

Example 4   Find the local extrema of the function $f(x, y) = 8y^3 + 12x^2 - 24xy$.

Solution   We begin by looking for critical points:

$$f_x(x, y) = 24x - 24y,$$
$$f_y(x, y) = 24y^2 - 24x.$$

Setting these expressions equal to zero gives the system of equations

$$x = y, \qquad x = y^2,$$

which has two solutions, $(0, 0)$ and $(1, 1)$. Are these local maxima, local minima or neither? Figure 15.11 shows contours of f near the points. Notice that $f(1, 1) = -4$ and that there is no other -4 contour. The contours near $(1, 1)$ suggests that f has a local minimum at the point $(1, 1)$.

We have $f(0, 0) = 0$ and the contours near $(0, 0)$ show that f takes both positive and negative values nearby. This suggests that $(0, 0)$ is a critical point which is neither a local maximum nor a local minimum.

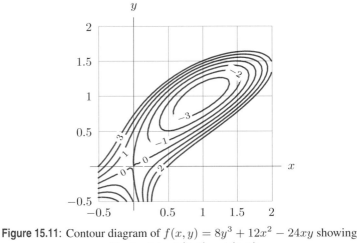

Figure 15.11: Contour diagram of $f(x, y) = 8y^3 + 12x^2 - 24xy$ showing critical points at $(0, 0)$ and $(1, 1)$

Classifying Critical Points

We can see whether a critical point of a function, f, is a local maximum, local minimum, or neither by looking at the contour diagram. There is also an analytic method for making this distinction.

Quadratic Functions of the form $f(x, y) = ax^2 + bxy + cy^2$

Near most critical points, a function has the same behavior as its quadratic Taylor approximation about that point. Thus, we start by investigating critical points of quadratic functions of the form $f(x, y) = ax^2 + bxy + cy^2$, where a, b and c are constants.

Example 5   Find and analyze the local extrema of the function $f(x, y) = x^2 + xy + y^2$.

Solution   To find critical points, we set

$$f_x(x, y) = 2x + y = 0,$$
$$f_y(x, y) = x + 2y = 0.$$

The only critical point is $(0, 0)$, and the value of the function there is $f(0, 0) = 0$. If f is always positive or zero near $(0, 0)$, then $(0, 0)$ is a local minimum; if f is always negative or zero near $(0, 0)$, it is a local maximum; if f takes both positive and negative values it is neither. The graph in Figure 15.12 suggests that $(0, 0)$ is a local minimum.

How can we be sure that $(0, 0)$ is a local minimum? We complete the square. Writing

$$f(x, y) = x^2 + xy + y^2 = \left(x + \frac{1}{2}y\right)^2 + \frac{3}{4}y^2,$$

shows that $f(x, y)$ is a sum of two squares, so it is always greater than or equal to zero. Thus, the critical point is both a local and a global minimum.

Local minimum

Figure 15.12: Graph of $f(x, y) = x^2 + xy + y^2 = (x + \frac{1}{2}y)^2 + \frac{3}{4}y^2$ showing local minimum at the origin

The Shape of the Graph of $f(x, y) = ax^2 + bxy + cy^2$

In general, a function of the form $f(x, y) = ax^2 + bxy + cy^2$ has one critical point at $(0, 0)$. Assuming $a \neq 0$, we complete the square and write

$$ax^2 + bxy + cy^2 = a\left[x^2 + \frac{b}{a}xy + \frac{c}{a}y^2\right] = a\left[\left(x + \frac{b}{2a}y\right)^2 + \left(\frac{c}{a} - \frac{b^2}{4a^2}\right)y^2\right]$$

$$= a\left[\left(x + \frac{b}{2a}y\right)^2 + \left(\frac{4ac - b^2}{4a^2}\right)y^2\right].$$

The shape of the graph of f depends on whether the coefficient of y^2 is positive, negative, or zero. The sign of the *discriminant*, $D = 4ac - b^2$, determines the sign of the coefficient of y^2.

- If $D > 0$, then the expression inside the square brackets is positive or zero, so the function has a local maximum or a local minimum.
  - If $a > 0$, the function has a local minimum, since the graph is a right-side-up paraboloid, like $z = x^2 + y^2$. (See Figure 15.13.)
  - If $a < 0$, the function has a local maximum, since the graph is an upside-down paraboloid, like $z = -x^2 - y^2$. (See Figure 15.14.)
- If $D < 0$, then the function goes up in some directions and goes down in others, like $z = x^2 - y^2$. We call this a *saddle point*. (See Figure 15.15.)
- If $D = 0$, then the quadratic function is $a(x + by/2a)^2$, whose graph is a parabolic cylinder. (See Figure 15.16.)

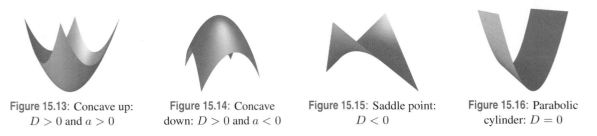

Figure 15.13: Concave up: $D > 0$ and $a > 0$

Figure 15.14: Concave down: $D > 0$ and $a < 0$

Figure 15.15: Saddle point: $D < 0$

Figure 15.16: Parabolic cylinder: $D = 0$

More generally, the graph of $g(x, y) = a(x - x_0)^2 + b(x - x_0)(y - y_0) + c(y - y_0)^2$ has the same shape as the graph of $f(x, y) = ax^2 + bxy + cy^2$, except that the critical point is at (x_0, y_0) rather than $(0, 0)$.[1]

Classifying the Critical Points of a Function

Suppose that f is any function with $\operatorname{grad} f(0, 0) = \vec{0}$. Its quadratic Taylor polynomial near $(0, 0)$,

$$f(x, y) \approx f(0, 0) + f_x(0, 0)x + f_y(0, 0)y$$
$$+ \frac{1}{2}f_{xx}(0, 0)x^2 + f_{xy}(0, 0)xy + \frac{1}{2}f_{yy}(0, 0)y^2,$$

can be simplified using $f_x(0, 0) = f_y(0, 0) = 0$, which gives

$$f(x, y) - f(0, 0) \approx \frac{1}{2}f_{xx}(0, 0)x^2 + f_{xy}(0, 0)xy + \frac{1}{2}f_{yy}(0, 0)y^2.$$

The discriminant of this quadratic polynomial is

$$D = 4ac - b^2 = 4\left(\frac{1}{2}f_{xx}(0, 0)\right)\left(\frac{1}{2}f_{yy}(0, 0)\right) - (f_{xy}(0, 0))^2,$$

which simplifies to

$$D = f_{xx}(0, 0)f_{yy}(0, 0) - (f_{xy}(0, 0))^2.$$

There is a similar formula for D if the critical point is at (x_0, y_0). An analogy with quadratic functions suggests the following test for classifying a critical point of a function of two variables:

Second Derivative Test for Functions of Two Variables

Suppose (x_0, y_0) is a point where $\operatorname{grad} f(x_0, y_0) = \vec{0}$. Let

$$D = f_{xx}(x_0, y_0)f_{yy}(x_0, y_0) - (f_{xy}(x_0, y_0))^2.$$

- If $D > 0$ and $f_{xx}(x_0, y_0) > 0$, then f has a local minimum at (x_0, y_0).
- If $D > 0$ and $f_{xx}(x_0, y_0) < 0$, then f has a local maximum at (x_0, y_0).
- If $D < 0$, then f has a saddle point at (x_0, y_0).
- If $D = 0$, anything can happen: f can have a local maximum, or a local minimum, or a saddle point, or none of these, at (x_0, y_0).

Example 6   Find the local maxima, minima, and saddle points of $f(x, y) = \frac{1}{2}x^2 + 3y^3 + 9y^2 - 3xy + 9y - 9x$.

Solution   The partial derivatives of f give

$$f_x(x, y) = x - 3y - 9 = 0,$$
$$f_y(x, y) = 9y^2 + 18y + 9 - 3x = 0.$$

---

[1]We assumed that $a \neq 0$. If $a = 0$ and $c \neq 0$, the same argument works. If both $a = 0$ and $c = 0$, then $f(x, y) = bxy$, which has a saddle point.

Eliminating x gives $9y^2 + 9y - 18 = 0$, with solutions $y = -2$ and $y = 1$. The corresponding values of x are $x = 3$ and $x = 12$, so the critical points of f are $(3, -2)$ and $(12, 1)$. The discriminant is

$$D(x, y) = f_{xx} f_{yy} - f_{xy}^2 = (1)(18y + 18) - (-3)^2 = 18y + 9.$$

Since $D(3, -2) = -36 + 9 < 0$, we know that $(3, -2)$ is a saddle point of f. Since $D(12, 1) = 18 + 9 > 0$ and $f_{xx}(12, 1) = 1 > 0$, we know that $(12, 1)$ is a local minimum of f.

The second derivative test does not give any information if $D = 0$. However, as the following example illustrates, we may still be able to classify the critical points.

Example 7   Classify the critical points of $f(x, y) = x^4 + y^4$, and $g(x, y) = -x^4 - y^4$, and $h(x, y) = x^4 - y^4$.

Solution   Each of these functions has a critical point at $(0, 0)$. Since all the second partial derivatives are 0 there, each function has $D = 0$. Near the origin, the graphs of f, g and h look like the surfaces in Figures 15.13–15.15, respectively, so f has a local minimum at $(0, 0)$, and g has a local maximum at $(0, 0)$, and h is saddle-shaped at $(0, 0)$.

We can get the same results algebraically. Since $f(0, 0) = 0$ and $f(x, y) > 0$ elsewhere, f has a local minimum at the origin. Since $g(0, 0) = 0$ and $g(x, y) < 0$ elsewhere, g has a local maximum at the origin. Lastly, h is saddle-shaped at the origin since $h(0, 0) = 0$ and, away from the origin, $h(x, y) > 0$ on the x-axis and $h(x, y) < 0$ on the y-axis.

Exercises and Problems for Section 15.1

Exercises

1. Which of the points A, B, C in Figure 15.17 appear to be critical points? Classify those that are critical points.

Figure 15.17

For Exercises 2–11, find the critical points and classify them as local maxima, local minima, saddle points, or none of these.

2. $f(x, y) = x^2 - 2xy + 3y^2 - 8y$

3. $f(x, y) = 400 - 3x^2 - 4x + 2xy - 5y^2 + 48y$

4. $f(x, y) = x^3 + y^2 - 3x^2 + 10y + 6$

5. $f(x, y) = x^3 - 3x + y^3 - 3y$

6. $f(x, y) = x^3 + y^3 - 3x^2 - 3y + 10$

7. $f(x, y) = x^3 + y^3 - 6y^2 - 3x + 9$

8. $f(x, y) = (x + y)(xy + 1)$

9. $f(x, y) = 8xy - \frac{1}{4}(x + y)^4$

10. $f(x, y) = 5 + 6x - x^2 + xy - y^2$

11. $f(x, y) = e^{2x^2 + y^2}$

Each function in Exercises 12–15 has a critical point at $(0, 0)$. What sort of critical point is it?

12. $f(x, y) = x^6 + y^6$

13. $g(x, y) = x^4 + y^3$

14. $h(x, y) = \cos x \cos y$

15. $k(x, y) = \sin x \sin y$

Problems

For Problems 16–18, use the contours of f in Figure 15.18.

16. Decide whether you think each point is a local maximum, local minimum, saddle point, or none of these.

(a) P (b) Q (c) R (d) S

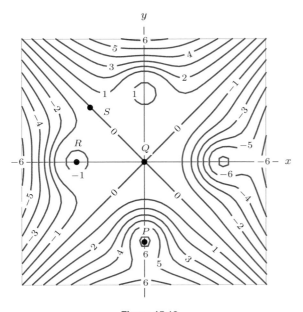

Figure 15.18

17. Sketch the direction of ∇f at several points around each of P, Q, and R.

18. Put arrows showing the direction of ∇f at the points where $\|\nabla f\|$ is largest.

For Problems 19–22, find critical points and classify them as local maxima, local minima, saddle points, or none of these.

19. $f(x, y) = x^3 + e^{-y^2}$

20. $E(x, y) = 1 - \cos x + y^2/2$

21. $f(x, y) = e^x(1 - \cos y)$

22. $f(x, y) = \sin x \sin y$

23. For $f(x, y) = A - (x^2 + Bx + y^2 + Cy)$, what values of A, B, and C give f a local maximum value of 15 at the point $(-2, 1)$?

24. At the point $(1, 3)$, suppose that $f_x = f_y = 0$ and $f_{xx} > 0$, $f_{yy} > 0$, $f_{xy} = 0$.

(a) What can you conclude about the behavior of the function near the point $(1, 3)$?

(b) Sketch a possible contour diagram.

25. At the point (a, b), suppose that $f_x = f_y = 0$, $f_{xx} > 0$, $f_{yy} = 0$, $f_{xy} > 0$.

(a) What can you conclude about the shape of the graph of f near the point (a, b)?

(b) Sketch a possible contour diagram.

26. Draw a possible contour diagram of f such that $f_x(-1, 0) = 0$, $f_y(-1, 0) < 0$, $f_x(3, 3) > 0$, $f_y(3, 3) > 0$, and f has a local maximum at $(3, -3)$.

27. Draw a possible contour diagram of a function with a saddle point at $(2, 1)$, a local minimum at $(2, 4)$, and no other critical points. Label the contours.

28. (a) Find the critical point of $f(x, y) = (x^2 - y)(x^2 + y)$.

(b) Show that at the critical point, the discriminant $D = 0$, so the second derivative test gives no information about the nature of the critical point.

(c) Sketch contours near the critical point to determine whether it is a local maximum, a local minimum, a saddle point, or none of these.

29. Let $f(x, y) = kx^2 + y^2 - 4xy$. Determine the values of k (if any) for which the critical point at $(0, 0)$ is:

(a) A saddle point

(b) A local maximum

(c) A local minimum

30. On a computer, draw contour diagrams for functions

$$f(x, y) = k(x^2 + y^2) - 2xy$$

for $k = -2, -1, 0, 1, 2$. Use these figures to classify the critical point at $(0, 0)$ for each value of k. Explain your observations using the discriminant, D.

31. The behavior of a function can be complicated near a critical point where $D = 0$. Suppose that

$$f(x, y) = x^3 - 3xy^2.$$

Show that there is one critical point at $(0, 0)$ and that $D = 0$ there. Show that the contour for $f(x, y) = 0$ consists of three lines intersecting at the origin and that these lines divide the plane into six regions around the origin where f alternates from positive to negative. Sketch a contour diagram for f near $(0, 0)$. The graph of this function is called a *monkey saddle*.

15.2 OPTIMIZATION

Suppose we want to find the highest and the lowest points in Colorado. A contour map is shown in Figure 15.19. The highest point is the top of a mountain peak (point A on the map, Mt. Elbert,

Figure 15.19: The highest and lowest points in the state of Colorado

14,431 feet high). What about the lowest point? Colorado does not have large pits without drainage, like Death Valley in California. A drop of rain falling at any point in Colorado will eventually flow out of the state. If there is no local minimum inside the state, where is the lowest point? It must be on the state boundary at a point where a river is flowing out of the state (point B where the Arkansas River leaves the state, 3,400 feet high). The highest point in Colorado is a global maximum for the elevation function in Colorado and the lowest point is the global minimum.

In general, if we are given a function f defined on a region R, we say:

> - f has a **global maximum on** R at the point P_0 if $f(P_0) \geq f(P)$ for all points P in R.
> - f has a **global minimum on** R at the point P_0 if $f(P_0) \leq f(P)$ for all points P in R.

The process of finding a global maximum or minimum for a function f on a region R is called *optimization*. If the region R is the entire xy-plane, we have *unconstrained optimization*; if the region R is not the entire xy-plane, that is, if x or y is restricted in some way, then we have *constrained optimization*. If the region R is not stated explicitly, it is understood to be the whole xy-plane.

How Do We Find Global Maxima and Minima?

As the Colorado example illustrates, a global extremum can occur either at a critical point inside the region or at a point on the boundary of the region. This is analogous to single-variable calculus, where a function achieves its global extrema on an interval either at a critical point inside the interval or at an endpoint of the interval. Optimization for functions of more than one variable, however, is more difficult because regions in 2-space can have very complicated boundaries.

> ### For an Unconstrained Optimization Problem
> - Find the critical points.
> - Investigate whether the critical points give global maxima or minima.

Not all functions have a global maximum or minimum: it depends on the function and the region. First, we consider applications in which global extrema are expected from practical considerations. At the end of this section, we examine the conditions that lead to global extrema. In general, the fact that a function has a single local maximum or minimum does not guarantee that the point is the global maximum or minimum. (See Problem 32.) An exception is if the function is quadratic, in which case the local maximum or minimum is the global maximum or minimum. (See Example 1 on page 749 and Example 5 on page 752.)

Maximizing Profit and Minimizing Cost

In planning production of an item, a company often chooses the combination of price and quantity that maximizes its profit. We use

$$\text{Profit} \ = \ \text{Revenue} \ - \ \text{Cost},$$

and, provided the price is constant,

$$\text{Revenue} \ = \ \text{Price} \ \cdot \ \text{Quantity} \ = pq.$$

In addition, we need to know how the cost and price depend on quantity.

Example 1   A company manufactures two items which are sold in two separate markets. The quantities, q_1 and q_2, demanded by consumers, and the prices, p_1 and p_2 (in dollars), of each item are related by

$$p_1 = 600 - 0.3q_1 \quad \text{and} \quad p_2 = 500 - 0.2q_2.$$

Thus, if the price for either item increases, the demand for it decreases. The company's total production cost is given by

$$C = 16 + 1.2q_1 + 1.5q_2 + 0.2q_1q_2.$$

To maximize its total profit, how much of each product should be produced? What is the maximum profit? [2]

Solution   The total revenue, R, is the sum of the revenues, p_1q_1 and p_2q_2, from each market. Substituting for p_1 and p_2, we get

$$R = p_1q_1 + p_2q_2 = (600 - 0.3q_1)q_1 + (500 - 0.2q_2)q_2$$
$$= 600q_1 - 0.3q_1^2 + 500q_2 - 0.2q_2^2.$$

Thus, the total profit P is given by

$$P = R - C = 600q_1 - 0.3q_1^2 + 500q_2 - 0.2q_2^2 - (16 + 1.2q_1 + 1.5q_2 + 0.2q_1q_2)$$
$$= -16 + 598.8q_1 - 0.3q_1^2 + 498.5q_2 - 0.2q_2^2 - 0.2q_1q_2.$$

To maximize P, we compute partial derivatives and set them equal to 0:

$$\frac{\partial P}{\partial q_1} = 598.8 - 0.6q_1 - 0.2q_2 = 0,$$

$$\frac{\partial P}{\partial q_2} = 498.5 - 0.4q_2 - 0.2q_1 = 0.$$

Since $\text{grad}\, P$ is defined everywhere, the only critical points of P are those where $\text{grad}\, P = \vec{0}$. Thus, solving for q_1, q_2, we find that

$$q_1 = 699.1 \quad \text{and} \quad q_2 = 896.7.$$

The corresponding prices are

$$p_1 = 390.27 \quad \text{and} \quad p_2 = 320.66.$$

To see whether or not we have found a maximum, we compute second partial derivatives:

$$\frac{\partial^2 P}{\partial q_1^2} = -0.6, \quad \frac{\partial^2 P}{\partial q_2^2} = -0.4, \quad \frac{\partial^2 P}{\partial q_1 \partial q_2} = -0.2,$$

---

[2] Adapted from M. Rosser, *Basic Mathematics for Economists*, p. 316 (New York: Routledge, 1993).

so,

$$D = \frac{\partial^2 P}{\partial q_1^2} \frac{\partial^2 P}{\partial q_2^2} - \left(\frac{\partial^2 P}{\partial q_1 \partial q_2}\right)^2 = (-0.6)(-0.4) - (-0.2)^2 = 0.2.$$

Therefore we have found a local maximum. The graph of P is an upside-down paraboloid, so $(699.1, 896.7)$ is a global maximum. The company should produce 699.1 units of the first item priced at \$390.27 per unit, and 896.7 units of the second item priced at \$320.66 per unit. The maximum profit $P(699.1, 896.7) \approx \$433,000$.

Example 2   Twenty cubic meters of gravel are to be delivered to a landfill. The trucker plans to purchase an open-top box in which to transport the gravel in numerous trips. The total cost to the trucker is the cost of the box plus \$2 per trip. The box must have height 0.5 m, but the trucker can choose the length and width. The cost of the box is \$20/m² for the ends and \$10/m² for the bottom and sides. Notice the tradeoff: A smaller box is cheaper to buy but requires more trips. What size box should the trucker buy to minimize the total cost? [3]

Solution   We first get an algebraic expression for the trucker's cost. Let the length of the box be x meters and the width be y meters and the height be 0.5 m (See Figure 15.20.)

Figure 15.20: The box for transporting gravel

Table 15.2   *Trucker's itemized cost*

$20/(0.5xy)$ at \$2/trip	$80/(xy)$
2 ends at \$20/m² · $0.5y$ m²	$20y$
2 sides at \$10/m² · $0.5x$ m²	$10x$
1 bottom at \$10/m² · xy m²	$10xy$
Total cost	$f(x, y)$

The volume of the box is $0.5xy$ m³, so delivery of 20 m³ of gravel requires $20/(0.5xy)$ trips. The trucker's cost is itemized in Table 15.2. The problem is to minimize

$$\text{Total cost} = f(x, y) = \frac{80}{xy} + 20y + 10x + 10xy.$$

The critical points of this function occur where

$$f_x(x, y) = -\frac{80}{x^2 y} + 10 + 10y = 0$$

$$f_y(x, y) = -\frac{80}{xy^2} + 10x + 20 = 0$$

We put the $80/(x^2 y)$ and $80/(xy^2)$ terms on the other side of the the equation, divide, and simplify:

$$\frac{80/(x^2 y)}{80/(xy^2)} = \frac{10 + 10y}{10x + 20} \quad \text{so} \quad \frac{y}{x} = \frac{1 + y}{x + 2} \quad \text{giving} \quad 2y = x.$$

---

[3] Adapted from Claude McMillan, Jr., *Mathematical Programming*, 2nd ed., p. 156-157 (New York: Wiley, 1978).

Substituting $x = 2y$ in the equation $f_y(x, y) = 0$ gives

$$-\frac{80}{2y \cdot y^2} + 10(2y) + 20 = 0$$

$$y^4 + y^3 - 2 = 0.$$

The only positive solution to this equation is $y = 1$, so the only critical point is $(2, 1)$.

To check that the critical point is a minimum, we use the second derivative test. Since

$$D(2, 1) = f_{xx}f_{yy} - (f_{xy})^2 = \frac{160}{2^3 \cdot 1} \cdot \frac{160}{2 \cdot 1^3} - \left(\frac{80}{2^2 \cdot 1^2} + 10\right)^2 = 700 > 0$$

and $f_{xx}(2, 1) > 0$, the point $(2, 1)$ is a local minimum. Since the value of f increases without bound as x or y increases without bound and as $x \to 0^+$ and $y \to 0^+$, it can be shown that $(2, 1)$ is a global minimum. (See Problem 33.) Thus, the optimal box is 2 meters long and 1 meter wide.

Fitting a Line to Data

Suppose we want to fit the "best" line to some data in the plane. We measure the distance from a line to the data points by adding the squares of the vertical distances from each point to the line. The smaller this sum of squares is, the better the line fits the data. The line with the minimum sum of square distances is called the *least squares line*, or the *regression line*. If the data is nearly linear, the least squares line is a good fit; otherwise it may not be. (See Figure 15.21.)

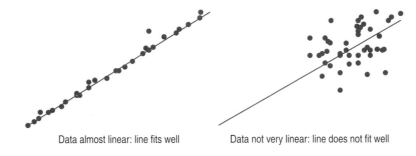

Data almost linear: line fits well   Data not very linear: line does not fit well

Figure 15.21: Fitting lines to data points

Example 3   Find a least squares line for the following data points: $(1, 1)$, $(2, 1)$, and $(3, 3)$.

Solution   Suppose the line has equation $y = b + mx$. If we find b and m then we have found the line. So, for this problem, b and m are the two variables. We want to minimize the function $f(b, m)$ that gives the sum of the three squared vertical distances from the points to the line in Figure 15.22.

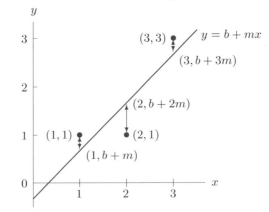

Figure 15.22: The least squares line minimizes the sum of the squares of these vertical distances

The vertical distance from the point $(1, 1)$ to the line is the difference in the y-coordinates $1 - (b + m)$; similarly for the other points. Thus, the sum of squares is

$$f(b, m) = (1 - (b + m))^2 + (1 - (b + 2m))^2 + (3 - (b + 3m))^2.$$

To minimize f we look for critical points. First we differentiate f with respect to b:

$$\frac{\partial f}{\partial b} = -2(1 - (b + m)) - 2(1 - (b + 2m)) - 2(3 - (b + 3m))$$
$$= -2 + 2b + 2m - 2 + 2b + 4m - 6 + 2b + 6m$$
$$= -10 + 6b + 12m.$$

Now we differentiate with respect to m:

$$\frac{\partial f}{\partial m} = 2(1 - (b + m))(-1) + 2(1 - (b + 2m))(-2) + 2(3 - (b + 3m))(-3)$$
$$= -2 + 2b + 2m - 4 + 4b + 8m - 18 + 6b + 18m$$
$$= -24 + 12b + 28m.$$

The equations $\dfrac{\partial f}{\partial b} = 0$ and $\dfrac{\partial f}{\partial m} = 0$ give a system of two linear equations in two unknowns:

$$-10 + 6b + 12m = 0,$$
$$-24 + 12b + 28m = 0.$$

The solution to this pair of equations is the critical point $b = -1/3$ and $m = 1$. Since

$$D = f_{bb}f_{mm} - (f_{mb})^2 = (6)(28) - 12^2 = 24 \quad \text{and} \quad f_{bb} = 6 > 0,$$

we have found a local minimum. The graph of $f(b, m)$ is a parabolic bowl, so the local minimum is the global minimum of f. Thus, the least squares line is

$$y = x - \frac{1}{3}.$$

As a check, notice that the line $y = x$ passes through the points $(1, 1)$ and $(3, 3)$. It is reasonable that introducing the point $(2, 1)$ moves the y-intercept down from 0 to $-1/3$.

The general formulas for the slope and y-intercept of a least squares line are in Project 2 on page 780. Many calculators have built in formulas for b and m, as well as for the *correlation coefficient*, which measures how well the data points fit the least squares line.

How Do We Know Whether a Function Has a Global Maximum or Minimum?

Under what circumstances does a function of two variables have a global maximum or minimum? The next example shows that a function may have both a global maximum and a global minimum on a region, or just one, or neither.

Example 4   Investigate the global maxima and minima of the following functions:
(a) $h(x, y) = 1 + x^2 + y^2$ on the disk $x^2 + y^2 \leq 1$.
(b) $f(x, y) = x^2 - 2x + y^2 - 4y + 5$ on the xy-plane.
(c) $g(x, y) = x^2 - y^2$ on the xy-plane.

Solution   (a) The graph of $h(x, y) = 1 + x^2 + y^2$ is a bowl shaped paraboloid with a global minimum of 1 at $(0, 0)$, and a global maximum of 2 on the edge of the region, $x^2 + y^2 = 1$.
(b) The graph of f in Figure 15.5 on page 749 shows that f has a global minimum at the point $(1, 2)$ and no global maximum (because the value of f increases without bound as $x \to \infty$, $y \to \infty$).

(c) The graph of g in Figure 15.7 on page 750 shows that g has no global maximum because $g(x, y) \to \infty$ as $x \to \infty$ if y is constant. Similarly, g has no global minimum because $g(x, y) \to -\infty$ as $y \to \infty$ if x is constant.

There are, however, conditions that guarantee that a function has a global maximum and minimum. For $h(x)$, a function of one variable, the function must be continuous on a closed interval $a \leq x \leq b$. If h is continuous on a non-closed interval, such as $a \leq x < b$ or $a < x < b$, or on an interval which is not bounded, such as $a < x < \infty$, then h need not have a maximum or minimum value. What is the situation for functions of two variables? As it turns out, a similar result is true for continuous functions defined on regions which are closed and bounded, analogous to the closed and bounded interval $a \leq x \leq b$. In everyday language we say

- A **closed** region is one which contains its boundary;
- A **bounded** region is one which does not stretch to infinity in any direction.

More precise definitions follow. Suppose R is a region in 2-space. A point (x_0, y_0) is a *boundary point* of R if, for every $r > 0$, the disk $(x - x_0)^2 + (y - y_0)^2 < r^2$ with center (x_0, y_0) and radius r contains both points which are in R and points which are not in R. See Figure 15.23. A point (x_0, y_0) can be a boundary point of the region R without belonging to R. A point (x_0, y_0) in R is an *interior point* if it is not a boundary point; thus, for small enough $r > 0$, the disk of radius r centered at (x_0, y_0) lies entirely in the region R. See Figure 15.24. The collection of all the boundary points is the *boundary of R* and the collection of all the interior points is the *interior* of R. The region R is *closed* if it contains its boundary; it is *open* if every point in R is an interior point.

A region R in 2-space is *bounded* if the distance between every point (x, y) in R and the origin is less than some constant K. Closed and bounded regions in 3-space are defined in the same way.

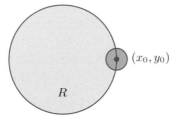

Figure 15.23: Boundary point (x_0, y_0) of R

Figure 15.24: Interior point (x_0, y_0) of R

Example 5

(a) The square $-1 \leq x \leq 1$, $-1 \leq y \leq 1$ is closed and bounded.
(b) The first quadrant $x \geq 0$, $y \geq 0$ is closed but is not bounded.
(c) The disk $x^2 + y^2 < 1$ is open and bounded, but is not closed.
(d) The half-plane $y > 0$ is open, but is neither closed nor bounded.

The reason that closed and bounded regions are useful is the following theorem, which is also true for functions of three or more variables:[4]

Theorem 15.1: Extreme Value Theorem for Multivariable Functions

If f is a continuous function on a closed and bounded region R, then f has a global maximum at some point (x_0, y_0) in R and a global minimum at some point (x_1, y_1) in R.

[4] For a proof, see W. Rudin, *Principles of Mathematical Analysis*, 2nd ed., p. 89, (New York: McGraw-Hill, 1976).

If f is not continuous or the region R is not closed and bounded, there is no guarantee that f achieves a global maximum or global minimum on R. In Example 4, the function g is continuous but does not achieve a global maximum or minimum in 2-space, a region which is closed but not bounded. Example 6 illustrates what can go wrong when the region is bounded but not closed.

Example 6   Does the function f have a global maximum or minimum on the region R given by $0 < x^2 + y^2 \leq 1$?

$$f(x, y) = \frac{1}{x^2 + y^2}$$

Solution   The region R is bounded, but it is not closed since it does not contain the boundary point $(0, 0)$. We see from the graph of $z = f(x, y)$ in Figure 15.25 that f has a global minimum on the circle $x^2 + y^2 = 1$. However, $f(x, y) \to \infty$ as $(x, y) \to (0, 0)$, so f has no global maximum.

Figure 15.25: Graph showing $f(x, y) = \frac{1}{x^2 + y^2}$ has no global maximum on $0 < x^2 + y^2 \leq 1$

Exercises and Problems for Section 15.2

Exercises

1. By looking at the weather map in Figure 12.1 on page 604, find the maximum and minimum daily high temperatures in the states of Mississippi, Alabama, Pennsylvania, New York, California, Arizona, and Massachusetts.

2. Compute the regression line for the points $(-1, 2)$, $(0, -1)$, $(1, 1)$ using least squares.

Do the functions in Exercises 3–7 have global maxima and minima?

3. $f(x, y) = x^2 - 2y^2$

4. $g(x, y) = x^2 y^2$

5. $h(x, y) = x^3 + y^3$

6. $f(x, y) = -2x^2 - 7y^2$

7. $f(x, y) = x^2/2 + 3y^3 + 9y^2 - 3x$

In Exercises 8–10, find the global maximum and minimum of the function on $-1 \leq x \leq 1$, $-1 \leq y \leq 1$, and say whether it occurs on the boundary of the square. [Hint: Use graphs.]

8. $z = x^2 + y^2$ 9. $z = -x^2 - y^2$ 10. $z = x^2 - y^2$

In Exercises 11–13, estimate the position and approximate value of the global maxima and minima on the region shown.

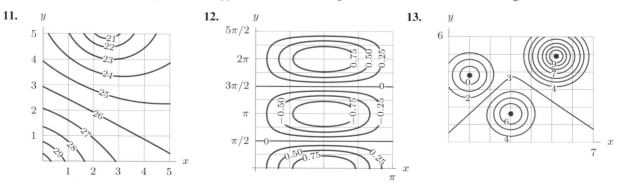

Problems

14. If $a > 0$ then the solution of the linear equation $ax = b$ minimizes the quadratic function $f(x) = (1/2)ax^2 - bx$. This problem shows the two variable analogue.

Suppose that $A > 0$ and $AC - B^2 > 0$. Show that the solution (x, y) of the simultaneous linear equations

$$Ax + By = D$$
$$Bx + Cy = E$$

minimizes the quadratic function $g(x, y) = (1/2)(Ax^2 + 2Bxy + Cy^2) - Dx - Ey$.

15. Find the parabola of the form $y = ax^2 + b$ which best fits the points $(1, 0)$, $(2, 2)$, $(3, 4)$ by minimizing the sum of squares, S, given by

$$S = (a + b)^2 + (4a + b - 2)^2 + (9a + b - 4)^2.$$

A *linear least squares approximation* of a function $f(x)$ on an interval $p \leq x \leq q$ is the linear function $L(x) = b + mx$ for which the point (b, m) minimizes the function $g(b, m) = \int_p^q (f(x) - (b + mx))^2 dx$. Find the linear least squares approximation on the interval $[0, 1]$ for the functions in Problems 16–17.

16. $f(x) = x^2$

17. $f(x) = x^3$

18. A missile has a guidance device which is sensitive to both temperature, $t°$C, and humidity, h. The range in km over which the missile can be controlled is given by

$$\text{Range} = 27{,}800 - 5t^2 - 6ht - 3h^2 + 400t + 300h.$$

What are the optimal atmospheric conditions for controlling the missile?

19. The quantity of a product demanded by consumers is a function of its price. The quantity of one product demanded may also depend on the price of other products. For example, the demand for tea is affected by the price of coffee; the demand for cars is affected by the price of gas. The quantities demanded, q_1 and q_2, of two products depend on their prices, p_1 and p_2, as follows

$$q_1 = 150 - 2p_1 - p_2$$

$$q_2 = 200 - p_1 - 3p_2.$$

(a) What does the fact that the coefficients of p_1 and p_2 are negative tell you? Give an example of two products that might be related this way.

(b) If one manufacturer sells both products, how should the prices be set to generate the maximum possible revenue? What is that maximum possible revenue?

20. A company operates two plants which manufacture the same item and whose total cost functions are

$$C_1 = 8.5 + 0.03q_1^2 \quad \text{and} \quad C_2 = 5.2 + 0.04q_2^2,$$

where q_1 and q_2 are the quantities produced by each plant. The total quantity demanded, $q = q_1 + q_2$, is related to the price, p, by

$$p = 60 - 0.04q.$$

How much should each plant produce in order to maximize the company's profit? [5]

21. Two products are manufactured in quantities q_1 and q_2 and sold at prices of p_1 and p_2, respectively. The cost of producing them is given by

$$C = 2q_1^2 + 2q_2^2 + 10.$$

(a) Find the maximum profit that can be made, assuming the prices are fixed.

(b) Find the rate of change of that maximum profit as p_1 increases.

22. Some items are sold at a discount to senior citizens or children. The reason is that these groups are more sensitive to price, so a discount has greater impact on their purchasing decisions. The seller faces an optimization problem: How large a discount to offer in order to maximize profits? Suppose a theater can sell q_c child tickets and q_a adult tickets at prices p_c and p_a, according to the demand functions:

$$q_c = rp_c^{-4} \quad \text{and} \quad q_a = sp_a^{-2},$$

and has operating costs proportional to the total number of tickets sold. What should be the relative price of children's and adults' tickets?

23. A closed rectangular box has volume 32 cm^3. What are the lengths of the edges giving the minimum surface area?

24. An open rectangular box has volume 32 cm^3. What are the lengths of the edges giving the minimum surface area?

25. A closed rectangular box with faces parallel to the coordinate planes has one bottom corner at the origin and the opposite top corner in the first octant on the plane $3x + 2y + z = 1$. What is the maximum volume of such a box?

26. An international airline has a regulation that each passenger can carry a suitcase having the sum of its width, length and height less than or equal to 135 cm. Find the dimensions of the suitcase of maximum volume that a passenger may carry under this regulation.

---

[5] Adapted from M. Rosser, *Basic Mathematics for Economists*, p. 318 (New York: Routledge, 1993).

27. Design a rectangular milk carton box of width w, length l, and height h which holds 512 cm^3 of milk. The sides of the box cost 1 cent/cm^2 and the top and bottom cost 2 cent/cm^2. Find the dimensions of the box that minimize the total cost of materials used.

28. Find the point on the plane $3x+2y+z=1$ that is closest to the origin by minimizing the square of the distance.

29. What is the shortest distance from the surface $xy+3x+z^2=9$ to the origin?

30. The population of the US was about 180 million in 1960, grew to 206 million in 1970, and 226 million in 1980.

(a) Assuming that the population was growing exponentially, use logarithms and the method of least squares to estimate the population in 1990.
(b) According to the national census, the 1990 population was 249 million. What does this say about the assumption of exponential growth?
(c) Predict the population in the year 2010.

31. A company manufactures a product which requires capital and labor to produce. The quantity, Q, of the product manufactured is given by the Cobb-Douglas function

$$Q = AK^aL^b,$$

where K is the quantity of capital and L is the quantity of labor used and A, a, and b are positive constants with $0 < a < 1$ and $0 < b < 1$. One unit of capital costs $\$k$ and one unit of labor costs $\$\ell$. The price of the product is fixed at $\$p$ per unit.

(a) If $a + b < 1$, how much capital and labor should the company use to maximize its profit?
(b) Is there a maximum profit in the case $a + b = 1$? What about $a + b \geq 1$? Explain.

32. Let $f(x,y) = x^2(y+1)^3 + y^2$. Show that f has only one critical point, namely $(0,0)$, and that point is a local minimum but not a global minimum. Contrast this with the case of a function with a single local minimum in one-variable calculus.

33. Let $f(x,y) = 80/(xy)+20y+10x+10xy$ in the region R where $x, y > 0$.

(a) Explain why $f(x,y) > f(2,1)$ at every point in R where
  (i) $x > 20$ (ii) $y > 20$
  (iii) $x < 0.01$ and $y \leq 20$
  (iv) $y < 0.01$ and $x \leq 20$
(b) Explain why f must have a global minimum at a critical point in R.
(c) Explain why f must have a global minimum in R at the point $(2,1)$.

34. Let $f(x,y) = 2/x + 3/y + 4x + 5y$ in the region R where $x, y > 0$.

(a) Explain why f must have a global minimum at some point in R.
(b) Find the global minimum.

35. This problem concerns optimization with a parameter. Let $f(x,a)$ give a family of functions of x with parameter a. Suppose that for each a the function $g(x) = f(x,a)$ in the family has a unique maximum $M(a)$ that is reached at a unique critical point $x = h(a)$ of g. Thus $M(a) = f(h(a),a)$. In good situations both M and h are differentiable.

(a) Show that $M'(a) = f_a(h(a),a)$
(b) Show that the graph of M is tangent at each point to the graph of one of the cross-sections of f with x fixed. Specifically, show that at the point $(a, M(a))$ the graph of M and the graph of the cross-section of f with $x = h(a)$ are tangent.
(c) Let $f(x,a) = -(1/2)x^2 + 2ax - a^2$. Find $h(a)$ and $M(a)$. Graph $M(a)$ and the cross-sections $f(x,a)$ with x fixed for $x = -2, 1, 0, 1, 2$, and 3 on the same axes. Looking at the graphs, why do you think that the equation in part (a) is called the *envelope theorem*?

15.3 CONSTRAINED OPTIMIZATION: LAGRANGE MULTIPLIERS

Many, perhaps most, real optimization problems are constrained by external circumstances. For example, a city wanting to build a public transportation system has only a limited number of tax dollars it can spend on the project. In this section, we see how to find an optimum value under such constraints.

Graphical Approach: Maximizing Production Subject to a Budget Constraint

Suppose we want to maximize the production under a budget constraint. Suppose production, f, is a function of two variables, x and y, which are quantities of two raw materials, and that

$$f(x,y) = x^{2/3}y^{1/3}.$$

If x and y are purchased at prices of p_1 and p_2 thousands of dollars per unit, what is the maximum production f that can be obtained with a budget of c thousand dollars?

To maximize f without regard to the budget, we simply increase x and y. However, the budget constraint prevents us from increasing x and y beyond a certain point. Exactly how does the budget

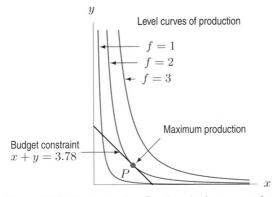

Figure 15.26: Optimal point, P, where budget constraint
is tangent to a level of production function

constrain us? With prices of p_1 and p_2, the amount spent on x is p_1x and the amount spent on y is p_2y, so we must have

$$g(x, y) = p_1x + p_2y \leq c,$$

where $g(x, y)$ is the total cost of the raw materials and c is the budget in thousands of dollars.

Let's look at the case when $p_1 = p_2 = 1$ and $c = 3.78$. Then

$$x + y \leq 3.78.$$

Figure 15.26 shows some contours of f and the budget constraint represented by the line $x+y = 3.78$. Any point on or below the line represents a pair of values of x and y that we can afford. A point on the line completely exhausts the budget, while a point below the line represents values of x and y which can be bought without using up the budget. Any point above the line represents a pair of values that we cannot afford.

To maximize f, we find the point which lies on the level curve with the largest possible value of f *and* which lies within the budget. The point must lie on the budget constraint because we should spend all the available money. Unless we are at the point where the budget constraint is tangent to the contour $f = 2$, we can increase f by moving along the line representing the budget constraint in Figure 15.26. For example, if we are on the line to the left of the point of tangency, moving right on the constraint will increase f; if we are on the line to the right of the point of tangency, moving left will increase f. Thus, the maximum value of f on the budget constraint occurs at the point where the budget constraint is tangent to the contour $f = 2$.

Analytical Solution: Lagrange Multipliers

Figure 15.26 suggests that maximum production is achieved at the point where the budget constraint is tangent to a level curve of the production function. The method of Lagrange multipliers uses this fact in algebraic form. Figure 15.27 shows that at the optimum point, P, the gradient of f and the normal to the budget line $g(x, y) = 3.78$ are parallel. Thus, at P, grad f and grad g are parallel, so for some scalar λ, called the *Lagrange multiplier*

$$\text{grad } f = \lambda \text{ grad } g.$$

Since grad $f = \left(\frac{2}{3}x^{-1/3}y^{1/3}\right)\vec{i} + \left(\frac{1}{3}x^{2/3}y^{-2/3}\right)\vec{j}$ and grad $g = \vec{i} + \vec{j}$, equating components gives

$$\frac{2}{3}x^{-1/3}y^{1/3} = \lambda \quad \text{and} \quad \frac{1}{3}x^{2/3}y^{-2/3} = \lambda.$$

Eliminating λ gives

$$\frac{2}{3}x^{-1/3}y^{1/3} = \frac{1}{3}x^{2/3}y^{-2/3}, \quad \text{which leads to} \quad 2y = x.$$

Since the constraint $x + y = 3.78$ must be satisfied, we have $x = 2.52$ and $y = 1.26$. Then

$$f(2.52, 1.26) = (2.52)^{2/3}(1.26)^{1/3} \approx 2.$$

As before, we see that the maximum value of f is approximately 2. Thus, to maximize production on a budget of \$3780, we should use 2.52 units of one raw material and 1.26 units of the other.

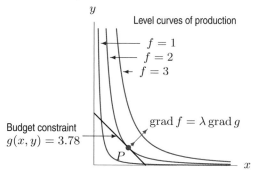

Figure 15.27: At the point, P, of maximum production, the vectors $\operatorname{grad} f$ and $\operatorname{grad} g$ are parallel

Lagrange Multipliers in General

Suppose we want to optimize an *objective function* $f(x, y)$ subject to a *constraint* $g(x, y) = c$. We look for extrema among the points which satisfy the constraint. We make the following definition.

> Suppose P_0 is a point satisfying the constraint $g(x, y) = c$.
> - f has a **local maximum** at P_0 **subject to the constraint** if $f(P_0) \geq f(P)$ for all points P near P_0 satisfying the constraint.
> - f has a **global maximum** at P_0 **subject to the constraint** if $f(P_0) \geq f(P)$ for all points P satisfying the constraint.
>
> Local and global minima are defined similarly.

As we saw in the production example, constrained extrema occur at points where the contours of f are tangent to the contours of g; they can also occur at endpoints of the constraint. In general, at any point, $\operatorname{grad} f$ is perpendicular to the directions in which the directional derivative of f is zero (provided $\operatorname{grad} f \neq \vec{0}$ there). Thus if $\operatorname{grad} f$ is not perpendicular to the constraint, it is possible to increase and decrease f by moving along the constraint. Therefore, at a constrained extremum, $\operatorname{grad} f$ must be perpendicular to the constraint and parallel to $\operatorname{grad} g$. (See Figure 15.28.) Thus, provided $\operatorname{grad} g \neq \vec{0}$ at P_0, there is a constant λ such that $\operatorname{grad} f = \lambda \operatorname{grad} g$ at P_0.

> **Optimizing f Subject to the Constraint $g = c$:**
> If a smooth function, f, has a maximum or minimum subject to a smooth constraint $g = c$ at a point P_0, then either P_0 satisfies the equations
>
> $$\operatorname{grad} f = \lambda \operatorname{grad} g \quad \text{and} \quad g = c,$$
>
> or P_0 is an endpoint of the constraint, or $\operatorname{grad} g(P_0) = \vec{0}$. To find P_0, compare values of f at these points. The number λ is called the **Lagrange multiplier**.

If the set of points satisfying the constraint is closed and bounded, such as a circle or line segment, then there must be a global maximum and minimum of f subject to the constraint. If the constraint is not closed and bounded, such as a line or hyperbola, then there may or may not be a global maximum and minimum.

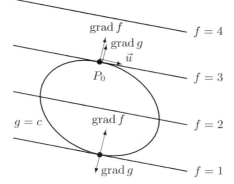

Figure 15.28: Maximum and minimum values of $f(x, y)$ on $g(x, y) = c$ are at points where grad f is parallel to grad g

Example 1 Find the maximum and minimum values of $x + y$ on the circle $x^2 + y^2 = 4$.

Solution The objective function is
$$f(x, y) = x + y,$$
and the constraint is
$$g(x, y) = x^2 + y^2 = 4.$$
Since grad $f = f_x \vec{i} + f_y \vec{j} = \vec{i} + \vec{j}$ and grad $g = g_x \vec{i} + g_y \vec{j} = 2x\vec{i} + 2y\vec{j}$, then grad $f = \lambda$ grad g gives
$$1 = 2\lambda x \quad \text{and} \quad 1 = 2\lambda y,$$
so
$$x = y.$$
We also know that
$$x^2 + y^2 = 4,$$
giving $x = y = \sqrt{2}$ or $x = y = -\sqrt{2}$. The constraint has no endpoints (it's a circle) and grad $g \neq \vec{0}$ on the circle, so we compare values of f at $(\sqrt{2}, \sqrt{2})$ and $(-\sqrt{2}, -\sqrt{2})$. Since $f(x, y) = x + y$, the maximum value of f is $f(\sqrt{2}, \sqrt{2}) = 2\sqrt{2}$; the minimum value is $f(-\sqrt{2}, -\sqrt{2}) = -2\sqrt{2}$. (See Figure 15.29.)

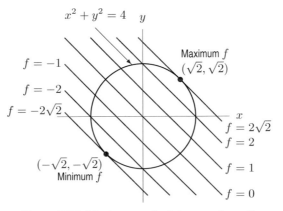

Figure 15.29: Maximum and minimum values of $f(x, y) = x + y$ on the circle $x^2 + y^2 = 4$ are at points where contours of f are tangent to the circle

How to Distinguish Maxima from Minima

There is a second derivative test[6] for classifying the critical points of constrained optimization problems, but it is more complicated than the test in Section 15.1. However, a graph of the constraint and some contours usually shows which points are maxima, which points are minima, and which are neither.

Optimization with Inequality Constraints

The production problem that we looked at first was to maximize production $f(x, y)$ subject to a budget constraint

$$g(x, y) = p_1 x + p_2 y \le c.$$

Since the inputs are nonnegative, $x \ge 0$ and $y \ge 0$, we have three inequality constraints, which restrict (x, y) to a region of the plane rather than to a curve in the plane. In principle, we should first check to see whether or not $f(x, y)$ has any critical points in the interior:

$$p_1 x + p_2 y < c, \qquad x > 0 \quad y > 0.$$

However, in the case of a budget constraint, we can see that the maximum of f must occur when the budget is exhausted, so we look for the maximum value of f on the boundary line:

$$p_1 x + p_2 y = c, \qquad x \ge 0 \quad y \ge 0.$$

Strategy for Optimizing $f(x, y)$ Subject to the Constraint $g(x, y) \le c$

- Find all points in the interior $g(x, y) < c$ where grad f is zero or undefined.
- Use Lagrange multipliers to find the local extrema of f on the boundary $g(x, y) = c$.
- Evaluate f at the points found in the previous two steps and compare the values.

From Section 15.2 we know that if f is continuous on a closed and bounded region, R, then f is guaranteed to attain its global maximum and minimum values on R.

Example 2   Find the maximum and minimum values of $f(x, y) = (x - 1)^2 + (y - 2)^2$ subject to the constraint $x^2 + y^2 \le 45$.

Solution   First, we look for all critical points for f in the interior of the region. Setting

$$f_x(x, y) = 2(x - 1) = 0$$
$$f_y(x, y) = 2(y - 2) = 0$$

we find f has exactly one critical point at $x = 1$, $y = 2$. Since $1^2 + 2^2 < 45$, that critical point is in the interior of the region.

Next, we find the local extrema of f on the boundary curve $x^2 + y^2 = 45$. To do this, we use Lagrange multipliers with constraint $g(x, y) = x^2 + y^2 = 45$. Setting grad $f = \lambda$ grad g, we get

$$2(x - 1) = \lambda \cdot 2x,$$
$$2(y - 2) = \lambda \cdot 2y.$$

We can't have $x = 0$ since the first equation would become $-2 = 0$. Similarly, $y \ne 0$. So we can solve each equation for λ by dividing by x and y. Setting the expressions for λ equal gives

$$\frac{x - 1}{x} = \frac{y - 2}{y},$$

---

[6]See J. E. Marsden and A. J. Tromba, *Vector Calculus*, 4th ed., p.218 (New York: W.H. Freeman, 1996).

so
$$y = 2x.$$

Combining this with the constraint $x^2 + y^2 = 45$, we get
$$5x^2 = 45$$

so
$$x = \pm 3.$$

Since $y = 2x$, we have possible local extrema at $x = 3$, $y = 6$ and $x = -3$, $y = -6$.

We conclude that the only candidates for the maximum and minimum values of f in the region occur at $(1, 2)$, $(3, 6)$, and $(-3, -6)$. Evaluating f at these three points we find
$$f(1, 2) = 0, \qquad f(3, 6) = 20, \qquad f(-3, -6) = 80.$$

Therefore, the minimum value of f is 0 at $(1, 2)$ and the maximum value is 80 at $(-3, -6)$.

The Meaning of λ

In the uses of Lagrange multipliers so far, we never found (or needed) the value of λ. However, λ does have a practical interpretation. In the production example, we wanted to maximize
$$f(x, y) = x^{2/3} y^{1/3}$$

subject to the constraint
$$g(x, y) = x + y = 3.78.$$

We solved the equations
$$\frac{2}{3} x^{-1/3} y^{1/3} = \lambda,$$
$$\frac{1}{3} x^{2/3} y^{-2/3} = \lambda,$$
$$x + y = 3.78,$$

to get $x = 2.52, y = 1.26$ and $f(2.52, 1.26) \approx 2$. Continuing to find λ gives us
$$\lambda \approx 0.53.$$

Now we do another, apparently unrelated, calculation. Suppose our budget is increased slightly, from 3.78 to 4.78, giving a new budget constraint of $x + y = 4.78$. Then the corresponding solution is at $x = 3.19$ and $y = 1.59$ and the new maximum value (instead of $f = 2$) is
$$f = (3.19)^{2/3}(1.59)^{1/3} \approx 2.53.$$

Notice that the amount by which f has increased is 0.53, the value of λ. Thus, in this example, the value of λ represents the extra production achieved by increasing the budget by one—in other words, the extra "bang" you get for an extra "buck" of budget. In fact, this is true in general:
- The value of λ is approximately the increase in the optimum value of f when the budget is increased by 1 unit.

More precisely:
- The value of λ represents the rate of change of the optimum value of f as the budget increases.

An Expression for λ

To interpret λ, we look at how the optimum value of the objective function f changes as the value c of the constraint function g is varied. In general, the optimum point (x_0, y_0) depends on the constraint value c. So, provided x_0 and y_0 are differentiable functions of c, we can use the chain rule to differentiate the optimum value $f(x_0(c), y_0(c))$ with respect to c:
$$\frac{df}{dc} = \frac{\partial f}{\partial x} \frac{dx_0}{dc} + \frac{\partial f}{\partial y} \frac{dy_0}{dc}.$$

At the optimum point (x_0, y_0), we have $f_x = \lambda g_x$ and $f_y = \lambda g_y$, and therefore

$$\frac{df}{dc} = \lambda \left(\frac{\partial g}{\partial x} \frac{dx_0}{dc} + \frac{\partial g}{\partial y} \frac{dy_0}{dc} \right) = \lambda \frac{dg}{dc}.$$

But, as $g(x_0(c), y_0(c)) = c$, we see that $dg/dc = 1$, so $df/dc = \lambda$. Thus, we have the following interpretation of the Lagrange multiplier λ:

The value of λ is the rate of change of the optimum value of f as c increases (where $g(x, y) = c$). If the optimum value of f is written as $f(x_0(c), y_0(c))$, then

$$\frac{d}{dc} f(x_0(c), y_0(c)) = \lambda.$$

Example 3   The quantity of goods produced according to the function $f(x, y) = x^{2/3} y^{1/3}$ is maximized subject to the budget constraint $x + y \leq 3.78$. The budget is increased to allow for a small increase in production. What is the price of the product if the sale of the additional goods covers the budget increase?

Solution   We know that $\lambda = 0.53$, which tells us that $df/dc = 0.53$. The constraint corresponds to a budget of 3.78 thousand. Therefore increasing the budget by $1000 increases production by about 0.53 units. In order to make the increase in budget profitable, the extra goods produced must sell for more than $1000. Thus, if p is the price of each unit of the good, then $0.53p$ is the revenue from the extra 0.53 units sold. Thus, we need $0.53p \geq 1000$ so $p \geq 1000/0.53 = \$1890$.

The Lagrangian Function

Constrained optimization problems are frequently solved using a *Lagrangian function*, $\mathcal{L}$. For example, to optimize $f(x, y)$ subject to the constraint $g(x, y) = c$, we use the Lagrangian function

$$\mathcal{L}(x, y, \lambda) = f(x, y) - \lambda(g(x, y) - c).$$

To see how the function $\mathcal{L}$ is used, compute the partial derivatives of $\mathcal{L}$:

$$\frac{\partial \mathcal{L}}{\partial x} = \frac{\partial f}{\partial x} - \lambda \frac{\partial g}{\partial x},$$

$$\frac{\partial \mathcal{L}}{\partial y} = \frac{\partial f}{\partial y} - \lambda \frac{\partial g}{\partial y},$$

$$\frac{\partial \mathcal{L}}{\partial \lambda} = -(g(x, y) - c).$$

Notice that if (x_0, y_0) is an extreme point of $f(x, y)$ subject to the constraint $g(x, y) = c$ and λ_0 is the corresponding Lagrange multiplier, then at the point (x_0, y_0, λ_0) we have

$$\frac{\partial \mathcal{L}}{\partial x} = 0 \quad \text{and} \quad \frac{\partial \mathcal{L}}{\partial y} = 0 \quad \text{and} \quad \frac{\partial \mathcal{L}}{\partial \lambda} = 0.$$

In other words, (x_0, y_0, λ_0) is a critical point for the unconstrained Lagrangian function, $\mathcal{L}(x, y, \lambda)$. Thus, the Lagrangian converts a constrained optimization problem to an unconstrained problem.

Example 4   A company has a production function with three inputs x, y, and z given by

$$f(x, y, z) = 50x^{2/5} y^{1/5} z^{1/5}.$$

The total budget is $24,000 and the company can buy x, y, and z at $80, $12, and $10 per unit, respectively. What combination of inputs will maximize production? [7]

---

[7] Adapted from M. Rosser, *Basic Mathematics for Economists*, p. 363 (New York: Routledge, 1993).

Solution   We need to maximize the objective function

$$f(x, y, z) = 50x^{2/5}y^{1/5}z^{1/5},$$

subject to the constraint

$$g(x, y, z) = 80x + 12y + 10z = 24,000.$$

Therefore, the Lagrangian function is

$$\mathcal{L}(x, y, z, \lambda) = 50x^{2/5}y^{1/5}z^{1/5} - \lambda(80x + 12y + 10z - 24,000),$$

so we solve the system of equations we get from grad $\mathcal{L} = \vec{0}$:

$$\frac{\partial \mathcal{L}}{\partial x} = 20x^{-3/5}y^{1/5}z^{1/5} - 80\lambda = 0,$$

$$\frac{\partial \mathcal{L}}{\partial y} = 10x^{2/5}y^{-4/5}z^{1/5} - 12\lambda = 0,$$

$$\frac{\partial \mathcal{L}}{\partial z} = 10x^{2/5}y^{1/5}z^{-4/5} - 10\lambda = 0,$$

$$\frac{\partial \mathcal{L}}{\partial \lambda} = -(80x + 12y + 10z - 24,000) = 0.$$

We simplify this system to give

$$\lambda = \frac{1}{4}x^{-3/5}y^{1/5}z^{1/5},$$

$$\lambda = \frac{5}{6}x^{2/5}y^{-4/5}z^{1/5},$$

$$\lambda = x^{2/5}y^{1/5}z^{-4/5},$$

$$80x + 12y + 10z = 24,000.$$

Eliminating z from the first two equations gives $x = 0.3y$. Eliminating x from the second and third equations gives $z = 1.2y$. Substituting for x and z into $80x + 12y + 10z = 24,000$ gives

$$80(0.3y) + 12y + 10(1.2y) = 24,000,$$

so $y = 500$. Then $x = 150$ and $z = 600$, and $f(150, 500, 600) = 4,622$ units.

The graph of the constraint, $80x + 12y + 10z = 24,000$, is a plane. Since the inputs x, y, z must be nonnegative, the graph is a triangle in the first quadrant, with edges on the coordinate planes. On the boundary of the triangle, one (or more) of the variables x, y, z is zero, so the function f is zero. Thus production is maximized within the budget using $x = 150$, $y = 500$, and $z = 600$.

Exercises and Problems for Section 15.3

Exercises

In Exercises 1–15, use Lagrange multipliers to find the maximum and minimum values of f subject to the given constraint, if such values exist.

1. $f(x, y) = x + y$, $x^2 + y^2 = 1$

2. $f(x, y) = 3x - 2y$, $x^2 + 2y^2 = 44$

3. $f(x, y) = xy$, $4x^2 + y^2 = 8$

4. $f(x, y) = x^2 + y^2$, $x^4 + y^4 = 2$

5. $f(x, y) = x^2 + y^2$, $4x - 2y = 15$

6. $f(x, y) = x^2 + y$, $x^2 - y^2 = 1$

7. $f(x, y) = x^2 - xy + y^2$, $x^2 - y^2 = 1$

8. $f(x, y, z) = x + 3y + 5z$, $x^2 + y^2 + z^2 = 1$

9. $f(x, y, z) = x^2 - 2y + 2z^2$, $x^2 + y^2 + z^2 = 1$

10. $f(x, y, z) = 2x + y + 4z$, $x^2 + y + z^2 = 16$

11. $f(x, y) = x^2 + 2y^2$, $x^2 + y^2 \le 4$

12. $f(x, y) = x + 3y$, $x^2 + y^2 \le 2$

13. $f(x, y) = xy$, $x^2 + 2y^2 \le 1$

14. $f(x, y) = x^3 - y^2$, $x^2 + y^2 \le 1$

15. $f(x, y) = x^3 + y$, $x + y \ge 1$

Problems

16. Figure 15.30 shows contours of f. Does f have a maximum value subject to the constraint $g(x,y) = c$ for $x \geq 0$, $y \geq 0$? If so, approximately where is it and what is its value? Does f have a minimum value subject to the constraint? If so, approximately where and what?

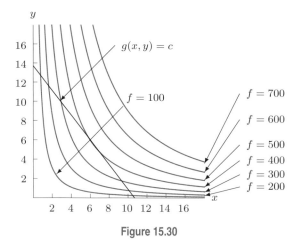

Figure 15.30

17. **(a)** Draw contours of $f(x,y) = 2x + y$ for $z = -7, -5, -3, -1, 1, 3, 5, 7$.
(b) On the same axes, graph the constraint $x^2 + y^2 = 5$.
(c) Use the graph to approximate the points at which f has a maximum or a minimum value subject to the constraint $x^2 + y^2 = 5$.
(d) Use Lagrange multipliers to find the maximum and minimum values of $f(x,y) = 2x + y$ subject to $x^2 + y^2 = 5$.

18. A firm manufactures a commodity at two different factories. The total cost of manufacturing depends on the quantities, q_1 and q_2, supplied by each factory, and is expressed by the *joint cost function*,

$$C = f(q_1, q_2) = 2q_1^2 + q_1 q_2 + q_2^2 + 500.$$

The company's objective is to produce 200 units, while minimizing production costs. How many units should be supplied by each factory?

19. Design a closed cylindrical container which holds $100\,\text{cm}^3$ and has the minimal possible surface area. What should its dimensions be?

20. An industry manufactures a product from two raw materials. The quantity produced, Q, can be given by the Cobb-Douglas function:

$$Q = cx^a y^b,$$

where x and y are quantities of each of the two raw materials used and a, b, and c are positive constants. The first raw material costs $\$P_1$ per unit and the second costs $\$P_2$ per unit. Find the maximum production possible if no more than $\$K$ can be spent on raw materials.

21. The function $P(x,y)$ gives the number of units produced and $C(x,y)$ gives the cost of production.

(a) A company wishes to maximize production at a fixed cost of $\$50,000$. What is the objective function f? What is the constraint equation? What is the meaning of λ in this situation?
(b) A company wishes to minimize costs at a fixed production level of 2000 units. What is the objective function f? What is the constraint equation? What is the meaning of λ in this situation?

22. A chemical manufacturing plant can produce z units of chemical Z given x units of chemical X and y units of chemical Y, where $z = 500x^{0.6}y^{0.3}$. Chemical X costs $\$10$ a unit and chemical Y costs $\$25$ a unit. The company wishes to produce as many units of chemical Z as possible given a fixed budget of $\$2000$.

(a) How many units each of chemical X and chemical Y should the company purchase? How many units of chemical Z can be produced given these purchases?
(b) What is the value of λ? Interpret it in this situation.

23. The quantity, q, of a product manufactured depends on the number of workers, W, and the amount of capital invested, K, and is given by

$$q = 6W^{3/4}K^{1/4}.$$

Labor costs are $\$10$ per worker and capital costs are $\$20$ per unit, and the budget is $\$3000$.

(a) What are the optimum number of workers and the optimum number of units of capital?
(b) Show that at the optimum values of W and K, the ratio of the marginal productivity of labor $(\partial q/\partial W)$ to the marginal productivity of capital $(\partial q/\partial K)$ is the same as the ratio of the cost of a unit of labor to the cost of a unit of capital.
(c) Recompute the optimum values of W and K when the budget is increased by one dollar. Check that increasing the budget by $\$1$ allows the production of λ extra units of the good, where λ is the Lagrange multiplier.

24. The director of a neighborhood health clinic has an annual budget of $\$600,000$. He wants to allocate his budget so as to maximize the number of patient visits, V, which is given as a function of the number of doctors, D, and the number of nurses, N, by

$$V = 1000D^{0.6}N^{0.3}.$$

A doctor's salary is $\$40,000$; nurses get $\$10,000$.

(a) Set up the director's constrained optimization problem.

(b) Describe, in words, the conditions which must be satisfied by $\partial V/\partial D$ and $\partial V/\partial N$ for V to have an optimum value.

(c) Solve the problem formulated in part (a).

(d) Find the value of the Lagrange multiplier and interpret its meaning in this problem.

(e) At the optimum point, what is the marginal cost of a patient visit (that is, the cost of an additional visit)? Will that marginal cost rise or fall with the number of visits? Why?

25. (a) In Problem 23, does the value of λ change if the budget changes from \$3000 to \$4000?

(b) In Problem 24, does the value of λ change if the budget changes from \$600,000 to \$700,000?

(c) What condition must a Cobb-Douglas production function, $Q = cK^aL^b$, satisfy to ensure that the marginal increase of production (that is, the rate of increase of production with budget) is not affected by the size of the budget?

26. Each person tries to balance his or her time between leisure and work. The tradeoff is that as you work less your income falls. Therefore each person has *indifference curves* which connect the number of hours of leisure, l, and income, s. If, for example, you are indifferent between 0 hours of leisure and an income of \$1125 a week on the one hand, and 10 hours of leisure and an income of \$750 a week on the other hand, then the points $l = 0$, $s = 1125$, and $l = 10$, $s = 750$ both lie on the same indifference curve. Table 15.3 gives information on three indifference curves, I, II, and III.

Table 15.3

Weekly income			Weekly leisure hours		
I	II	III	I	II	III
1125	1250	1375	0	20	40
750	875	1000	10	30	50
500	625	750	20	40	60
375	500	625	30	50	70
250	375	500	50	70	90

(a) Graph the three indifference curves.

(b) You have 100 hours a week available for work and leisure combined, and you earn \$10/hour. Write an equation in terms of l and s which represents this constraint.

(c) On the same axes, graph this constraint.

(d) Estimate from the graph what combination of leisure hours and income you would choose under these circumstances. Give the corresponding number of hours per week you would work.

27. Let $f(x, y) = ax^2 + bxy + cy^2$. Show that the maximum value of $f(x, y)$ subject to the constraint $x^2 + y^2 = 1$ is equal to λ, the Lagrange multiplier.

28. (a) If $\sum_{i=1}^{3} x_i = 1$, find the values of x_1, x_2, x_3 making $\sum_{i=1}^{3} x_i^2$ minimum.

(b) Generalize the result of part (a) to find the minimum value of $\sum_{i=1}^{n} x_i^2$ subject to $\sum_{i=1}^{n} x_i = 1$.

29. Figure 15.31 shows ∇f for a function $f(x, y)$ and two curves $g(x, y) = 1$ and $g(x, y) = 2$. Mark the following:

(a) The point(s) A where f has a local maximum.

(b) The point(s) B where f has a saddle point.

(c) The point C where f has a maximum on $g = 1$.

(d) The point D where f has a minimum on $g = 1$.

(e) If you used Lagrange multipliers to find C, what would the sign of λ be? Why?

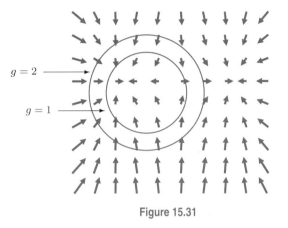

$g = 2$

$g = 1$

Figure 15.31

30. A mountain climber at the summit of a mountain wants to descend to a lower altitude as fast as possible. The altitude of the mountain is given approximately by

$$h(x, y) = 3000 - \frac{1}{10,000}(5x^2 + 4xy + 2y^2) \text{ meters,}$$

where x, y are horizontal coordinates on the earth (in meters), with the mountain summit located above the origin. In thirty minutes, the climber can reach any point (x, y) on a circle of radius 1000 m. In which direction should she travel in order to descend as far as possible?

31. Let $f(x, y, z) = \sqrt{(x-a)^2 + (y-b)^2 + (z-c)^2}$. Minimize f subject to $Ax + By + Cz + D = 0$. What is the geometric meaning of your solution?

32. An electrical current I flows through a circuit containing the resistors R_1 and R_2 in Figure 15.32. The currents i_1 and i_2 through the individual resistors minimize energy loss $i_1^2 R_1 + i_2^2 R_2$ subject to the constraint $i_1 + i_2 = I$ given by Kirchoff's current law.

(a) Find the currents i_1 and i_2 by the method of Lagrange multipliers.

(b) If you are familiar with Ohm's law, find the meaning of λ.

Figure 15.32

33. For constants a, b, c, let $f(x, y) = ax + by + c$ be a linear function, and let R be a region in the xy-plane.

(a) If R is any disk, show that the maximum and minimum values of f on R occur on the boundary of the disk.

(b) If R is any rectangle, show that the maximum and minimum values of f on R occur at the corners of the rectangle. They may occur at other points of the rectangle as well.

(c) Use a graph of the plane $z = f(x, y)$ to explain your answers in parts (a) and (b).

34. The function $f(x, y)$ is defined for all x, y and the origin does not lie on the surface representing $z = f(x, y)$. There is a unique point $P = (a, b, c)$ on the surface which is closest to the origin. Explain why the position vector from the origin to P must be perpendicular to the surface at that point.

35. A company produces one product from two inputs (for example capital and labor). Its production function $g(x, y)$ gives the quantity of the product that can be produced with x units of the first input and y units of the second. The *cost function* (or *expenditure function*) is the three variable function $C(p, q, u)$ where p and q are the unit prices of the two inputs. For fixed p, q, and u, the value $C(p, q, u)$ is the minimum of $f(x, y) = px + qy$ subject to the constraint $g(x, y) = u$.

(a) What is the practical meaning of $C(p, q, u)$?

(b) Find a formula for $C(p, q, u)$ if $g(x, y) = xy$.

36. A *utility function* $U(x, y)$ for two items gives the utility (benefit) to a consumer of x units of item 1 and y units of item 2. The *indirect utility function* is the three variable function $V(p, q, I)$ where p and q are the unit prices of the two items. For fixed p, q, and I, the value $V(p, q, I)$ is the maximum of $U(x, y)$ subject to the constraint $px + qy = I$.

(a) What is the practical meaning of $V(p, q, I)$?

(b) The Lagrange multiplier λ that arises in the maximization defining V is called the *marginal utility of money*. What is its practical meaning?

(c) Find formulas for $V(p, q, I)$ and λ if $U(x, y) = xy$.

37. For each value of λ the function $h(x, y) = x^2 + y^2 - \lambda(2x + 4y - 15)$ has a minimum value $m(\lambda)$.

(a) Find $m(\lambda)$.

(b) For which value of λ is $m(\lambda)$ the largest and what is that maximum value?

(c) Find the minimum value of $f(x, y) = x^2 + y^2$ subject to the constraint $2x + 4y = 15$ using the method of Lagrange multipliers and evaluate λ.

(d) Compare your answers to parts (b) and (c).

38. In a curious game you and your opponent will choose three real numbers. The rules say that you must first choose a value λ, then your opponent is free to choose any values he likes for x and y. Your goal is to make the value of $\mathcal{L}(x, y, \lambda) = 10 - x^2 - y^2 - 2x - \lambda(2x + 2y)$ as small as you can, and your opponent's goal is to make it as large as possible. What value of λ should you choose (assuming you have a brilliant opponent who never makes mistakes)?

CHAPTER SUMMARY (see also Ready Reference at the end of the book)

- **Critical Points**
  Definitions of local extrema, critical points, saddle points, finding critical points algebraically, behavior of contours near critical points, discriminant, second derivative test to classify critical points.

- **Unconstrained Optimization**
  Definitions of global extrema, method of least squares, closed and bounded regions, existence of global

  extrema.

- **Constrained Optimization**
  Objective function and constraint, definitions of extrema subject to a constraint, geometric interpretation of Lagrange multiplier method, solving Lagrange multiplier problems algebraically, inequality constraints, meaning of the Lagrange multiplier λ, the Lagrangian function.

REVIEW EXERCISES AND PROBLEMS FOR CHAPTER FIFTEEN

Exercises

For Exercises 1–4, find the critical points of the given function and classify them as local maxima, local minima, saddle points, or none of these.

1. $f(x, y) = 2xy^2 - x^2 - 2y^2 + 1$

2. $f(x, y) = x^2y + 2y^2 - 2xy + 6$

3. $f(x, y) = 2x^3 - 3x^2y + 6x^2 - 6y^2$

4. $f(x, y) = \sin x + \sin y + \sin(x + y)$, $\quad 0 < x < \pi$, $0 < y < \pi$.

For Exercises 5–8, find the local maxima, local minima, and saddle points of the function. Decide if the local maxima or minima are global maxima or minima. Explain.

5. $f(x, y) = 10 + 12x + 6y - 3x^2 - y^2$

6. $f(x, y) = x^2 + y^3 - 3xy$

7. $f(x, y) = xy + \ln x + y^2 - 10$, $\quad x > 0$

8. $f(x, y) = x + y + \dfrac{1}{x} + \dfrac{4}{y}$

For Exercises 9–15, use Lagrange multipliers to find the maximum and minimum values of f subject to the constraint.

9. $f(x, y) = 3x - 4y$, $\quad x^2 + y^2 = 5$

10. $f(x, y) = x^2 + 2y^2$, $\quad 3x + 5y = 200$

11. $f(x, y) = 2xy$, $\quad 5x + 4y = 100$

12. $f(x, y) = -3x^2 - 2y^2 + 20xy$, $\quad x + y = 100$?

13. $f(x, y, z) = x^2 - y^2 - 2z$, $\quad x^2 + y^2 = z$

14. $z = 4x^2 - xy + 4y^2$, $\quad x^2 + y^2 \leq 2$

15. $f(x, y) = x^2 - y^2$, $\quad x^2 \geq y$

16. Find the least squares line for the data points $(0, 4)$, $(1, 3)$, $(2, 1)$.

Problems

17. At the point $(1, 3)$, suppose $f_x = f_y = 0$ and $f_{xx} < 0$, $f_{yy} < 0$, $f_{xy} = 0$. Draw a possible contour diagram.

18. A company sells two products which are partial substitutes for each other, such as coffee and tea. If the price of one product rises, then the demand for the other product rises. The quantities demanded, q_1 and q_2, are given as a function of the prices, p_1 and p_2, by

$$q_1 = 517 - 3.5p_1 + 0.8p_2, \quad q_2 = 770 - 4.4p_2 + 1.4p_1.$$

What prices should the company charge in order to maximize the total sales revenue? [8]

19. An international organization must decide how to spend the $2000 they have been allotted for famine relief in a remote area. They expect to divide the money between buying rice at $5/sack and beans at $10/sack. The number, P, of people who would be fed if they buy x sacks of rice and y sacks of beans is given by

$$P = x + 2y + \frac{x^2y^2}{2 \cdot 10^8}.$$

What is the maximum number of people that can be fed, and how should the organization allocate its money?

20. The quantity, Q, of a certain product manufactured depends on the quantity of labor, L, and of capital, K, used according to the function

$$Q = 900L^{1/2}K^{2/3}.$$

Labor costs $100 per unit and capital costs $200 per unit. What combination of labor and capital should be used to produce 36,000 units of the goods at minimum cost? What is that minimum cost?

21. A company manufactures a product using inputs x, y, and z according to the production function

$$Q(x, y, z) = 20x^{1/2}y^{1/4}z^{2/5}.$$

The prices per unit are $20 for x, and $10 for y, and $5 for z. What quantity of each input should be used in order to manufacture 1,200 units at minimum cost? [9]

22. The quantity q of goods produced is given by

$$q = 80x^{0.75}y^{0.25},$$

where x is the number of units of labor used and y is the number of units of capital used. Labor costs are $6 per unit, capital costs are $4 per unit, and the budget is $8000. Find the optimal values of x and y in order to maximize the quantity produced.

23. A company manufactures x units of one item and y units of another. The total cost in dollars, C, of producing these two items is approximated by the function

$$C = 5x^2 + 2xy + 3y^2 + 800.$$

(a) If the production quota for the total number of items (both types combined) is 39, find the minimum production cost.

(b) Estimate the additional production cost or savings if the production quota is raised to 40 or lowered to 38.

[8] Adapted from M. Rosser, *Basic Mathematics for Economists*, p. 318 (New York: Routledge, 1993).
[9] Adapted from M. Rosser, *Basic Mathematics for Economists*, p.363 (New York: Routledge, 1993).

24. The quantity, Q, of a product manufactured by a company is given by

$$Q = aK^{0.6}L^{0.4},$$

where a is a positive constant, K is the quantity of capital and L is the quantity of labor used. Capital costs are $20 per unit, labor costs are $10 per unit, and the company wants costs for capital and labor combined to be no higher than $150. Suppose you are asked to consult for the company, and learn that 5 units each of capital and labor are being used.

(a) What do you advise? Should the company use more or less labor? More or less capital? If so, by how much?

(b) Write a one sentence summary that could be used to sell your advice to the board of directors.

25. The Cobb-Douglas function models the quantity, q, of a commodity produced as a function of the number of workers, W, and the amount of capital invested, K:

$$q = cW^{1-a}K^a,$$

where a and c are positive constants. Labor costs are $\$p_1$ per worker, capital costs are $\$p_2$ per unit, and there is a fixed budget of $\$b$. Show that when W and K are at their optimal levels, the ratio of marginal productivity of labor to marginal productivity of capital equals the ratio of the cost of one unit of labor to one unit of capital.

26. Figure 15.33 shows contours labeled with values of $f(x, y)$ and a constraint $g(x, y) = c$. Mark the approximate points at which:

(a) $\operatorname{grad} f = \lambda \operatorname{grad} g$

(b) f has a maximum

(c) f has a maximum on the constraint $g = c$.

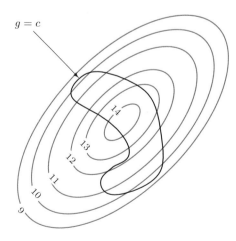

Figure 15.33

27. Let f be differentiable and $\operatorname{grad} f(2, 1) = -3\vec{i} + 4\vec{j}$. You want to see if $(2, 1)$ is a candidate for the maximum and minimum values of f subject to a constraint satisfied by the point $(2, 1)$.

(a) Show $(2, 1)$ is not a candidate if the constraint is $x^2 + y^2 = 5$.

(b) Show $(2, 1)$ is a candidate if the constraint is $(x - 5)^2 + (y + 3)^2 = 25$. From a sketch of the contours for f near $(2, 1)$ and the constraint, decide whether $(2, 1)$ is a candidate for a maximum or minimum.

(c) Do the same as part (b), but using the constraint $(x + 1)^2 + (y - 5)^2 = 25$.

28. A biological rule of thumb states that as the area A of an island increases tenfold, the number of animal species, N, living on it doubles. The table contains data for islands in the West Indies. Assume that N is a power function of A.

(a) Use the biological rule of thumb to find

(i) N as a function of A

(ii) $\ln N$ as a function of $\ln A$

(b) Using the data given, tabulate $\ln N$ against $\ln A$ and find the line of best fit. Does your answer agree with the biological rule of thumb?

Island	Area (sq km)	Number of species
Redonda	3	5
Saba	20	9
Montserrat	192	15
Puerto Rico	8858	75
Jamaica	10854	70
Hispaniola	75571	130
Cuba	113715	125

29. Figure 15.34 shows two weightless springs with spring constants k_1 and k_2 attached between a ceiling and floor without tension or compression. A mass m is placed between the springs which settle into equilibrium as in Figure 15.35. The magnitudes f_1 and f_2 of the forces of the springs on the mass minimize the complementary energy

$$\frac{f_1^2}{2k_1} + \frac{f_2^2}{2k_2}$$

subject to the force balance constraint $f_1 + f_2 = mg$.

(a) Determine f_1 and f_2 by the method of Lagrange multipliers.

(b) If you are familiar with Hooke's law, find the meaning of λ.

Figure 15.34   **Figure 15.35**

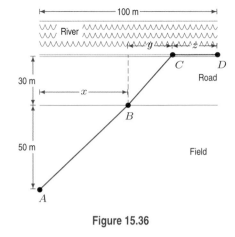

Figure 15.36

30. Find the point on the surface $z = xy + 1$ which is closest to the origin. Justify your answer.

31. A doctor wants to schedule visits for two patients who have been operated on for tumors so as to minimize the expected delay in detecting a new tumor. Visits for patients 1 and 2 are scheduled at intervals of x_1 and x_2 weeks, respectively. A total of m visits per week is available for both patients combined.

    The recurrence rates for tumors for patients 1 and 2 are judged to be v_1 and v_2 tumors per week, respectively. Thus, $v_1/(v_1 + v_2)$ and $v_2/(v_1 + v_2)$ are the probabilities that patient 1 and patient 2, respectively, will have the next tumor. It is known that the expected delay in detecting a tumor for a patient checked every x weeks is $x/2$. Hence, the expected detection delay for both patients combined is given by[10]

$$f(x_1, x_2) = \frac{v_1}{v_1 + v_2} \cdot \frac{x_1}{2} + \frac{v_2}{v_1 + v_2} \cdot \frac{x_2}{2}.$$

Find the values of x_1 and x_2 in terms of v_1 and v_2 that minimize $f(x_1, x_2)$ subject to the fact that m, the number of visits per week, is fixed.

32. What is the value of the Lagrange multiplier in Problem 31? What are the units of λ? What is its practical significance to the doctor?

33. An underground cable is to be laid from point A to point D in Figure 15.36. The cable will go under a field between points A and B, where the cost to lay it is $500/meter. From points B to C, the cable goes under a road, where the cost to lay it is $2000/meter. Between points C and D, it goes alongside a stream where the cost is $300/meter. Find the values of x, y, z in Figure 15.36 minimizing the total cost of laying the cable.

34. An irrigation canal has a trapezoidal cross-section of area 50 m^2, as in Figure 15.37. The average flow rate in the canal is inversely proportional to the wetted perimeter, p, of the canal, that is, to the perimeter of the trapezoid in Figure 15.37, excluding the top. Thus, to maximize the flow rate we must minimize p. Find the depth d, base width w, and angle θ that give the maximum flow rate.[11]

Figure 15.37

35. A light ray crossing the boundary between two different media (for example, vacuum and glass, or air and water) undergoes a change in direction or is *refracted* by an amount which depends on the properties of the media. Suppose that a light ray travels from point A to point B in Figure 15.38, with velocity v_1 in medium 1 and velocity v_2 in medium 2.

(a) Find the time $T(\theta_1, \theta_2)$ taken for the ray to travel from A to B in terms of the angles θ_1, θ_2 and the constants a, b, v_1, and v_2.

(b) Show that the angles θ_1, θ_2 satisfy the constraint

$$a \tan \theta_1 + b \tan \theta_2 = d.$$

(c) What is the effect on the time, T, of letting $\theta_1 \to -\pi/2$ (that is, moving R far to the left of A') or of letting $\theta_1 \to \pi/2$ (that is, moving R far to the right of B')?

[10]Adapted from Daniel Kent, Ross Shachter, *et al.,* Efficient Scheduling of Cystoscopies in Monitoring for Recurrent Bladder Cancer in *Medical Decision Making* (Philadelphia: Hanley and Belfus, 1989).

[11]Adapted from Robert M. Stark and Robert L. Nichols, *Mathematical Foundations of Design: Civil Engineering Systems,* (New York: McGraw-Hill, 1972).

(d) *Fermat's Principle* states that the light ray follows a path such that the time taken T is minimized. Use Lagrange multipliers to show that $T(\theta_1, \theta_2)$ is minimized when *Snell's Law of Refraction* holds:

$$\frac{\sin\theta_1}{\sin\theta_2} = \frac{v_1}{v_2}.$$

(The constant v_1/v_2 is called the *index of refraction* of medium 2 with respect to medium 1. For example, the indices of air, water, and glass with respect to a vacuum are approximately 1.0003, 1.33, and 1.52, respectively. The lenses of modern reading "glasses" are made of plastics with a high index of refraction in order to reduce weight and thickness when the prescription is strong.)

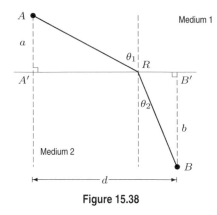

Figure 15.38

CAS Challenge Problems

36. Let $f(x, y) = \dfrac{\sqrt{a + x + y}}{1 + y + \sqrt{a + x}}$, for $x, y > 0$, where a is a positive constant.

    (a) Find the critical points of f and classify them as local maximum, local minimum, saddle point, or none of these.

    (b) Describe how the position and type of the critical points changes with respect to a, and explain this in terms of the graph of f.

37. Students are asked to find the global maximum of $f(x, y) = x^2 + y$ subject to the constraint $g(x, y) = x^2 + 2xy + y^2 - 9 = 0$. Student A uses the method of Lagrange multipliers with the help of a computer algebra system, and says that the global maximum is 11/4. Student B looks at a contour diagram of f and a graph of $g = 0$ and says there is no global maximum. Which student is correct and what mistake is the other one making?

38. Let $f(x, y) = 3x + 2y + 5$, $g(x, y) = 2x^2 - 4xy + 5y^2$.

    (a) Find the maximum of f subject to the constraint $g = 20$.

    (b) Using the value of λ in part (a), estimate the maximum of f subject to each of the constraints $g = 20.5$ and $g = 20.2$.

    (c) Use Lagrange multipliers to find the two maxima in part (b) exactly. Compare them with the estimates.

CHECK YOUR UNDERSTANDING

Are the statements in Problems 1–25 true or false? Give reasons for your answer.

1. If $f_x(P_0) = f_y(P_0) = 0$, then P_0 is a critical point of f.

2. If $f_x(P_0) = f_y(P_0) = 0$, then P_0 is a local maximum or local minimum of f.

3. If P_0 is a critical point of f, then P_0 is either a local maximum or local minimum of f.

4. If P_0 is a local maximum or local minimum of f, and not on the boundary of the domain of f, then P_0 is a critical point of f.

5. The function $f(x, y) = \sqrt{x^2 + y^2}$ has a local minimum at the origin.

6. The function $f(x, y) = x^2 - y^2$ has a local minimum at the origin.

7. If f has a local minimum at P_0 then so does the function $g(x, y) = f(x, y) + 5$.

8. If f has a local minimum at P_0 then the function $g(x, y) = -f(x, y)$ has a local maximum at P_0.

9. Every function has at least one local maximum.

10. If P_0 is a local maximum for f then $f(a, b) \leq f(P_0)$ for all points (a, b) in 2-space.

11. If P_0 is a local maximum of f, then P_0 is also a global maximum of f.

12. If P_0 is a global maximum of f, where f is defined on all of 2-space, then P_0 is also a local maximum of f.

13. Every function has a global maximum.

14. The region consisting of all points (x, y) satisfying $x^2 + y^2 < 1$ is bounded.

15. The region consisting of all points (x, y) satisfying $x^2 + y^2 < 1$ is closed.

16. The function $f(x, y) = x^2 + y^2$ has a global minimum on the region $x^2 + y^2 < 1$.

17. The function $f(x, y) = x^2 + y^2$ has a global maximum on the region $x^2 + y^2 < 1$.

18. If P and Q are two distinct points in 2-space, and f has a global maximum at P, then f cannot have a global maximum at Q.

19. The function $f(x, y) = \sin(1 + e^{xy})$ must have a global minimum in the square region $0 \leq x \leq 1, 0 \leq y \leq 1$.

20. If P_0 is a global minimum of f on a closed and bounded region, then P_0 need not be a critical point of f.

21. If $f(x, y)$ has a local maximum at (a, b) subject to the constraint $g(x, y) = c$, then $g(a, b) = c$.

22. If $f(x, y)$ has a local maximum at (a, b) subject to the constraint $g(x, y) = c$, then $\text{grad} f(a, b) = \vec{0}$.

23. The function $f(x, y) = x + y$ has no global maximum subject to the constraint $x - y = 0$.

24. The point $(2, -1)$ is a local minimum of $f(x, y) = x^2 + y^2$ subject to the constraint $x + 2y = 0$.

25. If $\text{grad} f(a, b)$ and $\text{grad} g(a, b)$ point in opposite directions, then (a, b) is a local minimum of $f(x, y)$ constrained by $g(x, y) = c$.

In Problems 26–33, suppose that M and m are the maximum and minimum values of $f(x, y)$ subject to the constraint $g(x, y) = c$ and that (a, b) satisfies $g(a, b) = c$. Decide whether the statements are true or false. Give an explanation for your answer.

26. If $f(a, b) = M$, then $f_x(a, b) = f_y(a, b) = 0$.

27. If $f(a, b) = M$, then $f(a, b) = \lambda g(a, b)$ for some value of λ.

28. If $\text{grad} f(a, b) = \lambda \, \text{grad} g(a, b)$, then $f(a, b) = M$ or $f(a, b) = m$.

29. If $f(a, b) = M$ and $f_x(a, b)/f_y(a, b) = 5$, then $g_x(a, b)/g_y(a, b) = 5$.

30. If $f(a, b) = m$ and $g_x(a, b) = 0$, then $f_x(a, b) = 0$.

31. Increasing the value of c will increase the value of M.

32. Suppose that $f(a, b) = M$ and that $\text{grad} f(a, b) = 3 \, \text{grad} g(a, b)$. Then increasing the value of c by 0.02 increases the value of M by about 0.06.

33. Suppose that $f(a, b) = m$ and that $\text{grad} f(a, b) = 3 \, \text{grad} g(a, b)$. Then increasing the value of c by 0.02 decreases the value of m by about 0.06.

PROJECTS FOR CHAPTER FIFTEEN

1. Hockey and Entropy

Thirty teams compete for the Stanley Cup in the National Hockey League (after expansion in 2000). At the beginning of the season an experienced fan estimates that the probability that team i will win is some number p_i, where $0 \leq p_i \leq 1$ and

$$\sum_{i=1}^{30} p_i = 1.$$

Exactly one team will actually win, so the probabilities have to add to 1. If one of the teams, say team i, is certain to win then p_i is equal to 1 and all the other p_j are zero. Another extreme case occurs if all the teams are equally likely to win, so all the p_i are equal to $1/30$, and the outcome of the hockey season is completely unpredictable. Thus, the *uncertainty* in the outcome of the hockey season depends on the probabilities $p_1, \ldots, p_{30}$. In this problem we measure this uncertainty quantitatively using the following function:

$$S(p_1, \ldots, p_{30}) = - \sum_{i=1}^{30} p_i \frac{\ln p_i}{\ln 2}.$$

Note that as $p_i \leq 1$, we have $-\ln p_i \geq 0$ and hence $S \geq 0$.

(a) Show that $\lim_{p \to 0} p \ln p = 0$. (This means that S is a continuous function of the p_i, where $0 \leq p_i \leq 1$ and $1 \leq i \leq 30$, if we set $p \ln p|_{p=0}$ equal to zero. Since S is then a continuous function on a closed and bounded region, it attains a maximum and a minimum value on this region.)

(b) Find the maximum value of $S(p_1, \ldots, p_{30})$ subject to the constraint $p_1 + \cdots + p_{30} = 1$. What are the values of p_i in this case? What does your answer mean in terms of the uncertainty in the outcome of the hockey season?

(c) Find the minimum value of $S(p_1, \ldots, p_{30})$, subject to the constraint $p_1 + \cdots + p_{30} = 1$. What are the values of p_i in this case? What does your answer mean in terms of the uncertainty in the outcome of the hockey season?

[Note: The function S is an example of an *entropy* function; the concept of entropy is used in information theory, statistical mechanics, and thermodynamics when measuring the uncertainty in an experiment (the hockey season in this problem) or physical system.]

2. Fitting a Line to Data Using Least Squares

In this problem you will derive the general formulas for the slope and y-intercept of a least squares line. Assume that you have n data points $(x_1, y_1), (x_2, y_2), \cdots, (x_n, y_n)$. Let the equation of the least squares line be $y = b + mx$.

(a) For each data point (x_i, y_i), show that the corresponding point directly above or below it on the least squares line has y-coordinate $b + mx_i$.

(b) For each data point (x_i, y_i), show that the square of the vertical distance from it to the point found in part (a) is $(y_i - (b + mx_i))^2$.

(c) Form the function $f(b, m)$ which is the sum of all of the n squared distances found in part (b). That is,

$$f(b, m) = \sum_{i=1}^{n} (y_i - (b + mx_i))^2.$$

Show that the partial derivatives $\dfrac{\partial f}{\partial b}$ and $\dfrac{\partial f}{\partial m}$ are given by

$$\frac{\partial f}{\partial b} = -2 \sum_{i=1}^{n} (y_i - (b + mx_i))$$

and

$$\frac{\partial f}{\partial m} = -2 \sum_{i=1}^{n} (y_i - (b + mx_i)) \cdot x_i.$$

(d) Show that the critical point equations $\dfrac{\partial f}{\partial b} = 0$ and $\dfrac{\partial f}{\partial m} = 0$ lead to a pair of simultaneous linear equations in b and m:

$$nb + \left(\sum x_i\right)m = \sum y_i$$

$$\left(\sum x_i\right)b + \left(\sum x_i^2\right)m = \sum x_i y_i$$

(e) Solve the equations in part (d) for b and m, getting

$$b = \left(\sum_{i=1}^{n} x_i^2 \sum_{i=1}^{n} y_i - \sum_{i=1}^{n} x_i \sum_{i=1}^{n} x_i y_i\right) \Big/ \left(n \sum_{i=1}^{n} x_i^2 - \left(\sum_{i=1}^{n} x_i\right)^2\right)$$

$$m = \left(n \sum_{i=1}^{n} x_i y_i - \sum_{i=1}^{n} x_i \sum_{i=1}^{n} y_i\right) \Big/ \left(n \sum_{i=1}^{n} x_i^2 - \left(\sum_{i=1}^{n} x_i\right)^2\right)$$

(f) Apply the formulas of part (e) to the data points $(1, 1), (2, 1), (3, 3)$ to check that you get the same result as in Example 3 on page 759.

Chapter Sixteen

INTEGRATING FUNCTIONS OF SEVERAL VARIABLES

A definite integral is the limit of a sum. We use a definite integral to calculate the total population of a region given the population density as a function of position. If the population density is a function of one variable only (for example, the distance from the center of a city), we have an ordinary one-variable definite integral. If the density depends on more than one variable, we need a multivariable integral to calculate the total population. In this chapter we develop double integrals in Cartesian and polar coordinates and triple integrals in Cartesian, cylindrical, and spherical coordinates. We see how to change variables using a Jacobian and apply integration to probability.

16.1 THE DEFINITE INTEGRAL OF A FUNCTION OF TWO VARIABLES

The definite integral of a continuous one-variable function, f, is a limit of Riemann sums:

$$\int_a^b f(x)\,dx = \lim_{\Delta x \to 0} \sum_i f(x_i)\,\Delta x,$$

where x_i is a point in the ith subdivision of the interval $[a, b]$. In this section we extend this definition to functions of two variables. We start by considering how to estimate total population from a two-variable population density.

Population Density of Foxes in England

The fox population in parts of England is important to public health officials concerned about the disease rabies, which is spread by animals. The contour diagram in Figure 16.1 shows the population density $D = f(x, y)$ of foxes in the southwestern corner of England, where x and y are in kilometers from the southwest corner of the map and D is in foxes per square kilometer.[1] The bold contour is the coastline, which may be thought of as the $D = 0$ contour; clearly the density is zero outside it.

Figure 16.1: Population density of foxes in southwestern England

---

Example 1   Estimate the total fox population in the region represented by the map in Figure 16.1.

Solution   We subdivide the map into the rectangles shown in Figure 16.1 and estimate the population in each rectangle. For simplicity, we use the population density at the northeast corner of each rectangle. For example, in the bottom left rectangle, the density is 0 at the northeast corner, in the next rectangle to the east (right), the density in the northeast corner is 1. Continuing in this way, we get the estimates in Table 16.1. To estimate the population in a rectangle, we multiply the density by the area of the rectangle, $30 \cdot 25 = 750 \text{ km}^2$. Adding the results, we obtain

$$\begin{aligned}
\text{Estimate of population} = {}&(0.2 + 0.7 + 1.2 + 1.2 + 0.1 + 1.6 + 0.5 + 1.4 \\
&+ 1.1 + 1.6 + 1.5 + 1.8 + 1.5 + 1.3 + 1.1 + 2.0 \\
&+ 1.4 + 1.0 + 1.0 + 0.6 + 1.2)750 = 18{,}000 \text{ foxes.}
\end{aligned}$$

Taking the upper and lower bounds for the population density on each rectangle enables us to find upper and lower estimates for the population. Using the same rectangles, the upper estimate is approximately 35,000 and the lower estimate is 4,000. There is a wide discrepancy between the upper and lower estimates; we could make them closer by taking finer subdivisions.

---

[1] Adapted from J. D. Murray, *Mathematical Biology*, Springer-Verlag, 1989.

Table 16.1 *Estimates of population density (northeast corner)*

0.0	0.0	0.2	0.7	1.2	1.2
0.0	0.0	0.0	0.0	0.1	1.6
0.0	0.0	0.5	1.4	1.1	1.6
0.0	0.0	1.5	1.8	1.5	1.3
0.0	1.1	2.0	1.4	1.0	0.0
0.0	1.0	0.6	1.2	0.0	0.0

Definition of the Definite Integral

The sums used to approximate the fox population are Riemann sums. We now define the definite integral for a function f of two variables on a rectangular region. Given a continuous function $f(x, y)$ defined on a region $a \leq x \leq b$ and $c \leq y \leq d$, we subdivide each of the intervals $a \leq x \leq b$ and $c \leq y \leq d$ into n and m equal subintervals respectively, giving nm subrectangles. (See Figure 16.2.)

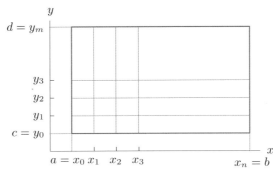

Figure 16.2: Subdivision of a rectangle into nm subrectangles

The area of each subrectangle is $\Delta A = \Delta x \, \Delta y$, where $\Delta x = (b - a)/n$ is the width of each subdivision on the x-axis, and $\Delta y = (d - c)/m$ is the width of each subdivision on the y-axis. To compute the Riemann sum, we multiply the area of each subrectangle by the value of the function at a point in the rectangle and add the resulting numbers. Choosing the point which gives the maximum value, M_{ij}, of the function on each rectangle, we get the *upper sum*, $\sum_{i,j} M_{ij} \Delta x \Delta y$.

The *lower sum*, $\sum_{i,j} L_{ij} \Delta x \Delta y$, is obtained by taking the minimum value on each rectangle. If (u_{ij}, v_{ij}) is any point in the ij-th subrectangle, any other Riemann sum satisfies

$$\sum_{i,j} L_{ij} \Delta x \Delta y \leq \sum_{i,j} f(u_{ij}, v_{ij}) \, \Delta x \, \Delta y \leq \sum_{i,j} M_{ij} \Delta x \Delta y.$$

We define the definite integral by taking the limit as the numbers of subdivisions, n and m, tend to infinity. By comparing upper and lower sums, as we did for the fox population, it can be shown that the limit exists when the function, f, is continuous. We get the same limit by letting Δx and Δy tend to 0. Thus, we have the following definition:

Suppose the function f is continuous on R, the rectangle $a \leq x \leq b$, $c \leq y \leq d$. If (u_{ij}, v_{ij}) is any point in the ij-th subrectangle, we define the **definite integral** of f over R

$$\int_R f \, dA = \lim_{\Delta x, \Delta y \to 0} \sum_{i,j} f(u_{ij}, v_{ij}) \Delta x \Delta y.$$

Such an integral is called a **double integral**.

The case when R is not rectangular is considered on page 786. Sometimes we think of dA as being the area of an infinitesimal rectangle of length dx and height dy, so that $dA = dx\, dy$. Then we use the notation[2]

$$\int_R f\, dA = \int_R f(x, y)\, dx\, dy.$$

For this definition, we used a particular type of Riemann sum with equal-sized rectangular subdivisions. In a general Riemann sum, the subdivisions do not all have to be the same size.

Interpretation of the Double Integral as Volume

Just as the definite integral of a positive one-variable function can be interpreted as an area, so the double integral of a positive two-variable function can be interpreted as a volume. In the one-variable case we visualize the Riemann sums as the total area of rectangles above the subdivisions. In the two-variable case we get solid bars instead of rectangles. As the number of subdivisions grows, the tops of the bars approximate the surface better, and the volume of the bars gets closer to the volume under the graph of the function. (See Figure 16.3.)

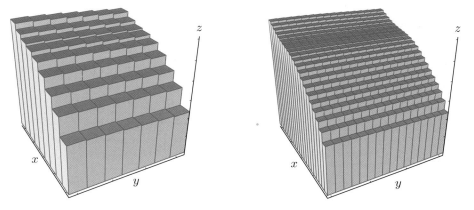

Figure 16.3: Approximating volume under a graph with finer and finer Riemann sums

Thus we have the following result:

If x, y, z represent length and f is positive, then

$$\text{Volume under graph of } f \text{ above region } R = \int_R f\, dA$$

Example 2   Let R be the rectangle $0 \le x \le 1$ and $0 \le y \le 1$. Use Riemann sums to make upper and lower estimates of the volume of the region above R and under the graph of $z = e^{-(x^2+y^2)}$.

Solution   If R is the rectangle $0 \le x \le 1, 0 \le y \le 1$, the volume we want is given by

$$\text{Volume} = \int_R e^{-(x^2+y^2)}\, dA.$$

We divide R into 16 subrectangles by dividing each edge into 4 parts. Figure 16.4 shows that $f(x, y) = e^{-(x^2+y^2)}$ decreases as we move away from the origin. Thus, to get an upper sum we evaluate f on each subrectangle at the corner nearest the origin. For example, in the rectangle $0 \le x \le 0.25, 0 \le y \le 0.25$, we evaluate f at $(0, 0)$. Using Table 16.2 we find that

[2]Another common notation for the double integral is $\int \int_R f\, dA$.

Figure 16.4: Graph of $e^{-(x^2+y^2)}$ above the rectangle R

Upper sum $= (\ 1 + 0.9394 + 0.7788 + 0.5698$

$+ 0.9394 + 0.8825 + 0.7316 + 0.5353$

$+ 0.7788 + 0.7316 + 0.6065 + 0.4437$

$+ 0.5698 + 0.5353 + 0.4437 + 0.3247)(0.0625) = 0.68.$

To get a lower sum, we evaluate f at the opposite corner of each rectangle because the surface slopes down in both the x and y directions. This yields a lower sum of 0.44. Thus,

$$0.44 \leq \int_R e^{-(x^2+y^2)}\, dA \leq 0.68.$$

To get a better approximation, we use more subdivisions. See Table 16.3.

Table 16.2   *Values of $f(x,y) = e^{-(x^2+y^2)}$ on the rectangle R*

		y				

		0.0	0.25	0.50	0.75	1.00
	0.0	1	0.9394	0.7788	0.5698	0.3679
	0.25	0.9394	0.8825	0.7316	0.5353	0.3456
x	0.50	0.7788	0.7316	0.6065	0.4437	0.2865
	0.75	0.5698	0.5353	0.4437	0.3247	0.2096
	1.00	0.3679	0.3456	0.2865	0.2096	0.1353

Table 16.3   *Riemann sum approximations to $\int_R e^{-(x^2+y^2)}\, dA$*

	Number of subdivisions in x and y directions			
	8	16	32	64
Upper	0.6168	0.5873	0.5725	0.5651
Lower	0.4989	0.5283	0.5430	0.5504

The true value of the double integral, $0.5577\ldots$, is trapped between the lower and upper sums. Notice that the lower sum increases and the upper sum decreases as the number of subdivisions increases. (Why?) However, even with 64 subdivisions, the lower and upper sums agree with the true value of the integral only in the first decimal place. Hence, we should look for better ways to approximate the integral.

Interpretation of the Double Integral as Area

In the special case that $f(x,y) = 1$ for all points (x,y) in the region R, each term in the Riemann sum is of the form $1 \cdot \Delta A = \Delta A$ and the double integral gives the area of the region R:

$$\text{Area}(R) = \int_R 1 \, dA = \int_R dA$$

Interpretation of the Double Integral as Average Value

As in the one-variable case, the definite integral can be used to compute the average value of a function:

$$\frac{\text{Average value of } f}{\text{on the region } R} = \frac{1}{\text{Area of } R} \int_R f \, dA$$

We can rewrite this as

$$\text{Average value} \times \text{Area of } R = \int_R f \, dA.$$

If we interpret the integral as the volume under the graph of f, then we can think of the average value of f as the height of the box with the same volume that is on the same base. (See Figure 16.5.) Imagine that the volume under the graph is made out of wax. If the wax melted within the perimeter of R, then it would end up box-shaped with height equal to the average value of f.

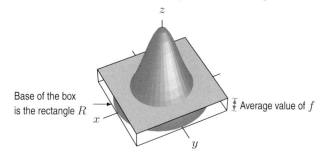

Figure 16.5: Volume and average value

Integral over Regions That are Not Rectangles

We defined the definite integral $\int_R f(x, y) \, dA$, for a rectangular region R. Now we extend the definition to regions of other shapes, including triangles, circles, and regions bounded by the graphs of piecewise continuous functions.

To approximate the definite integral over a region, R, which is not rectangular, we use a grid of rectangles approximating the region. We obtain this grid by enclosing R in a large rectangle and subdividing that rectangle; we consider just the subrectangles which are inside R.

As before, we pick a point (u_{ij}, v_{ij}) in each subrectangle and form a Riemann sum

$$\sum_{i,j} f(u_{ij}, v_{ij}) \Delta x \Delta y.$$

This time, however, the sum is over only those subrectangles within R. For example, in the case of the fox population we can use the rectangles which are entirely on land. As the subdivisions become finer, the grid resembles the region R more closely. For a function, f, which is continuous on R, we define the definite integral as follows:

$$\int_R f \, dA = \lim_{\Delta x, \Delta y \to 0} \sum_{i,j} f(u_{ij}, v_{ij}) \Delta x \Delta y$$

where the Riemann sum is taken over the subrectangles inside R.

You may wonder why we can leave out the rectangles which cover the edge of R—if we included them might we get a different value for the integral? The answer is that for any region that we are likely to meet, the area of the subrectangles covering the edge tends to 0 as the grid becomes finer. Therefore, omitting these rectangles does not affect the limit.

Convergence of Upper and Lower Sums to Same Limit

We have said that if f is continuous on the rectangle R, then the difference between upper and lower sums for f converges to 0 as Δx and Δy approach 0. In the following example, we show this in a particular case. The ideas in this example can be used in a general proof.

Example 3   Let $f(x, y) = x^2 y$ and let R be the rectangle $0 \le x \le 1$, $0 \le y \le 1$. Show that the difference between upper and lower Riemann sums for f on R converges to 0, as Δx and Δy approach 0.

Solution   The difference between the sums is

$$\sum M_{ij} \Delta x \Delta y - \sum L_{ij} \Delta x \Delta y = \sum (M_{ij} - L_{ij}) \Delta x \Delta y.$$

where M_{ij} and L_{ij} are the maximum and minimum of f on the ij-th subrectangle. Since f increases in both the x and y directions, M_{ij} occurs at the corner of the subrectangle farthest from the origin and L_{ij} at the closest. Moreover, since the slopes in the x and y directions don't decrease as x and y increase, the difference $M_{ij} - L_{ij}$ is largest in the subrectangle R_{nm}, which is farthest from the origin. Thus,

$$\sum (M_{ij} - L_{ij}) \Delta x \Delta y \le (M_{nm} - L_{nm}) \sum \Delta x \Delta y = (M_{nm} - L_{nm}) \text{Area}(R).$$

Thus, the difference converges to 0 as long as $(M_{nm} - L_{nm})$ does. The maximum M_{nm} of f on the nm-th subrectangle occurs at $(1, 1)$, the subrectangle's top right corner, and the minimum L_{nm} occurs at the opposite corner, $(1 - 1/n, 1 - 1/m)$. Substituting into $f(x, y) = x^2 y$ gives

$$M_{nm} - L_{nm} = (1)^2(1) - \left(1 - \frac{1}{n}\right)^2 \left(1 - \frac{1}{m}\right) = \frac{2}{n} - \frac{1}{n^2} + \frac{1}{m} - \frac{2}{nm} + \frac{1}{n^2 m}.$$

The right hand side converges to 0 as $n, m \to \infty$, that is, as $\Delta x, \Delta y \to 0$.

Exercises and Problems for Section 16.1

Exercises

1. Values of $f(x, y)$ are in Table 16.4. Let R be the rectangle $1 \le x \le 1.2$, $2 \le y \le 2.4$. Find Riemann sums which are reasonable over- and under-estimates for $\int_R f(x, y) \, dA$ with $\Delta x = 0.1$ and $\Delta y = 0.2$.

Table 16.4

	\multicolumn{3}{c}{x}	
	1.0	1.1	1.2
2.0	5	7	10
y 2.2	4	6	8
2.4	3	5	4

2. Figure 16.6 shows contours of $f(x, y)$ on the rectangle R with $0 \le x \le 30$ and $0 \le y \le 15$. Using $\Delta x = 10$ and $\Delta y = 5$, find an overestimate and an underestimate for $\int_R f(x, y) dA$.

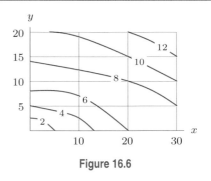

Figure 16.6

3. Let R be the rectangle with vertices $(0, 0)$, $(4, 0)$, $(4, 4)$, and $(0, 4)$ and let $f(x, y) = \sqrt{xy}$.

   (a) Find reasonable upper and lower bounds for $\int_R f \, dA$ without subdividing R.

   (b) Estimate $\int_R f \, dA$ by partitioning R into four subrectangles and evaluating f at its maximum and minimum values on each subrectangle.

4. Table 16.5 gives values of $z = f(x, y)$ on the rectangle R with $0 \le x \le 6$ and $0 \le y \le 8$.

   (a) Estimate $\int_R f(x, y)dA$ as accurately as possible.
   (b) Estimate the average value of $f(x, y)$ on R.

Table 16.5

		x		
		0	3	6
	0	100	90	81
y	4	85	79	68
	8	65	61	55

5. Using Riemann sums with four subdivisions in each direction, find upper and lower bounds for the volume under the graph of $f(x, y) = 2 + xy$ above the rectangle R with $0 \le x \le 2$, $0 \le y \le 4$.

6. If $f(x, y)$ gives pollution density, in micrograms per square meter, and x and y are in meters, give the units and a practical interpretation of $\int_R f(x, y)dA$.

In Exercises 7–8, use a computer or calculator program for two-dimensional Riemann sums to estimate the integral.

7. $\int_R (x^2 + y^2)\, dA$, if R is $1 \le x \le 2, 1 \le y \le 3$.

8. $\int_R \sin(xy)\, dA$, if R is $-\pi \le x \le 0, 0 \le y \le \pi/2$.

Problems

In Problems 9–24, decide (without calculation) whether the integrals are positive, negative, or zero. Let D be the region inside the unit circle centered at the origin, let R be the right half of D and let B be the bottom half of D.

9. $\int_D dA$

10. $\int_B dA$

11. $\int_R 5x\, dA$

12. $\int_B 5x\, dA$

13. $\int_D 5x\, dA$

14. $\int_D (y^3 + y^5)\, dA$

15. $\int_B (y^3 + y^5)\, dA$

16. $\int_R (y^3 + y^5)\, dA$

17. $\int_B (y - y^3)\, dA$

18. $\int_D (y - y^3)\, dA$

19. $\int_D \sin y\, dA$

20. $\int_D \cos y\, dA$

21. $\int_D e^x\, dA$

22. $\int_D xe^x\, dA$

23. $\int_D xy^2\, dA$

24. $\int_B x \cos y\, dA$

25. Table 16.6 gives values of $f(x, y)$, the number of milligrams of mosquito larvae per square meter in a swamp. If x and y are in meters and R is the rectangle $0 \le x \le 8$, $0 \le y \le 6$, estimate $\int_R f(x, y)dA$. Give units and interpret your answer.

Table 16.6

		x		
		0	4	8
	0	1	3	6
y	3	2	5	9
	6	4	9	15

26. A biologist studying insect populations measures the population density of flies and mosquitos at various points in a rectangular study region. The graphs of the two population densities for the region are shown in Figures 16.7 and 16.8. Assuming that the units along the corresponding axes are the same in the two graphs, are there more flies or more mosquitos in the region?

flies

Figure 16.7

mosquitos

Figure 16.8

27. Figure 16.9 shows the distribution of temperature, in °C, in a 5 meter by 5 meter heated room. Using Riemann sums, estimate the average temperature in the room.

Figure 16.9

28. Figure 16.10 shows contours of $f(x, y)$. Let R be the square $-0.5 \leq x \leq 1$, $-0.5 \leq y \leq 1$. Is the integral $\int_R f \, dA$ positive or negative? Explain your reasoning.

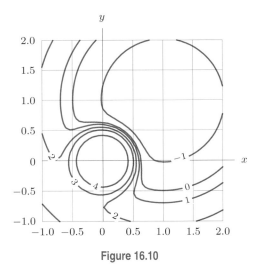

Figure 16.10

29. Figure 16.11 shows contours for the annual frequency of tornados per 10,000 square miles in the U.S.[3] Each grid square is 100 miles on a side. Use the map to estimate the total number of tornados per year in

(a) Texas   (b) Florida   (c) Arizona.

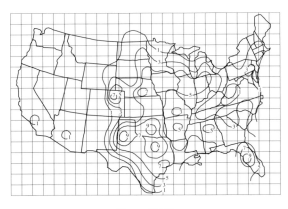

Figure 16.11

30. The hull of a boat has width $w(x, y)$ feet at a point x feet from the front and y feet below the water line. Values of w are in Table 16.7. Set up a definite integral that gives the volume of the hull below the waterline, and then estimate the value of the integral.

Table 16.7

		0	10	20	30	40	50	60
	0	2	8	13	16	17	16	10
Depth	2	1	4	8	10	11	10	8
below	4	0	3	4	6	7	6	4
waterline	6	0	1	2	3	4	3	2
(in feet)	8	0	0	1	1	1	1	1

Front of boat → Back of boat

31. Let $f(x, y)$ be a function of x and y which is independent of y, that is, $f(x, y) = g(x)$ for some one-variable function g.

(a) What does the graph of f look like?

(b) Let R be the rectangle $a \leq x \leq b$, $c \leq y \leq d$. By interpreting the integral as a volume, and using your answer to part (a), express $\int_R f \, dA$ in terms of a one-variable integral.

32. A pile of earth standing on flat ground by an abandoned mine has height 15 meters. The ground is the xy-plane; the origin is directly below the top of the pile and the z-axis is upward. The cross-section at height z is $x^2 + y^2 = 15 - z$ for $0 \leq z \leq 15$, with x, y, z in meters

(a) What equation gives the edge of the base of the pile?

(b) What is the area of the base of the pile?

(c) What equation gives the cross-section of the pile with the plane $z = 10$?

(d) What is the area of the cross-section in part (c)?

(e) Imagine the pile sliced horizontally. What is the area, $A(z)$, of the slice at height z? What are its units?

(f) If Δz is the thickness of the slice at height z, what is represented by $A(z) \Delta z$? What are its units?

(g) Evaluate and interpret $\int_0^{15} A(z) \, dz$. What are its units?

33. For any region, R, in the plane, does the value of the following integral increase or decrease if the positive constant a is increased? Justify your answer with Riemann sums.

$$\int_R \frac{1}{1 + ax^2 + y^2} \, dA.$$

34. For any numbers a and b, assume that $|a + b| \leq |a| + |b|$. Use this to explain why

$$\left| \int_R f \, dA \right| \leq \int_R |f| \, dA.$$

---

[3]From *Modern Physical Geography*, Alan H. Strahler and Arthur H. Strahler, Fourth Edition, John Wiley & Sons, New York, 1992, p. 128

16.2   ITERATED INTEGRALS

In Section 16.1 we approximated double integrals using Riemann sums. In this section we see how to compute double integrals exactly using one-variable integrals.

The Fox Population Again: Expressing a Double Integral as an Iterated Integral

To estimate the fox population, we computed a sum of the form

$$\text{Total population} \approx \sum_{i,j} f(u_{ij}, v_{ij}) \Delta x \, \Delta y,$$

where $1 \le i \le n$ and $1 \le j \le m$ and the values $f(u_{ij}, v_{ij})$ can be arranged as in Table 16.8.

Table 16.8   *Estimates for fox population densities for $n = m = 6$*

0.0	0.0	0.2	0.7	1.2	1.2
0.0	0.0	0.0	0.0	0.1	1.6
0.0	0.0	0.5	1.4	1.1	1.6
0.0	0.0	1.5	1.8	1.5	1.3
0.0	1.1	2.0	1.4	1.0	0.0
0.0	1.0	0.6	1.2	0.0	0.0

For any values of n and m, we can either add across the rows first, or we can add down the columns first. If we add rows first, we can write the sum in the form

$$\text{Total population} \approx \sum_{j=1}^{m} \left(\sum_{i=1}^{n} f(u_{ij}, v_{ij}) \Delta x \right) \Delta y.$$

The inner sum, $\sum_{i=1}^{n} f(u_{ij}, v_{ij}) \Delta x$, approximates the integral $\int_0^{180} f(x, v_{ij}) \, dx$. Thus, we have

$$\text{Total population} \approx \sum_{j=1}^{m} \left(\int_0^{180} f(x, v_{ij}) \, dx \right) \Delta y.$$

The outer Riemann sum approximates another integral, this time with integrand $\int_0^{180} f(x, y) \, dx$, which is a function of y. Thus, we can write the total population in terms of nested, or *iterated*, one-variable integrals:

$$\text{Total population} = \int_0^{150} \left(\int_0^{180} f(x, y) \, dx \right) dy.$$

Since the total population is represented by $\int_R f \, dA$, this suggests the method of computing double integrals in the following theorem:[4]

Theorem 16.1: Writing a Double Integral as an Iterated Integral

If R is the rectangle $a \le x \le b$, $c \le y \le d$ and f is a continuous function on R, then the integral of f over R exists and is equal to the **iterated integral**

$$\int_R f \, dA = \int_{y=c}^{y=d} \left(\int_{x=a}^{x=b} f(x, y) \, dx \right) dy.$$

The expression $\int_{y=c}^{y=d} \left(\int_{x=a}^{x=b} f(x, y) \, dx \right) dy$ can be written $\int_c^d \int_a^b f(x, y) \, dx \, dy$.

---

[4]For a proof, see M. Spivak, *Calculus on Manifolds*, pp. 53 and 58 (New York: Benjamin, 1965).

To evaluate the iterated integral, first perform the inside integral with respect to x, holding y constant; then integrate the result with respect to y.

The Parallel Between Repeated Summation and Repeated Integration

It is helpful to notice how the summation and integral notation parallel each other. Viewing a Riemann sum as a sum of sums,

$$\sum_{i,j} f(u_{ij}, v_{ij}) \, \Delta x \, \Delta y = \sum_j \left(\sum_i f(u_{ij}, v_{ij}) \, \Delta x \right) \Delta y$$

leads us to see a double integral as an integral of integrals:

$$\int_R f(x,y) \, dA = \int_c^d \left(\int_a^b f(x,y) \, dx \right) dy,$$

where R is the rectangle $a \le x \le b$ and $c \le y \le d$.

Example 1 A building is 8 meters wide and 16 meters long. It has a flat roof that is 12 meters high at one corner, and 10 meters high at each of the adjacent corners. What is the volume of the building?

Solution If we put the high corner on the z-axis, the long side along the y-axis, and the short side along the x-axis, as in Figure 16.12, then the roof is a plane with z-intercept 12, and x slope $(-2)/8 = -1/4$, and y slope $(-2)/16 = -1/8$. Hence, the equation of the roof is

$$z = 12 - \tfrac{1}{4}x - \tfrac{1}{8}y.$$

The volume is given by the double integral

$$\text{Volume} = \int_R (12 - \tfrac{1}{4}x - \tfrac{1}{8}y) \, dA,$$

where R is the rectangle $0 \le x \le 8$, $0 \le y \le 16$. Setting up an iterated integral, we get

$$\text{Volume} = \int_0^{16} \int_0^8 (12 - \tfrac{1}{4}x - \tfrac{1}{8}y) \, dx \, dy.$$

The inside integral is

$$\int_0^8 (12 - \tfrac{1}{4}x - \tfrac{1}{8}y) \, dx = \left(12x - \tfrac{1}{8}x^2 - \tfrac{1}{8}xy \right) \Big|_{x=0}^{x=8} = 88 - y.$$

Then the outside integral gives

$$\text{Volume} = \int_0^{16} (88 - y) \, dy = (88y - \tfrac{1}{2}y^2) \Big|_0^{16} = 1280.$$

The volume of the building is 1280 cubic meters.

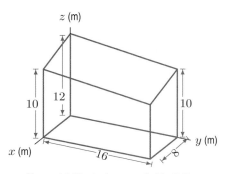

Figure 16.12: A slant-roofed building

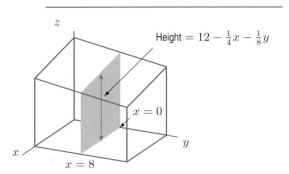

Figure 16.13: Cross-section of a building

Notice that the inner integral $\int_0^8 (12 - \frac{1}{4}x - \frac{1}{8}y)\,dx$ in Example 1 gives the area of the cross section of the building perpendicular to the y-axis in Figure 16.13.

The iterated integral $\int_0^{16} \int_0^8 (12 - \frac{1}{4}x - \frac{1}{8}y)\,dx\,dy$ thus calculates the volume by adding the volumes of thin cross-sectional slabs.

The Order of Integration

In computing the fox population, we could have chosen to add columns (fixed x) first, instead of the rows. This leads to an iterated integral where x is constant in the inner integral instead of y. Thus,

$$\int_R f(x, y)\,dA = \int_a^b \left(\int_c^d f(x, y)\,dy \right) dx$$

where R is the rectangle $a \leq x \leq b$ and $c \leq y \leq d$.

For any function we are likely to meet, it does not matter in which order we integrate over a rectangular region R; we get the same value for the double integral either way.

$$\int_R f\,dA = \int_c^d \left(\int_a^b f(x, y)\,dx \right) dy = \int_a^b \left(\int_c^d f(x, y)\,dy \right) dx$$

Example 2   Compute the volume of Example 1 as an iterated integral with y fixed in the inner integral.

Solution   Rewriting the integral, we have

$$\text{Volume} = \int_0^8 \left(\int_0^{16} (12 - \tfrac{1}{4}x - \tfrac{1}{8}y)\,dy \right) dx = \int_0^8 \left((12y - \tfrac{1}{4}xy - \tfrac{1}{16}y^2) \Big|_{y=0}^{y=16} \right) dx$$

$$= \int_0^8 (176 - 4x)\,dx = (176x - 2x^2) \Big|_0^8 = 1280 \text{ meter}^3.$$

Iterated Integrals Over Non-Rectangular Regions

Example 3   The density at the point (x, y) of a triangular metal plate, as shown in Figure 16.14, is $\delta(x, y)$. Express its mass as an iterated integral.

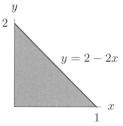

Figure 16.14: A triangular metal plate with density $\delta(x, y)$ at the point (x, y)

Solution   Approximate the triangular region using a grid of small rectangles of sides Δx and Δy. The mass of one rectangle is given by

$$\text{Mass of rectangle} \approx \text{Density} \cdot \text{Area} \approx \delta(x, y)\Delta x \Delta y.$$

Summing over all rectangles gives a Riemann sum which approximates the double integral

$$\text{Mass} = \int_R \delta(x, y)\,dA,$$

where R is the triangle. We want to compute this integral using an iterated integral.

Think about how the iterated integral over the rectangle $a \leq x \leq b$, $c \leq y \leq d$ works:

$$\int_a^b \int_c^d f(x,y) \, dy \, dx.$$

The inside integral with respect to y is along vertical strips which begin at the horizontal line $y = c$ and end at the line $y = d$. There is one such strip for each x between $x = a$ and $x = b$. (See Figure 16.15).

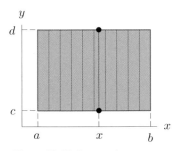

Figure 16.15: Integrating over a rectangle using vertical strips

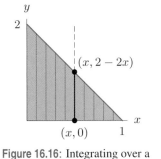

Figure 16.16: Integrating over a triangle using vertical strips

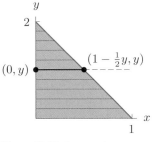

Figure 16.17: Integrating over a triangle using horizontal strips

For the triangular region in Figure 16.14, the idea is the same. The only difference is that the individual vertical strips no longer all go from $y = c$ to $y = d$. The vertical strip that starts at the point $(x, 0)$ ends at the point $(x, 2 - 2x)$, because the top edge of the triangle is the line $y = 2 - 2x$. See Figure 16.16. On this vertical strip, y goes from 0 to $2 - 2x$. Hence, the inside integral is

$$\int_0^{2-2x} \delta(x,y) \, dy.$$

Finally, since there is a vertical strip for each x between 0 and 1, the outside integral goes from $x = 0$ to $x = 1$. Thus, the iterated integral we want is

$$\text{Mass} = \int_0^1 \int_0^{2-2x} \delta(x,y) \, dy \, dx.$$

We could have chosen to integrate in the opposite order, keeping y fixed in the inner integral instead of x. The limits are formed by looking at horizontal strips instead of vertical ones, and expressing the x-values at the end points in terms of y. To find the right endpoint of the strip, we use the equation of the top edge of the triangle in the form $x = 1 - \frac{1}{2}y$. Thus, a horizontal strip goes from $x = 0$ to $x = 1 - \frac{1}{2}y$. Since there is a strip for every y from 0 and 2, the iterated integral is

$$\text{Mass} = \int_0^2 \int_0^{1-\frac{1}{2}y} \delta(x,y) \, dx \, dy.$$

Limits on Iterated Integrals

- The limits on the outer integral must be constants.
- The limits on the inner integral can involve only the variable in the outer integral. For example, if the inner integral is with respect to x, its limits can be functions of y.

Example 4   Find the mass M of a metal plate R bounded by $y = x$ and $y = x^2$, with density given by $\delta(x, y) = 1 + xy$ kg/meter^2. (See Figure 16.18.)

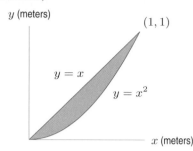

Figure 16.18: A metal plate with density $\delta(x, y)$

Solution   The mass is given by

$$M = \int_R \delta(x, y) \, dA.$$

We integrate along vertical strips first; this means we do the y integral first, which goes from the bottom boundary $y = x^2$ to the top boundary $y = x$. The left edge of the region is at $x = 0$ and the right edge is at the intersection point of $y = x$ and $y = x^2$, which is $(1, 1)$. Thus, the x-coordinate of the vertical strips can vary from $x = 0$ to $x = 1$, and so the mass is given by

$$M = \int_0^1 \int_{x^2}^{x} \delta(x, y) \, dy \, dx = \int_0^1 \int_{x^2}^{x} (1 + xy) \, dy \, dx.$$

Calculating the inner integral first gives

$$M = \int_0^1 \int_{x^2}^{x} (1 + xy) \, dy \, dx = \int_0^1 \left(y + x\frac{y^2}{2} \right) \Bigg|_{y=x^2}^{y=x} dx$$

$$= \int_0^1 \left(x - x^2 + \frac{x^3}{2} - \frac{x^5}{2} \right) dx = \left(\frac{x^2}{2} - \frac{x^3}{3} + \frac{x^4}{8} - \frac{x^6}{12} \right) \Bigg|_0^1 = \frac{5}{24} \approx 0.21 \text{ kg.}$$

Example 5   A city occupies a semicircular region of radius 3 km bordering on the ocean. Find the average distance from points in the city to the ocean.

Solution   Think of the ocean as everything below the x-axis in the xy-plane and think of the city as the upper half of the circular disk of radius 3 bounded by $x^2 + y^2 = 9$. (See Figure 16.19).

The distance from any point (x, y) in the city to the ocean is the vertical distance to the x-axis, namely y. Thus, we want to compute

$$\text{Average distance} = \frac{1}{\text{Area}(R)} \int_R y \, dA,$$

where R is the region between the upper half of the circle $x^2 + y^2 = 9$ and the x-axis. The area of R is $\pi 3^2 / 2 = 9\pi / 2$.

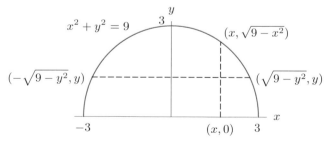

Figure 16.19: The city by the ocean showing a typical vertical strip and a typical horizontal strip

To compute the integral, let's take the inner integral with respect to y. Then a vertical strip goes from the x-axis, namely $y = 0$, to the semicircle. The upper limit must be expressed in terms of x so we solve $x^2 + y^2 = 9$ to get $y = \sqrt{9 - x^2}$. Since there is a strip for every x from -3 to 3, the integral is:

$$\int_R y \, dA = \int_{-3}^{3} \left(\int_0^{\sqrt{9-x^2}} y \, dy \right) dx = \int_{-3}^{3} \left(\frac{y^2}{2} \Big|_0^{\sqrt{9-x^2}} \right) dx$$

$$= \int_{-3}^{3} \frac{1}{2}(9 - x^2) \, dx = \frac{1}{2} \left(9x - \frac{x^3}{3} \right) \Big|_{-3}^{3} = \frac{1}{2}(18 - (-18)) = 18.$$

Therefore, the average distance is $18/(9\pi/2) = 4/\pi \approx 1.3$ km.

What if we choose the inner integral with respect to x? Then we get the limits by looking at horizontal strips, not vertical, and we solve $x^2 + y^2 = 9$ for x in terms of y. We get $x = -\sqrt{9 - y^2}$ at the left end of the strip and $x = \sqrt{9 - y^2}$ at the right. There is a strip for every y from 0 to 3, so

$$\int_R y \, dA = \int_0^3 \left(\int_{-\sqrt{9-y^2}}^{\sqrt{9-y^2}} y \, dx \right) dy = \int_0^3 \left(yx \Big|_{x=-\sqrt{9-y^2}}^{x=\sqrt{9-y^2}} \right) dy = \int_0^3 2y\sqrt{9 - y^2} \, dy$$

$$= -\frac{2}{3}(9 - y^2)^{3/2} \Big|_0^3 = -\frac{2}{3}(0 - 27) = 18.$$

We get the same result as before. The average distance to the ocean is $(2/(9\pi))18 = 4/\pi \approx 1.3$ km.

In the examples so far, a region was given and the problem was to determine the limits for an iterated integral. Sometimes the limits are known and we want to determine the region.

Example 6   Sketch the region of integration for the iterated integral $\displaystyle\int_0^6 \int_{x/3}^2 x\sqrt{y^3 + 1} \, dy \, dx$.

Solution   The inner integral is with respect to y, so we imagine the region built of vertical strips. The bottom of each strip is on the line $y = x/3$, and the top is on the horizontal line $y = 2$. Since the limits of the outer integral are 0 and 6, the whole region is contained between the vertical lines $x = 0$ and $x = 6$. Notice that the lines $y = 2$ and $y = x/3$ meet where $x = 6$. See Figure 16.20.

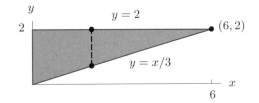

Figure 16.20: The region of integration for Example 6, showing the vertical strip

Reversing the Order of Integration

It is sometimes helpful to reverse the order of integration in an iterated integral. An integral which is difficult or impossible with the integration in one order can be quite straightforward in the other. The next example is such a case.

Example 7   Evaluate $\displaystyle\int_0^6 \int_{x/3}^2 x\sqrt{y^3 + 1} \, dy \, dx$ using the region sketched in Figure 16.20.

Solution   Since $\sqrt{y^3 + 1}$ has no elementary antiderivative, we cannot calculate the inner integral symbolically. We try reversing the order of integration. From Figure 16.20, we see that horizontal strips go from $x = 0$ to $x = 3y$ and that there is a strip for every y from 0 to 2. Thus, when we change the order of integration we get

$$\int_0^6 \int_{x/3}^2 x\sqrt{y^3 + 1}\, dy\, dx = \int_0^2 \int_0^{3y} x\sqrt{y^3 + 1}\, dx\, dy.$$

Now we can at least do the inner integral because we know the antiderivative of x. What about the outer integral?

$$\int_0^2 \int_0^{3y} x\sqrt{y^3 + 1}\, dx\, dy = \int_0^2 \left(\frac{x^2}{2}\sqrt{y^3 + 1}\right)\Big|_{x=0}^{x=3y} dy = \int_0^2 \frac{9y^2}{2}(y^3 + 1)^{1/2}\, dy$$

$$= (y^3 + 1)^{3/2}\Big|_0^2 = 27 - 1 = 26.$$

Thus, reversing the order of integration made the integral in the previous problem much easier. Notice that to reverse the order it is essential first to sketch the region over which the integration is being performed.

Exercises and Problems for Section 16.2

Exercises

For Exercises 1–4, evaluate the integral.

1. $\int_0^3 \int_0^4 (4x + 3y)\, dx\, dy$

2. $\int_0^2 \int_0^3 (x^2 + y^2)\, dy\, dx$

3. $\int_0^3 \int_0^2 (6xy)\, dy\, dx$

4. $\int_0^1 \int_0^2 (x^2 y)\, dy\, dx$

For each of the shaded regions R in Exercises 5–10, write $\int_R f\, dA$ as an iterated integral.

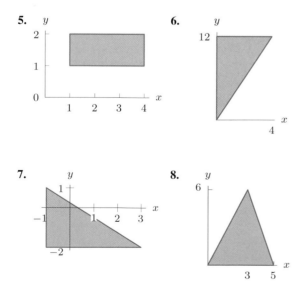

5.

6.

7.

8.

9.

10.

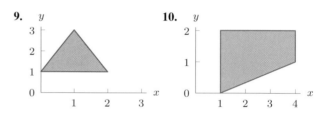

For Exercises 11–16, sketch the region of integration and evaluate the integral.

11. $\int_1^3 \int_0^4 e^{x+y}\, dy\, dx$

12. $\int_0^2 \int_0^x e^{x^2}\, dy\, dx$

13. $\int_1^5 \int_x^{2x} \sin x\, dy\, dx$

14. $\int_1^4 \int_{\sqrt{y}}^y x^2 y^3\, dx\, dy$

15. $\int_0^3 \int_0^{2x} (x^2 + y^2)\, dy\, dx$

16. $\int_{-2}^0 \int_{-\sqrt{9-x^2}}^0 2xy\, dy\, dx$

For Exercises 17–21, evaluate the integral.

17. $\int_R \sqrt{x + y}\, dA$, where R is the rectangle $0 \le x \le 1$, $0 \le y \le 2$.

18. Calculate the integral in Exercise 17 using the other order of integration.

19. $\int_R (5x^2 + 1)\sin 3y\, dA$, where R is the rectangle $-1 \le x \le 1$, $0 \le y \le \pi/3$.

20. $\int_R xy\, dA$, where R is the triangle $x + y \le 1$, $x \ge 0$, $y \ge 0$.

21. $\int_R (2x + 3y)^2\, dA$, where R is the triangle with vertices at $(-1, 0)$, $(0, 1)$, and $(1, 0)$.

22. Find the volume under the graph of the function $f(x, y) = 6x^2 y$ over the region shown in Figure 16.21.

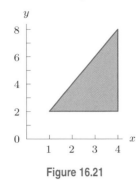

Figure 16.21

23. Find the average value of $f(x, y) = x^2 + 4y$ on the rectangle $0 \leq x \leq 3$ and $0 \leq y \leq 6$.

24. Find the average value of $f(x, y) = xy^2$ on the rectangle $0 \leq x \leq 4, 0 \leq y \leq 3$.

25. (a) Sketch the region in the xy-plane bounded by the x-axis, $y = x$, $x + y = 1$.
(b) Express the integral of $f(x, y)$ over this region in terms of iterated integrals in two ways. (In one, use $dx\,dy$; in the other, use $dy\,dx$.)
(c) Using one of your answers to part (b), evaluate the integral exactly with $f(x, y) = x$.

26. Consider the integral $\int_0^4 \int_0^{-(y-4)/2} g(x, y)\, dx\, dy$.

(a) Sketch the region over which the integration is being performed.
(b) Write the integral with the order of the integration reversed.

Problems

Evaluate the integrals in Problems 27–32 by reversing the order of integration.

27. $\displaystyle\int_0^1 \int_y^1 e^{x^2}\, dx\, dy$

28. $\displaystyle\int_0^1 \int_y^1 \sin\left(x^2\right)\, dx\, dy$

29. $\displaystyle\int_0^1 \int_{\sqrt{y}}^1 \sqrt{2 + x^3}\, dx\, dy$

30. $\displaystyle\int_0^3 \int_{y^2}^9 y \sin(x^2)\, dx\, dy$

31. $\displaystyle\int_0^1 \int_{e^y}^e \frac{x}{\ln x}\, dx\, dy$

32. $\displaystyle\int_{-4}^0 \int_0^{2x+8} f(x, y)\, dy\, dx + \int_0^4 \int_0^{-2x+8} f(x, y)\, dy\, dx$

33. (a) Use four subrectangles to approximate the volume of the object whose base is the region $0 \leq x \leq 4$ and $0 \leq y \leq 6$, and whose height is given by $f(x, y) = xy$. Find an overestimate and an underestimate and average the two.
(b) Integrate to find the exact volume of the three-dimensional object described in part (a).

34. The region W lies below the surface $f(x, y) = 2e^{-(x-1)^2 - y^2}$ and above the disk $x^2 + y^2 \leq 4$ in the xy-plane.

(a) Describe in words the contours of f, using $f(x, y) = 1$ as an example.
(b) Write an integral giving the area of the cross-section of W in the plane $x = 1$.
(c) Write an iterated double integral giving the volume of W.

For Problems 35–37 set up, but do not evaluate, an iterated integral for the volume of the solid.

35. Under the graph of $f(x, y) = 25 - x^2 - y^2$ and above the xy-plane.

36. Below the graph of $f(x, y) = 25 - x^2 - y^2$ and above the plane $z = 16$.

37. The three-sided pyramid whose base is on the xy-plane and whose three sides are the vertical planes $y = 0$ and $y - x = 4$, and the slanted plane $2x + y + z = 4$.

For Problems 38–43, find the volume of the region.

38. Under the graph of $f(x, y) = xy$ and above the square $0 \leq x \leq 2, 0 \leq y \leq 2$ in the xy-plane.

39. Under the graph of $f(x, y) = x^2 + y^2$ and above the triangle $y \leq x, 0 \leq x \leq 1$.

40. Under the graph of $f(x, y) = x + y$ and above the region $y^2 \leq x, 0 \leq x \leq 9$.

41. Under the graph of $2x + y + z = 4$ in the first octant.

42. The solid between the planes $z = 3x + 2y + 1$ and $z = x + y$, and above the triangle with vertices $(1, 0, 0)$, $(2, 2, 0)$, and $(0, 1, 0)$ in the xy-plane. See Figure 16.22.

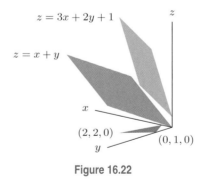

$z = 3x + 2y + 1$

$z = x + y$

$(2, 2, 0)$ $(0, 1, 0)$

Figure 16.22

43. The region R bounded by the graph of $ax + by + cz = 1$ and the coordinate planes. Assume a, b, and $c > 0$.

44. Find the average distance to the x-axis for points in the region bounded by the x-axis and the graph of $y = x - x^2$.

45. Show that for a right triangle the average distance from any point in the triangle to one of the legs is one-third the length of the other leg. (The legs of a right triangle are the two sides that are not the hypotenuse.)

46. The function $f(x, y) = ax^2 + bxy + cy^2$ has an average value of 20 on the square $0 \leq x \leq 2, 0 \leq y \leq 2$.

   (a) What can you say about the constants a, b, and c?
   (b) Find two different choices for f that have average value 20 on the square, and give their contour diagrams on the square.

47. The function $f(x, y) = ax + by$ has an average value of 20 on the rectangle $0 \leq x \leq 2, 0 \leq y \leq 3$.

   (a) What can you say about the constants a and b?

   (b) Find two different choices for f that have average value 20 on the rectangle, and give their contour diagrams on the rectangle.

48. Give the contour diagram of a function f whose average value on the square $0 \leq x \leq 1, 0 \leq y \leq 1$ is

   (a) Greater than the average of the values of f at the four corners of the square.
   (b) Less than the average of the values of f at the four corners of the square.

49. A rectangular plate of sides a and b is subjected to a normal force (that is, perpendicular to the plate). The pressure, p, at any point on the plate is proportional to the square of the distance of that point from one corner. Find the total force on the plate. [Note that pressure is force per unit area.]

16.3 TRIPLE INTEGRALS

A continuous function of three variables can be integrated over a solid region W in 3-space in the same way as a function of two variables is integrated over a flat region in 2-space. Again, we start with a Riemann sum. First we subdivide W into smaller regions, then we multiply the volume of each region by a value of the function in that region, and then we add the results. For example, if W is the box $a \leq x \leq b, c \leq y \leq d, p \leq z \leq q$, then we subdivide each side into n, m, and l pieces, thereby chopping W into nml smaller boxes, as shown in Figure 16.23.

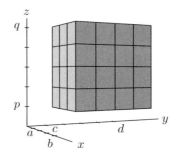

Figure 16.23: Subdividing a three-dimensional box

The volume of each smaller box is

$$\Delta V = \Delta x \Delta y \Delta z,$$

where $\Delta x = (b - a)/n$, and $\Delta y = (d - c)/m$, and $\Delta z = (q - p)/l$. Using this subdivision, we pick a point $(u_{ijk}, v_{ijk}, w_{ijk})$ in the ijk-th small box and construct a Riemann sum

$$\sum_{i,j,k} f(u_{ijk}, v_{ijk}, w_{ijk}) \Delta V.$$

If f is continuous, as Δx, Δy, and Δz approach 0, this Riemann sum approaches the definite integral, $\int_W f \, dV$, called a *triple integral*, which is defined as

$$\int_W f \, dV = \lim_{n,m,l \to \infty} \sum_{i,j,k} f(u_{ijk}, v_{ijk}, w_{ijk}) \Delta V.$$

As in the case of a double integral, we can evaluate this integral as an iterated integral:

Triple integral as an iterated integral

$$\int_W f\, dV = \int_p^q \left(\int_c^d \left(\int_a^b f(x,y,z)\, dx \right) dy \right) dz,$$

where y and z are treated as constants in the innermost (dx) integral, and z is treated as a constant in the middle (dy) integral. Other orders of integration are possible.

Example 1   A cube C has sides of length 4 cm and is made of a material of variable density. If one corner is at the origin and the adjacent corners are on the positive x, y, and z axes, then the density at the point (x,y,z) is $\delta(x,y,z) = 1 + xyz$ gm/cm^3. Find the mass of the cube.

Solution   Consider a small piece ΔV of the cube, small enough so that the density remains close to constant over the piece. Then

$$\text{Mass of small piece} = \text{Density} \cdot \text{Volume} \approx \delta(x,y,z)\, \Delta V.$$

To get the total mass, we add the masses of the small pieces and take the limit as $\Delta V \to 0$. Thus, the mass is the triple integral

$$M = \int_C \delta\, dV = \int_0^4 \int_0^4 \int_0^4 (1+xyz)\, dx\, dy\, dz = \int_0^4 \int_0^4 \left(x + \frac{1}{2}x^2 yz \right)\Big|_{x=0}^{x=4} dy\, dz$$

$$= \int_0^4 \int_0^4 (4 + 8yz)\, dy\, dz = \int_0^4 \left(4y + 4y^2 z \right)\Big|_{y=0}^{y=4} dz = \int_0^4 (16 + 64z)\, dz = 576\,\text{gm}.$$

Example 2   Express the volume of the building described in Example 1 on page 791 as a triple integral.

Solution   The building is given by $0 \le x \le 8$, $0 \le y \le 16$, and $0 \le z \le 12 - x/4 - y/8$. (See Figure 16.24 on page 800.) To find its volume, divide it into small cubes of volume $\Delta V = \Delta x\, \Delta y\, \Delta z$ and add. First, make a vertical stack of cubes above the point $(x,y,0)$. This stack goes from $z=0$ to $z = 12 - x/4 - y/8$, so

$$\text{Volume of vertical stack} \approx \sum \Delta V = \sum_z \Delta x\, \Delta y\, \Delta z = \left(\sum_z \Delta z \right) \Delta x\, \Delta y.$$

Next, line up these stacks parallel to the y-axis to form a slice from $y=0$ to $y=16$. So

$$\text{Volume of slice} \approx \left(\sum_y \sum_z \Delta z\, \Delta y \right) \Delta x.$$

Finally, line up the slices along the x-axis from $x=0$ to $x=8$ and add up their volumes, to get

$$\text{Volume of building} \approx \sum_x \sum_y \sum_z \Delta z\, \Delta y\, \Delta x.$$

Thus, in the limit,

$$\text{Volume of building} = \int_0^8 \int_0^{16} \int_0^{12-x/4-y/8} 1\, dz\, dy\, dx.$$

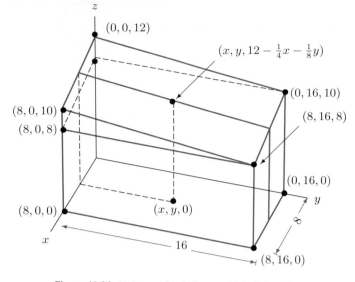

Figure 16.24: Volume of building as triple integral

Example 3   Set up an iterated integral to compute the mass of the solid cone bounded by $z = \sqrt{x^2 + y^2}$ and $z = 3$, if the density is given by $\delta(x, y, z) = z$.

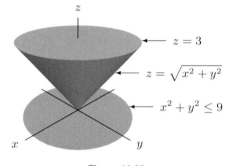

Figure 16.25

Solution   We break the cone in Figure 16.25 into small cubes of volume $\Delta V = \Delta x \, \Delta y \, \Delta z$, on which the density is approximately constant, and approximate the mass of each cube by $\delta(x, y, z) \, \Delta x \, \Delta y \, \Delta z$. Stacking the cubes vertically above the point $(x, y, 0)$, starting on the cone at height $z = \sqrt{x^2 + y^2}$ and going up to $z = 3$, tells us that the inner integral is

$$\int_{\sqrt{x^2+y^2}}^{3} \delta(x, y, z) \, dz = \int_{\sqrt{x^2+y^2}}^{3} z \, dz.$$

There is a stack for every point in the xy-plane in the shadow of the cone. The cone $z = \sqrt{x^2 + y^2}$ intersects the horizontal plane $z = 3$ in the circle $x^2 + y^2 = 9$, so there is a stack for all (x, y) in the region $x^2 + y^2 \leq 9$. Lining up the stacks parallel to the y-axis gives a slice from $y = -\sqrt{9 - x^2}$ to $y = \sqrt{9 - x^2}$, for each fixed value of x. Thus, the limits on the middle integral are

$$\int_{-\sqrt{9-x^2}}^{\sqrt{9-x^2}} \int_{\sqrt{x^2+y^2}}^{3} z \, dz \, dy.$$

Finally, there is a slice for each x between -3 and 3, so the integral we want is

$$\text{Mass} = \int_{-3}^{3} \int_{-\sqrt{9-x^2}}^{\sqrt{9-x^2}} \int_{\sqrt{x^2+y^2}}^{3} z \, dz \, dy \, dx.$$

Notice that setting up the limits on the two outer integrals was just like setting up the limits for a double integral over the region $x^2 + y^2 \leq 9$.

As the previous example illustrates, for a region W contained between two surfaces, the innermost limits correspond to these surfaces. The middle and outer limits ensure that we integrate over the "shadow" of W in the xy-plane.

Limits on Triple Integrals

- The limits for the outer integral are constants.
- The limits for the middle integral can involve only one variable (that in the outer integral).
- The limits for the inner integral can involve two variables (those on the two outer integrals).

Exercises and Problems for Section 16.3

Exercises

In Exercises 1–4, find the triple integrals of the function over the region W.

1. $f(x, y, z) = x^2 + 5y^2 - z$, W is the rectangular box $0 \le x \le 2, -1 \le y \le 1, 2 \le z \le 3$.

2. $h(x, y, z) = ax + by + cz$, W is the rectangular box $0 \le x \le 1, 0 \le y \le 1, 0 \le z \le 2$.

3. $f(x, y, z) = e^{-x-y-z}$, W is the rectangular box with corners at $(0, 0, 0)$, $(a, 0, 0)$, $(0, b, 0)$, and $(0, 0, c)$.

4. $f(x, y, z) = \sin x \cos(y + z)$, W is the cube $0 \le x \le \pi$, $0 \le y \le \pi, 0 \le z \le \pi$.

Sketch the region of integration in Exercises 5–13.

5. $\displaystyle\int_0^1 \int_{-1}^1 \int_0^{\sqrt{1-z^2}} f(x, y, z)\, dy\, dz\, dx$

6. $\displaystyle\int_0^1 \int_{-1}^1 \int_0^{\sqrt{1-x^2}} f(x, y, z)\, dz\, dx\, dy$

7. $\displaystyle\int_0^1 \int_{-\sqrt{1-z^2}}^{\sqrt{1-z^2}} \int_0^{\sqrt{1-x^2-z^2}} f(x, y, z)\, dy\, dx\, dz$

8. $\displaystyle\int_{-1}^1 \int_0^1 \int_{-\sqrt{1-z^2}}^{\sqrt{1-z^2}} f(x, y, z)\, dy\, dz\, dx$

9. $\displaystyle\int_0^1 \int_{-1}^1 \int_{-\sqrt{1-x^2}}^{\sqrt{1-x^2}} f(x, y, z)\, dz\, dx\, dy$

10. $\displaystyle\int_{-1}^1 \int_{-\sqrt{1-x^2}}^{\sqrt{1-x^2}} \int_0^{\sqrt{1-x^2-z^2}} f(x, y, z)\, dy\, dz\, dx$

11. $\displaystyle\int_0^1 \int_{-\sqrt{1-z^2}}^{\sqrt{1-z^2}} \int_{-\sqrt{1-y^2-z^2}}^{\sqrt{1-y^2-z^2}} f(x, y, z)\, dx\, dy\, dz$

12. $\displaystyle\int_0^1 \int_0^{\sqrt{1-z^2}} \int_{-\sqrt{1-x^2-z^2}}^{\sqrt{1-x^2-z^2}} f(x, y, z)\, dy\, dx\, dz$

13. $\displaystyle\int_0^1 \int_0^{\sqrt{1-y^2}} \int_{-\sqrt{1-x^2-y^2}}^{\sqrt{1-x^2-y^2}} f(x, y, z)\, dz\, dx\, dy$

Problems

In Problems 14–17, write limits of integration for the integral $\int_W f(x, y, z)\, dV$ where W is the quarter or half sphere or cylinder shown.

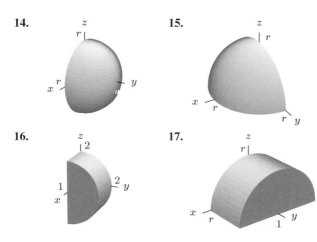

14.

15.

16.

17.

18. Find the volume of the region bounded by $z = x + y, z = 10$, and the planes $x = 0, y = 0$.

19. Find the volume of the pyramid with base in the plane $z = -6$ and sides formed by the three planes $y = 0$ and $y - x = 4$ and $2x + y + z = 4$.

20. Find the volume of the region bounded by the planes $z = 3y, z = y, y = 1, x = 1$, and $x = 2$.

21. Find the volume of the region between the plane $z = x$ and the surface $z = x^2$, and the planes $y = 0$, and $y = 3$.

22. Find the volume of the region bounded by $z = x + y$, $0 \le x \le 5, 0 \le y \le 5$, and the planes $x = 0, y = 0$, and $z = 0$.

23. Find the volume of the region bounded by $z = x^2$, $0 \le x \le 5$, and the planes $y = 0, y = 3$, and $z = 0$.

24. **(a)** What is the equation of the plane passing through the points $(1, 0, 0)$, $(0, 1, 0)$, and $(0, 0, 1)$?

(b) Find the volume of the region bounded by this plane and the planes $x = 0$, $y = 0$, and $z = 0$.

25. A solid region D is a half cylinder of radius 1 lying horizontally with its rectangular base in the xy-plane and its axis along the y-axis from $y = 0$ to $y = 10$. (The region is above the xy-plane.)

(a) What is the equation of the curved surface of this half cylinder?

(b) Write the limits of integration of the integral $\int_D f(x, y, z)\, dV$ in Cartesian coordinates.

26. Find the mass of the solid bounded by the xy-plane, yz-plane, xz-plane, and the plane $(x/3) + (y/2) + (z/6) = 1$, if the density of the solid is given by $\delta(x, y, z) = x + y$.

27. Find the mass of the pyramid with base in the plane $z = -6$ and sides formed by the three planes $y = 0$ and $y - x = 4$ and $2x + y + z = 4$, if the density of the solid is given by $\delta(x, y, z) = y$.

28. Find the average value of the sum of the squares of three numbers x, y, z, where each number is between 0 and 2.

Let W be the solid cone bounded by $z = \sqrt{x^2 + y^2}$ and $z = 2$. For Problems 29–37, decide (without calculating its value) whether the integral is positive, negative, or zero.

29. $\int_W \sqrt{x^2 + y^2}\, dV$

30. $\int_W (z - \sqrt{x^2 + y^2})\, dV$

31. $\int_W x\, dV$

32. $\int_W y\, dV$

33. $\int_W z\, dV$

34. $\int_W xy\, dV$

35. $\int_W xyz\, dV$

36. $\int_W (z - 2)\, dV$

37. $\int_W e^{-xyz}\, dV$

Let W be the solid half-cone bounded by $z = \sqrt{x^2 + y^2}$, $z = 2$ and the yz-plane with $x \geq 0$. For Problems 38–46, decide (without calculating its value) whether the integral is positive, negative, or zero.

38. $\int_W \sqrt{x^2 + y^2}\, dV$

39. $\int_W (z - \sqrt{x^2 + y^2})\, dV$

40. $\int_W x\, dV$

41. $\int_W y\, dV$

42. $\int_W z\, dV$

43. $\int_W xy\, dV$

44. $\int_W xyz\, dV$

45. $\int_W (z - 2)\, dV$

46. $\int_W e^{-xyz}\, dV$

47. Figure 16.26 shows part of a spherical ball of radius 5 cm. Write an iterated triple integral which represents the volume of this region.

2 cm

Figure 16.26

48. Set up, but do not evaluate, an iterated integral for the volume of the solid formed by the intersections of the cylinders $x^2 + z^2 = 1$ and $y^2 + z^2 = 1$.

49. A cube of side length 2 m is composed of two materials, separated by a plane slanting down from the top corner to cut the bottom face in a diagonal, as in Figure 16.27. The material below the plane has density δ_1 gm/m^3 and the material above the plane has density δ_2 gm/m^3. Find the mass of the cube.

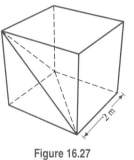

2 m

Figure 16.27

The motion of a solid object can be analyzed by thinking of the mass as concentrated at a single point, the *center of mass*. If the object has density $\rho(x, y, z)$ at the point (x, y, z) and occupies a region W, then the coordinates $(\bar{x}, \bar{y}, \bar{z})$ of the center of mass are given by

$$\bar{x} = \frac{1}{m} \int_W x\rho\, dV \quad \bar{y} = \frac{1}{m} \int_W y\rho\, dV \quad \bar{z} = \frac{1}{m} \int_W z\rho\, dV$$

where $m = \int_W \rho\, dV$ is the total mass of the body. Use these definitions for Problems 50–51.

50. A solid is bounded below by the square $z = 0$, $0 \leq x \leq 1$, $0 \leq y \leq 1$ and above by the surface $z = x + y + 1$. Find the total mass and the coordinates of the center of mass if the density is 1 gm/cm^3 and x, y, z are measured in centimeters.

51. Find the center of mass of the tetrahedron that is bounded by the xy, yz, xz planes and the plane $x + 2y + 3z = 1$. Assume the density is 1 gm/cm^3 and x, y, z are in centimeters.

The *moment of inertia* of a solid body about an axis in 3-space relates the angular acceleration about this axis to torque (force twisting the body). The moments of inertia about the coordinate axes of a body of constant density and mass m occupying a region W of volume V are defined to be

$$I_x = \frac{m}{V} \int_W (y^2 + z^2)\, dV \qquad I_y = \frac{m}{V} \int_W (x^2 + z^2)\, dV$$

$$I_z = \frac{m}{V} \int_W (x^2 + y^2)\, dV$$

Use these definitions for Problems 52–54.

52. Find the moment of inertia about the z-axis of the rectangular solid of mass m given by $0 \le x \le 1$, $0 \le y \le 2$, $0 \le z \le 3$.

53. Find the moment of inertia about the x-axis of the rectangular solid $-a \le x \le a$, $-b \le y \le b$ and $-c \le z \le c$ of mass m.

54. Let a, b, and c denote the moments of inertia of a homogeneous solid object about the x, y and z-axes respectively. Explain why $a + b > c$.

16.4   DOUBLE INTEGRALS IN POLAR COORDINATES

Integration in Polar Coordinates

We started this chapter by putting a rectangular grid on the fox population density map, to estimate the total population using a Riemann sum. However, sometimes a polar grid is more appropriate.

Example 1   A biologist studying insect populations around a circular lake divides the area into the polar sectors in Figure 16.28. The population density in each sector is shown in millions per square km. Estimate the total insect population around the lake.

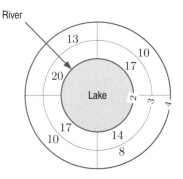

Figure 16.28: An insect infested lake showing the insect population density by sector

Solution   To get the estimate, we multiply the population density in each sector by the area of that sector. Unlike the rectangles in a rectangular grid, the sectors in this grid do not all have the same area. The inner sectors have area

$$\frac{1}{4}(\pi 3^2 - \pi 2^2) = \frac{5\pi}{4} \approx 3.93 \text{ km}^2,$$

and the outer sectors have area

$$\frac{1}{4}(\pi 4^2 - \pi 3^2) = \frac{7\pi}{4} \approx 5.50 \text{ km}^2,$$

so we estimate

$$\text{Population} \approx (20)(3.93) + (17)(3.93) + (14)(3.93) + (17)(3.93) +$$
$$(13)(5.50) + (10)(5.50) + (8)(5.50) + (10)(5.50)$$
$$\approx 493 \text{ million insects.}$$

What is dA in Polar Coordinates?

The previous example used a polar grid rather than a rectangular grid. A rectangular grid is constructed from vertical and horizontal lines of the form $x = k$ (a constant) and $y = l$ (another constant). In polar coordinates, $r = k$ gives a circle of radius k centered at the origin and $\theta = l$ gives a ray emanating from the origin (at angle l with the x-axis). A polar grid is built out of these circles and rays. See Figure 16.29.

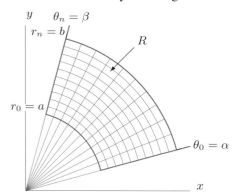

Figure 16.29: Dividing up a region using a polar grid

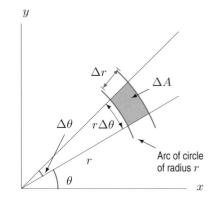

Figure 16.30: Calculating area ΔA in polar coordinates

Choosing (r_{ij}, θ_{ij}) in the ij-th bent rectangle in Figure 16.29 gives a Riemann sum:

$$\sum_{i,j} f(r_{ij}, \theta_{ij}) \, \Delta A.$$

To calculate the area ΔA, look at Figure 16.30. If Δr and $\Delta \theta$ are small, the shaded region is approximately a rectangle with sides $r \, \Delta \theta$ and Δr, so

$$\Delta A \approx r \Delta \theta \Delta r.$$

Thus, the Riemann sum is approximately

$$\sum_{i,j} f(r_{ij}, \theta_{ij}) \, r_{ij} \, \Delta \theta \, \Delta r.$$

If we take the limit as Δr and $\Delta \theta$ approach 0, we obtain

$$\int_R f \, dA = \int_\alpha^\beta \int_a^b f(r, \theta) \, r \, dr \, d\theta.$$

When computing integrals in polar coordinates, use $x = r \cos \theta, y = r \sin \theta, x^2 + y^2 = r^2$. Put $dA = r \, dr \, d\theta$ or $dA = r \, d\theta \, dr$.

Example 2 Compute the integral of $f(x, y) = 1/(x^2 + y^2)^{3/2}$ over the region R shown in Figure 16.31.

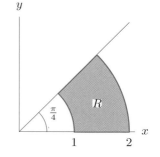

Figure 16.31: Integrate f over the polar region

Solution   The region R is described by the inequalities $1 \leq r \leq 2$, $0 \leq \theta \leq \pi/4$. In polar coordinates, $r = \sqrt{x^2 + y^2}$, so we can write f as

$$f(x, y) = \frac{1}{(x^2 + y^2)^{3/2}} = \frac{1}{(r^2)^{3/2}} = \frac{1}{r^3}.$$

Then

$$\int_R f \, dA = \int_0^{\pi/4} \int_1^2 \frac{1}{r^3} r \, dr \, d\theta = \int_0^{\pi/4} \left(\int_1^2 r^{-2} \, dr \right) d\theta$$

$$= \int_0^{\pi/4} -\frac{1}{r} \bigg|_{r=1}^{r=2} d\theta = \int_0^{\pi/4} \frac{1}{2} \, d\theta = \frac{\pi}{8}.$$

Example 3   For each region in Figure 16.32, decide whether to integrate using polar or Cartesian coordinates. On the basis of its shape, write an iterated integral of an arbitrary function $f(x, y)$ over the region.

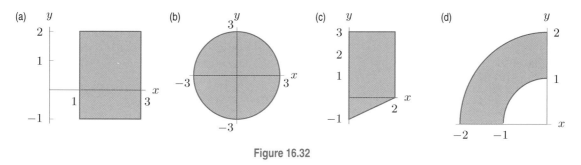

Figure 16.32

Solution   (a) Since this is a rectangular region, Cartesian coordinates are likely to be a better choice. The rectangle is described by the inequalities $1 \leq x \leq 3$ and $-1 \leq y \leq 2$, so the integral is

$$\int_{-1}^2 \int_1^3 f(x, y) \, dx \, dy.$$

(b) A circle is best described in polar coordinates. The radius is 3, so r goes from 0 to 3, and to describe the whole circle, θ goes from 0 to 2π. The integral is

$$\int_0^{2\pi} \int_0^3 f(r \cos \theta, r \sin \theta) \, r \, dr \, d\theta.$$

(c) The bottom boundary of this trapezoid is the line $y = (x/2) - 1$ and the top is the line $y = 3$, so we use Cartesian coordinates. If we integrate with respect to y first, the lower limit of the integral is $(x/2) - 1$ and the upper limit is 3. The x limits are $x = 0$ to $x = 2$. So the integral is

$$\int_0^2 \int_{(x/2)-1}^3 f(x, y) \, dy \, dx.$$

(d) This is another polar region: it is a piece of a ring in which r goes from 1 to 2. Since it is in the second quadrant, θ goes from $\pi/2$ to π. The integral is

$$\int_{\pi/2}^{\pi} \int_1^2 f(r \cos \theta, r \sin \theta) \, r \, dr \, d\theta.$$

Exercises and Problems for Section 16.4

Exercises

For the regions R in Exercises 1–4, write $\int_R f \, dA$ as an iterated integral in polar coordinates.

1.

2.

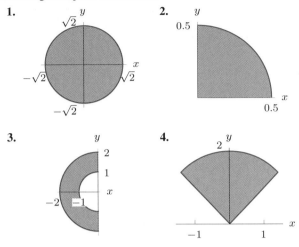

3.

4.

Sketch the region of integration in Exercises 5–11.

5. $\displaystyle\int_0^4 \int_{-\pi/2}^{\pi/2} f(r, \theta) \, r \, d\theta \, dr$

6. $\displaystyle\int_{\pi/2}^{\pi} \int_0^1 f(r, \theta) \, r \, dr \, d\theta$

7. $\displaystyle\int_0^{2\pi} \int_1^2 f(r, \theta) \, r \, dr \, d\theta$

8. $\displaystyle\int_{\pi/6}^{\pi/3} \int_0^1 f(r, \theta) \, r \, dr \, d\theta$

9. $\displaystyle\int_3^4 \int_{3\pi/4}^{3\pi/2} f(r, \theta) \, r \, d\theta \, dr$

10. $\displaystyle\int_0^{\pi/4} \int_0^{1/\cos\theta} f(r, \theta) \, r \, dr \, d\theta$

11. $\displaystyle\int_{\pi/4}^{\pi/2} \int_0^{2/\sin\theta} f(r, \theta) \, r \, dr \, d\theta$

In Exercises 12–14, evaluate the integral.

12. $\int_R \sin(x^2 + y^2) \, dA$, where R is the disk of radius 2 centered at the origin.

13. $\int_R (x^2 - y^2) \, dA$, where R is the first quadrant region between the circles of radius 1 and radius 2.

14. $\int_R \sqrt{x^2 + y^2} \, dxdy$ where R is $4 \le x^2 + y^2 \le 9$.

15. Consider the integral $\int_0^3 \int_{x/3}^1 f(x, y) \, dy \, dx$.

   (a) Sketch the region R over which the integration is being performed.
   (b) Rewrite the integral with the order of integration reversed.
   (c) Rewrite the integral in polar coordinates.

Problems

Convert the integrals in Problems 16–18 to polar coordinates and evaluate.

16. $\displaystyle\int_{-1}^0 \int_{-\sqrt{1-x^2}}^{\sqrt{1-x^2}} x \, dy \, dx$ **17.** $\displaystyle\int_0^{\sqrt6} \int_{-x}^x dy \, dx$

18. $\displaystyle\int_0^{\sqrt2} \int_y^{\sqrt{4-y^2}} xy \, dx \, dy$

19. Find the volume of the region between the graph of $f(x, y) = 25 - x^2 - y^2$ and the xy plane.

20. Find the volume of an ice cream cone bounded by the hemisphere $z = \sqrt{8 - x^2 - y^2}$ and the cone $z = \sqrt{x^2 + y^2}$.

21. A city surrounds a bay as shown in Figure 16.33. The population density of the city (in thousands of people per square km) is $\delta(r, \theta)$, where r and θ are polar coordinates and distances are in km.

   (a) Set up an iterated integral in polar coordinates giving the total population of the city.

   (b) The population density decreases the farther you live from the shoreline of the bay; it also decreases the farther you live from the ocean. Which of the following functions best describes this situation?
     (i) $\delta(r, \theta) = (4 - r)(2 + \cos\theta)$
     (ii) $\delta(r, \theta) = (4 - r)(2 + \sin\theta)$
     (iii) $\delta(r, \theta) = (r + 4)(2 + \cos\theta)$
   (c) Estimate the population using your answers to parts (a) and (b).

Figure 16.33

22. **(a)** For $a > 0$, find the volume under the graph of $z = e^{-(x^2+y^2)}$ above the disk $x^2 + y^2 \leq a^2$.
(b) What happens to the volume as $a \to \infty$?

23. A disk of radius 5 cm has density 10 gm/cm^2 at its center, density 0 at its edge, and its density is a linear function of the distance from the center. Find the mass of the disk.

24. A watch spring lies flat on the table. It is made of a coiled steel strip standing a height of 0.2 inches above the table. The inner edge is the spiral $r = 0.25 + 0.04\theta$, where $0 \leq \theta \leq 4\pi$ (so the spiral makes two complete turns). The outer edge is given by $r = 0.26 + 0.04\theta$. Find the volume of the spring.

25. Electric charge is distributed over the xy-plane, with density inversely proportional to the distance from the origin. Show that the total charge inside a circle of radius R centered at the origin is proportional to R. What is the constant of proportionality?

26. Two circular disks, each of radius 1, have centers which are 1 unit apart. Write, but do not evaluate, a double integral, including limits of integration, giving the area of overlap of the disks in

(a) Cartesian coordinates **(b)** Polar coordinates

16.5 INTEGRALS IN CYLINDRICAL AND SPHERICAL COORDINATES

Some double integrals are easier to evaluate in polar, rather than Cartesian, coordinates. Similarly, some triple integrals are easier in non-Cartesian coordinates.

Cylindrical Coordinates

The cylindrical coordinates of a point (x, y, z) in 3-space are obtained by representing the x and y coordinates in polar coordinates and letting the z-coordinate be the z-coordinate of the Cartesian coordinate system. (See Figure 16.34.)

Relation between Cartesian and Cylindrical Coordinates

Each point in 3-space is represented using $0 \leq r < \infty, 0 \leq \theta \leq 2\pi, -\infty < z < \infty$.

$$x = r\cos\theta,$$
$$y = r\sin\theta,$$
$$z = z.$$

As with polar coordinates in the plane, note that $x^2 + y^2 = r^2$.

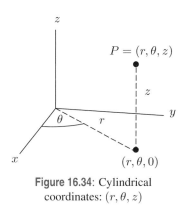

Figure 16.34: Cylindrical coordinates: (r, θ, z)

A useful way to visualize cylindrical coordinates is to sketch the surfaces obtained by setting one of the coordinates equal to a constant. See Figures 16.35–16.37.

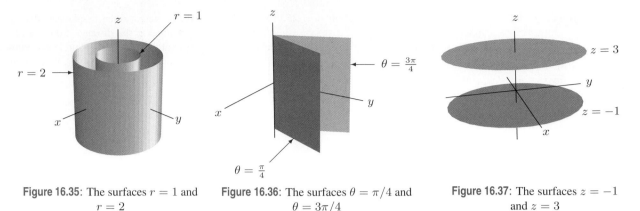

Figure 16.35: The surfaces $r = 1$ and $r = 2$

Figure 16.36: The surfaces $\theta = \pi/4$ and $\theta = 3\pi/4$

Figure 16.37: The surfaces $z = -1$ and $z = 3$

Setting $r = c$ (where c is constant) gives a cylinder around the z-axis whose radius is c. Setting $\theta = c$ gives a half-plane perpendicular to the xy plane, with one edge along the z-axis, making an angle c with the x-axis. Setting $z = c$ gives a horizontal plane $|c|$ units from the xy-plane. We call these *fundamental surfaces*.

The regions that can most easily be described in cylindrical coordinates are those regions whose boundaries are such fundamental surfaces. (For example, vertical cylinders, or wedge-shaped parts of vertical cylinders.)

Example 1   Describe in cylindrical coordinates a wedge of cheese cut from a cylinder 4 cm high and 6 cm in radius; this wedge subtends an angle of $\pi/6$ at the center. (See Figure 16.38.)

Solution   The wedge is described by the inequalities $0 \leq r \leq 6$, and $0 \leq z \leq 4$, and $0 \leq \theta \leq \pi/6$.

Figure 16.38: A wedge of cheese

Integration in Cylindrical Coordinates

To integrate in polar coordinates, we had to express the area element dA in terms of polar coordinates: $dA = r\,dr\,d\theta$. To evaluate a triple integral $\int_W f\,dV$ in cylindrical coordinates, we need to express the volume element dV in cylindrical coordinates.

Consider the volume element ΔV, shown in Figure 16.39. It is a wedge bounded by fundamental surfaces. The area of the base is $\Delta A \approx r\Delta r\Delta\theta$. Since the height is Δz, the volume element is given approximately by $\Delta V \approx r\,\Delta r\,\Delta\theta\,\Delta z$.

> When computing integrals in cylindrical coordinates, put $dV = r\,dr\,d\theta\,dz$. Other orders of integration are also possible.

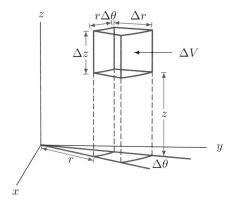

Figure 16.39: Volume element in cylindrical coordinates

Example 2   Find the mass of the wedge of cheese in Example 1, if its density is 1.2 grams/cm^3.

Solution   If the wedge is W, its mass is

$$\int_W 1.2 \, dV.$$

In cylindrical coordinates this integral is

$$\int_0^4 \int_0^{\pi/6} \int_0^6 1.2 \, r \, dr \, d\theta \, dz = \int_0^4 \int_0^{\pi/6} 0.6 r^2 \Big|_0^6 \, d\theta \, dz = 21.6 \int_0^4 \int_0^{\pi/6} d\theta \, dz$$

$$= 21.6 \left(\frac{\pi}{6}\right) 4 \approx 45.24 \text{ grams.}$$

Example 3   A water tank in the shape of a hemisphere has radius a; its base is its plane face. Find the volume, V, of water in the tank as a function of h, the depth of the water.

Solution   In Cartesian coordinates a sphere of radius a has the equation $x^2 + y^2 + z^2 = a^2$. (See Figure 16.40.) In cylindrical coordinates, $r^2 = x^2 + y^2$, so this becomes

$$r^2 + z^2 = a^2.$$

Thus, if we want to describe the amount of water in the tank in cylindrical coordinates, we let r go from 0 to $\sqrt{a^2 - z^2}$, we let θ go from 0 to 2π, and we let z go from 0 to h, giving

$$\begin{aligned} \text{Volume of water} &= \int_W dV = \int_0^{2\pi} \int_0^h \int_0^{\sqrt{a^2-z^2}} r \, dr \, dz \, d\theta = \int_0^{2\pi} \int_0^h \frac{r^2}{2} \Big|_{r=0}^{r=\sqrt{a^2-z^2}} dz \, d\theta \\ &= \int_0^{2\pi} \int_0^h \frac{1}{2}(a^2 - z^2) \, dz \, d\theta = \int_0^{2\pi} \frac{1}{2}\left(a^2 z - \frac{z^3}{3}\right) \Big|_{z=0}^{z=h} d\theta \\ &= \int_0^{2\pi} \frac{1}{2}\left(a^2 h - \frac{h^3}{3}\right) d\theta = \pi\left(a^2 h - \frac{h^3}{3}\right). \end{aligned}$$

Figure 16.40: Hemispherical water tank with radius a and water of depth h

Spherical Coordinates

In Figure 16.41, the point P has coordinates (x, y, z) in the Cartesian coordinate system. We define spherical coordinates ρ, ϕ, and θ for P as follows: $\rho = \sqrt{x^2 + y^2 + z^2}$ is the distance of P from the origin; ϕ is the angle between the positive z-axis and the line through the origin and the point P; and θ is the same as in cylindrical coordinates.

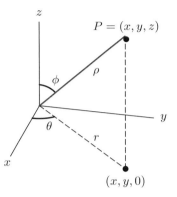

Figure 16.41: Spherical coordinates

In cylindrical coordinates,

$$x = r\cos\theta, \quad \text{and} \quad y = r\sin\theta, \quad \text{and} \quad z = z.$$

From Figure 16.41 we have $z = \rho\cos\phi$ and $r = \rho\sin\phi$, giving the following relationship:

Relation between Cartesian and Spherical Coordinates

Each point in 3-space is represented using $0 \le \rho < \infty, 0 \le \phi \le \pi$, and $0 \le \theta \le 2\pi$.

$$x = \rho\sin\phi\cos\theta$$
$$y = \rho\sin\phi\sin\theta$$
$$z = \rho\cos\phi.$$

Also, $\rho^2 = x^2 + y^2 + z^2$.

This system of coordinates is useful when there is spherical symmetry with respect to the origin, either in the region of integration or in the integrand. The fundamental surfaces in spherical coordinates are $\rho = k$ (a constant), which is a sphere of radius k centered at the origin, $\theta = k$ (a constant), which is the half-plane with its edge along the z-axis, and $\phi = k$ (a constant), which is a cone if $k \ne \pi/2$ and the xy-plane if $k = \pi/2$. (See Figures 16.42–16.44.)

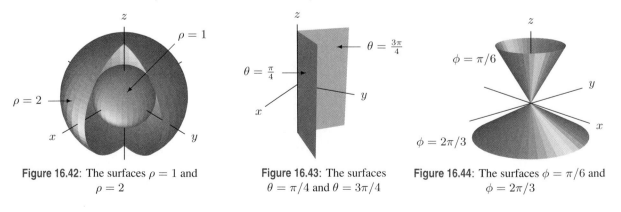

Figure 16.42: The surfaces $\rho = 1$ and $\rho = 2$

Figure 16.43: The surfaces $\theta = \pi/4$ and $\theta = 3\pi/4$

Figure 16.44: The surfaces $\phi = \pi/6$ and $\phi = 2\pi/3$

Integration in Spherical Coordinates

To use spherical coordinates in triple integrals we need to express the volume element, dV, in spherical coordinates. From Figure 16.45, we see that the volume element can be approximated by a box with curved edges. One edge has length $\Delta\rho$. The edge parallel to the xy-plane is an arc of a circle made from rotating the cylindrical radius r $(= \rho\sin\phi)$ through an angle $\Delta\theta$, and so has length $\rho\sin\phi\,\Delta\theta$. The remaining side comes from rotating the radius ρ through an angle $\Delta\phi$, and so has length $\rho\,\Delta\phi$. Therefore, $\Delta V \approx \Delta\rho(\rho\,\Delta\phi)(\rho\sin\phi\,\Delta\theta) = \rho^2\sin\phi\,\Delta\rho\,\Delta\phi\,\Delta\theta$.

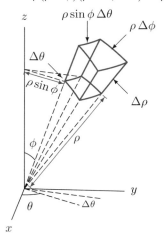

Figure 16.45: Volume element in spherical coordinates

Thus,

> When computing integrals in spherical coordinates, put $dV = \rho^2\sin\phi\,d\rho\,d\phi\,d\theta$. Other orders of integration are also possible.

Example 4   Use spherical coordinates to derive the formula for the volume of a solid sphere of radius a.

Solution   In spherical coordinates, a sphere of radius a is described by the inequalities $0 \le \rho \le a$, $0 \le \theta \le 2\pi$, and $0 \le \phi \le \pi$. Note that θ goes all the way around the circle, whereas ϕ only goes from 0 to π. We find the volume by integrating the constant density function 1 over the sphere:

$$\text{Volume} = \int_R 1\,dV = \int_0^{2\pi}\int_0^{\pi}\int_0^a \rho^2\sin\phi\,d\rho\,d\phi\,d\theta = \int_0^{2\pi}\int_0^{\pi}\frac{1}{3}a^3\sin\phi\,d\phi\,d\theta$$

$$= \frac{1}{3}a^3\int_0^{2\pi} -\cos\phi\Big|_0^{\pi}\,d\theta = \frac{2}{3}a^3\int_0^{2\pi}d\theta = \frac{4\pi a^3}{3}.$$

Example 5   Find the magnitude of the gravitational force exerted by a solid hemisphere of radius a and constant density δ on a unit mass located at the center of the base of the hemisphere.

Solution   Assume the base of the hemisphere rests on the xy-plane with center at the origin. (See Figure 16.46.) Newton's law of gravitation says that the force between two masses m_1 and m_2 at a distance r apart is $F = Gm_1m_2/r^2$. In this example, symmetry shows that the net component of the force on the particle at the origin due to the hemisphere is in the z direction only. Any force in the x or y direction from some part of the hemisphere will be canceled by the force from another part of the hemisphere directly opposite the first. To compute the net z component of the gravitational force, we imagine a small piece of the hemisphere with volume ΔV, located at spherical coordinates (ρ, θ, ϕ). This piece has mass $\delta\Delta V$, and exerts a force of magnitude F on the unit mass at the origin. The z-component of this force is given by its projection onto the z-axis, which can be seen from the figure to be $F \cos \phi$. The distance from the mass $\delta \Delta V$ to the unit mass at the origin is the spherical coordinate ρ. Therefore the z-component of the force due to the small piece ΔV is

$$\begin{matrix} z\text{-component} \\ \text{of force} \end{matrix} = \frac{G(\delta \Delta V)(1)}{\rho^2} \cos \phi.$$

Adding the contributions of the small pieces, we get a vertical force with magnitude

$$F = \int_0^{2\pi} \int_0^{\pi/2} \int_0^a \left(\frac{G\delta}{\rho^2}\right) (\cos \phi) \rho^2 \sin \phi \, d\rho \, d\phi \, d\theta = \int_0^{2\pi} \int_0^{\pi/2} G\delta(\cos \phi \sin \phi)\rho \Big|_{\rho=0}^{\rho=a} \, d\phi \, d\theta$$

$$= \int_0^{2\pi} \int_0^{\pi/2} G\delta a \cos \phi \sin \phi \, d\phi \, d\theta = \int_0^{2\pi} G\delta a \left(-\frac{(\cos \phi)^2}{2}\right) \Big|_{\phi=0}^{\phi=\pi/2} \, d\theta$$

$$= \int_0^{2\pi} G\delta a(\tfrac{1}{2}) \, d\theta = G\delta a\pi.$$

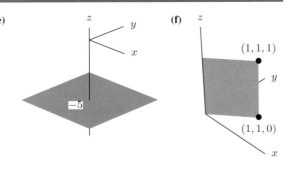

z-component of force

Force, F, due to mass δdV

Unit mass

Figure 16.46: Gravitational force of hemisphere on mass at origin

Exercises and Problems for Section 16.5

Exercises

1. Using rectangular, cylindrical, or spherical coordinates, write an equation for the following surfaces. Each equation should be of the form "Coordinate = Constant."

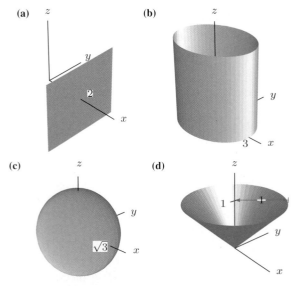

In Exercises 2–3, evaluate the triple integrals in cylindrical coordinates over the region W.

2. $f(x, y, z) = x^2 + y^2 + z^2$, W is the region $0 \leq r \leq 4$, $\pi/4 \leq \theta \leq 3\pi/4$, $-1 \leq z \leq 1$.

3. $f(x, y, z) = \sin(x^2 + y^2)$, W is the solid cylinder with height 4 and with base of radius 1 centered on the z axis at $z = -1$.

In Exercises 4–5, evaluate the triple integrals in spherical coordinates.

4. $f(x, y, z) = 1/(x^2 + y^2 + z^2)^{1/2}$ over the bottom half of the sphere of radius 5 centered at the origin.

5. $f(\rho, \theta, \phi) = \sin\phi$, over the region $0 \leq \theta \leq 2\pi$, $0 \leq \phi \leq \pi/4$, $1 \leq \rho \leq 2$.

For Exercises 6–12, choose coordinates and set up a triple integral, including limits of integration, for a density function f over the given region.

6.

7.

8.

9.

10. A piece of a sphere; angle at the center is $\pi/3$.

11.

12.

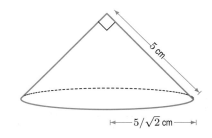

13. Sketch the region R over which the integration is being performed:

$$\int_0^{\pi/2} \int_{\pi/2}^{\pi} \int_0^1 f(\rho, \phi, \theta)\rho^2 \sin\phi \, d\rho \, d\phi \, d\theta.$$

14. (a) Convert the following triple integral to spherical coordinates:

$$\int_0^{2\pi} \int_0^3 \int_0^r r \, dz dr d\theta.$$

(b) Evaluate either the original integral or your answer to part (a).

Problems

If W is the region in Figure 16.47, what are the limits of integration in Problems 15–17?

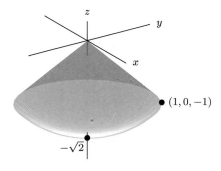

Figure 16.47: Cone with spherical cap

15. $\displaystyle\int_?^? \int_?^? \int_?^? f(\rho, \phi, \theta)\rho^2 \sin\phi \, d\rho \, d\phi \, d\theta$

16. $\displaystyle\int_?^? \int_?^? \int_?^? g(r, \theta, z)r \, dz \, dr \, d\theta$

17. $\displaystyle\int_?^? \int_?^? \int_?^? h(x, y, z) \, dz \, dy \, dx$

18. Write a triple integral representing the volume of the cone in Figure 16.48 and evaluate it.

Figure 16.48

For the regions W shown in Problems 19–21, write the limits of integration for $\int_W dV$ in the following coordinates:

(a) Cartesian   (b) Cylindrical   (c) Spherical

19.

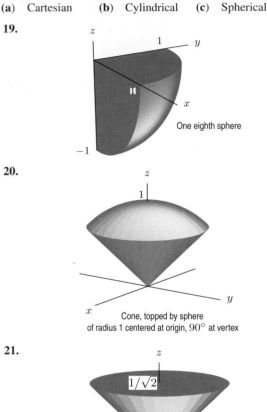

One eighth sphere

20.

Cone, topped by sphere
of radius 1 centered at origin, $90°$ at vertex

21.

Cone, flat on top,
$\pi/2$ at vertex

Evaluate the integrals in Problems 22–23.

22. $\displaystyle\int_0^1 \int_{-1}^1 \int_{-\sqrt{1-x^2}}^{\sqrt{1-x^2}} \frac{1}{(x^2+y^2)^{1/2}}\, dy\, dx\, dz$

23. $\displaystyle\int_0^1 \int_{-\sqrt{1-x^2}}^{\sqrt{1-x^2}} \int_{-\sqrt{1-x^2-z^2}}^{\sqrt{1-x^2-z^2}} \frac{1}{(x^2+y^2+z^2)^{1/2}}\, dy\, dz\, dx$

24. Suppose W is the region outside the cylinder $x^2+y^2=1$ and inside the sphere $x^2+y^2+z^2=2$. Calculate

$$\int_W (x^2+y^2)\, dV.$$

Without performing the integration, decide whether each of the integrals in Problems 25–26 is positive, negative, or zero. Give reasons for your decision.

25. W_1 is the unit ball, $x^2+y^2+z^2 \le 1$.

(a) $\int_{W_1} \sin\phi\, dV$ (b) $\int_{W_1} \cos\phi\, dV$

26. W_2 is $0 \le z \le \sqrt{1-x^2-y^2}$, the top half of the unit ball.

(a) $\int_{W_2} (z^2 - z)\, dV$ (b) $\int_{W_2} (-xz)\, dV$

27. Write a triple integral representing the volume of a slice of the cylindrical cake of height 2 and radius 5 between the planes $\theta = \pi/6$ and $\theta = \pi/3$. Evaluate this integral.

28. Write a triple integral representing the volume above the cone $z = \sqrt{x^2+y^2}$ and below the sphere of radius 2 centered at the origin. Include limits of integration but do not evaluate. Use:

(a) Cylindrical coordinates
(b) Spherical coordinates

29. Write a triple integral representing the volume of the region between spheres of radius 1 and 2, both centered at the origin. Include limits of integration but do not evaluate. Use:

(a) Spherical coordinates.
(b) Cylindrical coordinates. Write your answer as the difference of two integrals.

30. Figure 16.49 shows part of a spherical ball of radius 5 cm. Write an integral in cylindrical coordinates representing the volume of this region and evaluate it.

2 cm

Figure 16.49

31. Figure 16.49 shows part of a spherical ball of radius 5 cm. Write an integral in spherical coordinates representing the volume of this region and evaluate it.

32. Find the mass M of the solid region W given in spherical coordinates by $0 \le \rho \le 3, 0 \le \theta < 2\pi, 0 \le \phi \le \pi/4$. The density, $\delta(P)$, at any point P is given by the distance of P from the origin.

33. A spherical cloud of gas of radius 3 km is more dense at the center than toward the edge. At a distance ρ km from the center, the density is $\delta(\rho) = 3 - \rho$. Write an integral representing the total mass of the cloud of gas, and evaluate it.

34. A spherical shell centered at the origin has an inner radius of 6 cm and an outer radius of 7 cm. The density, δ, of the material increases linearly with the distance from the center. At the inner surface, $\delta = 9$ gm/cm^3; at the outer surface, $\delta = 11$ gm/cm^3.

(a) Using spherical coordinates, write the density, δ, as a function of radius, ρ.
(b) Write an integral giving the mass of the shell.
(c) Find the mass of the shell.

35. Use appropriate coordinates to find the average distance to the origin for points in the ice cream cone region bounded by the hemisphere $z = \sqrt{8 - x^2 - y^2}$ and the cone $z = \sqrt{x^2 + y^2}$. [Hint: The volume of this region is computed in Problem 20 on page 806.]

36. **(a)** Write an iterated integral which represents the mass of a solid ball of radius a. The density at each point in the ball is k times the distance from that point to a fixed plane passing through the center of the ball.
   (b) Evaluate the integral.

37. A bead is made by drilling a cylindrical hole of radius 1 mm through a sphere of radius 5 mm. See Figure 16.50.

   (a) Set up a triple integral in cylindrical coordinates representing the volume of the bead.
   (b) Evaluate the integral.

1 mm

5 mm

Figure 16.50

38. A half-melon is approximated by the region between two concentric spheres, one of radius a and the other of radius b, with $0 < a < b$. Write a triple integral, including limits of integration, giving the volume of the half-melon. Evaluate the integral.

For Problems 39–42, use the definition of center of mass given on page 802.

39. Let C be a solid cone with both height and radius 1 and contained between the surfaces $z = \sqrt{x^2 + y^2}$ and $z = 1$. If C has constant mass density of 1 gm/cm^3, find the z-coordinate of C's center of mass.

40. The density of the cone C in Problem 39 is given by $\delta(z) = z^2$ gm/cm^3. Find
   (a) The mass of C.
   (b) The z-coordinate of C's center of mass.

41. For $a > 0$, consider the family of solids bounded below by the paraboloid $z = a(x^2 + y^2)$ and above by the plane $z = 1$. If the solids all have constant mass density 1 gm/cm^3, show that the z-coordinate of the center of mass is $2/3$ and so independent of the parameter a.

42. Find the location of the center of mass of a hemisphere of radius a and density b gm/cm^3.

For Problems 43–44, use the definition of moment of inertia given on page 803.

43. The moment of inertia of a solid homogeneous ball B of mass 1 and radius a centered at the origin is the same about any of the coordinate axes (due to the symmetry of the ball). It is easier to evaluate the sum of the three integrals involved in computing the moment of inertia about each of the axes than to compute them individually. Find the sum of the moments of inertia about the x, y and z-axes and thus find the individual moments of inertia.

44. Find the moment of inertia about the z axis of the solid "fat ice cream cone" given in spherical coordinates by $0 \le \rho \le a, 0 \le \phi \le \frac{\pi}{3}$ and $0 \le \theta \le 2\pi$. Assume that the solid is homogeneous with mass m.

45. Compute the force of gravity exerted by a solid cylinder of radius R, height H, and constant density δ on a unit mass at the center of the base of the cylinder.

46. Figure 16.51 shows an alternative notation for spherical coordinates, used often in electrical engineering. Write the volume element dV in this coordinate system.

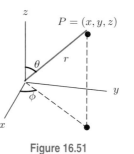

Figure 16.51

47. Electric charge is distributed throughout 3-space, with density proportional to the distance from the xy-plane. Show that the total charge inside a cylinder of radius R and height h, sitting on the xy-plane and centered along the z-axis, is proportional to $R^2 h^2$.

48. Electric charge is distributed throughout 3-space with density inversely proportional to the distance from the origin. Show that the total charge inside a sphere of radius R is proportional to R^2.

Problems 49–50 deal with the energy stored in an electric field. If a region of space W contains an electric field whose magnitude at a point (x, y, z) is $E(x, y, z)$, then

$$\text{Energy stored by field} = \frac{1}{2} \int_W \epsilon E^2 \, dV,$$

where ϵ is a property of the material called the *permittivity*.

49. The region between two concentric spheres, with radii $a < b$, contains an electric field with magnitude $E = q/(4\pi\epsilon\rho^2)$, where ρ is the distance from the center of the spheres and q is the charge on the inner sphere. Assuming the permittivity, ϵ, is constant, find the total energy stored in the region between the two spheres.

50. Figure 16.52 shows a coaxial cable consisting of two cylindrical conductors centered on the same axis, of radii $a < b$. The electric field between the conductors has magnitude $E = q/(2\pi\epsilon r)$, where r is the distance from the axis and q is the charge per unit length on the cable. The permittivity of the material between the conductors is constant.[5] Show that the stored energy per unit length is proportional to $\ln(b/a)$.

Figure 16.52

16.6   APPLICATIONS OF INTEGRATION TO PROBABILITY

To represent how a quantity such as height or weight is distributed throughout a population, we use a density function. To study two or more quantities at the same time and see how they are related, we use a multivariable density function.

Density Functions

Distribution of Weight and Height in Expectant Mothers

Table 16.9 shows the distribution of weight and height in a survey of expectant mothers. The histogram in Figure 16.53 is constructed in such a way that the volume of each bar represents the percentage in the corresponding weight and height range. For example, the bar representing the mothers who weighed 60–70 kg and were 160–165 cm tall has base of area $10 \text{ kg} \cdot 5 \text{ cm} = 50 \text{ kg cm}$. The volume of this bar is 12%, so its height is $12\%/50 \text{ kg cm} = 0.24\%/\text{ kg cm}$. Notice that the units on the vertical axis are percent/ kg cm. Thus, volumes under the histogram are in units of %. The total volume is 100% = 1.

Table 16.9   *Distribution of weight and height in a survey of expectant mothers*

	45-50 kg	50-60 kg	60-70 kg	70-80 kg	80-105 kg	Totals by height
150-155 cm	2	4	4	2	1	13
155-160 cm	0	12	8	2	1	23
160-165 cm	1	7	12	4	3	27
165-170 cm	0	8	12	6	2	28
170-180 cm	0	1	3	4	1	9
Totals by weight	3	32	39	18	8	100

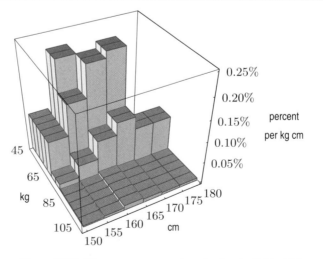

Figure 16.53: Histogram representing the data in Table 16.9.

[5]See C. R. Paul and S. A. Nasar, *Introduction to Electromagnetic Fields*, 2nd ed. (New York: McGraw-Hill, 1987).

Example 1   Find the percentage of mothers in the survey with height between 170 and 180 cm.

Solution   We add the percentages across the row corresponding to the 170–180 cm height range; this is equivalent to adding the volumes of the corresponding rectangular solids in the histogram.

$$\text{Percentage of mothers} = 0 + 1 + 3 + 4 + 1 = 9\%.$$

Smoothing the Histogram

If we have smaller weight and height groups (and a larger sample), we can draw a smoother histogram and get finer estimates. In the limit, we replace the histogram with a smooth surface, in such a way that the volume under the surface above a rectangle is the percentage of mothers in that rectangle. We define a *density function*, $p(w, h)$, to be the function whose graph is the smooth surface. It has the property that

$$
\begin{array}{ccc}
\text{Fraction of sample with} & & \text{Volume under graph of } p \\
\text{weight between } a \text{ and } b \text{ and} & = & \text{over the rectangle} & = \int_a^b \int_c^d p(w, h)\, dh\, dw. \\
\text{height between } c \text{ and } d & & a \le w \le b, c \le h \le d
\end{array}
$$

Joint Density Functions

We generalize this idea to represent any two characteristics, x and y, distributed throughout a population.

A function $p(x, y)$ is called a **joint density function** for x and y if

$$
\begin{array}{ccc}
\text{Fraction of population with} & & \text{Volume under graph of } p \\
x \text{ between } a \text{ and } b \text{ and} & = & \text{above the rectangle} & = \int_a^b \int_c^d p(x, y)\, dy\, dx \\
y \text{ between } c \text{ and } d & & a \le x \le b, c \le y \le d
\end{array}
$$

where

$$\int_{-\infty}^{\infty} \int_{-\infty}^{\infty} p(x, y)\, dy\, dx = 1 \quad \text{and} \quad p(x, y) \ge 0 \text{ for all } x \text{ and } y.$$

A joint density function need not be continuous, as in Example 2 which follows. In addition, as in Example 4, the integrals involved may be improper and must be computed by methods similar to those used for improper one-variable integrals.

Example 2   Let $p(x, y)$ be defined on the square $0 \le x \le 1, 0 \le y \le 1$ by $p(x, y) = x + y$; let $p(x, y) = 0$ if (x, y) is outside this square. Verify that p is a joint density function. In terms of the distribution of x and y in the population, what does it mean that $p(x, y) = 0$ outside the square?

Solution   First, we have $p(x, y) \ge 0$ for all x and y. To verify that p is a joint density function, we check that the total volume under the graph is 1:

$$\int_{-\infty}^{\infty} \int_{-\infty}^{\infty} p(x, y)\, dy\, dx = \int_0^1 \int_0^1 (x + y)\, dy\, dx$$

$$= \int_0^1 \left(xy + \frac{y^2}{2} \right) \Bigg|_0^1 dx = \int_0^1 \left(x + \frac{1}{2} \right) dx = \left(\frac{x^2}{2} + \frac{x}{2} \right) \Bigg|_0^1 = 1$$

The fact that $p(x, y) = 0$ outside the square means that the variables x and y never take values outside the interval $[0, 1]$, that is, the value of x and y for any individual in the population is always between 0 and 1.

Example 3   Suppose two variables x and y are distributed in a population according to the density function of Example 2. Find the fraction of the population with $x \leq 1/2$, the fraction with $y \leq 1/2$, and the fraction with both $x \leq 1/2$ and $y \leq 1/2$.

Solution   The fraction with $x \leq 1/2$ is the volume under the graph to the left of the line $x = 1/2$:

$$\int_0^{1/2} \int_0^1 (x+y)\, dy\, dx = \int_0^{1/2} \left(xy + \frac{y^2}{2}\right)\Big|_0^1 dx = \int_0^{1/2} \left(x + \frac{1}{2}\right) dx$$

$$= \left(\frac{x^2}{2} + \frac{x}{2}\right)\Big|_0^{1/2} = \frac{1}{8} + \frac{1}{4} = \frac{3}{8}.$$

Since the function is symmetric in x and y, the fraction with $y \leq 1/2$ is also $3/8$. Finally, the fraction with both $x \leq 1/2$ and $y \leq 1/2$ is

$$\int_0^{1/2} \int_0^{1/2} (x+y)\, dy\, dx = \int_0^{1/2} \left(xy + \frac{y^2}{2}\right)\Big|_0^{1/2} dx = \int_0^{1/2} \left(\frac{1}{2}x + \frac{1}{8}\right) dx$$

$$= \left(\frac{1}{4}x^2 + \frac{1}{8}x\right)\Big|_0^{1/2} = \frac{1}{16} + \frac{1}{16} = \frac{1}{8}$$

Recall that a one-variable density function $p(x)$ is a function such that $p(x) \geq 0$ for all x, and $\int_{-\infty}^\infty p(x)\, dx = 1$.

Example 4   Let p_1 and p_2 be one-variable density functions for x and y, respectively. Verify that $p(x,y) = p_1(x)p_2(y)$ is a joint density function.

Solution   Since both p_1 and p_2 are density functions, they are nonnegative everywhere. Thus, their product $p_1(x)p_2(x) = p(x,y)$ is nonnegative everywhere. Now we must check that the volume under the graph of p is 1. Since $\int_{-\infty}^\infty p_2(y)\, dy = 1$ and $\int_{-\infty}^\infty p_1(x)\, dx = 1$, we have

$$\int_{-\infty}^\infty \int_{-\infty}^\infty p(x,y)\, dy\, dx = \int_{-\infty}^\infty \int_{-\infty}^\infty p_1(x)p_2(y)\, dy\, dx = \int_{-\infty}^\infty p_1(x)\left(\int_{-\infty}^\infty p_2(y)\, dy\right) dx$$

$$= \int_{-\infty}^\infty p_1(x)(1)\, dx = \int_{-\infty}^\infty p_1(x)\, dx = 1.$$

Density Functions and Probability

What is the probability that an expectant mother weighs 60–70 kg and is 155–160 cm tall? Table 16.9 shows that 8% of mothers fall in this group, so the probability that a randomly chosen mother falls in this group is 0.08.

$$\begin{array}{ccc} \text{Probability that a mother} & \text{Volume under graph of } p & \\ \text{has weight between } a \text{ and } b & = \quad \text{above the rectangle} & = \int_a^b \int_c^d p(w,h)\, dh\, dw. \\ \text{and height between } c \text{ and } d & a \leq w \leq b, c \leq h \leq d & \end{array}$$

For a joint density function $p(x,y)$, the probability that x falls in an interval of width Δx around x_0 and y falls in an interval of width Δy around y_0 is approximately $p(x_0, y_0)\Delta x \Delta y$. Thus, p is often called a *probability density function*.

Example 5   A machine in a factory is set to produce components 10 cm long and 5 cm in diameter. In fact, there is a slight variation from one component to the next. A component is usable if its length and diameter deviate from the correct values by less than 0.1 cm. If the length is x cm and the diameter is y cm, the probability density function for the variation in x and y is

$$p(x,y) = \frac{50\sqrt{2}}{\pi} e^{-100(x-10)^2} e^{-50(y-5)^2}.$$

What is the probability that a component is usable? (See Figure 16.54.)

Figure 16.54: The density function $p(x,y) = \frac{50\sqrt{2}}{\pi} e^{-100(x-10)^2} e^{-50(y-5)^2}$.

Solution   We know that

Probability that x and y satisfy
$x_0 - \Delta x \le x \le x_0 + \Delta x$
$y_0 - \Delta y \le y \le y_0 + \Delta y$
$$= \frac{50\sqrt{2}}{\pi} \int_{y_0-\Delta y}^{y_0+\Delta y} \int_{x_0-\Delta x}^{x_0+\Delta x} e^{-100(x-10)^2} e^{-50(y-5)^2}\, dx\, dy.$$

Thus,

Probability that component is usable
$$= \frac{50\sqrt{2}}{\pi} \int_{4.9}^{5.1} \int_{9.9}^{10.1} e^{-100(x-10)^2} e^{-50(y-5)^2}\, dx\, dy.$$

The double integral must be evaluated numerically. This yields

Probability that component is usable
$$= \frac{50\sqrt{2}}{\pi}(0.02556) \approx 0.57530.$$

Thus, there is a 57.5% chance that the component is usable.

Exercises and Problems for Section 16.6

Exercises

In Exercises 1–8, let p be the joint density function such that $p(x,y) = xy$ in R, the rectangle $0 \le x \le 2, 0 \le y \le 1$, and $p(x,y) = 0$ outside R. Find the fraction of the population satisfying the given constraints.

1. $x \ge 3$

2. $x = 1$

3. $x + y \le 3$

4. $-1 \le x \le 1$

5. $x \ge y$

6. $x + y \le 1$

7. $0 \le x \le 1, 0 \le y \le 1/2$

8. Within a distance 1 from the origin

9. Table 16.10 gives values of a joint density function for x and y. Assume x takes the values 1, 2, 3, 4 and y takes the values 1, 2, 3.

(a) Explain why this table defines a joint density function.
(b) What is the probability that $x = 2$?
(c) Find the probability that $y \le 2$.
(d) Find the probability that $x \le 3$ and $y \le 2$.

Table 16.10

		y		
		1	2	3
x	1	0.3	0.2	0.1
	2	0.2	0.1	0
	3	0.1	0	0
	4	0	0	0

In Exercises 10–15, check whether p is a joint density function. Assume $p(x, y) = 0$ outside the region R.

10. $p(x, y) = 1$, where R is $0 \le x \le 1, 0 \le y \le 2$

11. $p(x, y) = 1/2$, where R is $4 \le x \le 5, -2 \le y \le 0$

12. $p(x, y) = x + y$, where R is $-1 \le x \le 1, 0 \le y \le 1$

13. $p(x, y) = 6(y - x)$, where R is $0 \le x \le y \le 2$

14. $p(x, y) = (2/\pi)(1 - x^2 - y^2)$, where R is $x^2 + y^2 \le 1$

15. $p(x, y) = xye^{-x-y}$, where R is $x \ge 0, y \ge 0$

16. Let x and y have joint density function

$$p(x, y) = \begin{cases} \frac{2}{3}(x + 2y) & \text{for } 0 \le x \le 1, 0 \le y \le 1, \\ 0 & \text{otherwise.} \end{cases}$$

Find the probability that

(a) $x > 1/3$.

(b) $x < (1/3) + y$.

Problems

17. The joint density function for x, y is given by

$$f(x, y) = \begin{cases} kxy & \text{for } 0 \le x \le y \le 1, \\ 0 & \text{otherwise.} \end{cases}$$

(a) Determine the value of k.

(b) Find the probability that (x, y) lies in the shaded region in Figure 16.55.

Figure 16.55

18. A joint density function is given by

$$f(x, y) = \begin{cases} kx^2 & \text{for } 0 \le x \le 2 \text{ and } 0 \le y \le 1, \\ 0 & \text{otherwise.} \end{cases}$$

(a) Find the value of the constant k.

(b) Find the probability that (x, y) satisfies $x + y \le 2$.

(c) Find the probability that (x, y) satisfies $x \le 1$ and $y \le 1/2$.

19. A health insurance company wants to know what proportion of its policies are going to cost them a lot of money because the insured people are over 65 and sick. In order to compute this proportion, the company defines a *disability index*, x, with $0 \le x \le 1$, where $x = 0$ represents perfect health and $x = 1$ represents total disability. In addition, the company uses a density function, $f(x, y)$, defined in such a way that the quantity

$$f(x, y) \, \Delta x \, \Delta y$$

approximates the fraction of the population with disability index between x and $x + \Delta x$, and aged between y and $y + \Delta y$. The company knows from experience that a policy no longer covers its costs if the insured person is over 65 and has a disability index exceeding 0.8. Write an expression for the fraction of the company's policies held by people meeting these criteria.

20. A point is chosen at random from the region S in the xy-plane containing all points (x, y) such that $-1 \le x \le 1, -2 \le y \le 2$ and $x - y \ge 0$ (at random means that the density function is constant on S).

(a) Determine the joint density function for x and y.

(b) If T is a subset of S with area α, then find the probability that a point (x, y) is in T.

21. The probability that a radioactive substance will decay at time t is modeled by the density function

$$p(t) = \lambda e^{-\lambda t}$$

for $t \ge 0$, and $p(t) = 0$ for $t < 0$. The positive constant λ depends on the material, and is called the decay rate.

(a) Check that p is a density function.

(b) Consider two materials with decay rates λ and μ, which decay independently of each other. Write the joint density function for the probability that the first material decays at time t and the second at time s.

(c) Find the probability that the first substance decays before the second.

22. Figure 16.56 represents a baseball field, with the bases at $(1, 0)$, $(1, 1)$, $(0, 1)$, and home plate at $(0, 0)$. The outer bound of the outfield is a piece of a circle about the origin with radius 4. When a ball is hit by a batter we record the spot on the field where the ball is caught. Let $p(r, \theta)$ be a function in the plane that gives the density of the distribution of such spots. Write an expression that represents the probability that a hit is caught in

(a) The right field (region R).

(b) The center field (region C).

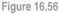

Figure 16.56

23. Two independent random numbers x and y between 0 and 1 have joint density function

$$p(x, y) = \begin{cases} 1 & \text{if } 0 \leq x, y \leq 1 \\ 0 & \text{otherwise.} \end{cases}$$

This problem concerns the average $z = (x+y)/2$, which has a one-variable probability density function of its own.

(a) Find $F(t)$, the probability that $z \leq t$. Treat separately the cases $t \leq 0, 0 < t \leq 1/2, 1/2 < t \leq 1$, $1 < t$. Note that $F(t)$ is the cumulative distribution function of z.

(b) Find and graph the probability density function of z.

(c) Are x and y more likely to be near 0, 1/2, or 1? What about z?

16.7 CHANGE OF VARIABLES IN A MULTIPLE INTEGRAL

In the previous sections, we used polar, cylindrical, and spherical coordinates to simplify iterated integrals. In this section, we discuss more general changes of variable. In the process, we will see where the extra factor of r comes from when we change from Cartesian to polar coordinates and the factor $\rho^2 \sin\phi$ when we change from Cartesian to spherical coordinates.

Polar Change of Variables Revisited

Consider the integral $\int_R (x+y)\, dA$ where R is the region in the first quadrant bounded by the circle $x^2 + y^2 = 16$ and the x and y-axes. Writing the integral in Cartesian and polar coordinates we have

$$\int_R (x+y)\, dA = \int_0^4 \int_0^{\sqrt{16-x^2}} (x+y)\, dy\, dx = \int_0^{\pi/2} \int_0^4 (r\cos\theta + r\sin\theta)r\, dr d\theta.$$

This is an integral over the rectangle in the $r\theta$-space given by $0 \leq r \leq 4$, $0 \leq \theta \leq \pi/2$. The conversion from polar to Cartesian coordinates changes this rectangle into a quarter-disk. Figure 16.57 shows how a typical rectangle (shaded) in the $r\theta$-plane with sides of length Δr and $\Delta\theta$ corresponds to a curved rectangle in the xy-plane with sides of length Δr and $r\Delta\theta$. The extra r is needed because the correspondence between r, θ and x, y not only curves the lines $r = 1, 2, 3 \ldots$ into circles, it also stretches those lines around larger and larger circles.

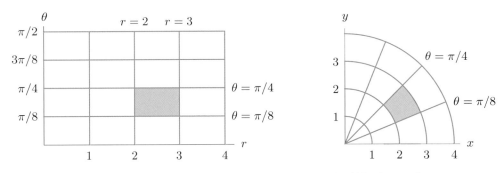

Figure 16.57: A grid in the $r\theta$-plane and the corresponding curved grid in the xy-plane

General Change of Variables

We now consider a general change of variables, where x, y coordinates are related to s, t coordinates by the differentiable functions

$$x = x(s, t) \quad y = y(s, t).$$

Just as a rectangular region in the $r\theta$-plane corresponds to a region in the xy-plane, a rectangular region, T, in the st-plane corresponds to a region, R, in the xy-plane. We assume that the change of coordinates is one-to-one, that is, that each point R corresponds to one point in T.

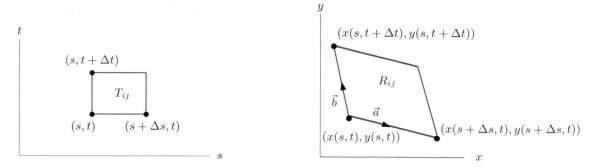

Figure 16.58: A small rectangle T_{ij} in the st-plane and the corresponding region R_{ij} of the xy-plane

We divide T into small rectangles T_{ij} with sides of length Δs and Δt. (See Figure 16.58.) The corresponding piece R_{ij} of the xy-plane is a quadrilateral with curved sides. If we choose Δs and Δt small, then by local linearity of $x(s,t)$ and $y(s,t)$, we know R_{ij} is approximately a parallelogram.

Recall from Chapter 13 that the area of the parallelogram with sides $\vec{a}$ and $\vec{b}$ is $\|\vec{a} \times \vec{b}\|$. Thus, we need to find the sides of R_{ij} as vectors. The side of R_{ij} corresponding to the bottom side of T_{ij} has endpoints $(x(s,t), y(s,t))$ and $(x(s + \Delta s, t), y(s + \Delta s, t))$, so in vector form that side is

$$\vec{a} = (x(s + \Delta s, t) - x(s, t))\vec{i} + (y(s + \Delta s, t) - y(s, t))\vec{j} \approx \left(\frac{\partial x}{\partial s}\Delta s\right)\vec{i} + \left(\frac{\partial y}{\partial s}\Delta s\right)\vec{j}.$$

Similarly, the side of R_{ij} corresponding to the left edge of T_{ij} is given by

$$\vec{b} \approx \left(\frac{\partial x}{\partial t}\Delta t\right)\vec{i} + \left(\frac{\partial y}{\partial t}\Delta t\right)\vec{j}.$$

Computing the cross product, we get

$$\text{Area } R_{ij} \approx \|\vec{a} \times \vec{b}\| \approx \left|\left(\frac{\partial x}{\partial s}\Delta s\right)\left(\frac{\partial y}{\partial t}\Delta t\right) - \left(\frac{\partial x}{\partial t}\Delta t\right)\left(\frac{\partial y}{\partial s}\Delta s\right)\right|$$

$$= \left|\frac{\partial x}{\partial s} \cdot \frac{\partial y}{\partial t} - \frac{\partial x}{\partial t} \cdot \frac{\partial y}{\partial s}\right|\Delta s\Delta t.$$

Using determinant notation,[6] we define the *Jacobian*, $\dfrac{\partial(x, y)}{\partial(s, t)}$, as follows

$$\frac{\partial(x, y)}{\partial(s, t)} = \frac{\partial x}{\partial s} \cdot \frac{\partial y}{\partial t} - \frac{\partial x}{\partial t} \cdot \frac{\partial y}{\partial s} = \begin{vmatrix} \dfrac{\partial x}{\partial s} & \dfrac{\partial x}{\partial t} \\[2mm] \dfrac{\partial y}{\partial s} & \dfrac{\partial y}{\partial t} \end{vmatrix}.$$

Thus, we can write

$$\text{Area } R_{ij} \approx \left|\frac{\partial(x, y)}{\partial(s, t)}\right|\Delta s\,\Delta t.$$

To compute $\int_R f(x, y)\,dA$, where f is a continuous function, we look at the Riemann sum obtained by dividing the region R into the small curved regions R_{ij}, giving

$$\int_R f(x, y)\,dA \approx \sum_{i,j} f(u_{ij}, v_{ij}) \cdot \text{Area of } R_{ij} \approx \sum_{i,j} f(u_{ij}, v_{ij})\left|\frac{\partial(x, y)}{\partial(s, t)}\right|\Delta s\,\Delta t.$$

---

[6]See Appendix D.

Each point (u_{ij}, v_{ij}) in R_{ij} corresponds to a point (s_{ij}, t_{ij}) in T_{ij}, so the sum can be written in terms of s and t:

$$\sum_{i,j} f(x(s_{ij}, t_{ij}), y(s_{ij}, t_{ij})) \left| \frac{\partial(x,y)}{\partial(s,t)} \right| \Delta s \, \Delta t.$$

This is a Riemann sum in terms of s and t, so as Δs and Δt approach 0, we get

$$\int_R f(x,y) \, dA = \int_T f(x(s,t), y(s,t)) \left| \frac{\partial(x,y)}{\partial(s,t)} \right| ds \, dt.$$

To convert an integral from x, y to s, t coordinates we make three changes:
1. Substitute for x and y in the integrand in terms of s and t.
2. Change the xy region R into an st region T.
3. Use the absolute value of the Jacobian to change the area element by making the substitution $dx dy = \left| \dfrac{\partial(x,y)}{\partial(s,t)} \right| ds dt.$

Example 1   Check that the Jacobian $\dfrac{\partial(x,y)}{\partial(r,\theta)} = r$ for polar coordinates $x = r\cos\theta,\ y = r\sin\theta$.

Solution   We have $\dfrac{\partial(x,y)}{\partial(r,\theta)} = \begin{vmatrix} \frac{\partial x}{\partial r} & \frac{\partial x}{\partial \theta} \\ \frac{\partial y}{\partial r} & \frac{\partial y}{\partial \theta} \end{vmatrix} = \begin{vmatrix} \cos\theta & \sin\theta \\ -r\sin\theta & r\cos\theta \end{vmatrix} = r\cos^2\theta + r\sin^2\theta = r.$

Example 2   Find the area of the ellipse $\dfrac{x^2}{a^2} + \dfrac{y^2}{b^2} = 1$.

Solution   Let $x = as,\ y = bt$. Then the ellipse $x^2/a^2 + y^2/b^2 = 1$ in the xy-plane corresponds to the circle $s^2 + t^2 = 1$ in the st-plane. The Jacobian is $\begin{vmatrix} a & 0 \\ 0 & b \end{vmatrix} = ab$. Thus, if R is the ellipse in the xy-plane and T is the unit circle in the st-plane, we get

$$\text{Area of } xy\text{-ellipse} = \int_R 1 \, dA = \int_T 1ab \, ds \, dt = ab \int_T ds \, dt = ab \cdot \text{Area of } st\text{-circle} = \pi ab.$$

Change of Variables in Triple Integrals

For triple integrals, there is a similar formula. Suppose the differentiable functions

$$x = x(s,t,u), \quad y = y(s,t,u), \quad z = z(s,t,u)$$

define a change of variables from a region S in stu-space to a region W in xyz-space. Then, the Jacobian of this change of variables is given by the determinant[7]

$$\frac{\partial(x,y,z)}{\partial(s,t,u)} = \begin{vmatrix} \frac{\partial x}{\partial s} & \frac{\partial x}{\partial t} & \frac{\partial x}{\partial u} \\ \frac{\partial y}{\partial s} & \frac{\partial y}{\partial t} & \frac{\partial y}{\partial u} \\ \frac{\partial z}{\partial s} & \frac{\partial z}{\partial t} & \frac{\partial z}{\partial u} \end{vmatrix}.$$

---

[7] See Appendix D.

Just as the Jacobian in two dimensions gives us the change in the area element, the Jacobian in three dimensions represents the change in the volume element. Thus, we have

$$\int_W f(x, y, z) \, dx \, dy \, dz = \int_S f(x(s, t, u), y(s, t, u), z(s, t, u)) \left| \frac{\partial(x, y, z)}{\partial(s, t, u)} \right| ds \, dt \, du.$$

Problem 17 at the end of this section asks you to check that the Jacobian for the change of variables to spherical coordinates is $\rho^2 \sin \phi$. The next example generalizes Example 2 to ellipsoids.

Example 3   Find the volume of the ellipsoid $\dfrac{x^2}{a^2} + \dfrac{y^2}{b^2} + \dfrac{z^2}{c^2} = 1$.

Solution   Let $x = as$, $y = bt$, $z = cu$. The Jacobian is computed to be abc. The xyz-ellipsoid corresponds to the stu-sphere $s^2 + t^2 + u^2 = 1$. Thus, as in Example 2,

$$\text{Volume of } xyz\text{-ellipsoid} = abc \cdot \text{Volume of } stu\text{-sphere} = abc\frac{4}{3}\pi = \frac{4}{3}\pi abc.$$

Exercises and Problems for Section 16.7

Exercises

In Exercises 1–4, find the absolute value of the Jacobian, $\left| \frac{\partial(x,y)}{\partial(s,t)} \right|$, for the given change of variables.

1. $x = 5s + 2t, y = 3s + t$

2. $x = s^2 - t^2, y = 2st$

3. $x = e^s \cos t, y = e^s \sin t$

4. $x = s^3 - 3st^2, y = 3s^2t - t^3$

In Exercises 5–6, find the Jacobian.

5. $\dfrac{\partial(x, y, z)}{\partial(s, t, u)}$, where $x = 3s + t + 2u, y = s + 5t - u, z = 2s - t + u$.

6. $\dfrac{\partial(x, y, z)}{\partial(r, \theta, z)}$, where $x = r \cos \theta, y = r \sin \theta, z = z$.

In Exercises 7–11, find positive numbers a and b so that the change of variables $s = ax, t = by$ transforms the integral $\int \int_R dx \, dy$ into

$$\int \int_T \left| \frac{\partial(x, y)}{\partial(s, t)} \right| ds \, dt$$

for the given regions R and T.

7. R is the rectangle $0 \le x \le 10, 0 \le y \le 1$ and T is the square $0 \le s, t \le 1$.

8. R is the rectangle $0 \le x \le 1, 0 \le y \le 1/4$ and T is the square $0 \le s, t \le 1$.

9. R is the rectangle $0 \le x \le 50, 0 \le y \le 10$ and T is the square $0 \le s, t \le 1$.

10. R is the circular disc of radius 15 centered at the origin and T is the circular disc $s^2 + t^2 \le 1$.

11. R is the elliptical region $x^2/4 + y^2/9 \le 1$ and T is the circular disc $s^2 + t^2 \le 1$.

In Exercises 12–14, find a number a so that the change of variables $s = x + ay, t = y$ transforms the integral $\int \int_R dx \, dy$ over the parallelogram R in the xy-plane into an integral

$$\int \int_T \left| \frac{\partial(x, y)}{\partial(s, t)} \right| ds \, dt$$

over a rectangle T in the st-plane.

12. R has vertices $(0, 0), (10, 0), (12, 3), (22, 3)$

13. R has vertices $(0, 0), (10, 0), (-15, 5), (-5, 5)$

14. R has vertices $(10, 15), (30, 15), (20, 35), (40, 35)$

Problems

15. Find the region R in the xy-plane corresponding to the region $T = \{(s, t) \mid 0 \le s \le 3, 0 \le t \le 2\}$ under the change of variables $x = 2s - 3t, y = s - 2t$. Check that

$$\int_R dx \, dy = \int_T \left| \frac{\partial(x, y)}{\partial(s, t)} \right| ds \, dt.$$

16. Find the region R in the xy-plane corresponding to the region $T = \{(s, t) \mid 0 \le s \le 2, s \le t \le 2\}$ under the change of variables $x = s^2, y = t$. Check that

$$\int_R dx \, dy = \int_T \left| \frac{\partial(x, y)}{\partial(s, t)} \right| ds \, dt.$$

17. Compute the Jacobian for the change of variables into spherical coordinates:

$$x = \rho \sin \phi \cos \theta, \quad y = \rho \sin \phi \sin \theta, \quad z = \rho \cos \phi.$$

18. For the change of variables $x = 3s - 4t$, $y = 5s + 2t$, show that

$$\frac{\partial(x,y)}{\partial(s,t)} \cdot \frac{\partial(s,t)}{\partial(x,y)} = 1$$

19. Use the change of variables $x = 2s + t$, $y = s - t$ to compute the integral $\int_R (x+y)\, dA$, where R is the parallelogram formed by $(0,0)$, $(3,-3)$, $(5,-2)$, and $(2,1)$.

20. Use the change of variables $x = \frac{1}{2}s$, $y = \frac{1}{3}t$ to compute the integral $\int_R (x^2 + y^2)\, dA$, where R is the region bounded by the curve $4x^2 + 9y^2 = 36$.

21. Use the change of variables $s = x + y$, $t = y$ to find the area of the ellipse $x^2 + 2xy + 2y^2 \le 1$.

22. Use the change of variables $s = x - y$, $t = x + y$ to evaluate $\int \int_R \sin(x+y)\, dx\, dy$ over the disc $x^2 + y^2 \le 1$.

23. Use the change of variables $s = y$, $t = y - x^2$ to evaluate $\int \int_R x\, dx\, dy$ over the region R in the first quadrant bounded by $y = 0$, $y = 16$, $y = x^2$, and $y = x^2 - 9$.

24. Use the change of variables $s = 1/r$, $t = \theta$ to transform the polar coordinate integral $\int_R (1/r^3) r\, dr\, d\theta$ over the infinite region $R : r \ge 1, 0 \le \theta \le 2\pi$ to an iterated integral over a finite region.

25. Use the change of variables $s = xy$, $t = xy^2$ to compute $\int_R xy^2\, dA$, where R is the region bounded by $xy = 1$, $xy = 4$, $xy^2 = 1$, $xy^2 = 4$.

26. If R is the triangle bounded by $x + y = 1$, $x = 0$, and $y = 0$, evaluate the integral $\int_R \cos\left(\frac{x-y}{x+y}\right)\, dx\, dy$.

CHAPTER SUMMARY (see also Ready Reference at the end of the book)

- **Double Integral**

  Definition as a limit of Riemann sum, interpretation as volume under graph, as area, as average value, or as total mass from density, estimating from contour diagrams or tables, evaluating using iterated integrals, setting up in polar coordinates.

- **Triple Integral**

  Definition as a limit of Riemann sum, interpretation as volume of solid, as total mass, or as average value, evaluating using iterated integrals, setting up in cylindrical or spherical coordinates.

- **Change of Variable** Polar coordinates, spherical and cylindrical coordinates, general change of variables and Jacobians

- **Probability**

  Joint density functions, using integrals to calculate probability .

REVIEW EXERCISES AND PROBLEMS FOR CHAPTER SIXTEEN

Exercises

Sketch the regions of integration in Exercises 1–5.

1. $\displaystyle\int_{-1}^{1} \int_{-\sqrt{1-x^2}}^{\sqrt{1-x^2}} f(x,y)\, dy\, dx$

2. $\displaystyle\int_{0}^{2} \int_{-\sqrt{4-y^2}}^{0} f(x,y)\, dx\, dy$

3. $\displaystyle\int_{1}^{4} \int_{-\sqrt{y}}^{\sqrt{y}} f(x,y)\, dx\, dy$

4. $\displaystyle\int_{0}^{1} \int_{0}^{\sin^{-1} y} f(x,y)\, dx\, dy$

5. $\displaystyle\int_{-1}^{1} \int_{-1}^{1} \int_{0}^{\sqrt{1-z^2}} f(x,y,z)\, dy\, dz\, dx$

Choose coordinates and write a triple integral, including limits of integration, for a function over the regions in Exercises 6–9.

6.

7.

8.

9.

10. Write $\int_R f(x, y)\, dA$ as an iterated integral if R is the region in Figure 16.59.

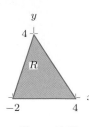

Figure 16.59

11. Evaluate $\int_R \sqrt{x^2 + y^2}\, dA$ where R is the region in Figure 16.60.

Figure 16.60

Calculate exactly the integrals in Exercises 12–17.

12. $\int_0^{10} \int_0^{0.1} xe^{xy}\, dy\, dx$

13. $\int_0^1 \int_3^4 (\sin(2 - y)) \cos(3x - 7)\, dx\, dy$

14. $\int_0^1 \int_0^y (\sin^3 x)(\cos x)(\cos y)\, dx\, dy$

15. $\int_3^4 \int_0^1 x^2 y \cos(xy)\, dy\, dx$

16. $\int_0^1 \int_0^z \int_0^2 (y + z)^7\, dx\, dy\, dz$

17. $\int_0^1 \int_{-\sqrt{1-x^2}}^{\sqrt{1-x^2}} e^{-(x^2+y^2)}\, dy\, dx$

If W is the region in Figure 16.61, what are the limits of integration in Exercises 18–20?

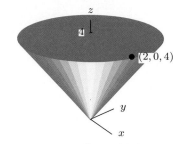

Figure 16.61: Cone with flat top, symmetric about z-axis

18. $\int_?^? \int_?^? \int_?^? f(r, \theta, z) r\, dz\, dr\, d\theta$

19. $\int_?^? \int_?^? \int_?^? g(\rho, \phi, \theta) \rho^2 \sin \phi\, d\rho\, d\phi\, d\theta$

20. $\int_?^? \int_?^? \int_?^? h(x, y, z)\, dz\, dy\, dx$

21. Set up $\int_R f\, dV$ as an iterated integral in all six possible orders of integration, where R is the hemisphere bounded by the upper half of $x^2 + y^2 + z^2 = 1$ and the xy-plane.

Evaluate the integrals in Exercises 22–25 by changing them to cylindrical or spherical coordinates.

22. $\int_{-\sqrt{3}}^{\sqrt{3}} \int_{-\sqrt{3-x^2}}^{\sqrt{3-x^2}} \int_1^{4-x^2-y^2} \frac{1}{z^2}\, dz\, dy\, dx$

23. $\int_0^3 \int_{-\sqrt{9-z^2}}^{\sqrt{9-z^2}} \int_{-\sqrt{9-y^2-z^2}}^{\sqrt{9-y^2-z^2}} x^2\, dx\, dy\, dz$

24. $\int_0^1 \int_0^{\sqrt{1-x^2}} \int_0^{\sqrt{x^2+y^2}} (z + \sqrt{x^2+y^2})\, dz\, dy\, dx$

25. $\int_W \frac{z}{(x^2+y^2)^{3/2}}\, dV$, if W is $1 \leq x^2 + y^2 \leq 4$,
$0 \leq z \leq 4$

Problems

The finite regions T, B, R, L are in the xy-plane.

T lies in the region where $y > 0$,
R lies in the region where $x > 0$,
B lies in the region where $y < 0$,
L lies in the region where $x < 0$.

Are the double integrals in Problems 26–30 positive, negative, or is it impossible to tell?

26. $\int_T e^{-x}\, dA$

27. $\int_B y^3\, dA$

28. $\int_R (x + y^2)\, dA$

29. $\int_L y^3\, dA$

30. $\int_L (x + y^2)\, dA$

Let W be the solid sphere bounded by $x^2 + y^2 + z^2 = 1$. For Problems 31–39, decide (without calculating its value) whether the integral is positive, negative, or zero.

31. $\int_W z\, dV$

32. $\int_W x\, dV$

33. $\int_W xy\, dV$

34. $\int_W xyz\, dV$

35. $\int_W \sin\left(\frac{\pi}{2}xy\right) dV$

36. $\int_W \cos\left(\frac{\pi}{2}xy\right) dV$

37. $\int_W (z^2 - 1)\, dV$

38. $\int_W e^{-xyz}\, dV$

39. $\int_W \sqrt{x^2 + y^2 + z^2}\, dV$

40. A solid region D is a half cylinder with radius 1 lying horizontally with its rectangular base in the xy-plane and its axis along the line $y = 1$ from $x = 0$ to $x = 10$. (The region is above the xy-plane.)

   (a) What is the equation of the curved surface of this half cylinder?
   (b) Write the limits of integration of the integral $\int_D f(x, y, z)\, dV$ in Cartesian coordinates.

41. Find the mass of the solid bounded by the xy-plane, yz-plane, xz-plane, and the plane $4x + 3y + z = 12$, if the density of the solid is given by $\delta(x, y, z) = x^2$.

42. Write an integral representing the mass of a sphere of radius 3 if the density of the sphere at any point is twice the distance of that point from the center of the sphere.

43. A forest next to a road has the shape in Figure 16.62. The population density of rabbits is proportional to the distance from the road. It is 0 at the road, and 10 rabbits per square mile at the opposite edge of the forest. Find the total rabbit population in the forest.

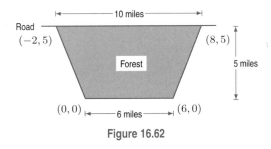

Figure 16.62

44. The insulation surrounding a pipe is the region between two cylinders of length l and with the same axis. The inner cylinder has radius a, the outer radius of the pipe, and the insulation has thickness h. Write a triple integral, including limits of integration, giving the volume of the insulation. Evaluate the integral.

45. Assume p, q, r are positive constants. Find the volume contained between the coordinates planes and the plane
$$\frac{x}{p} + \frac{y}{q} + \frac{z}{r} = 1.$$

46. A cone stands with its flat base on a table. The cone's circular base has radius a; the vertex (tip) is at a height of h above the center of the base. Write a triple integral, including limits of integration, representing the volume of the cone. Evaluate the integral.

47. A solid hemisphere of radius 2 cm has density, in gm/cm^3, at each point equal to the distance in centimeters from the point to the center of the base. Write a triple integral representing the total mass of the hemisphere. Evaluate the integral.

48. Find the volume that remains after a cylindrical hole of radius R is bored through a sphere of radius a, where $0 < R < a$, passing through the center of the sphere along the pole.

49. Two spheres, one of radius 1, one of radius $\sqrt{2}$, have centers that are 1 unit apart. Write a triple integral, including limits of integration, giving the volume of the smaller region that is outside one sphere and inside the other. Evaluate the integral.

50. Give the joint density of x and y where x and y are independent, x has a normal distribution with mean 5 and standard deviation $1/10$, and y has a normal distribution with mean 15 and standard deviation $1/6$.

For Problems 51–52, use the definition of moment of inertia on page 803.

51. Consider a rectangular brick with length 5, width 3, and height 1, and of uniform density 1. Compute the moment of inertia about each of the three axes passing through the center of the brick, perpendicular to one of the sides.

52. Compute the moment of inertia of a ball of radius R about an axis passing through its center. Assume that the ball has a constant density of 1.

53. A particle of mass m is placed at the center of one base of a circular cylindrical shell of inner radius r_1, outer radius r_2, height h, and constant density δ. Find the force of gravity exerted by the cylinder on the particle.

54. Find the area of the half-moon shape with circular arcs as edges and the dimensions shown in Figure 16.63.

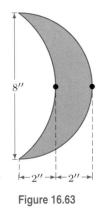

Figure 16.63

55. A river follows the path $y = f(x)$ where x, y are in kilometers. Near the sea, it widens into a lagoon, then narrows again at its mouth. See Figure 16.64. At the point (x, y), the depth, $d(x, y)$, of the lagoon is given by

$$d(x, y) = 40 - 160(y - f(x))^2 - 40x^2 \text{ meters.}$$

The lagoon itself is described by $d(x, y) \geq 0$. What is the volume of the lagoon in cubic meters? [Hint: Use new coordinates $u = x/2$, $v = y - f(x)$ and Jacobians.]

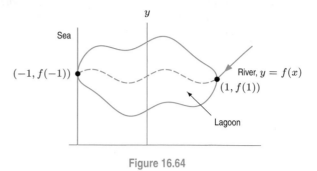

Figure 16.64

CAS Challenge Problems

56. Let D be the region inside the triangle with vertices $(0,0)$, $(1,1)$ and $(0,1)$. Express the double integral $\int_D e^{y^2} \, dA$ as an iterated integral in two different ways. Calculate whichever of the two you can do by hand, and calculate the other with a computer algebra system if possible. Compare the answers.

57. Let D be the region inside the circle $x^2 + y^2 = 1$. Express the integral $\int_D \sqrt[3]{x^2 + y^2} \, dA$ as an iterated integral in both Cartesian and polar coordinates. Calculate whichever of the two you can do by hand, and calculate the other with a computer algebra system if possible. Compare the answers.

58. Compute the iterated integrals $\int_0^1 \int_{-1}^0 \dfrac{x+y}{(x-y)^3} \, dy dx$

and $\int_{-1}^0 \int_0^1 \dfrac{x+y}{(x-y)^3} \, dx dy$. Explain why your answers do not contradict Theorem 16.1 on page 790

59. For each of the following functions, find its average value over the square $-h \leq x \leq h$, $-h \leq y \leq h$, calculate the limit of your answer as $h \to 0$, and compare with the value of the function at $(0,0)$. Assume a, b, c, d, e, and k are constants.

$$F(x, y) = a + bx^4 + cy^4 + dx^2 y^2 + ex^3 y^3$$
$$G(x, y) = a\sin(kx) + b\cos(ky) + c$$
$$H(x, y) = ax^2 e^{x+y} + by^2 e^{x-y}$$

Formulate a conjecture from your results and explain why it makes sense.

CHECK YOUR UNDERSTANDING

Are the statements in Problems 1–30 true or false? Give reasons for your answer.

1. The double integral $\int_R f \, dA$ is always positive.

2. If $f(x, y) = k$ for all points (x, y) in a region R then $\int_R f \, dA = k \cdot \text{Area}(R)$.

3. If R is the rectangle $0 \leq x \leq 1, 0 \leq y \leq 1$ then $\int_R e^{xy} \, dA > 3$.

4. If R is the rectangle $0 \leq x \leq 2, 0 \leq y \leq 3$ and S is the rectangle $-2 \leq x \leq 0, -3 \leq y \leq 0$ then $\int_R f \, dA = -\int_S f \, dA$.

5. Let $\rho(x, y)$ be the population density of a city, in people per km^2. If R is a region in the city, then $\int_R \rho \, dA$ gives the total number of people in the region R.

6. If $\int_R f \, dA = 0$ then $f(x, y) = 0$ at all points of R.

7. If $g(x, y) = kf(x, y)$, where k is constant, then $\int_R g \, dA = k \int_R f \, dA$.

8. If f and g are two functions continuous on a region R, then $\int_R f \cdot g \, dA = \int_R f \, dA \cdot \int_R g \, dA$.

9. If R is the rectangle $0 \leq x \leq 1, 0 \leq y \leq 2$ and S is the square $0 \leq x \leq 1, 0 \leq y \leq 1$, then $\int_R f \, dA = 2 \int_S f \, dA$.

10. If R is the rectangle $2 \leq x \leq 4, 5 \leq y \leq 9$, $f(x, y) = 2x$ and $g(x, y) = x + y$, then the average value of f on R is less than the average value of g on R.

11. The iterated integral $\int_0^1 \int_5^{12} f \, dx dy$ is computed over the rectangle $0 \leq x \leq 1, 5 \leq y \leq 12$.

12. If R is the region inside the triangle with vertices $(0,0), (1,1)$ and $(0,2)$, then the double integral $\int_R f \, dA$ can be evaluated by an iterated integral of the form $\int_0^2 \int_0^1 f \, dx dy$.

13. The region of integration of the iterated integral $\int_1^2 \int_{x^2}^{x^3} f \, dy dx$ lies completely in the first quadrant (that is, $x \geq 0, y \geq 0$).

14. If the limits a, b, c and d in the iterated integral $\int_a^b \int_c^d f dy dx$ are all positive, then the value of $\int_a^b \int_c^d f dy dx$ is also positive.

15. The iterated integrals $\int_0^1 \int_0^x f \, dy dx$ and $\int_0^1 \int_0^y f \, dx dy$ give the same answer for any function f.

16. If $f(x, y)$ is a function of y only, then $\int_a^b \int_0^1 f \, dx dy = \int_a^b f \, dy$.

17. If R is the region inside a circle of radius a, centered at the origin, then $\int_R f \, dA = \int_{-a}^a \int_0^{\sqrt{a^2 - x^2}} f \, dy dx$.

18. There is a region R over which $\int_R f \, dA$ cannot be evaluated by a *single* iterated integral in the order $\int \int f \, dx dy$ or $\int \int f \, dy dx$.

19. If $f(x, y) = g(x) \cdot h(y)$, where g and h are single-variable functions, then

$$\int_a^b \int_c^d f \, dy dx = \left(\int_a^b g(x) \, dx \right) \cdot \left(\int_c^d h(y) \, dy \right).$$

20. If $f(x, y) = g(x) + h(y)$, where g and h are single-variable functions, then

$$\int_a^b \int_c^d f \, dx dy = \left(\int_a^b g(x) \, dx \right) + \left(\int_c^d h(y) \, dy \right).$$

21. If $\rho(x, y, z)$ is mass density of a material in 3-space, then $\int_W \rho(x, y, z) \, dV$ gives the volume of the solid region W.

22. The region of integration of the triple iterated integral $\int_0^1 \int_0^1 \int_0^x f \, dz dy dx$ lies above a square in the xy-plane and below a plane.

23. If W is the entire unit ball $x^2 + y^2 + z^2 \leq 1$ then an iterated integral over W has limits $\int_0^1 \int_0^{\sqrt{1-x^2}} \int_0^{\sqrt{1-x^2-y^2}} f \, dz dy dx$.

24. The iterated integrals $\int_0^1 \int_0^{1-x} \int_0^{1-x-y} f \, dz dy dx$ and $\int_0^1 \int_0^{1-z} \int_0^{1-y-z} f \, dx dy dz$ are equal.

25. The iterated integrals $\int_{-1}^1 \int_0^1 \int_0^{1-x^2} f \, dz dy dx$ and $\int_0^1 \int_0^1 \int_{-\sqrt{1-z}}^{\sqrt{1-z}} f \, dx dy dz$ are equal.

26. If W is a rectangular solid in 3-space, then $\int_W f \, dV = \int_a^b \int_c^d \int_e^k f \, dz \, dy \, dx$, where $a, b, c, d, e,$ and k are constants.

27. If W is the unit cube $0 \leq x \leq 1, 0 \leq y \leq 1, 0 \leq z \leq 1$ and $\int_W f \, dV = 0$, then $f = 0$ everywhere in the unit cube.

28. If $f > g$ at all points in the solid region W, then $\int_W f \, dV > \int_W g \, dV$.

29. If W_1 and W_2 are solid regions with volume$(W_1) >$ volume(W_2) then $\int_{W_1} f \, dV > \int_{W_2} f \, dV$.

30. Both double and triple integrals can be used to compute volume.

PROJECTS FOR CHAPTER SIXTEEN

1. A Connection Between e and π

In this problem you will derive one of the remarkable formulas of mathematics, namely that

$$\int_{-\infty}^{\infty} e^{-x^2} \, dx = \sqrt{\pi}.$$

(a) Change the following double integral into polar coordinates and evaluate it:

$$\int_{-\infty}^{\infty} \int_{-\infty}^{\infty} e^{-(x^2+y^2)} \, dx dy.$$

(b) Explain why

$$\int_{-\infty}^{\infty} \int_{-\infty}^{\infty} e^{-(x^2+y^2)} \, dx dy = \left(\int_{-\infty}^{\infty} e^{-x^2} \, dx \right)^2.$$

(c) Explain why the answers to parts (a) and (b) give the formula we want.

2. Average Distance Walked to an Airport Gate

At airports, departure gates are often lined up in a terminal like points along a line. If you arrive at one gate and proceed to another gate for a connecting flight, what proportion of the length of the terminal will you have to walk, on average?

(a) One way to model this situation is to randomly choose two numbers, $0 \leq x \leq 1$ and $0 \leq y \leq 1$, and calculate the average value of $|x - y|$. Use a double integral to show that, on average, you have to walk $1/3$ the length of the terminal.

(b) The terminal gates are not actually located continuously from 0 to 1, as we assumed in part (a). There are only a finite number of gates and they are likely to be equally spaced. Suppose there are $n + 1$ gates located $1/n$ units apart from one end of the terminal ($x_0 = 0$) to the other ($x_n = 1$). Assuming that all pairs (i, j) of arrival and departure gates are equally likely, show that

$$\text{Average distance between gates} = \frac{1}{(n+1)^2} \cdot \sum_{i=0}^{n} \sum_{j=0}^{n} \left| \frac{i}{n} - \frac{j}{n} \right|.$$

Identify this sum as approximately (but not exactly) a Riemann sum with n subdivisions for the integrand used in part (a). Compute this sum for $n = 5$ and $n = 10$ and compare to the answer of $1/3$ obtained in part (a).

Chapter Seventeen

PARAMETERIZATION AND VECTOR FIELDS

Some physical quantities (such as temperature) are best represented by scalars; others (such as velocity) are best represented by vectors. We have looked at functions whose values are scalars. For example, in single-variable calculus we represented the motion of an object thrown straight up into the air by a function $h(t)$, the height of the object above the ground at time t. In multivariable calculus we considered temperature as a function of position on a weather map. Such functions are called scalar-valued functions.

In this chapter we look at vector-valued functions. We represent the motion of a particle in space by giving its position vector $\vec{r}(t)$ at time t. This enables us to find the velocity and acceleration vectors of the particle. We represent a surface in 3-space by giving the position vector of a point on the surface as a function of two parameters.

Another example of a vector-valued function is a *vector field*, which assigns a vector to each point in space. We have already seen one important example of a vector field, namely the gradient of a scalar-valued function. In this chapter we look at other examples, such as velocity vector fields describing a fluid flow. We also look at the path followed by a particle moving with the flow, which is called a *flow line* of the vector field.

17.1 PARAMETERIZED CURVES

A curve in the plane may be parameterized by a pair of equations of the form $x = f(t)$, $y = g(t)$. As the parameter t changes, the point (x, y) traces out the curve. In this section we find parametric equations for curves in three dimensions, and we see how to write parametric equations using position vectors.

Parametric Equations in Three Dimensions

We describe motion in the plane by giving parametric equations for x and y in terms of t. To describe a motion in 3-dimensional space parametrically, we need a third equation giving z in terms of t.

Example 1 Find parametric equations for the curve $y = x^2$ in the xy-plane.

Solution A possible parameterization in two dimensions is $x = t$, $y = t^2$. Since the curve is in the xy-plane, the z-coordinate is zero, so a parameterization in three dimensions is

$$x = t, \quad y = t^2, \quad z = 0.$$

Example 2 Find parametric equations for a particle that starts at $(0, 3, 0)$ and moves around a circle as shown in Figure 17.1.

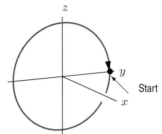

Figure 17.1: Circle of radius 3 in the yz-plane, centered at origin

Solution Since the motion is in the yz-plane, we have $x = 0$ at all times t. Looking at the yz-plane from the positive x-direction we see motion around a circle of radius 3 in the clockwise direction. Thus

$$x = 0, \quad y = 3\cos t, \quad z = -3\sin t.$$

Example 3 Describe in words the motion given parametrically by

$$x = \cos t, \quad y = \sin t, \quad z = t.$$

Solution The particle's x- and y-coordinates give circular motion in the xy-plane, while the z-coordinate increases steadily. Thus, the particle traces out a rising spiral, like a coiled spring. (See Figure 17.2.) This curve is called a *helix*.

Figure 17.2: The helix $x = \cos t, y = \sin t, z = t$

Example 4   Find parametric equations for the line parallel to the vector $2\vec{i} + 3\vec{j} + 4\vec{k}$ and through the point $(1, 5, 7)$.

Solution   Let's imagine a particle at the point $(1, 5, 7)$ at time $t = 0$ and moving through a displacement of $2\vec{i} + 3\vec{j} + 4\vec{k}$ for each unit of time, t. When $t = 0$, $x = 1$ and x increases by 2 units for every unit of time. Thus, at time t, the x-coordinate of the particle is given by

$$x = 1 + 2t.$$

Similarly, the y-coordinate starts at $y = 5$ and increases at a rate of 3 units for every unit of time. The z-coordinate starts at $y = 7$ and increases by 4 units for every unit of time. Thus, the parametric equations of the line are

$$x = 1 + 2t, \quad y = 5 + 3t, \quad z = 7 + 4t.$$

We can generalize the previous example as follows:

Parametric Equations of a Line through the point (x_0, y_0, z_0) and parallel to the vector $a\vec{i} + b\vec{j} + c\vec{k}$ are

$$x = x_0 + at, \quad y = y_0 + bt, \quad z = z_0 + ct.$$

Notice that the coordinates x, y, and z are linear functions of the parameter t.

Example 5   (a) Describe in words the curve given by the parametric equations $x = 3 + t$, $y = 2t$, $z = 1 - t$.
(b) Find parametric equations for the line through the points $(1, 2, -1)$ and $(3, 3, 4)$.

Solution   (a) The curve is a line through the point $(3, 0, 1)$ and parallel to the vector $\vec{i} + 2\vec{j} - \vec{k}$.
(b) The line is parallel to the vector between the points $P = (1, 2, -1)$ and $Q = (3, 3, 4)$.

$$\overrightarrow{PQ} = (3 - 1)\vec{i} + (3 - 2)\vec{j} + (4 - (-1))\vec{k} = 2\vec{i} + \vec{j} + 5\vec{k}.$$

Thus, using the point P, the parametric equations are

$$x = 1 + 2t, \quad y = 2 + t, \quad z = -1 + 5t.$$

Using the point Q gives the equations $x = 3 + 2t$, $y = 3 + t$, $z = 4 + 5t$, which represent the same line.

Using Position Vectors to Write Parameterized Curves as Vector-Valued Functions

A point in the plane with coordinates (x, y) can be represented by the position vector $\vec{r} = x\vec{i} + y\vec{j}$ in Figure 17.3. Similarly, in 3-space we write $\vec{r} = x\vec{i} + y\vec{j} + z\vec{k}$. (See Figure 17.4.)

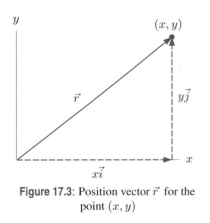

Figure 17.3: Position vector $\vec{r}$ for the point (x, y)

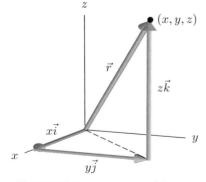

Figure 17.4: Position vector $\vec{r}$ for the point (x, y, z)

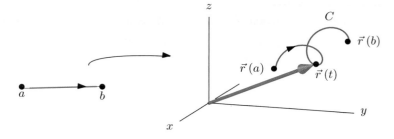

Figure 17.5: The parameterization sends the interval, $a \leq t \leq b$, to the curve, C, in 3-space

We can write the parametric equations $x = f(t), y = g(t), z = h(t)$ as a single vector equation

$$\vec{r}(t) = f(t)\vec{i} + g(t)\vec{j} + h(t)\vec{k}$$

called a *parameterization*. As the parameter t varies, the point with position vector $\vec{r}(t)$ traces out a curve in 3-space. For example, the circular motion in the plane

$$x = \cos t, y = \sin t \quad \text{can be written as} \quad \vec{r} = (\cos t)\vec{i} + (\sin t)\vec{j}$$

and the helix in 3-space

$$x = \cos t, y = \sin t, z = t \quad \text{can be written as} \quad \vec{r} = (\cos t)\vec{i} + (\sin t)\vec{j} + t\vec{k}.$$

See Figure 17.5.

Example 6   Use vectors to give a parameterization for the circle of radius $\frac{1}{2}$ centered at the point $(-1, 2)$.

Solution   The circle of radius 1 centered at the origin is parameterized by the vector-valued function

$$\vec{r}_1(t) = \cos t\vec{i} + \sin t\vec{j}, \quad 0 \leq t \leq 2\pi.$$

The point $(-1, 2)$ has position vector $\vec{r}_0 = -\vec{i} + 2\vec{j}$. The position vector, $\vec{r}(t)$, of a point on the circle of radius $\frac{1}{2}$ centered at $(-1, 2)$ is found by adding $\frac{1}{2}\vec{r}_1$ to $\vec{r}_0$. (See Figures 17.6 and 17.7.) Thus,

$$\vec{r}(t) = \vec{r}_0 + \tfrac{1}{2}\vec{r}_1(t) = -\vec{i} + 2\vec{j} + \tfrac{1}{2}(\cos t\vec{i} + \sin t\vec{j}) = (-1 + \tfrac{1}{2}\cos t)\vec{i} + (2 + \tfrac{1}{2}\sin t)\vec{j},$$

or, equivalently,

$$x = -1 + \tfrac{1}{2}\cos t, \quad y = 2 + \tfrac{1}{2}\sin t, \quad 0 \leq t \leq 2\pi.$$

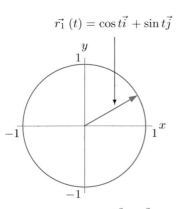

Figure 17.6: The circle $x^2 + y^2 = 1$ parameterized by $\vec{r}_1(t) = \cos t\vec{i} + \sin t\vec{j}$

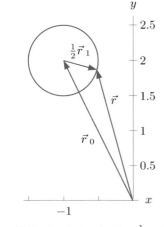

Figure 17.7: The circle of radius $\frac{1}{2}$ and center $(-1, 2)$ parameterized by $\vec{r}(t) = \vec{r}_0 + \frac{1}{2}\vec{r}_1(t)$

Parametric Equation of a Line

Consider a straight line in the direction of a vector $\vec{v}$ passing through the point (x_0, y_0, z_0) with position vector $\vec{r}_0$. We start at $\vec{r}_0$ and move up and down the line, adding different multiples of $\vec{v}$ to $\vec{r}_0$. (See Figure 17.8.)

Figure 17.8: The line $\vec{r}(t) = \vec{r}_0 + t\vec{v}$

In this way, every point on the line can be written as $\vec{r}_0 + t\vec{v}$, which yields the following:

Parametric Equation of a Line in Vector Form

The line through the point with position vector $\vec{r_0} = x_0\vec{i} + y_0\vec{j} + z_0\vec{k}$ in the direction of the vector $\vec{v} = a\vec{i} + b\vec{j} + c\vec{k}$ has parametric equation

$$\vec{r}(t) = \vec{r}_0 + t\vec{v}.$$

Example 7

(a) Find parametric equations for the line passing through the points $(2, -1, 3)$ and $(-1, 5, 4)$.

(b) Represent the line segment from $(2, -1, 3)$ to $(-1, 5, 4)$ parametrically.

Solution

(a) The line passes through $(2, -1, 3)$ and is parallel to the displacement vector $\vec{v} = -3\vec{i} + 6\vec{j} + \vec{k}$ from $(2, -1, 3)$ to $(-1, 5, 4)$. Thus the parametric equation is

$$\vec{r}(t) = 2\vec{i} - \vec{j} + 3\vec{k} + t(-3\vec{i} + 6\vec{j} + \vec{k}).$$

(b) In the parameterization in part (a), $t = 0$ corresponds to the point $(2, -1, 3)$ and $t = 1$ corresponds to the point $(-1, 5, 4)$. So the parameterization of the segment is

$$\vec{r}(t) = 2\vec{i} - \vec{j} + 3\vec{k} + t(-3\vec{i} + 6\vec{j} + \vec{k}), \qquad 0 \le t \le 1.$$

Intersection of Curves and Surfaces

Parametric equations for a curve enable us to find where a curve intersects a surface.

Example 8

Find the points at which the line $x = t$, $y = 2t$, $z = 1 + t$ pierces the sphere of radius 10 centered at the origin.

Solution

The equation for the sphere of radius 10 and center at the origin is

$$x^2 + y^2 + z^2 = 100.$$

To find the intersection points of the line and the sphere, substitute the parametric equations of the line into the equation of the sphere, giving

$$t^2 + 4t^2 + (1+t)^2 = 100,$$

so

$$6t^2 + 2t - 99 = 0,$$

which has the two solutions at approximately $t = -4.23$ and $t = 3.90$. Using the parametric equation for the line, $(x, y, z) = (t, 2t, 1 + t)$, we see that the line cuts the sphere at the two points

$$(x, y, z) = (-4.23, 2(-4.23), 1 + (-4.23)) = (-4.23, -8.46, -3.23),$$

and

$$(x, y, z) = (3.90, 2(3.90), 1 + 3.90) = (3.90, 7.80, 4.90).$$

We can also use parametric equations to find the intersection of two curves.

Example 9   Two particles move through space, with equations $\vec{r}_1(t) = t\vec{i} + (1 + 2t)\vec{j} + (3 - 2t)\vec{k}$ and $\vec{r}_2(t) = (-2 - 2t)\vec{i} + (1 - 2t)\vec{j} + (1 + t)\vec{k}$. Do the particles ever collide? Do their paths cross?

Solution   To see if the particles collide, we must find out if they pass through the same point at the same time t. So we must find a solution to the vector equation $\vec{r}_1(t) = \vec{r}_2(t)$, which is the same as finding a common solution to the three scalar equations

$$t = -2 - 2t, \qquad 1 + 2t = 1 - 2t, \qquad 3 - 2t = 1 + t.$$

Separately, the solutions are $t = -2/3$, $t = 0$, and $t = 2/3$, so there is no common solution, and the particles don't collide. To see if their paths cross, we find out if they pass through the same point at two possibly different times, t_1 and t_2. So we solve the equations

$$t_1 = -2 - 2t_2, \qquad 1 + 2t_1 = 1 - 2t_2, \qquad 3 - 2t_1 = 1 + t_2.$$

We solve the first two equations simultaneously and get $t_1 = 2$, $t_2 = -2$. Since these values also satisfy the third equation, the paths cross. The position of the first particle at time $t = 2$ is the same as the position of the second particle at time $t = -2$, namely the point $(2, 5, -1)$.

Example 10   Are the lines $x = -1 + t$, $y = 1 + 2t$, $z = 5 - t$ and $x = 2 + 2t$, $y = 4 + t$, $z = 3 + t$ parallel? Do they intersect?

Solution   In vector form the lines are parameterized by

$$\vec{r} = -\vec{i} + \vec{j} + 5\vec{k} + t(\vec{i} + 2\vec{j} - \vec{k})$$
$$\vec{r} = 2\vec{i} + 4\vec{j} + 3\vec{k} + t(2\vec{i} + \vec{j} + \vec{k})$$

Their direction vectors $\vec{i} + 2\vec{j} - \vec{k}$ and $2\vec{i} + \vec{j} + \vec{k}$ are not multiplies of each other, so the lines are not parallel. To find out if they intersect, we see if they pass through same point at two possibly different times, t_1 and t_2:

$$-1 + t_1 = 2 + 2t_2, \qquad 1 + 2t_1 = 4 + t_2, \qquad 5 - t_1 = 3 + t_2.$$

The first two equations give $t_1 = 1$, $t_2 = -1$. Since these values do not satisfy the third equation, the paths do not cross, and so the lines do not intersect.

The next example shows how to tell if two different parameterizations give the same line.

Example 11   Show that the following two lines are the same:

$$\vec{r} = -\vec{i} - \vec{j} + \vec{k} + t(3\vec{i} + 6\vec{j} - 3\vec{k})$$
$$\vec{r} = \vec{i} + 3\vec{j} - \vec{k} + t(-\vec{i} - 2\vec{j} + \vec{k})$$

Solution   The direction vectors of the two lines, $3\vec{i} + 6\vec{j} - 3\vec{k}$ and $-\vec{i} - 2\vec{j} + \vec{k}$, are multiples of each other, so the lines are parallel. To see if they are the same, we pick a point on the first line and see if it is on the second line. For example, the point on the first line with $t = 0$ has position vector $-\vec{i} - \vec{j} + \vec{k}$. Solving

$$\vec{i} + 3\vec{j} - \vec{k} + t(-\vec{i} - 2\vec{j} + \vec{k}) = -\vec{i} - \vec{j} + \vec{k}$$

we get $t = 2$, so the two lines have a point in common. Thus they are the same line, parameterized in two different ways.

Exercises and Problems for Section 17.1

Exercises

In Exercises 1–11, find parametric equations for the line.

1. The line in the direction of the vector $\vec{i} + 2\vec{j} - \vec{k}$ and through the point $(3, 0, -4)$.

2. The line in the direction of the vector $3\vec{i} - 3\vec{j} + \vec{k}$ and through the point $(1, 2, 3)$.

3. The line in the direction of the vector $2\vec{i} + 2\vec{j} - 3\vec{k}$ and through the point $(-3, 4, -2)$.

4. The line in the direction of the vector $5\vec{j} + 2\vec{k}$ and through the point $(5, -1, 1)$.

5. The line in the direction of the vector $\vec{i} - \vec{k}$ and through the point $(0, 1, 0)$.

6. The line parallel to the z-axis passing through the point $(1, 0, 0)$.

7. The line through the points $(1, 5, 2)$ and $(5, 0, -1)$.

8. The line through the points $(2, 3, -1)$ and $(5, 2, 0)$.

9. The line through $(-3, -2, 1)$ and $(-1, -3, -1)$.

10. The line through $(3, -2, 2)$ and intersecting the y-axis at $y = 2$.

11. The line intersecting the x-axis at $x = 3$ and the z-axis at $z = -5$.

12. Do the lines in Exercises 2 and 8 intersect?

In Exercises 13–25, find a parameterization for the curve.

13. The circle of radius 3 in the xy-plane, centered at the origin, counterclockwise.

14. The circle of radius 2 in the xy-plane, centered at the origin, clockwise.

15. The circle of radius 2 in the xz-plane, centered at the origin.

16. The circle of radius 3 parallel to the xy-plane, centered at the point $(0, 0, 2)$.

17. The circle of radius 3 in the yz-plane, centered at the point $(0, 0, 2)$.

18. The circle of radius 5 parallel to the yz-plane, centered at the point $(-1, 0, -2)$.

19. The curve $y = x^3$ in the xy-plane.

20. The curve $x = y^2$ in the xy-plane.

21. The curve $x = -3z^2$ in the xz-plane.

22. The curve $y = 4 - 5x^4$ through the point $(0, 4, 4)$, parallel to the xy-plane.

23. The ellipse of major diameter 6 along the x-axis and minor diameter 4 along the y-axis, centered at the origin.

24. The ellipse of major diameter 5 parallel to the y-axis and minor diameter 2 parallel to the z-axis, centered at $(0, 1, -2)$.

25. The ellipse of major diameter 3 parallel to the x-axis and minor diameter 2 parallel to the z-axis, centered at $(0, 1, -2)$.

In Exercises 26–29, find parametric equations for the curve segment.

26. Line from $P_0 = (1, -3, 2)$ to $P_1 = (4, 1, -3)$.

27. Line from $P_0 = (-1, -3)$ to $P_1 = (5, 2)$.

28. Semicircle from $(1, 0, 0)$ to $(-1, 0, 0)$ in the xy-plane with $y \geq 0$.

29. Semicircle from $(0, 0, 5)$ to $(0, 0, -5)$ in the yz-plane with $y \geq 0$.

Are the lines in Exercises 30–33 parallel? Do they intersect? If so, where?

30. l_1 : $x = 5 + 2t$, $y = -t$, $z = -2 + 3t$
    l_2 : $x = 4 - 3t$, $y = 1 + 5t$, $z = 4$

31. l_1 : $x = 3 - 2t$, $y = 5 + t$, $z = -4 + 2t$
    l_2 : $x = 7 + 6t$, $y = 1 - 3t$, $z = 5 - 6t$

32. l_1 : $x = 1 + 4t$, $y = 1 - 3t$, $z = 2t$
    l_2 : $x = 3 - 10t$, $y = 7$, $z = -6 + 2t$

33. l_1 : $x = 3 + 2t$, $y = 5 - 2t$, $z = 1 + t$
    l_2 : $x = -1 - t$, $y = 4t$, $z = 4 - 2t$

Problems

In Problems 34–37, find parametric equations for the curve segment.

34. The graph of $y = \sqrt{x}$ from $(1, 1)$ to $(16, 4)$.

35. The quarter-ellipse from $(4, 0, 3)$ to $(0, -3, 3)$ in the plane $z = 3$.

36. A circular arc from $P = (0, 0)$ to $Q = (10, 0)$ if the radius of the circle is 5.

37. A circular arc from $P = (0, 0)$ to $Q = (10, 0)$ if the radius of the circle is 10.

In Problems 38–42, parameterize the line through $P = (2, 5)$ and $Q = (12, 9)$ so that the points P and Q correspond to the given parameter values.

38. $t = 0$ and 1

39. $t = 0$ and 5

40. $t = 10$ and 11

41. $t = 20$ and 30

42. $t = 0$ and -1

43. Consider two points P_0 and P_1 in 3-space.

  (a) Show that the line segment from P_0 to P_1 can be parametrized by
  $$\vec{r}(t) = (1 - t)\overrightarrow{OP_0} + t\overrightarrow{OP_1}, \quad 0 \le t \le 1.$$

  (b) What is represented by the parametric equation
  $$\vec{r}(t) = t\overrightarrow{OP_0} + (1 - t)\overrightarrow{OP_1}, \quad 0 \le t \le 1?$$

44. (a) Explain why the line of intersection of two planes must be parallel to the cross product of a normal vector to the first plane and a normal vector to the second.

  (b) Find a vector parallel to the line of intersection of the two planes $x + 2y - 3z = 7$ and $3x - y + z = 0$.

  (c) Find parametric equations for the line in part (b).

45. (a) Find a vector parallel to the line of intersection of the planes $2x - y - 3z = 0$ and $x + y + z = 1$.

  (b) Show that the point $(1, -1, 1)$ lies on both planes.

  (c) Find parametric equations for the line of intersection.

In Problems 46–49, find parametric equations for the line.

46. The line of intersection of the planes $x + y + z = 3$ and $x - y + 2z = 2$.

47. The line of intersection of the planes $x - y + z = 3$ and $2x + y - z = 5$.

48. The line perpendicular to the surface $z = x^2 + y^2$ at the point $(1, 2, 5)$.

49. The line through the point $(-4, 2, 3)$ and parallel to a line in the yz-plane which makes a $45°$ angle with the positive y-axis and the positive z-axis.

50. If it exists, find the value of c for which the lines $l(t) = (c+t, 1+t, 5+t)$ and $m(s) = (s, 1-s, 3+s)$ intersect.

51. Show that the equations $x = 3 + t$, $y = 2t$, $z = 1 - t$ satisfy the equations $x + y + 3z = 6$ and $x - y - z = 2$. What does this tell you about the curve parameterized by these equations?

52. Two particles are traveling through space. At time t the first particle is at the point $(-1 + t, 4 - t, -1 + 2t)$ and the second particle is at $(-7 + 2t, -6 + 2t, -1 + t)$.

  (a) Describe the two paths in words.

  (b) Do the two particles collide? If so, when and where?

  (c) Do the paths of the two particles cross? If so, where?

In Problems 53–55, describe in words the motion represented by the equations.

53. $x = 3 + t^3$, $\quad y = 5 - t^3$, $\quad z = 7 + 2t^3$

54. $x = 5 \cos t$, $\quad y = 5 \sin t$, $\quad z = t$

55. $x = \cos t$, $\quad y = t^2$, $\quad z = \sin t$

56. For $t > 0$, a particle moves along the curve $x = a + b\sin kt$, $y = a + b\cos kt$, where a, b, k are positive constants.

  (a) Describe the motion in words.

  (b) What is the effect on the curve of the following changes?

  (i) Increasing b

  (ii) Increasing a

  (iii) Increasing k

  (iv) Setting a and b equal.

57. Is the point $(-3, -4, 2)$ visible from the point $(4, 5, 0)$ if there is an opaque ball of radius 1 centered at the origin?

58. A light shines on the helix of Example 3 on page 832 from far down each axis. Sketch the shadow the helix casts on each of the coordinate planes: xy, xz, and yz.

59. (a) Find a parametric equation for the line through the point $(2, 1, 3)$ and in the direction of $a\vec{i} + b\vec{j} + c\vec{k}$.

  (b) Find conditions on a, b, c so that the line you found in part (a) goes through the origin. Give a reason for your answer.

60. Consider the line $x = 5 - 2t$, $y = 3 + 7t$, $z = 4t$ and the plane $ax + by + cz = d$. All the following questions have many possible answers. Find values of a, b, c, d such that:

  (a) The plane is perpendicular to the line.

  (b) The plane is perpendicular to the line and through the point $(5, 3, 0)$.

  (c) The line lies in the plane.

61. A plane from Denver, Colorado, (altitude 1650 meters) flies to Bismark, North Dakota (altitude 550 meters). It travels at 650 km/hour along a horizontal line at 8,000 meters above the line joining Denver and Bismark. Bismark is about 850 km in the direction $60°$ north of east from Denver. Find parametric equations describing the plane's motion. Assume the origin is at sea level beneath Denver, that the x-axis points east and the y-axis points north, and that the earth is flat. Measure distances in kilometers and time in hours.

62. A line has equation $\vec{r} = \vec{a} + t\vec{b}$ where $\vec{r} = x\vec{i} + y\vec{j} + z\vec{k}$ and $\vec{a}$ and $\vec{b}$ are constant vectors such that $\vec{a} \ne \vec{0}, \vec{b} \ne \vec{0}, \vec{a} \ne \vec{b}, \vec{b}$ not perpendicular to $\vec{a}$. For each of the planes (a)–(c), pick the equation (i)–(ix) which represents it. Explain your choice.

  (a) A plane perpendicular to the line and through the origin.

(b) A plane perpendicular to the line and not through the origin.

(c) A plane containing the line.

(i) $\vec{a} \cdot \vec{r} = ||\vec{b}||$ (ii) $\vec{b} \cdot \vec{r} = ||\vec{a}||$

(iii) $\vec{a} \cdot \vec{r} = \vec{b} \cdot \vec{r}$ (iv) $(\vec{a} \times \vec{b}) \cdot (\vec{r} - \vec{a}) = 0$

(v) $(\vec{a} \times \vec{b}) \cdot (\vec{r} - \vec{b}) = 0$ (vi) $\vec{a} \cdot \vec{r} = 0$

(vii) $\vec{b} \cdot \vec{r} = 0$ (viii) $\vec{a} + \vec{r} = \vec{b}$

(ix) $\vec{r} - \vec{a} = \vec{b}$

63. The plane $x + 3y - 2z = 6$ is colored blue and the plane $2x + y + z = 3$ is colored yellow. The planes intersect in a line, which is colored green. You are at the point $P = (1, -2, -1)$.

(a) You look in the direction $\vec{v} = \vec{i} + 2\vec{j} + \vec{k}$. Do you see the blue plane or the yellow plane?

(b) In what direction(s) are you looking directly at the green line?

(c) In what direction(s) should you look to see the yellow plane? The blue plane?

64. (a) Find a parametric equation for the line through the point $(1, 5, 2)$ and in the direction of the vector $2\vec{i} + 3\vec{j} - \vec{k}$.

(b) By minimizing the square of the distance from a point on the line to the origin, find the exact point on the line which is closest to the origin.

17.2 MOTION, VELOCITY, AND ACCELERATION

In this section we see how to find the vector quantities of velocity and acceleration from a parametric equation for the motion of an object.

The Velocity Vector

The velocity of a moving particle can be represented by a vector with the following properties:

The **velocity vector** of a moving object is a vector $\vec{v}$ such that:
- The magnitude of $\vec{v}$ is the speed of the object.
- The direction of $\vec{v}$ is the direction of motion.

Thus the speed of the object is $||\vec{v}||$ and the velocity vector is tangent to the object's path.

Example 1   A child is sitting on a ferris wheel of diameter 10 meters, making one revolution every 2 minutes. Find the speed of the child and draw velocity vectors at two different times.

Solution   The child moves at a constant speed around a circle of radius 5 meters, completing one revolution every 2 minutes. One revolution around a circle of radius 5 is a distance of 10π, so the child's speed is $10\pi/2 = 5\pi \approx 15.7$ m/min. Hence, the magnitude of the velocity vector is 15.7 m/min. The direction of motion is tangent to the circle, and hence perpendicular to the radius at that point. Figure 17.9 shows the direction of the vector at two different times.

Figure 17.9: Velocity vectors of a child on a ferris wheel (Note that vectors would be in opposite direction if viewed from the other side)

Computing the Velocity

We find the velocity, as in one-variable calculus, by taking a limit. If the position vector of the particle is $\vec{r}(t)$ at time t, then the displacement vector between its positions at times t and $t + \Delta t$ is $\Delta \vec{r} = \vec{r}(t + \Delta t) - \vec{r}(t)$. (See Figure 17.10.) Over this interval,

$$\text{Average velocity} = \frac{\Delta \vec{r}}{\Delta t}.$$

In the limit as Δt goes to zero we have the instantaneous velocity at time t:

The **velocity vector**, $\vec{v}(t)$, of a moving object with position vector $\vec{r}(t)$ at time t is

$$\vec{v}(t) = \lim_{\Delta t \to 0} \frac{\Delta \vec{r}}{\Delta t} = \lim_{\Delta t \to 0} \frac{\vec{r}(t + \Delta t) - \vec{r}(t)}{\Delta t},$$

whenever the limit exists. We use the notation $\vec{v} = \dfrac{d\vec{r}}{dt} = \vec{r}'(t)$.

Notice that the direction of the velocity vector $\vec{r}'(t)$ in Figure 17.10 is approximated by the direction of the vector $\Delta \vec{r}$ and that the approximation gets better as $\Delta t \to 0$.

Figure 17.10: The change, $\Delta \vec{r}$, in the position vector for a particle moving on a curve and the velocity vector $\vec{v} = \vec{r}'(t)$

The Components of the Velocity Vector

If we represent a curve parametrically by $x = f(t), y = g(t), z = h(t)$, then we can write its position vector as: $\vec{r}(t) = f(t)\vec{i} + g(t)\vec{j} + h(t)\vec{k}$. Now we can compute the velocity vector:

$$\begin{aligned}
\vec{v}(t) &= \lim_{\Delta t \to 0} \frac{\vec{r}(t + \Delta t) - \vec{r}(t)}{\Delta t} \\
&= \lim_{\Delta t \to 0} \frac{(f(t + \Delta t)\vec{i} + g(t + \Delta t)\vec{j} + h(t + \Delta t)\vec{k}) - (f(t)\vec{i} + g(t)\vec{j} + h(t)\vec{k})}{\Delta t} \\
&= \lim_{\Delta t \to 0} \left(\frac{f(t + \Delta t) - f(t)}{\Delta t}\vec{i} + \frac{g(t + \Delta t) - g(t)}{\Delta t}\vec{j} + \frac{h(t + \Delta t) - h(t)}{\Delta t}\vec{k} \right) \\
&= f'(t)\vec{i} + g'(t)\vec{j} + h'(t)\vec{k} \\
&= \frac{dx}{dt}\vec{i} + \frac{dy}{dt}\vec{j} + \frac{dz}{dt}\vec{k}.
\end{aligned}$$

Thus we have the following result:

The **components of the velocity vector** of a particle moving in space with position vector $\vec{r}(t) = f(t)\vec{i} + g(t)\vec{j} + h(t)\vec{k}$ at time t are given by

$$\vec{v}(t) = f'(t)\vec{i} + g'(t)\vec{j} + h'(t)\vec{k} = \frac{dx}{dt}\vec{i} + \frac{dy}{dt}\vec{j} + \frac{dz}{dt}\vec{k}.$$

Example 2   Find the components of the velocity vector for the child on the ferris wheel in Example 1 using a coordinate system which has its origin at the center of the ferris wheel and which makes the rotation counterclockwise.

Solution   The ferris wheel has radius 5 meters and completes 1 revolution counterclockwise every 2 minutes. The motion is parameterized by an equation of the form

$$\vec{r}(t) = 5\cos(\omega t)\vec{i} + 5\sin(\omega t)\vec{j},$$

where ω is chosen to make the period 2 minutes. Since the period of $\cos(\omega t)$ and $\sin(\omega t)$ is $2\pi/\omega$, we must have

$$\frac{2\pi}{\omega} = 2, \quad \text{so} \quad \omega = \pi.$$

Thus, the motion is described by the equation

$$\vec{r}(t) = 5\cos(\pi t)\vec{i} + 5\sin(\pi t)\vec{j},$$

where t is in minutes. The velocity is given by

$$\vec{v} = \frac{dx}{dt}\vec{i} + \frac{dy}{dt}\vec{j} = -5\pi\sin(\pi t)\vec{i} + 5\pi\cos(\pi t)\vec{j}.$$

To check, we calculate the magnitude of $\vec{v}$,

$$\|\vec{v}\| = \sqrt{(-5\pi)^2\sin^2(\pi t) + (5\pi)^2\cos^2(\pi t)} = 5\pi\sqrt{\sin^2(\pi t) + \cos^2(\pi t)} = 5\pi \approx 15.7,$$

which agrees with the speed we calculated in Example 1. To see that the direction is correct, we must show that the vector $\vec{v}$ at any time t is perpendicular to the position vector of the particle at time t. To do this, we compute the dot product of $\vec{v}$ and $\vec{r}$:

$$\vec{v} \cdot \vec{r} = (-5\pi\sin(\pi t)\vec{i} + 5\pi\cos(\pi t)\vec{j}) \cdot (5\cos(\pi t)\vec{i} + 5\sin(\pi t)\vec{j})$$
$$= -25\pi\,\sin(\pi t)\,\cos(\pi t) + 25\pi\cos(\pi t)\,\sin(\pi t) = 0.$$

So the velocity vector, $\vec{v}$, is perpendicular to $\vec{r}$ and hence tangent to the circle. (See Figure 17.11.)

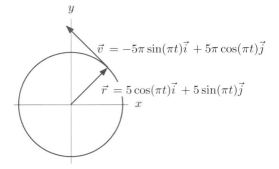

Figure 17.11: Velocity and radius vector of motion around a circle

Velocity Vectors and Tangent Lines

Since the velocity vector is tangent to the path of motion, it can be used to find parametric equations for the tangent line, if there is one.

Example 3   Find the tangent line at the point $(1, 1, 2)$ to the curve defined by the parametric equation

$$\vec{r}(t) = t^2\vec{i} + t^3\vec{j} + 2t\vec{k}.$$

Solution   At time $t = 1$ the particle is at the point $(1, 1, 2)$ with position vector $\vec{r}_0 = \vec{i} + \vec{j} + 2\vec{k}$. The velocity vector at time t is $\vec{r}'(t) = 2t\vec{i} + 3t^2\vec{j} + 2\vec{k}$, so at time $t = 1$ the velocity is $\vec{v} = \vec{r}'(1) = 2\vec{i} + 3\vec{j} + 2\vec{k}$. The tangent line passes through $(1, 1, 2)$ in the direction of $\vec{v}$, so it has the parametric equation

$$\vec{r}(t) = \vec{r}_0 + t\vec{v} = (\vec{i} + \vec{j} + 2\vec{k}) + t(2\vec{i} + 3\vec{j} + 2\vec{k}).$$

The Acceleration Vector

Just as the velocity of a particle moving in 2-space or 3-space is a vector quantity, so is the rate of change of the velocity of the particle, namely its acceleration. Figure 17.12 shows a particle at time t with velocity vector $\vec{v}(t)$ and then a little later at time $t + \Delta t$. The vector $\Delta \vec{v} = \vec{v}(t + \Delta t) - \vec{v}(t)$ is the change in velocity and points approximately in the direction of the acceleration. So,

$$\text{Average acceleration} = \frac{\Delta \vec{v}}{\Delta t}.$$

In the limit as $\Delta t \to 0$, we have the instantaneous acceleration at time t:

The **acceleration vector** of an object moving with velocity $\vec{v}(t)$ at time t is

$$\vec{a}(t) = \lim_{\Delta t \to 0} \frac{\Delta \vec{v}}{\Delta t} = \lim_{\Delta t \to 0} \frac{\vec{v}(t + \Delta t) - \vec{v}(t)}{\Delta t},$$

if the limit exists. We use the notation $\vec{a} = \dfrac{d\vec{v}}{dt} = \dfrac{d^2 \vec{r}}{dt^2} = \vec{r}''(t)$.

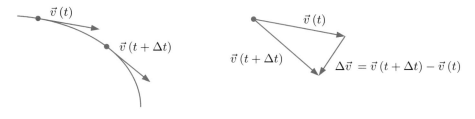

Figure 17.12: Computing the difference between two velocity vectors

Components of the Acceleration Vector

If we represent a curve in space parametrically by $x = f(t)$, $y = g(t)$, $z = h(t)$, we can express the acceleration in components. The velocity vector $\vec{v}(t)$ is given by

$$\vec{v}(t) = f'(t)\vec{i} + g'(t)\vec{j} + h'(t)\vec{k}.$$

From the definition of the acceleration vector, we have

$$\vec{a}(t) = \lim_{\Delta t \to 0} \frac{\vec{v}(t + \Delta t) - \vec{v}(t)}{\Delta t} = \frac{d\vec{v}}{dt}.$$

Using the same method to compute $d\vec{v}/dt$ as we used to compute $d\vec{r}/dt$ on page 840, we obtain

---

The **components of the acceleration vector**, $\vec{a}(t)$, at time t of a particle moving in space with position vector $\vec{r}(t) = f(t)\vec{i} + g(t)\vec{j} + h(t)\vec{k}$ at time t are given by

$$\vec{a}(t) = f''(t)\vec{i} + g''(t)\vec{j} + h''(t)\vec{k} = \frac{d^2 x}{dt^2}\vec{i} + \frac{d^2 y}{dt^2}\vec{j} + \frac{d^2 z}{dt^2}\vec{k}.$$

Motion In a Circle and Along a Line

We now consider the velocity and acceleration vectors for two basic motions: uniform motion around a circle, and motion along a straight line.

Example 4   Find the acceleration vector for the child on the ferris wheel in Examples 1 and 2.

Solution   The child's position vector is given by $\vec{r}(t) = 5\cos(\pi t)\vec{i} + 5\sin(\pi t)\vec{j}$. In Example 2 we saw that the velocity vector is

$$\vec{v}(t) = \frac{dx}{dt}\vec{i} + \frac{dy}{dt}\vec{j} = -5\pi\sin(\pi t)\vec{i} + 5\pi\cos(\pi t)\vec{j}.$$

Thus, the acceleration vector is

$$\vec{a}(t) = \frac{d^2x}{dt^2}\vec{i} + \frac{d^2y}{dt^2}\vec{j} = -(5\pi)\cdot\pi\cos(\pi t)\vec{i} - (5\pi)\cdot\pi\sin(\pi t)\vec{j}$$
$$= -5\pi^2\cos(\pi t)\vec{i} - 5\pi^2\sin(\pi t)\vec{j}.$$

Notice that $\vec{a}(t) = -\pi^2\vec{r}(t)$. Thus, the acceleration vector is a multiple of $\vec{r}(t)$ and points toward the origin.

The motion of the child on the ferris wheel is an example of uniform circular motion, whose properties follow. (See Problem 30.)

Uniform Circular Motion: For a particle whose motion is described by

$$\vec{r}(t) = R\cos(\omega t)\vec{i} + R\sin(\omega t)\vec{j}$$

- Motion is in a circle of radius R with period $2\pi/\omega$.
- Velocity, $\vec{v}$, is tangent to the circle and speed is constant $\|\vec{v}\| = \omega R$.
- Acceleration, $\vec{a}$, points toward the center of the circle with $\|\vec{a}\| = \|\vec{v}\|^2/R$.

In uniform circular motion, the acceleration vector is perpendicular to the velocity vector because the velocity vector does not change in magnitude, only in direction. We now look at straight line motion in which the velocity vector always has the same direction but its magnitude changes. In straight line motion, the acceleration vector points in the same direction as the velocity vector if the speed is increasing and in the opposite direction to the velocity vector if the speed is decreasing.

Example 5   Consider the motion given by the vector equation

$$\vec{r}(t) = 2\vec{i} + 6\vec{j} + (t^3 + t)(4\vec{i} + 3\vec{j} + \vec{k}).$$

Show that this is straight line motion in the direction of the vector $4\vec{i} + 3\vec{j} + \vec{k}$ and relate the acceleration vector to the velocity vector.

Solution   The velocity vector is
$$\vec{v} = (3t^2 + 1)(4\vec{i} + 3\vec{j} + \vec{k}).$$

Since $(3t^2 + 1)$ is a positive scalar, the velocity vector $\vec{v}$ always points in the direction of the vector $4\vec{i} + 3\vec{j} + \vec{k}$. In addition,

$$\text{Speed} = \|\vec{v}\| = (3t^2 + 1)\sqrt{4^2 + 3^2 + 1^2} = \sqrt{26}(3t^2 + 1).$$

Notice that the speed is decreasing until $t = 0$, then starts increasing. The acceleration vector is
$$\vec{a} = 6t(4\vec{i} + 3\vec{j} + \vec{k}).$$

For $t > 0$, the acceleration vector points in the same direction as $4\vec{i} + 3\vec{j} + \vec{k}$, which is the same direction as $\vec{v}$. This makes sense because the object is speeding up. For $t < 0$, the acceleration vector $6t(4\vec{i} + 3\vec{j} + \vec{k})$ points in the opposite direction to $\vec{v}$ because the object is slowing down.

Motion in a Straight Line: For a particle whose motion is described by

$$\vec{r}(t) = \vec{r}_0 + f(t)\vec{v}_0$$

- Motion is along a straight line through the point with position vector $\vec{r}_0$ parallel to $\vec{v}_0$.
- Velocity, $\vec{v}$, and acceleration, $\vec{a}$, are parallel to the line.

The Length of a Curve

The speed of a particle is the magnitude of its velocity vector:

$$\text{Speed} = \|\vec{v}\| = \sqrt{\left(\frac{dx}{dt}\right)^2 + \left(\frac{dy}{dt}\right)^2 + \left(\frac{dz}{dt}\right)^2}.$$

As in one dimension, we can find the distance traveled by a particle along a curve by integrating its speed. Thus,

$$\text{Distance traveled} = \int_a^b \|\vec{v}(t)\|\, dt.$$

If the particle never stops or reverses its direction as it moves along the curve, the distance it travels will be the same as the length of the curve. This suggests the following formula, which is justified in Problem 37:

If the curve C is given parametrically for $a \leq t \leq b$ by smooth functions and if the velocity vector $\vec{v}$ is not $\vec{0}$ for $a < t < b$, then

$$\text{Length of } C = \int_a^b \|\vec{v}\|\, dt.$$

Example 6   Find the circumference of the ellipse given by the parametric equations

$$x = 2\cos t, \quad y = \sin t, \quad 0 \leq t \leq 2\pi.$$

Solution   The circumference of this curve is given by an integral which must be calculated numerically:

$$\text{Circumference} = \int_0^{2\pi} \sqrt{\left(\frac{dx}{dt}\right)^2 + \left(\frac{dy}{dt}\right)^2}\, dt = \int_0^{2\pi} \sqrt{(-2\sin t)^2 + (\cos t)^2}\, dt$$

$$= \int_0^{2\pi} \sqrt{4\sin^2 t + \cos^2 t}\, dt = 9.69.$$

Since the ellipse is inscribed in a circle of radius 2 and circumscribes a circle of radius 1, we would expect the length of the ellipse to be between $2\pi(2) \approx 12.57$ and $2\pi(1) \approx 6.28$, so the value of 9.69 is reasonable.

Exercises and Problems for Section 17.2

Exercises

In Exercises 1–4, find the velocity and acceleration vectors.

1. $x = t, y = t^3 - t$

2. $x = 2 + 3t, y = 4 + t, z = 1 - t$

3. $x = 3\cos t, y = 4\sin t$

4. $x = 3\cos(t^2), y = 3\sin(t^2), z = t^2$

For Exercises 5–8, find the velocity $\vec{v}(t)$ and the speed $\|\vec{v}(t)\|$. Find any times at which the particle stops.

5. $x = 3t^2$, $y = t^3 + 1$

6. $x = (t-1)^2$, $y = 2$, $z = 2t^3 - 3t^2$

7. $x = 3\sin(t^2) - 1$, $y = 3\cos(t^2)$

8. $x = 3\sin^2 t$, $y = \cos t - 1$, $z = t^2$

Find the length of the curves in Exercises 9–11.

9. $x = 3 + 5t$, $y = 1 + 4t$, $z = 3 - t$ for $1 \le t \le 2$. Explain your answer.

10. $x = \cos(e^t)$, $y = \sin(e^t)$ for $0 \le t \le 1$. Explain why your answer is reasonable.

11. $x = \cos 3t$, $y = \sin 5t$ for $0 \le t \le 2\pi$.

In Exercises 12–13, find the velocity and acceleration vectors of the uniform circular motion and check that they are perpendicular. Check that the speed and magnitude of the acceleration are constant.

12. $x = 3\cos(2\pi t)$, $y = 3\sin(2\pi t)$, $z = 0$

13. $x = 2\pi$, $y = 2\sin(3t)$, $z = 2\cos(3t)$

In Exercises 14–15, find the velocity and acceleration vectors of the straight line motion. Check that the acceleration vector points in the same direction as the velocity vector if the speed is increasing and in the opposite direction if the speed is decreasing.

14. $x = 2 + t^2$, $y = 3 - 2t^2$, $z = 5 - t^2$

15. $x = -2t^3 - 3t + 1$, $y = 4t^3 + 6t - 5$, $z = 6t^3 + 9t - 2$

16. Find parametric equations for the tangent line at $t = 2$ for Exercise 6.

Problems

17. A particle moves with $x = 3\cos 2t$, $y = 3\sin 2t$. Describe the motion in words and sketch the particle's path. Find the particle's position, velocity, and acceleration vectors at $t = 0$, and add these vectors to your sketch.

18. Using time increments of 0.01, give a table of values near $t = 1$ for the position vector of the circular motion

$$\vec{r}(t) = (\cos t)\vec{i} + (\sin t)\vec{j}.$$

Use the table to approximate the velocity vector, $\vec{v}$, at time $t = 1$. Show that $\vec{v}$ is perpendicular to the radius from the origin at $t = 1$.

19. **(a)** Sketch the parameterized curve $x = t\cos t$, $y = t\sin t$ for $0 \le t \le 4\pi$.
(b) Use difference quotients to approximate the velocity vectors $\vec{v}(t)$ for $t = 2, 4$, and 6.
(c) Compute the velocity vectors $\vec{v}(t)$ for $t = 2, 4$, and 6, exactly and sketch them on the graph of the curve.

20. The table gives x and y coordinates of a particle in the plane at time t. Assuming the path is smooth, estimate the following quantities:

(a) The velocity vector and speed at time $t = 2$.
(b) Any times when the particle is moving parallel to the y-axis.
(c) Any times when the particle has come to a stop.

t	0	0.5	1.0	1.5	2.0	2.5	3.0	3.5	4.0
x	1	4	6	7	6	3	2	3	5
y	3	2	3	5	8	10	11	10	9

In Problems 21–22, find all values of t for which the particle is moving parallel to the x-axis and to the y-axis. Determine the end behavior and graph the particle's path.

21. $x = t^2 - 6t$, $y = t^3 - 3t$

22. $x = t^3 - 12t$, $y = t^2 + 10t$

23. A particle passes through the point $P = (5, 4, -2)$ at time $t = 4$, moving with constant velocity $\vec{v} = 2\vec{i} - 3\vec{j} + \vec{k}$. Find a parametric equation for its motion.

24. A particle starts at the point $P = (3, 2, -5)$ and moves along a straight line toward $Q = (5, 7, -2)$ at a speed of 5 cm/sec. Let x, y, z be measured in centimeters.

(a) Find the particle's velocity vector.
(b) Find parametric equations for the particle's motion.

25. A particle moves at a constant speed along a line from the point $P = (2, -1, 5)$ to the point $Q = (5, 3, -1)$. Find a parametric equation of the line if:

(a) The particle takes 5 seconds to move from P to Q.
(b) The speed of the particle is 5 units per second.

26. A stone is thrown from a rooftop at time $t = 0$ seconds. Its position at time t is given by

$$\vec{r}(t) = 10t\vec{i} - 5t\vec{j} + (6.4 - 4.9t^2)\vec{k}.$$

The origin is at the base of the building, which is standing on flat ground. Distance is measured in meters. The vector $\vec{i}$ points east, $\vec{j}$ points north, and $\vec{k}$ points up.

(a) How high is the rooftop above the ground?
(b) At what time does the stone hit the ground?
(c) How fast is the stone moving when it hits the ground?
(d) Where does the stone hit the ground?
(e) What is the stone's acceleration when it hits the ground?

27. The base of a 20-meter tower is at the origin; the base of a 20-meter tree is at $(0, 20, 0)$. The ground is flat and the z-axis points upward. The following parametric equations describe the motion of six projectiles each launched at time $t = 0$ in seconds.

(I) $\vec{r}(t) = (20 + t^2)\vec{k}$

(II) $\vec{r}(t) = 2t^2\vec{j} + 2t^2\vec{k}$

(III) $\vec{r}(t) = 20\vec{i} + 20\vec{j} + (20 - t^2)\vec{k}$

(IV) $\vec{r}(t) = 2t\vec{j} + (20 - t^2)\vec{k}$

(V) $\vec{r}(t) = (20 - 2t)\vec{i} + 2t\vec{j} + (20 - t)\vec{k}$

(VI) $\vec{r}(t) = t\vec{i} + t\vec{j} + t\vec{k}$

(a) Which projectile is launched from the top of the tower and goes downward? When and where does it hit the ground?

(b) Which projectile hits the top of the tree? When? From where is it launched?

(c) Which projectile is not launched from somewhere on the tower and hits the tree? Where and when does it hit the tree?

28. A particle moves on a circle of radius 5 cm, centered at the origin, in the xy-plane (x and y measured in centimeters). It starts at the point $(0, 5)$ and moves counterclockwise, going once around the circle in 8 seconds.

(a) Write a parameterization for the particle's motion.

(b) What is the particle's speed? Give units.

29. Determine the position vector $\vec{r}(t)$ for a rocket which is launched from the origin at time $t = 0$ seconds, reaches its highest point of $(x, y, z) = (1000, 3000, 10,000)$, where x, y, z are in meters, and after the launch is subject only to the acceleration due to gravity, 9.8 m/sec^2.

30. Suppose $\vec{r}(t) = \cos t\, \vec{i} + \sin t\, \vec{j} + 2t\, \vec{k}$ represents the position of a particle on a helix, where z is the height of the particle above the ground.

(a) Is the particle ever moving downward? When?

(b) When does the particle reach a point 10 units above the ground?

(c) What is the velocity of the particle when it is 10 units above the ground?

(d) When it is 10 units above the ground, the particle leaves the helix and moves along the tangent. Find parametric equations for this tangent line.

31. Emily is standing on the outer edge of a merry-go-round, 10 meters from the center. The merry-go-round completes one full revolution every 20 seconds. As Emily passes over a point P on the ground, she drops a ball from 3 meters above the ground.

(a) How fast is Emily going?

(b) How far from P does the ball hit the ground? (The acceleration due to gravity is 9.8 m/sec^2.)

(c) How far from Emily does the ball hit the ground?

32. A wheel of radius 1 meter rests on the x-axis with its center on the y-axis. There is a spot on the rim at the point

$(1, 1)$. See Figure 17.13. At time $t = 0$ the wheel starts rolling on the x-axis in the direction shown at a rate of 1 radian per second.

(a) Find parametric equations describing the motion of the center of the wheel.

(b) Find parametric equations describing the motion of the spot on the rim. Plot its path.

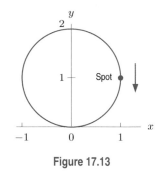

Figure 17.13

33. An ant crawls along the radius from the center to the edge of a circular disk of radius 1 meter, moving at a constant rate of 1 cm/sec. Meanwhile, the disk is turning counterclockwise about its center at 1 revolution per second.

(a) Parameterize the motion of the ant.

(b) Find the velocity and speed of the ant.

(c) Determine the acceleration and magnitude of the acceleration of the ant.

34. How do the motions of objects A and B differ, if A has position vector $\vec{r}_A(t)$ and B has position vector $\vec{r}_B(t) = \vec{r}_A(2t)$ for $t \geq 0$. Illustrate your answer with $\vec{r}_A(t) = t\vec{i} + t^2\vec{j}$.

35. At time $t = 0$ an object is moving with velocity vector $\vec{v} = 2\vec{i} + \vec{j}$ and acceleration vector $\vec{a} = \vec{i} + \vec{j}$. Can it be in uniform circular motion about some point in the plane?

36. Figure 17.14 shows the velocity and acceleration vectors of an object in uniform circular motion about a point in the plane at a particular moment. Is it moving round the circle in the clockwise or counterclockwise direction?

Figure 17.14

37. In this problem we justify the formula for the length of a curve given on page 844. Suppose the curve C is given by smooth parametric equations $x = x(t)$, $y = y(t)$, $z = z(t)$ for $a \leq t \leq b$. By dividing the parameter interval $a \leq t \leq b$ at points $t_1, \ldots, t_{n-1}$ into small segments of length $\Delta t = t_{i+1} - t_i$, we get a corresponding division of the curve C into small pieces. See Figure 17.15, where the points $P_i = (x(t_i), y(t_i), z(t_i))$ on the curve C correspond to parameter values $t = t_i$. Let C_i be the portion of the curve C between P_i and P_{i+1}.

(a) Use local linearity to show that

$$\text{Length of } C_i \approx \sqrt{x'(t_i)^2 + y'(t_i)^2 + z'(t_i)^2}\, \Delta t.$$

(b) Use part (a) and a Riemann sum to explain why

$$\text{Length of } C = \int_a^b \sqrt{x'(t)^2 + y'(t)^2 + z'(t)^2}\, dt.$$

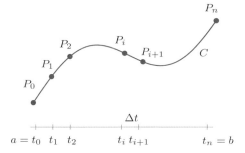

Figure 17.15: A subdivision of the parameter interval and the corresponding subdivision of the curve C

17.3 VECTOR FIELDS

Introduction to Vector Fields

A *vector field* is a function that assigns a vector to each point in the plane or in 3-space. One example of a vector field is the gradient of a function $f(x, y)$; at each point (x, y) the vector $\text{grad } f(x, y)$ points in the direction of maximum rate of increase of f. In this section we will look at other vector fields representing velocities and forces.

Velocity Vector Fields

Figure 17.16 shows the flow of a part of the Gulf stream, a current in the Atlantic Ocean.[1] It is an example of a *velocity vector field*: each vector shows the velocity of the current at that point. The current is fastest where the velocity vectors are longest in the middle of the stream. Beside the stream are eddies where the water flows round and round in circles.

Figure 17.16: The velocity vector field of the Gulf stream

[1]Based on data supplied by Avijit Gangopadhyay of the Jet Propulsion Laboratory.

Figure 17.17: The gravitational field of the earth

Force Fields

Another physical quantity represented by a vector is force. When we experience a force, sometimes it results from direct contact with the object that supplies the force (for example, a push). Many forces, however, can be felt at all points in space. For example, the earth exerts a gravitational pull on all other masses. Such forces can be represented by vector fields.

   Figure 17.17 shows the gravitational force exerted by the earth on a mass of one kilogram at different points in space. This is a sketch of the vector field in 3-space. You can see that the vectors all point toward the earth (which is not shown in the diagram) and that the vectors further from the earth are smaller in magnitude.

Definition of a Vector Field

Now that you have seen some examples of vector fields we give a more formal definition.

> A **vector field** in 2-space is a function $\vec{F}(x, y)$ whose value at a point (x, y) is a 2-dimensional vector. Similarly, a vector field in 3-space is a function $\vec{F}(x, y, z)$ whose values are 3-dimensional vectors.

   Notice the arrow over the function, $\vec{F}$, indicating that its value is a vector, not a scalar. We often represent the point (x, y) or (x, y, z) by its position vector $\vec{r}$ and write the vector field as $\vec{F}(\vec{r})$.

Visualizing a Vector Field Given by a Formula

Since a vector field is a function that assigns a vector to each point, a vector field can often be given by a formula.

Example 1   Sketch the vector field in 2-space given by $\vec{F}(x, y) = -y\vec{i} + x\vec{j}$.

Solution   Table 17.1 shows the value of the vector field at a few points. Notice that each value is a vector. To plot the vector field, we plot $\vec{F}(x, y)$ with its tail at (x, y). (See Figure 17.18.)

Table 17.1   Values of $\vec{F}(x, y) = -y\vec{i} + x\vec{j}$

		y		
		-1	0	1
	-1	$\vec{i} - \vec{j}$	$-\vec{j}$	$-\vec{i} - \vec{j}$
x	0	$\vec{i}$	$\vec{0}$	$-\vec{i}$
	1	$\vec{i} + \vec{j}$	$\vec{j}$	$-\vec{i} + \vec{j}$

   Now we look at the formula. The magnitude of the vector at (x, y) is $\|\vec{F}(x, y)\| = \| - y\vec{i} + x\vec{j}\| = \sqrt{x^2 + y^2}$, which is the distance from (x, y) to the origin. Therefore, all the vectors at a fixed distance from the origin (that is, on a circle centered at the origin) have the same magnitude. The magnitude gets larger as we move further from the origin. What about the direction? Figure 17.18 suggests that at each point (x, y) the vector $\vec{F}(x, y)$ is perpendicular to the position vector $\vec{r} = x\vec{i} + y\vec{j}$. We confirm this using the dot product: $\vec{r} \cdot \vec{F}(x, y) = (x\vec{i} + y\vec{j}) \cdot (-y\vec{i} + x\vec{j}) = 0$. Thus,

Figure 17.18: The value $\vec{F}(x, y)$ is placed at the point (x, y)

Figure 17.19: The vector field $\vec{F}(x, y) = -y\vec{i} + x\vec{j}$, vectors scaled smaller to fit in diagram

the vectors are tangent to circles centered at the origin and get longer as we go out. In Figure 17.19, the vectors have been scaled so that they do not obscure each other.

Example 2   Sketch the vector fields in 2-space given by (a) $\vec{F}(x, y) = x\vec{j}$ (b) $\vec{G}(x, y) = x\vec{i}$.

Solution   (a) The vector $x\vec{j}$ is parallel to the y-direction, pointing up when x is positive and down when x is negative. Also, the larger $|x|$ is, the longer the vector. The vectors in the field are constant along vertical lines since the vector field does not depend on y. (See Figure 17.20.)

Figure 17.20: The vector field $\vec{F}(x, y) = x\vec{j}$

Figure 17.21: The vector field $\vec{F}(x, y) = x\vec{i}$

(b) This is similar to the previous example, except that the vector $x\vec{i}$ is parallel to the x-direction, pointing to the right when x is positive and to the left when x is negative. Again, the larger $|x|$ is the longer the vector, and the vectors are constant along vertical lines, since the vector field does not depend on y. (See Figure 17.21.)

Example 3   Describe the vector field in 3-space given by $\vec{F}(\vec{r}) = \vec{r}$, where $\vec{r} = x\vec{i} + y\vec{j} + z\vec{k}$.

Solution   The notation $\vec{F}(\vec{r}) = \vec{r}$ means that the value of $\vec{F}$ at the point (x, y, z) with position vector $\vec{r}$ is the vector $\vec{r}$ with its tail at (x, y, z). Thus, the vector field points outward. See Figure 17.22. Note that the lengths of the vectors have been scaled down so as to fit into the diagram.

Figure 17.22: The vector field $\vec{F}(\vec{r}) = \vec{r}$

$M \bullet$

Figure 17.23: Force exerted on mass m by mass M

Finding a Formula for a Vector Field

Example 4   Newton's Law of Gravitation states that the magnitude of the gravitational force exerted by an object of mass M on an object of mass m is proportional to M and m and inversely proportional to the square of the distance between them. The direction of the force is from m to M along the line connecting them. (See Figure 17.23.) Find a formula for the vector field $\vec{F}(\vec{r})$ that represents the gravitational force, assuming M is located at the origin and m is located at the point with position vector $\vec{r}$.

Solution   Since the mass m is located at $\vec{r}$, Newton's law says that the magnitude of the force is given by

$$\|\vec{F}(\vec{r})\| = \frac{GMm}{\|\vec{r}\|^2},$$

where G is the universal gravitational constant. A unit vector in the direction of the force is $-\vec{r}/\|\vec{r}\|$, where the negative sign indicates that the direction of force is toward the origin (gravity is attractive). By taking the product of the magnitude of the force and a unit vector in the direction of the force, we obtain an expression for the force vector field:

$$\vec{F}(\vec{r}) = \frac{GMm}{\|\vec{r}\|^2}\left(-\frac{\vec{r}}{\|\vec{r}\|}\right) = \frac{-GMm\vec{r}}{\|\vec{r}\|^3}.$$

We have already seen a picture of this vector field in Figure 17.17.

Gradient Vector Fields

The gradient of a scalar function f is a function that assigns a vector to each point, and is therefore a vector field. It is called the *gradient field* of f. Many vector fields in physics are gradient fields.

Example 5   Sketch the gradient field of the functions in Figures 17.24–17.26.

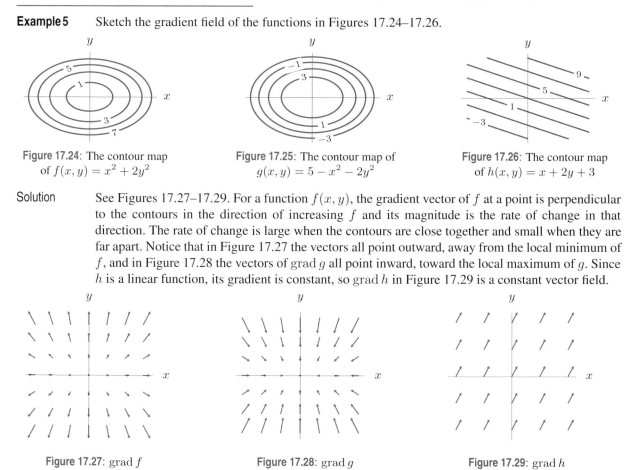

Figure 17.24: The contour map of $f(x, y) = x^2 + 2y^2$

Figure 17.25: The contour map of $g(x, y) = 5 - x^2 - 2y^2$

Figure 17.26: The contour map of $h(x, y) = x + 2y + 3$

Solution   See Figures 17.27–17.29. For a function $f(x, y)$, the gradient vector of f at a point is perpendicular to the contours in the direction of increasing f and its magnitude is the rate of change in that direction. The rate of change is large when the contours are close together and small when they are far apart. Notice that in Figure 17.27 the vectors all point outward, away from the local minimum of f, and in Figure 17.28 the vectors of grad g all point inward, toward the local maximum of g. Since h is a linear function, its gradient is constant, so grad h in Figure 17.29 is a constant vector field.

Figure 17.27: grad f

Figure 17.28: grad g

Figure 17.29: grad h

Exercises and Problems for Section 17.3

Exercises

For Exercises 1–6, find formulas for the vector fields. (There are many possible answers.)

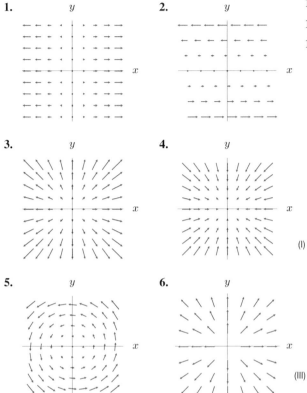

1.

2.

3.

4.

5.

6.

11. $\vec{F}(\vec{r}) = \vec{r}/\|\vec{r}\|$

12. $\vec{F}(\vec{r}) = -\vec{r}/\|\vec{r}\|^3$

13. $\vec{F}(x,y) = -y\vec{i} + x\vec{j}$

14. $\vec{F}(x,y) = 2x\vec{i} + x\vec{j}$

15. $\vec{F}(x,y) = (x+y)\vec{i} + (x-y)\vec{j}$

16. Each vector field in Figures (I)–(IV) represents the force on a particle at different points in space as a result of another particle at the origin. Match up the vector fields with the descriptions below.

   (a) A repulsive force whose magnitude decreases as distance increases, such as between electric charges of the same sign.

   (b) A repulsive force whose magnitude increases as distance increases.

   (c) An attractive force whose magnitude decreases as distance increases, such as gravity.

   (d) An attractive force whose magnitude increases as distance increases.

(I)

(II)

(III)

(IV)

Sketch the vector fields in Exercises 7–15 in the xy-plane.

7. $\vec{F}(x,y) = 2\vec{i} + 3\vec{j}$

8. $\vec{F}(x,y) = y\vec{i}$

9. $\vec{F}(x,y) = -y\vec{j}$

10. $\vec{F}(\vec{r}) = 2\vec{r}$

Problems

17. Match the level curves in (I)–(IV) with the gradient fields in (A)–(D). All figures have $-2 \le x \le 2$, $-2 \le y \le 2$.

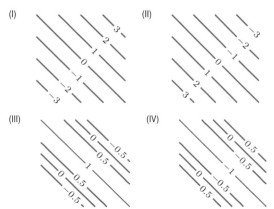

(I)

(II)

(III)

(IV)

(A)

(B)

(C)

(D)

Problems 18–19 concern the vector fields $\vec{F} = x\vec{i} + y\vec{j}$, $\vec{G} = -y\vec{i} + x\vec{j}$, and $\vec{H} = x\vec{i} - y\vec{j}$.

18. Match $\vec{F}, \vec{G}, \vec{H}$ with their sketches in (I)–(III).

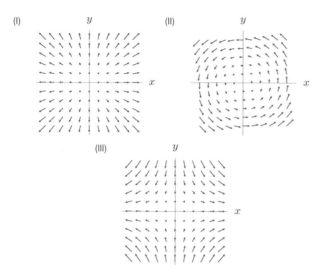

19. Match the vector fields with their sketches, (I)–(IV).

(a) $\vec{F} + \vec{G}$ (b) $\vec{F} + \vec{H}$ (c) $\vec{G} + \vec{H}$ (d) $-\vec{F} + \vec{G}$

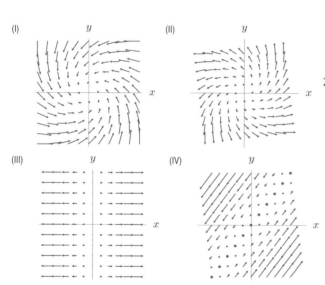

In Problems 20–22, write formulas for vector fields with the given properties.

20. All vectors are parallel to the x-axis; all vectors on a vertical line have the same magnitude.

21. All vectors point toward the origin and have constant length.

22. All vectors are of unit length and perpendicular to the position vector at that point.

23. Figures 17.30 and 17.31 show the gradient of the functions $z = f(x, y)$ and $z = g(x, y)$.

(a) For each function, draw a rough sketch of the level curves, showing possible z-values.

(b) The xz-plane cuts each of the surfaces $z = f(x, y)$ and $z = g(x, y)$ in a curve. Sketch each of these curves, making clear how they are similar and how they are different from one another.

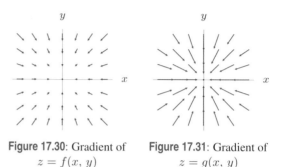

Figure 17.30: Gradient of $z = f(x, y)$ **Figure 17.31**: Gradient of $z = g(x, y)$

24. (a) Let $\vec{F} = x\vec{i} + (x + y)\vec{j} + (x - y + z)\vec{k}$. Find a point at which $\vec{F}$ is parallel to l, the line $x = 5 + t$, $y = 6 - 2t, z = 7 - 3t$.

(b) Find a point at which $\vec{F}$ and l are perpendicular.

(c) Give an equation for and describe in words the set of all points at which $\vec{F}$ and l are perpendicular.

25. In the middle of a wide, steadily flowing river there is a fountain that spouts water horizontally in all directions. The river flows in the $\vec{i}$-direction in the xy-plane and the fountain is at the origin.

(a) If $A > 0, K > 0$, explain why the following expression could represent the velocity field for the combined flow of the river and the fountain:

$$\vec{v} = A\vec{i} + K(x^2 + y^2)^{-1}(x\vec{i} + y\vec{j}).$$

(b) What is the significance of the constants A and K?

(c) Using a computer, sketch the vector field $\vec{v}$ for $K = 1$ and $A = 1$ and $A = 2$, and for $A = 0.2, K = 2$.

In Problems 26–27, let $\vec{F} = x\vec{i} + y\vec{j}$ and $\vec{G} = -y\vec{i} + x\vec{j}$.

26. Sketch the vector field $\vec{L} = a\vec{F} + \vec{G}$ if:

(a) $a = 0$ (b) $a > 0$ (c) $a < 0$

27. Sketch the vector field $\vec{L} = \vec{F} + b\vec{G}$ if:

(a) $b = 0$ (b) $b > 0$ (c) $b < 0$

28. Let $\vec{F} = u\vec{i} + v\vec{j}$ be a vector field in 2-space with magnitude $F = \|\vec{F}\|$.

(a) Let $\vec{T} = (1/F)\vec{F}$. Show that $\vec{T}$ is the unit vector in the direction of $\vec{F}$. See Figure 17.32.

(b) Let $\vec{N} = (1/F)(\vec{k} \times \vec{F}) = (1/F)(-v\vec{i} + u\vec{j})$. Show that $\vec{N}$ is the unit vector pointing to the left of and at right angles to $\vec{F}$. See Figure 17.32.

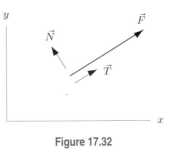

Figure 17.32

17.4 THE FLOW OF A VECTOR FIELD

When an iceberg is spotted in the North Atlantic, it is important to be able to predict where the iceberg is likely to be a day or a week later. To do this, one needs to know the velocity vector field of the ocean currents, that is, how fast and in what direction the water is moving at each point.

In this section we use differential equations to find the path of an object in a fluid flow. This path is called a *flow line*. Figure 17.33 shows several flow lines for the Gulf stream velocity vector field in Figure 17.16 on page 847. The arrows on each flow line indicate the direction of flow.

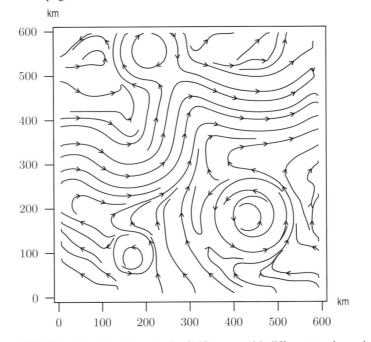

Figure 17.33: Flow lines for objects in the Gulf stream with different starting points

How Do We Find a Flow Line?

Suppose that $\vec{F}$ is the velocity vector field of water on the surface of a creek and imagine a seed being carried along by the current. We want to know the position vector $\vec{r}(t)$ of the seed at time t. We know

$$\begin{array}{c}\text{Velocity of seed}\\\text{at time } t\end{array} = \begin{array}{c}\text{Velocity of current at seed's position}\\\text{at time } t\end{array}$$

that is,

$$\vec{r}\,'(t) = \vec{F}\,(\vec{r}\,(t)).$$

We make the following definition:

A **flow line** of a vector field $\vec{v} = \vec{F}(\vec{r})$ is a path $\vec{r}(t)$ whose velocity vector equals $\vec{v}$. Thus

$$\vec{r}'(t) = \vec{v} = \vec{F}(\vec{r}(t)).$$

The **flow** of a vector field is the family of all of its flow lines.

A flow line is also called an *integral curve* or a *streamline*. We define flow lines for any vector field, as it turns out to be useful to study the flow of fields (for example, electric and magnetic) that are not velocity fields.

Resolving $\vec{F}$ and $\vec{r}$ into components, $\vec{F} = F_1\vec{i} + F_2\vec{j}$ and $\vec{r}(t) = x(t)\vec{i} + y(t)\vec{j}$, the definition of a flow line tells us that $x(t)$ and $y(t)$ satisfy the system of differential equations

$$x'(t) = F_1(x(t), y(t)) \quad \text{and} \quad y'(t) = F_2(x(t), y(t)).$$

Solving these differential equations gives a parameterization of the flow line.

Example 1   Find the flow line of the constant velocity field $\vec{v} = 3\vec{i} + 4\vec{j}$ cm/sec that passes through the point $(1, 2)$ at time $t = 0$.

Solution   Let $\vec{r}(t) = x(t)\vec{i} + y(t)\vec{j}$ be the position in cm of a particle at time t, where t is in seconds. We have

$$x'(t) = 3 \quad \text{and} \quad y'(t) = 4.$$

Thus,

$$x(t) = 3t + x_0 \quad \text{and} \quad y(t) = 4t + y_0.$$

Since the path passes the point $(1, 2)$ at $t = 0$, we have $x_0 = 1$ and $y_0 = 2$ and so

$$x(t) = 3t + 1 \quad \text{and} \quad y(t) = 4t + 2.$$

Thus, the path is the line given parametrically by

$$\vec{r}(t) = (3t + 1)\vec{i} + (4t + 2)\vec{j}.$$

(See Figure 17.34.) To find an explicit equation for the path, eliminate t between these expressions to get

$$\frac{x - 1}{3} = \frac{y - 2}{4} \quad \text{or} \quad y = \frac{4}{3}x + \frac{2}{3}.$$

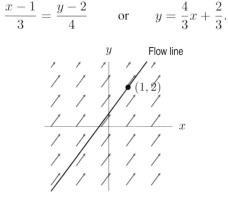

Figure 17.34: Vector field $\vec{F} = 3\vec{i} + 4\vec{j}$ with the flow line through $(1, 2)$

Example 2   The velocity of a flow at the point (x, y) is $\vec{F}(x, y) = \vec{i} + x\vec{j}$. Find the path of motion of an object in the flow that is at the point $(-2, 2)$ at time $t = 0$.

Solution

Figure 17.35 shows a sketch of this field. Since $\vec{r}\,'(t) = \vec{F}\,(\vec{r}\,(t))$, we are looking for the flow line that satisfies the system of differential equations

$$x'(t) = 1, \quad y'(t) = x(t).$$

Solving for $x(t)$ first, we get $x(t) = t + x_0$, where x_0 is a constant of integration. Thus, $y'(t) = t + x_0$, so $y(t) = \frac{1}{2}t^2 + x_0 t + y_0$, where y_0 is also a constant of integration. Since $x(0) = x_0 = -2$ and $y(0) = y_0 = 2$, the path of motion is given by

$$x(t) = t - 2, \quad y(t) = \tfrac{1}{2}t^2 - 2t + 2,$$

or, equivalently,

$$\vec{r}\,(t) = (t-2)\vec{i} + (\tfrac{1}{2}t^2 - 2t + 2)\vec{j}\,.$$

The graph of this flow line in Figure 17.36 looks like a parabola. We check this by seeing that an explicit equation for the path is $y = \frac{1}{2}x^2$.

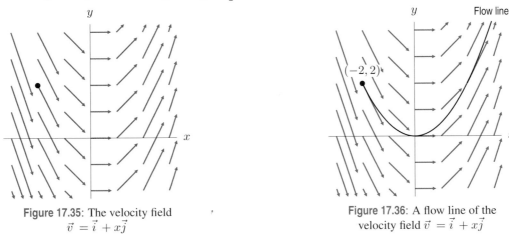

Figure 17.35: The velocity field $\vec{v} = \vec{i} + x\vec{j}$

Figure 17.36: A flow line of the velocity field $\vec{v} = \vec{i} + x\vec{j}$

Example 3   Determine the flow of the vector field $\vec{v} = -y\vec{i} + x\vec{j}$.

Solution

Figure 17.37 suggests that the flow consists of concentric counterclockwise circles, centered at the origin. The system of differential equations for the flow is

$$x'(t) = -y(t) \qquad y'(t) = x(t).$$

The equations $(x(t), y(t)) = (a\cos t, a\sin t)$ parameterize a family of counterclockwise circles of radius a, centered at the origin. We check that this family satisfies the system of differential equations:

$$x'(t) = -a\sin t = -y(t) \quad \text{and} \quad y'(t) = a\cos t = x(t).$$

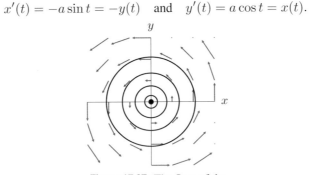

Figure 17.37: The flow of the vector field $\vec{v} = -y\vec{i} + x\vec{j}$

Approximating Flow Lines Numerically

Often it is not possible to find formulas for the flow lines of a vector field. However, we can approximate them numerically by Euler's method for solving differential equations. Since the flow lines $\vec{r}(t) = x(t)\vec{i} + y(t)\vec{j}$ of a vector field $\vec{v} = \vec{F}(x, y)$ satisfy the differential equation $\vec{r}'(t) = \vec{F}(\vec{r}(t))$, we have

$$\vec{r}(t + \Delta t) \approx \vec{r}(t) + (\Delta t)\vec{r}'(t)$$
$$= \vec{r}(t) + (\Delta t)\vec{F}(\vec{r}(t)) \quad \text{for } \Delta t \text{ near } 0.$$

To approximate a flow line, we start at a point $\vec{r}_0 = \vec{r}(0)$ and estimate the position $\vec{r}_1$ of a particle at time Δt later:

$$\vec{r}_1 = \vec{r}(\Delta t) \approx \vec{r}(0) + (\Delta t)\vec{F}(\vec{r}(0))$$
$$= \vec{r}_0 + (\Delta t)\vec{F}(\vec{r}_0).$$

We then repeat the same procedure starting at $\vec{r}_1$, and so on. The general formula for getting from one point to the next is

$$\vec{r}_{n+1} = \vec{r}_n + (\Delta t)\vec{F}(\vec{r}_n).$$

The points with position vectors $\vec{r}_0, \vec{r}_1, \ldots$ trace out the path, as shown in the next example.

Example 4 Use Euler's method to approximate the flow line through $(1, 2)$ for the vector field $\vec{v} = y^2\vec{i} + 2x^2\vec{j}$.

Solution The flow is determined by the differential equations $\vec{r}'(t) = \vec{v}$, or equivalently

$$x'(t) = y^2, \qquad y'(t) = 2x^2.$$

We use Euler's method with $\Delta t = 0.02$, giving

$$\vec{r}_{n+1} = \vec{r}_n + 0.02\,\vec{v}(x_n, y_n)$$
$$= x_n\vec{i} + y_n\vec{j} + 0.02(y_n^2\vec{i} + 2x_n^2\vec{j}),$$

or equivalently,

$$x_{n+1} = x_n + 0.02y_n^2, \qquad y_{n+1} = y_n + 0.02 \cdot 2x_n^2.$$

When $t = 0$, we have $(x_0, y_0) = (1, 2)$. Then

$$x_1 = x_0 + 0.02 \cdot y_0^2 = 1 + 0.02 \cdot 2^2 = 1.08,$$
$$y_1 = y_0 + 0.02 \cdot 2x_0^2 = 2 + 0.02 \cdot 2 \cdot 1^2 = 2.04.$$

So after one step $x(0.02) \approx 1.08$ and $y(0.02) \approx 2.04$. Similarly, $x(0.04) = x(2\Delta t) \approx 1.16$, $y(0.04) = y(2\Delta t) \approx 2.08$ and so on. Further values along the flow line are given in Table 17.2 and plotted in Figure 17.38.

Table 17.2 *Approximated flow line starting at $(1, 2)$ for the vector field $\vec{v} = y^2\vec{i} + 2x^2\vec{j}$*

t	0	0.02	0.04	0.06	0.08	0.1	0.12	0.14	0.16	0.18
x	1	1.08	1.16	1.25	1.34	1.44	1.54	1.65	1.77	1.90
y	2	2.04	2.08	2.14	2.20	2.28	2.36	2.45	2.56	2.69

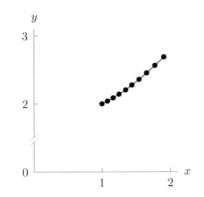

Figure 17.38: Euler's method solution to $x' = y^2$, $y' = 2x^2$

Exercises and Problems for Section 17.4

Exercises

In Exercises 1–3, sketch the vector field and its flow.

1. $\vec{v} = 3\vec{i}$ **2.** $\vec{v} = 2\vec{j}$ **3.** $\vec{v} = 3\vec{i} - 2\vec{j}$

In Exercises 4–9, sketch the vector field and the flow. Then find the system of differential equations associated with the vector field and check that the flow satisfies the system.

4. $\vec{v} = x\vec{i} + y\vec{j}$; $x(t) = ae^t$, $y(t) = be^t$.

5. $\vec{v} = x\vec{i}$; $x(t) = ae^t$, $y(t) = b$.

6. $\vec{v} = x\vec{j}$; $x(t) = a$, $y(t) = at + b$

7. $\vec{v} = x\vec{i} - y\vec{j}$; $x(t) = ae^t$, $y(t) = be^{-t}$.

8. $\vec{v} = y\vec{i} - x\vec{j}$; $x(t) = a \sin t$, $y(t) = a \cos t$.

9. $\vec{v} = y\vec{i} + x\vec{j}$; $x(t) = a(e^t + e^{-t})$, $y(t) = a(e^t - e^{-t})$.

10. Use a computer or calculator with Euler's method to approximate the flow line through $(1, 2)$ for the vector field $\vec{v} = y^2\vec{i} + 2x^2\vec{j}$ using 5 steps with $\Delta t = 0.1$.

Problems

For Problems 11–14, find the region of the Gulf stream velocity field in Figure 17.16 on page 847 represented by the given table of velocity vectors (in cm/sec).

11.

$35\vec{i} + 131\vec{j}$	$48\vec{i} + 92\vec{j}$	$47\vec{i} + \vec{j}$
$-32\vec{i} + 132\vec{j}$	$-44\vec{i} + 92\vec{j}$	$-42\vec{i} + \vec{j}$
$-51\vec{i} + 73\vec{j}$	$-119\vec{i} + 84\vec{j}$	$-128\vec{i} + 6\vec{j}$

12.

$10\vec{i} - 3\vec{j}$	$11\vec{i} + 16\vec{j}$	$20\vec{i} + 75\vec{j}$
$53\vec{i} - 7\vec{j}$	$58\vec{i} + 23\vec{j}$	$64\vec{i} + 80\vec{j}$
$119\vec{i} - 8\vec{j}$	$121\vec{i} + 31\vec{j}$	$114\vec{i} + 66\vec{j}$

13.

$97\vec{i} - 41\vec{j}$	$72\vec{i} - 24\vec{j}$	$54\vec{i} - 10\vec{j}$
$134\vec{i} - 49\vec{j}$	$131\vec{i} - 44\vec{j}$	$129\vec{i} - 18\vec{j}$
$103\vec{i} - 36\vec{j}$	$122\vec{i} - 30\vec{j}$	$131\vec{i} - 17\vec{j}$

14.

$-95\vec{i} - 60\vec{j}$	$18\vec{i} - 48\vec{j}$	$82\vec{i} - 22\vec{j}$
$-29\vec{i} + 48\vec{j}$	$76\vec{i} + 63\vec{j}$	$128\vec{i} - 16\vec{j}$
$26\vec{i} + 105\vec{j}$	$49\vec{i} + 119\vec{j}$	$88\vec{i} + 13\vec{j}$

15. $\vec{F}(x, y)$ and $\vec{G}(x, y) = 2\vec{F}(x, y)$ are two vector fields. Illustrating your answer with $\vec{F}(x, y) = -y\vec{i} + x\vec{j}$, describe the graphical difference between:

  (a) The vector fields   **(b)** Their flows

16. Show that the acceleration $\vec{a}$ of an object flowing in a velocity field $\vec{F}(x, y) = u(x, y)\vec{i} + v(x, y)\vec{j}$ is given by $\vec{a} = (u_x u + u_y v)\vec{i} + (v_x u + v_y v)\vec{j}$

17. A velocity vector field $\vec{v} = -H_y\vec{i} + H_x\vec{j}$ is based on the partial derivatives of a smooth function $H(x, y)$. Explain why

  (a) $\vec{v}$ is perpendicular to grad H.

  (b) the flow lines of $\vec{v}$ are along the level curves of H.

18. Match the vector fields (a)–(f) with their flow lines (I)–(VI). Put arrows on the flow lines indicating the direction of flow.

(a) $y\vec{i} + x\vec{j}$ (b) $-y\vec{i} + x\vec{j}$
(c) $x\vec{i} + y\vec{j}$ (d) $-y\vec{i} + (x + y/10)\vec{j}$
(e) $-y\vec{i} + (x - y/10)\vec{j}$ (f) $(x - y)\vec{i} + (x - y)\vec{j}$

19. We have a fixed set of axes and a solid metal ball whose center is at the origin. The ball rotates once every 24 hours around the z-axis. The direction of rotation is counterclockwise when viewed from above. Consider the point (x, y, z) in our coordinate system lying within the ball. Let $\vec{v}(x, y, z)$ be the velocity vector of the particle of metal at this point. Assume x, y, z are in meters and time is in hours.

(a) Find a formula for the vector field $\vec{v}$. Give units for your answer.

(b) Describe in words the flow lines of $\vec{v}$.

17.5 PARAMETERIZED SURFACES

How Do We Parameterize a Surface?

In Section 17.1 we parameterized a circle in 2-space using the equations

$$x = \cos t, \quad y = \sin t.$$

In 3-space, the same circle in the xy-plane has parametric equations

$$x = \cos t, \quad y = \sin t, \quad z = 0.$$

We add the equation $z = 0$ to specify that the circle is in the xy-plane. If we wanted a circle in the plane $z = 3$, we would use the equations

$$x = \cos t, \quad y = \sin t, \quad z = 3.$$

Suppose now we let z vary freely, as well as t. We get circles in every horizontal plane, forming the cylinder in Figure 17.39. Thus, we need two parameters, t and z, to parameterize the cylinder.

This is true in general. A curve, though it may live in two or three dimensions, is itself one-dimensional; if we move along it we can only move backward and forward in one direction. Thus, it only requires one parameter to trace out a curve.

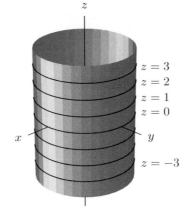

Figure 17.39: The cylinder $x = \cos t,\ y = \sin t,\ z = z$

A surface is 2-dimensional; at any given point there are two independent directions we can move in. For example, on the cylinder we can move vertically, or we can circle around the z-axis horizontally. So we need *two* parameters to describe it. We can think of the parameters as map coordinates, like longitude and latitude on the surface of the earth.

In the case of the cylinder our parameters are t and z, so

$$x = \cos t, \quad y = \sin t, \quad z = z, \qquad 0 \le t < 2\pi, \quad -\infty < z < \infty.$$

The last equation, $z = z$, looks strange, but it reminds us that we are in three dimensions, not two, and that the z-coordinate on our surface is allowed to vary freely.

In general, we express the coordinates, (x, y, z) of a point on a surface S in terms of two parameters, s and t:

$$x = f_1(s, t), \quad y = f_2(s, t), \quad z = f_3(s, t).$$

As the values of s and t vary, the corresponding point (x, y, z) sweeps out the surface, S. (See Figure 17.40.) The function which sends the point (s, t) to the point (x, y, z) is called the *parameterization of the surface.*

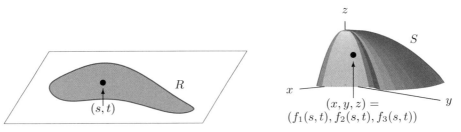

Figure 17.40: The parameterization sends each point (s, t) in the parameter region, R, to a point $(x, y, z) = (f_1(s, t), f_2(s, t), f_3(s, t))$ in the surface, S

Using Position Vectors

We can use the position vector $\vec{r} = x\vec{i} + y\vec{j} + z\vec{k}$ to combine the three parametric equations for a surface into a single vector equation. For example, the parameterization of the cylinder $x = \cos t, y = \sin t, z = z$ can be written as

$$\vec{r}(t, z) = \cos t\vec{i} + \sin t\vec{j} + z\vec{k} \qquad 0 \le t < 2\pi, \quad -\infty < z < \infty.$$

For a general parameterized surface S, we write

$$\vec{r}(s, t) = f_1(s, t)\vec{i} + f_2(s, t)\vec{j} + f_3(s, t)\vec{k}.$$

Parameterizing a Surface of the Form $z = f(x, y)$

The graph of a function $z = f(x, y)$ can be given parametrically simply by letting the parameters s and t be x and y:

$$x = s, \quad y = t, \quad z = f(s, t).$$

Example 1   Give a parametric description of the lower hemisphere of the sphere $x^2 + y^2 + z^2 = 1$.

Solution   The surface is the graph of the function $z = -\sqrt{1 - x^2 - y^2}$ over the region $x^2 + y^2 \le 1$ in the plane. Then parametric equations are $x = s$, $y = t$, $z = -\sqrt{1 - s^2 - t^2}$, where the parameters s and t vary inside the unit circle.

In practice we often think of x and y as parameters rather than introduce new variables s and t. Thus, we may write $x = x$, $y = y$, $z = f(x, y)$.

Parameterizing Planes

Consider a plane containing two nonparallel vectors $\vec{v}_1$ and $\vec{v}_2$ and a point P_0 with position vector $\vec{r}_0$. We can get to any point on the plane by starting at P_0 and moving parallel to $\vec{v}_1$ or $\vec{v}_2$, adding multiples of them to $\vec{r}_0$. (See Figure 17.41.)

Figure 17.41: The plane $\vec{r}(s, t) = \vec{r}_0 + s\vec{v}_1 + t\vec{v}_2$ and some points corresponding to various choices of s and t

Since $s\vec{v}_1$ is parallel to $\vec{v}_1$ and $t\vec{v}_2$ is parallel to $\vec{v}_2$, we have the following result:

Parametric Equations for a Plane

The plane through the point with position vector $\vec{r}_0$ and containing the two nonparallel vectors $\vec{v}_1$ and $\vec{v}_2$ has parametric equation

$$\vec{r}(s, t) = \vec{r}_0 + s\vec{v}_1 + t\vec{v}_2.$$

If $\vec{r}_0 = x_0\vec{i} + y_0\vec{j} + z_0\vec{k}$, and $\vec{v}_1 = a_1\vec{i} + a_2\vec{j} + a_3\vec{k}$, and $\vec{v}_2 = b_1\vec{i} + b_2\vec{j} + b_3\vec{k}$, then the parametric equations of the plane can be written in the form

$$x = x_0 + sa_1 + tb_1, \quad y = y_0 + sa_2 + tb_2, \quad z = z_0 + sa_3 + tb_3.$$

Notice that the parameterization of the plane expresses the coordinates x, y, and z as linear functions of the parameters s and t.

Example 2   Write a parametric equation for the plane through the point $(2, -1, 3)$ and containing the vectors $\vec{v}_1 = 2\vec{i} + 3\vec{j} - \vec{k}$ and $\vec{v}_2 = \vec{i} - 4\vec{j} + 5\vec{k}$.

Solution   The parametric equation is

$$\vec{r}(s, t) = \vec{r}_0 + s\vec{v}_1 + t\vec{v}_2 = 2\vec{i} - \vec{j} + 3\vec{k} + s(2\vec{i} + 3\vec{j} - \vec{k}) + t(\vec{i} - 4\vec{j} + 5\vec{k})$$

$$= (2 + 2s + t)\vec{i} + (-1 + 3s - 4t)\vec{j} + (3 - s + 5t)\vec{k},$$

or equivalently,

$$x = 2 + 2s + t, \quad y = -1 + 3s - 4t, \quad z = 3 - s + 5t.$$

Parameterizations Using Spherical Coordinates

Recall the spherical coordinates ρ, ϕ, and θ introduced on page 810 of Chapter 16. On a sphere of radius $\rho = a$ we can use ϕ and θ as coordinates, similar to latitude and longitude on the surface of the earth. (See Figure 17.42.) The latitude, however, is measured from the equator, whereas ϕ is measured from the north pole. If the positive x-axis passes through the Greenwich meridian, the longitude and θ are equal for $0 \leq \theta \leq \pi$.

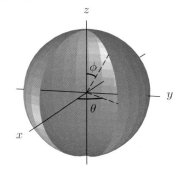

Figure 17.42: Parameterizing the sphere by ϕ and θ

Example 3   You are at a point on a sphere with $\phi = 3\pi/4$. Are you in the northern or southern hemisphere? If ϕ decreases, do you move closer to or farther from the equator?

Solution   The equator has $\phi = \pi/2$. Since $3\pi/4 > \pi/2$, you are in the southern hemisphere. If ϕ decreases, you move closer to the equator.

Example 4   On a sphere, you are standing at a point with coordinates θ_0 and ϕ_0. Your *antipodal* point is the point on the other side of the sphere on a line through you and the center. What are the θ, ϕ coordinates of your antipodal point?

Solution   Figure 17.43 shows that the coordinates are $\theta = \pi + \theta_0$ if $\theta_0 < \pi$ or $\theta = \theta_0 - \pi$ if $\pi \leq \theta_0 \leq 2\pi$, and $\phi = \pi - \phi_0$. Notice that if you are on the equator, then so is your antipodal point.

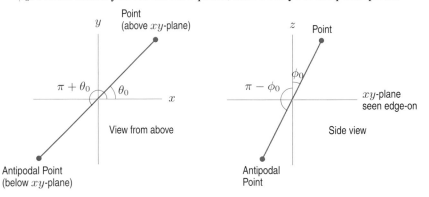

Figure 17.43: Two views of the xyz-coordinate system showing coordinates of antipodal points

Parameterizing a Sphere Using Spherical Coordinates

The sphere with radius 1 centered at the origin is parameterized by

$$x = \sin \phi \cos \theta, \qquad y = \sin \phi \sin \theta, \qquad z = \cos \phi.$$

Where $0 \leq \theta \leq 2\pi$ and $0 \leq \phi \leq \pi$. (See Figure 17.44.)

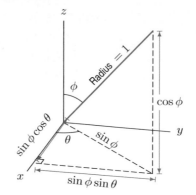

Figure 17.44: The relationship between x, y, z and ϕ, θ on a sphere of radius 1

We can also write these equations in vector form:

$$\vec{r}(\theta, \phi) = \sin\phi\cos\theta\,\vec{i} + \sin\phi\sin\theta\,\vec{j} + \cos\phi\,\vec{k}.$$

Since $x^2 + y^2 + z^2 = \sin^2\phi(\cos^2\theta + \sin^2\theta) + \cos^2\phi = \sin^2\phi + \cos^2\phi = 1$ this verifies that the point with position vector $\vec{r}(\theta, \phi)$ does lie on the sphere of radius 1. Notice that the z-coordinate depends only on the parameter ϕ. Geometrically, this means that all points on the same latitude have the same z-coordinate.

Example 5   Find parametric equations for the following spheres:
(a) Center at the origin and radius 2.
(b) Center at the point $(2, -1, 3)$ and radius 2.

Solution   (a) We must scale the distance from the origin by 2. Thus, we have

$$x = 2\sin\phi\cos\theta, \qquad y = 2\sin\phi\sin\theta, \qquad z = 2\cos\phi,$$

where $0 \leq \theta \leq 2\pi$ and $0 \leq \phi \leq \pi$. In vector form, this is written

$$\vec{r}(\theta, \phi) = 2\sin\phi\cos\theta\vec{i} + 2\sin\phi\sin\theta\vec{j} + 2\cos\phi\vec{k}.$$

(b) To shift the center of the sphere from the origin to the point $(2, -1, 3)$, we add the vector parameterization we found in part (a) to the position vector of $(2, -1, 3)$. (See Figure 17.45.) This gives

$$\vec{r}(\theta, \phi) = 2\vec{i} - \vec{j} + 3\vec{k} + (2\sin\phi\cos\theta\vec{i} + 2\sin\phi\sin\theta\vec{j} + 2\cos\phi\vec{k})$$
$$= (2 + 2\sin\phi\cos\theta)\vec{i} + (-1 + 2\sin\phi\sin\theta)\vec{j} + (3 + 2\cos\phi)\vec{k},$$

where $0 \leq \theta \leq 2\pi$ and $0 \leq \phi \leq \pi$. Alternatively

$$x = 2 + 2\sin\phi\cos\theta, \qquad y = -1 + 2\sin\phi\sin\theta, \qquad z = 3 + 2\cos\phi.$$

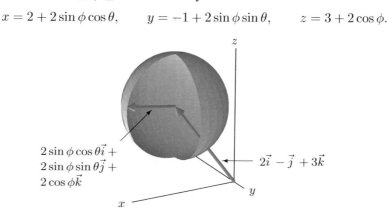

Figure 17.45: Sphere with center at the point $(2, -1, 3)$ and radius 2

Note that the same point can have more than one value for θ or ϕ. For example, points with $\theta = 0$ also have $\theta = 2\pi$, unless we restrict θ to the range $0 \leq \theta < 2\pi$. Also, the north pole, at $\phi = 0$, and the south pole, at $\phi = \pi$, can have any value of θ.

Parameterizing Surfaces of Revolution

Many surfaces have an axis of rotational symmetry and circular cross-sections perpendicular to that axis. These surfaces are referred to as *surfaces of revolution.*

Example 6   Find a parameterization of the cone whose base is the circle $x^2 + y^2 = a^2$ in the xy-plane and whose vertex is at a height h above the xy-plane. (See Figure 17.46.)

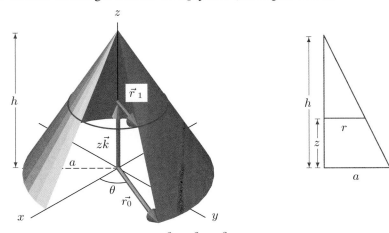

Figure 17.46: The cone whose base is the circle $x^2 + y^2 = a^2$ in the xy-plane and whose vertex is at the point $(0, 0, h)$ and the vertical cross-section through the cone

Solution   We use cylindrical coordinates, r, θ, z. (See Figure 17.46.) In the xy-plane, the radius vector, $\vec{r}_0$, from the z-axis to a point on the cone in the xy-plane is

$$\vec{r}_0 = a\cos\theta\vec{i} + a\sin\theta\vec{j}.$$

Above the xy-plane, the radius of the circular cross section, r, decreases linearly from $r = a$ when $z = 0$ to $r = 0$ when $z = h$. From the similar triangles in Figure 17.46,

$$\frac{a}{h} = \frac{r}{h - z}.$$

Solving for r, we have

$$r = \left(1 - \frac{z}{h}\right)a.$$

The horizontal radius vector, $\vec{r}_1$, at height z has components similar to $\vec{r}_0$, but with a replaced by r:

$$\vec{r}_1 = r\cos\theta\vec{i} + r\sin\theta\vec{j} = \left(1 - \frac{z}{h}\right)a\cos\theta\vec{i} + \left(1 - \frac{z}{h}\right)a\sin\theta\vec{j}.$$

As θ goes from 0 to 2π, the vector $\vec{r}_1$ traces out the horizontal circle in Figure 17.46. We get the position vector, $\vec{r}$, of a point on the cone by adding the vector $z\vec{k}$, so

$$\vec{r} = \vec{r}_1 + z\vec{k} = a\left(1 - \frac{z}{h}\right)\cos\theta\vec{i} + a\left(1 - \frac{z}{h}\right)\sin\theta\vec{j} + z\vec{k}, \quad \text{for } 0 \leq z \leq h \text{ and } 0 \leq \theta \leq 2\pi.$$

These equations can be written as

$$x = \left(1 - \frac{z}{h}\right)a\cos\theta, \quad y = \left(1 - \frac{z}{h}\right)a\sin\theta, \quad z = z.$$

The parameters are θ and z.

Example 7   Consider the bell of a trumpet. A model for the radius $z = f(x)$ of the horn (in cm) at a distance x cm from the large open end is given by the function

$$f(x) = \frac{6}{(x+1)^{0.7}}.$$

The bell is obtained by rotating the graph of f about the x-axis. Find a parameterization for the first 24 cm of the bell. (See Figure 17.47.)

Figure 17.47: The bell of a trumpet obtained by rotating the graph of $z = f(x)$ about the x-axis

Solution   At distance x from the large open end of the horn, the cross section parallel to the yz-plane is a circle of radius $f(x)$, with center on the x-axis. Such a circle can be parameterized by $y = f(x)\cos\theta$, $z = f(x)\sin\theta$. Thus we have the parameterization

$$x = x, \quad y = \left(\frac{6}{(x+1)^{0.7}}\right)\cos\theta, \quad z = \left(\frac{6}{(x+1)^{0.7}}\right)\sin\theta, \quad 0 \le x \le 24, \quad 0 \le \theta \le 2\pi.$$

The parameters are x and θ.

Parameter Curves

On a parameterized surface, the curve obtained by setting one of the parameters equal to a constant and letting the other vary is called a *parameter curve*. If the surface is parameterized by

$$\vec{r}(s,t) = f_1(s,t)\vec{i} + f_2(s,t)\vec{j} + f_3(s,t)\vec{k},$$

there are two families of parameter curves on the surface, one family with t constant and the other with s constant.

Example 8   Consider the vertical cylinder

$$x = \cos t, \quad y = \sin t, \quad z = z.$$

(a) Describe the two parameter curves through the point $(0, 1, 1)$.
(b) Describe the family of parameter curves with t constant and the family with z constant.

Solution   (a) Since the point $(0, 1, 1)$ corresponds to the parameter values $t = \pi/2$ and $z = 1$, there are two parameter curves, one with $t = \pi/2$ and the other with $z = 1$. The parameter curve with $t = \pi/2$ has the parametric equations

$$x = \cos\left(\frac{\pi}{2}\right) = 0, \quad y = \sin\left(\frac{\pi}{2}\right) = 1, \quad z = z,$$

with parameter z. This is a line through the point $(0, 1, 1)$ parallel to the z-axis.

The parameter curve with $z = 1$ has the parametric equations

$$x = \cos t, \quad y = \sin t, \quad z = 1,$$

with parameter t. This is a unit circle parallel to and one unit above the xy-plane centered on the z-axis.

(b) First, fix $t = t_0$ for t and let z vary. The curves parameterized by z have equation

$$x = \cos t_0, \quad y = \sin t_0, \quad z = z.$$

These are vertical lines on the cylinder parallel to the z-axis. (See Figure 17.48.)

The other family is obtained by fixing $z = z_0$ and varying t. Curves in this family are parameterized by t and have equation

$$x = \cos t, \quad y = \sin t, \quad z = z_0.$$

They are circles of radius 1 parallel to the xy-plane centered on the z-axis. (See Figure 17.49.)

Figure 17.48: The family of parameter curves with $t = t_0$ for the cylinder $x = \cos t, y = \sin t, z = z$

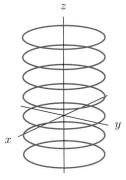

Figure 17.49: The family of parameter curves with $z = z_0$ for the cylinder $x = \cos t, y = \sin t, z = z$

Example 9   Describe the families of parameter curves with $\theta = \theta_0$ and $\phi = \phi_0$ for the sphere

$$x = \sin \phi \cos \theta, \qquad y = \sin \phi \sin \theta, \qquad z = \cos \phi,$$

where $0 \leq \theta \leq 2\pi, 0 \leq \phi \leq \pi$.

Solution   Since ϕ measures latitude, the family with ϕ constant consists of the circles of constant latitude. (See Figure 17.50.) Similarly, the family with θ constant consists of the meridians (semicircles) running between the north and south poles. (See Figure 17.51.)

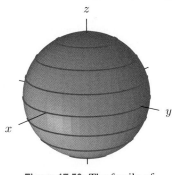

Figure 17.50: The family of parameter curves with $\phi = \phi_0$ for the sphere parameterized by (θ, ϕ)

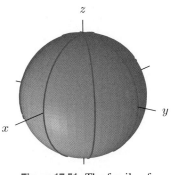

Figure 17.51: The family of parameter curves with $\theta = \theta_0$ for the sphere parameterized by (θ, ϕ)

We have seen parameter curves before in pages 612-614 of Section 12.2: The cross-sections with $x = a$ and $y = b$ on a surface $z = f(x, y)$ are examples of parameter curves. So are the grid lines on a computer sketch of a surface. The small regions shaped like parallelograms surrounded by nearby pairs of parameter curves are called *parameter rectangles*. See Figure 17.52.

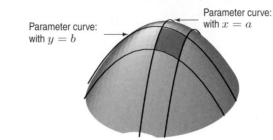

Parameter curve: with $y = b$

Parameter curve: with $x = a$

Figure 17.52: Parameter curves $x = a$ and $y = b$ on a surface $z = f(x, y)$; the darker region is a parameter rectangle

Exercises and Problems for Section 17.5

Exercises

Describe in words the objects parameterized by the equations in Exercises 1–8. (Note: r and θ are cylindrical coordinates.)

1. $x = r \cos \theta \qquad y = r \sin \theta \qquad z = 7$
$\quad 0 \le r \le 5 \qquad 0 \le \theta \le 2\pi$

2. $x = 5 \cos \theta \qquad y = 5 \sin \theta \qquad z = 7$
$\quad 0 \le \theta \le 2\pi$

3. $x = 5 \cos \theta \qquad y = 5 \sin \theta \qquad z = z$
$\quad 0 \le \theta \le 2\pi \qquad 0 \le z \le 7$

4. $x = 5 \cos \theta \qquad y = 5 \sin \theta \qquad z = 5\theta$
$\quad 0 \le \theta \le 2\pi$

5. $x = r \cos \theta \qquad y = r \sin \theta \qquad z = r$
$\quad 0 \le r \le 5 \qquad 0 \le \theta \le 2\pi$

6. $x = 2z \cos \theta \qquad y = 2z \sin \theta \qquad z = z$
$\quad 0 \le z \le 7 \qquad 0 \le \theta \le 2\pi$

7. $x = 3 \cos \theta \qquad y = 2 \sin \theta \qquad z = z$
$\quad 0 \le \theta \le 2\pi \qquad 0 \le z \le 7$

8. $x = x \qquad y = x^2 \qquad z = z$
$\quad -5 \le x \le 5 \qquad 0 \le z \le 7$

In Exercises 9–12, for a sphere parameterized using the spherical coordinates θ and ϕ, describe in words the part of the sphere given by the restrictions.

9. $0 \le \theta < 2\pi, \quad 0 \le \phi \le \pi/2$

10. $\pi \le \theta < 2\pi, \quad 0 \le \phi \le \pi$

11. $\pi/4 \le \theta < \pi/3, \quad 0 \le \phi \le \pi$

12. $0 \le \theta \le \pi, \quad \pi/4 \le \phi < \pi/3$

Problems

13. Find a parameterization of the circle of radius 2 centered at $(0, 1, 0)$ lying in the plane $x + z = 0$.

14. Find a parameterization for the plane through $(1, 3, 4)$ and orthogonal to $\vec{n} = 2\vec{i} + \vec{j} - \vec{k}$.

15. Does the plane $\vec{r}(s, t) = (2+s)\vec{i} + (3+s+t)\vec{j} + 4t\vec{k}$ contain the following points?
   (a) $(4, 8, 12)$ (b) $(1, 2, 3)$

16. Are the following two planes parallel?

$$x = 2 + s + t, \quad y = 4 + s - t, \quad z = 1 + 2s, \quad \text{and}$$

$$x = 2 + s + 2t, \quad y = t, \quad z = s - t.$$

17. A city is described parametrically by the equation

$$\vec{r} = (x_0\vec{i} + y_0\vec{j} + z_0\vec{k}) + s\vec{v_1} + t\vec{v_2}$$

where $\vec{v}_1 = 2\vec{i} - 3\vec{j} + 2\vec{k}$ and $\vec{v}_2 = \vec{i} + 4\vec{j} + 5\vec{k}$. A city block is a rectangle determined by $\vec{v}_1$ and $\vec{v}_2$. East

is in the direction of $\vec{v}_1$ and north is in the direction of $\vec{v}_2$. Starting at the point (x_0, y_0, z_0), you walk 5 blocks east, 4 blocks north, 1 block west and 2 blocks south. What are the parameters of the point where you end up? What are your x, y and z coordinates at that point?

18. You are at a point on the earth with longitude $80°$ West of Greenwich, England, and latitude $40°$ North of the equator.

   (a) If your latitude decreases have you moved nearer to or farther from the equator?
   (b) If your latitude decreases, have you moved nearer to or farther from the north pole?
   (c) If your longitude increases (say, to $90°$ West), have you moved nearer to or farther from Greenwich?

19. Describe in words the curve $\phi = \pi/4$ on the surface of the globe.

20. Describe in words the curve $\theta = \pi/4$ on the surface of the globe.

21. Find a parameterization of a circular cylinder of radius a whose axis is along the z-axis, from $z = 0$ to a height $z = h$. See Figure 17.53.

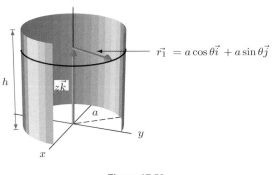

$$\vec{r_1} = a\cos\theta\vec{i} + a\sin\theta\vec{j}$$

Figure 17.53

22. Find parametric equations for the sphere centered at the origin and with radius 5.

23. Find parametric equations for the sphere $(x-a)^2 + (y-b)^2 + (z-c)^2 = d^2$.

24. Suppose you are standing at a point on the equator of a sphere, parameterized by spherical coordinates θ_0, and ϕ_0. If you go halfway around the equator and halfway up toward the north pole along a longitude, what are your new θ and ϕ coordinates?

25. Find parametric equations for the cone $x^2 + y^2 = z^2$.

26. Parameterize the cone in Example 6 on page 863 in terms of r and θ.

27. Parameterize the paraboloid $z = x^2 + y^2$ using cylindrical coordinates.

28. Parameterize a vase formed by rotating the curve $z = 10\sqrt{x-1}$, $1 \le x \le 2$, around the z-axis. Sketch the vase.

29. Give a parameterization of the circle of radius a centered at the point (x_0, y_0, z_0) and in the plane parallel to two given unit vectors $\vec{u}$ and $\vec{v}$ such that $\vec{u} \cdot \vec{v} = 0$.

For Problems 30–33

(a) Write an equation in x, y, z and identify the parametric surface.

(b) Draw a picture of the surface.

30. $x = 2s$ $y = s + t$ $z = 1 + s - t$
$0 \le s \le 1$ $0 \le t \le 1$

31. $x = s$ $y = t$ $z = \sqrt{1 - s^2 - t^2}$
$s^2 + t^2 \le 1$ $s, t \ge 0$

32. $x = s + t$ $y = s - t$ $z = s^2 + t^2$
$0 \le s \le 1$ $0 \le t \le 1$

33. $x = 3\sin s$ $y = 3\cos s$ $z = t + 1$
$0 \le s \le \pi$ $0 \le t \le 1$

A (not necessarily circular) cone S made from any curve C and point P is the surface consisting of all line segments from

P, the *vertex* of the cone, to points on C. If C is parameterized by $\vec{r}(u) = f(u)\vec{i} + g(u)\vec{j} + h(u)\vec{k}$, $a \le u \le b$, then S can be parameterized by $\vec{r}(u,t) = (1-t)\overrightarrow{OP} + t\vec{r}(u)$, $a \le u \le b$, $0 \le t \le 1$, where $\overrightarrow{OP}$ is the position vector of P. See Figure 17.54. Parameterize and graph the cones in Problems 34–36.

$P : t = 0$ $C : t = 1$

Figure 17.54

34. $C : x = 3$, $y = 5\cos u$, $z = 10\sin u$, $0 \le u \le 2\pi$; $P = (0,0,0)$

35. $C : x = u$, $y = u^2$, $z = u - u^2 + 5$, $0 \le u \le 3$; $P = (0,0,0)$

36. $C : x = 2\cos u$, $y = 2\sin u$, $z = -1$, $0 \le u \le \pi$; $P = (2,3,4)$

37. A torus (doughnut) is constructed by rotating a small circle of radius a in a large circle of radius b about the origin. The small circle is in a (rotating) vertical plane through the origin and the large circle is in the xy-plane. (See Figure 17.55.) Parameterize the torus as follows.

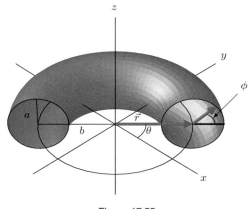

Figure 17.55

(a) Parameterize the large circle.

(b) For a typical point on the large circle, find two unit vectors which are perpendicular to one another and in the plane of the small circle at that point. Use these vectors to parameterize the small circle relative to its center.

(c) Combine your answers to parts (b) and (c) to parameterize the torus.

CHAPTER SUMMARY (see also Ready Reference at the end of the book)

- **Parameterized Curves**
  Parameterizations representing motion in 2- and 3-space, change of parameter, vector form of parametric equations, parametric equation of a line.
- **Velocity and Acceleration Vectors**
  Computing velocity and acceleration, uniform circular motion, the length of a parametric curve.
- **Vector Fields**

Definition of vector field, visualizing fields, gradient fields.
- **Flow Lines**
  Parametric equations of flow lines, approximating flow lines numerically.
- **Parameterized Surfaces** Parametric equations for planes, graphs of functions, spheres, and cylinders. Parameter curves.

REVIEW EXERCISES AND PROBLEMS FOR CHAPTER SEVENTEEN

Exercises

Write a parameterization for the curves in Exercises 1–10.

1. The horizontal line through the point $(0, 5)$.

2. The circle of radius 2 centered at the origin starting at the point $(0, 2)$ when $t = 0$.

3. The circle of radius 4 centered at the point $(4, 4)$ starting on the x-axis when $t = 0$.

4. The circle of radius 1 in the xy-plane centered at the origin, traversed counterclockwise when viewed from above.

5. The line through the points $(2, -1, 4)$ and $(1, 2, 5)$.

6. The line through the point $(1, 3, 2)$ perpendicular to the xz-plane.

7. The line through the point $(1, 1, 1)$ perpendicular to the plane $2x - 3y + 5z = 4$.

8. The circle of radius 2 parallel to the xy-plane, centered at the point $(0, 0, 1)$, and traversed counterclockwise when viewed from below.

9. The circle of radius 3 parallel to the xz-plane, centered at the point $(0, 5, 0)$, and traversed counterclockwise when viewed from $(0, 10, 0)$.

10. The line of intersection of the planes $z = 4 + 2x + 5y$ and $z = 3 + x + 3y$.

11. Are the lines $x = 3 + 2t, y = 5 - t, z = 7 + 3t$ and $x = 3 + t, y = 5 + 2t, z = 7 + 2t$ parallel?

12. Explain how you know the following equations parameterize the same line:

$$\vec{r} = (3 - t)\vec{i} + (3 + 4t)\vec{j} - (1 + 2t)\vec{k}$$
$$\vec{r} = (1 + 2t)\vec{i} + (11 - 8t)\vec{j} + (4t - 5)\vec{k}$$

13. A line is parameterized by $\vec{r} = 10\vec{k} + t(\vec{i} + 2\vec{j} + 3\vec{k})$.

   (a) Suppose we restrict ourselves to $t < 0$. What part of the line do we get?
   (b) Suppose we restrict ourselves to $0 \le t \le 1$. What part of the line do we get?

Sketch the vector fields in Exercises 14–16.

14. $\vec{F} = y\vec{i} - x\vec{j}$

15. $\vec{F} = \dfrac{y}{x^2 + y^2}\vec{i} - \dfrac{x}{x^2 + y^2}\vec{j}$

16. $\vec{F} = \dfrac{y}{\sqrt{x^2 + y^2}}\vec{i} - \dfrac{x}{\sqrt{x^2 + y^2}}\vec{j}$

Problems

17. (a) What is meant by a vector field?
    (b) Suppose $\vec{a} = a_1\vec{i} + a_2\vec{j} + a_3\vec{k}$ is a constant vector. Which of the following are vector fields? Explain.
    (i) $\vec{r} + \vec{a}$ (ii) $\vec{r} \cdot \vec{a}$
    (iii) $x^2\vec{i} + y^2\vec{j} + z^2\vec{k}$ (iv) $x^2 + y^2 + z^2$

18. A particle travels along a line, with position at time t given by $\vec{r}(t) = (2 + 5t)\vec{i} + (3 + t)\vec{j} + 2t\vec{k}$.

    (a) Where is the particle when $t = 0$?
    (b) At what time does the particle reach the point $(12, 5, 4)$?
    (c) Does the particle ever reach the point $(12, 4, 4)$? Explain.

19. Consider the parametric equations for $0 \le t \le \pi$:

    (I) $\vec{r} = \cos(2t)\vec{i} + \sin(2t)\vec{j}$
    (II) $\vec{r} = 2\cos t\vec{i} + 2\sin t\vec{j}$
    (III) $\vec{r} = \cos(t/2)\vec{i} + \sin(t/2)\vec{j}$
    (IV) $\vec{r} = 2\cos t\vec{i} - 2\sin t\vec{j}$

    (a) Match the equations above with four of the curves C_1, C_2, C_3, C_4, C_5 and C_6 in Figure 17.56. (Each curve is part of a circle.)
    (b) Give parametric equations for the curves which have not been matched, again assuming $0 \le t \le \pi$.

Figure 17.56

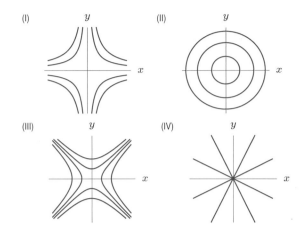

20. Match the level curves in (I)–(IV) with the gradient fields in (A)–(D). All figures have $-2 \leq x \leq 2, -2 \leq y \leq 2$.

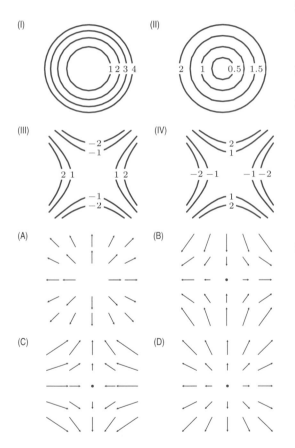

21. Each of the vector fields $\vec{E}, \vec{F}, \vec{G}, \vec{H}$ is tangent to one of the families of curves (I)–(IV). Match them.

$$\vec{E} = x\vec{i} + y\vec{j} \qquad \vec{F} = x\vec{i} - y\vec{j}$$
$$\vec{G} = y\vec{i} - x\vec{j} \qquad \vec{H} = y\vec{i} + x\vec{j}$$

22. A particle passes through the point $P = (5, 4, 3)$ at time $t = 7$, moving with constant velocity $\vec{v} = 3\vec{i} + \vec{j} + 2\vec{k}$. Find equations for its position at time t.

23. An object moving with constant velocity in 3-space, with coordinates in meters, passes through the point $(1, 1, 1)$, then through $(2, -1, 3)$ five seconds later. What is its velocity vector? What is its acceleration vector?

24. Find parametric equations for a particle moving along the line $y = -2x + 5$ with speed 3.

25. The temperature in $^\circ$C at (x, y) in the plane is $H = f(x, y)$, where x, y are in centimeters. A particle moves along the curve $x = g(t), y = k(t)$, with t in seconds.

(a) What does the quantity $\| \operatorname{grad} f \|$ represent in this context? What is its units?

(b) Write an expression for the speed of the particle. What is its units?

(c) Write an expression for the rate of change of the particle's temperature with time. What is its units?

26. A stone is swung around on a string at a constant speed with period 2π seconds in a horizontal circle centered at the point $(0, 0, 8)$. When $t = 0$, the stone is at the point $(0, 5, 8)$; it travels clockwise when viewed from above. When the stone is at the point $(5, 0, 8)$, the string breaks and it moves under gravity.

(a) Parameterize the stone's circular trajectory.

(b) Find the velocity and acceleration of the stone at the moment before the string breaks.

(c) Write, but do not solve, the differential equations (with initial conditions) satisfied by the coordinates x, y, z giving the position of the stone after it has left the circle.

27. Let $f(x, y) = \dfrac{x^2 - y^2}{x^2 + y^2}$.

(a) In which direction should you move from the point $(1, 1)$ to obtain the maximum rate of increase of f?

(b) Find a direction in which the directional derivative at the point $(1, 1)$ is equal to zero.

(c) Suppose you move along the curve $x = e^{2t}, y = 2t^3 + 6t + 1$. What is df/dt at $t = 0$?

28. Find parametric equations of the line passing through the points $(1, 2, 3)$, $(3, 5, 7)$ and calculate the shortest distance from the line to the origin.

29. An ant, starting at the origin, moves at 2 units/sec along the x-axis to the point $(1, 0)$. The ant then moves counterclockwise along the unit circle to $(0, 1)$ at a speed of $3\pi/2$ units/sec, then straight down to the origin at a speed of 2 units/sec along the y-axis.

    (a) Express the ant's coordinates as a function of time, t, in secs.
    (b) Express the reverse path as a function of time.

30. The motion of a particle is given by $\vec{r}(t) = R\cos(\omega t)\vec{i} + R\sin(\omega t)\vec{j}$, with $R > 0$, $\omega > 0$.

    (a) Show that the particle moves on a circle and find the radius, direction, and period.
    (b) Determine the velocity vector of the particle and its direction and speed.
    (c) What are the direction and magnitude of the acceleration vector of the particle?

31. On a calculator or a computer, plot $x = 2t/(t^2 + 1)$, $y = (t^2 - 1)/(t^2 + 1)$, first for $-50 \le t \le 50$ then for $-5 \le t \le 5$. Explain what you see. Is the curve really a circle?

32. (a) The ellipse $2x^2 + 3y^2 = 8$ intersects the line of slope t through the point $P = (-2, 0)$ in two points, one of which is P. Compute the coordinate of the other point Q. See Figure 17.57.
    (b) Give a parameterization of the ellipse $2x^2 + 3y^2 = 8$ by rational functions.

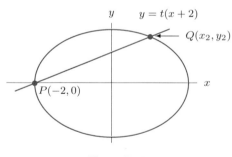

Figure 17.57

33. Let $x(t) = \cos^2 t$, $y(t) = \sin t \cos t$, $z(t) = \sin t$.

    (a) Show that this curve lies on the sphere of radius 1 centered at the origin.
    (b) Find a vector normal to the sphere and the curve at the point where $t = \pi/4$.
    (c) Find a tangent vector to the curve at the point where $t = \pi/4$.

34. A cheerleader has a 0.4 m long baton with a light on one end. She throws the baton in such a way that its center

moves along a parabola, and the baton rotates counterclockwise around the center with a constant angular velocity. The baton is initially horizontal and 1.5 m above the ground; its initial velocity is 8 m/sec horizontally and 10 m/sec vertically, and its angular velocity is 2 revolutions per second. Find parametric equations describing the following motions:

    (a) The center of the baton relative to the ground.
    (b) The end of the baton relative to its center.
    (c) The path traced out by the end of the baton relative to the ground.
    (d) Sketch a graph of the motion of the end of the baton.

35. For a and ω positive constants and $t \ge 0$, the position vector of a particle moving in a spiral counterclockwise outward from the origin is given by

$$\vec{r}(t) = at\cos(\omega t)\vec{i} + at\sin(\omega t)\vec{j}.$$

What is the significance of the parameters ω and a?

36. An object is moving on a straight line path. Can you conclude at all times that:

    (a) Its velocity vector is parallel to the line? Justify your answer.
    (b) Its acceleration vector is parallel to the line? Justify your answer.

37. An object is moving in the plane on a circular path. Can you conclude at all times that

    (a) Its velocity vector is tangent to the circle ? Justify your answer.
    (b) Its acceleration vector points toward the center of the circle? Justify your answer.

38. If $\vec{F} = \vec{r}/\|\vec{r}\|^3$, find the following quantities in terms of x, y, z, or t.

    (a) $\|\vec{F}\|$
    (b) $\vec{F} \cdot \vec{r}$
    (c) A unit vector parallel to $\vec{F}$ and pointing in the same direction.
    (d) A unit vector parallel to $\vec{F}$ and pointing in the opposite direction.
    (e) $\vec{F}$ if $\vec{r} = \cos t\vec{i} + \sin t\vec{j} + \vec{k}$
    (f) $\vec{F} \cdot \vec{r}$ if $\vec{r} = \cos t\vec{i} + \sin t\vec{j} + \vec{k}$

39. Find parametric equations for the sphere centered at the point $(2, -1, 3)$ and with radius 5.

40. Parameterize a cone of height h and maximum radius a with vertex at the origin and opening upward. Do this in two ways, giving the range of values for each parameter in each case: (a) Use r and θ.   (b) Use z and θ.

41. (a) Describe the surface given parametrically by the equations

$$x = \cos(s - t), \quad y = \sin(s - t), \quad z = s + t.$$

    (b) Describe the two families of parameter curves on the surface.

42. Each of the following vector fields represents an ocean current. Sketch the vector field, and sketch the path of an iceberg in this current. Determine the location of an iceberg at time $t = 7$ if it is at the point $(1, 3)$ at time $t = 0$.

    (a) The current everywhere is $\vec{i}$.
    (b) The current at (x, y) is $2x\vec{i} + y\vec{j}$.
    (c) The current at (x, y) is $-y\vec{i} + x\vec{j}$.

43. Adapt the parameterization for the sphere to find a parameterization for the ellipsoid

$$\frac{x^2}{a^2} + \frac{y^2}{b^2} + \frac{z^2}{c^2} = 1.$$

44. There is a famous way to parameterize a sphere called *stereographic projection*. We work with the sphere $x^2 +$ $y^2 + z^2 = 1$. Draw a line from a point (x, y) in the xy-plane to the north pole $(0, 0, 1)$. This line intersects the sphere in a point (x, y, z). This gives a parameterization of the sphere by points in the plane.

    (a) Which point corresponds to the south pole?
    (b) Which points correspond to the equator?
    (c) Do we get all the points of the sphere by this parameterization?
    (d) Which points correspond to the upper hemisphere?
    (e) Which points correspond to the lower hemisphere?

45. Wire is stretched taught from the point $P = (7, 12, -10)$ to the point $Q = (-2, -3, 2)$ and from the point $R = (-20, 17, 1)$ to the point $S = (37, 2, 25)$. Spherical balls of diameter 8 cm slide along each cable through holes along an axis through their centers. Can the beads pass each other without touching, regardless of their position?

CAS Challenge Problems

46. Let $\vec{r}_0 = x_0\vec{i} + y_0\vec{j} + z_0\vec{k}$, and let $\vec{e_1}$ and $\vec{e_2}$ be perpendicular unit vectors. A circle of radius R, centered at (x_0, y_0, z_0), and lying the plane parallel to $\vec{e}_1$ and $\vec{e}_2$, is parameterized by $\vec{r}(t) = \vec{r}_0 + R\cos t\vec{e}_1 + R\sin t\vec{e}_2$. We want to parameterize a circle in 3-space with radius 5, centered at $(1, 2, 3)$, and lying in the plane $x+y+z = 6$.

    (a) Let $\vec{e}_1 = a\vec{i} + b\vec{j}$ and $\vec{e}_2 = c\vec{i} + d\vec{j} + e\vec{k}$. Write down conditions on $\vec{e}_1$ and $\vec{e}_2$ that make them unit vectors, perpendicular to each other, and lie in the given plane.
    (b) Solve the equations in part (a) for a, b, c, d, and e and write a parameterization of the circle.

47. Let $\vec{F}(x, y) = -y(1 - y^2)\vec{i} + x(1 - y^2)\vec{j}$.

    (a) Show that $\vec{r} \cdot \vec{F} = 0$. What does this tell you about the shape of the flow lines?
    (b) Show that $\vec{r}(t) = \cos t\vec{i} + \sin t\vec{j}$ has velocity vector parallel to $\vec{F}$ at every point, but is not a flow line.
    (c) Show $\vec{r}(t) = (1/(\sqrt{1 + t^2}))\vec{i} + (t/(\sqrt{1 + t^2}))\vec{j}$ is a flow line for $\vec{F}$. What is the difference between this curve and the one in part (b)?

48. Let $\vec{F}(x, y) = (x + y)\vec{i} + (4x + y)\vec{j}$.

    (a) Show that $\vec{r}(t) = (ae^{3t} + be^{-t})\vec{i} + (2ae^{3t} - 2be^{-t})\vec{j}$, for constant a, b, is a flow line for $\vec{F}$.
    (b) Find the flow line passing through $(1, -2)$ at $t = 0$ and describe its behavior as $t \to \infty$. Do the same for the points $(1, -1.99)$ and $(1, -2.01)$. Compare the behavior of the three flow lines.

49. Two surfaces generally intersect in a curve. For each of the following pair of surfaces $f(x, y, z) = 0$ and $g(x, y, z) = 0$, find a parameterization for the curve of intersection by solving for two of the variables in terms of the third.

    (a) $3x - 5y + z = 5, 2x + y + z = 3,$
    (b) $3x^2 - 5y + z = 5, 2x + y + z = 3,$
    (c) $x^2 + y^2 = 2, 3x - y + z = 5.$

CHECK YOUR UNDERSTANDING

Are the statements in Problems 1–38 true or false? Give reasons for your answer.

1. The parametric curve $x = 3t+2, y = -2t$ for $0 \le t \le 5$ passes through the origin.

2. The parametric curve $x = t^2, y = t^4$ for $0 \le t \le 1$ is a parabola.

3. A parametric curve $x = g(t), y = h(t)$ for $a \le t \le b$ is always the graph of a function $y = f(x)$.

4. The parametric curve $x = (3t + 2)^2, y = (3t + 2)^2 - 1$ for $0 \le t \le 3$ is a line.

5. The parametric curve $x = -\sin t, y = -\cos t$ for $0 \le t \le 2\pi$ traces out a unit circle counterclockwise as t increases.

6. A parameterization of the graph of $y = \ln x$ for $x > 0$ is given by $x = e^t, y = t$ for $-\infty < t < \infty$.

7. Both $x = -t + 1, y = 2t$ and $x = 2s, y = -4s + 2$ describe the same line.

8. The line of intersection of the two planes $z = x + y$ and $z = 1 - x - y$ can be parameterized by $x = t, y = \frac{1}{2} - t, z = \frac{1}{2}$.

9. The two lines given by $x = t, y = 2 + t, z = 3 + t$ and $x = 2s, y = 1 - s, z = s$ do not intersect.

10. The line parameterized by $x = 1, y = 2t, z = 3 + t$ is parallel to the x-axis.

11. The equation $\vec{r}(t) = 3t\vec{i} + (6t + 1)\vec{j}$ parameterizes a line.

12. The lines parameterized by $\vec{r}_1(t) = t\vec{i} + (-2t+1)\vec{j}$ and $\vec{r}_2(t) = (2t+5)\vec{i} + (-t)\vec{j}$ are parallel.

13. A particle whose motion in the plane is given by $\vec{r}(t) = t^2\vec{i} + (1-t)\vec{j}$ has the same velocity at $t=1$ and $t=-1$.

14. A particle whose motion in the plane is given by $\vec{r}(t) = t^2\vec{i} + (1-t)\vec{j}$ has the same speed at $t=1$ and $t=-1$.

15. If a particle is moving along a parameterized curve $\vec{r}(t)$ then the acceleration vector at any point is always perpendicular to the velocity vector at that point.

16. If a particle is moving along a parameterized curve $\vec{r}(t)$ then the acceleration vector at a point cannot be parallel to the velocity vector at that point.

17. If $\vec{r}(t)$ for $a \le t \le b$ is a parameterized curve, then $\vec{r}(-t)$ for $a \le t \le b$ is the same curve traced backward.

18. If $\vec{r}(t)$ for $a \le t \le b$ is a parameterized curve C and the speed $\|\vec{v}(t)\| = 1$, then the length of C is $b-a$.

19. If a particle moves with motion $\vec{r}(t) = 3t\vec{i} + 2t\vec{j} + t\vec{k}$, then the particle stops at the origin.

20. If a particle moves with constant speed, the path of the particle must be a line.

21. The flow lines for $\vec{F}(x,y) = x\vec{j}$ are parallel to the y-axis.

22. The flow lines of $\vec{F}(x,y) = y\vec{i} - x\vec{j}$ are hyperbolas.

23. The flow lines of $\vec{F}(x,y) = x\vec{i}$ are parabolas.

24. The vector field in Figure 17.58 has a flow line which lies in the first and third quadrants.

Figure 17.58

25. The vector field in Figure 17.58 has a flow line on which both x and y tend to infinity.

26. If $\vec{F}$ is a gradient vector field, $\vec{F}(x,y) = \nabla f(x,y)$, then the flow lines for $\vec{F}$ are the contours for f.

27. If the flow lines for the vector field $\vec{F}(\vec{r})$ are all concentric circles centered at the origin, then $\vec{F}(\vec{r}) \cdot \vec{r} = 0$ for all $\vec{r}$.

28. If the flow lines for the vector field $\vec{F}(x,y)$ are all straight lines parallel to the constant vector $\vec{v} = 3\vec{i} + 5\vec{j}$, then $\vec{F}(x,y) = \vec{v}$.

29. No flow line for the vector field $\vec{F}(x,y) = x\vec{i} + 2\vec{j}$ has a point where the y-coordinate reaches a relative maximum.

30. The vector field $\vec{F}(x,y) = e^x\vec{i} + y\vec{j}$ has a flow line that crosses the x-axis.

31. The vector field $\vec{F}(x,y) = y\vec{i} + \vec{j}$ is a gradient vector field.

32. The equations $x = s+1$, $y = t-2$, $z = 3$ parameterize a plane.

33. The equations $x = 2s-1$, $y = -s+3$, $z = 4+s$ parameterize a plane.

34. If $\vec{r} = \vec{r}(s,t)$ parameterizes the upper hemisphere $x^2 + y^2 + z^2 = 1$, $z \ge 0$, then $\vec{r} = -\vec{r}(s,t)$ parameterizes the lower hemisphere $x^2 + y^2 + z^2 = 1$, $z \le 0$.

35. If $\vec{r} = \vec{r}(s,t)$ parameterizes the upper hemisphere $x^2 + y^2 + z^2 = 1$, $z \ge 0$, then $\vec{r} = \vec{r}(-s,-t)$ parameterizes the lower hemisphere $x^2 + y^2 + z^2 = 1$, $z \le 0$.

36. If $\vec{r}_1(s,t)$ parameterizes a plane then $\vec{r}_2(s,t) = \vec{r}_1(s,t) + 2\vec{i} - 3\vec{j} + \vec{k}$ parameterizes a parallel plane.

37. Every point on a parameterized surface has a parameter curve passing through it.

38. If $s_0 \ne s_1$, then the parameter curves $\vec{r}(s_0,t)$ and $\vec{r}(s_1,t)$ do not intersect.

For Problems 39–42, decide if the statement is true or false for all smooth parameterized curves $\vec{r}(t)$ and all values of t for which $\vec{r}'(t) \ne \vec{0}$.

39. The vector $\vec{r}'(t)$ is tangent to the curve at the point with position vector $\vec{r}(t)$.

40. $\vec{r}'(t) \times \vec{r}(t) = \vec{0}$

41. $\vec{r}'(t) \cdot \vec{r}(t) = 0$

42. $\vec{r}''(t) = -\omega^2 \vec{r}(t)$

PROJECTS FOR CHAPTER SEVENTEEN

1. **Shooting a Basketball**

   A basketball player shoots the ball from 6 feet above the ground toward a basket that is 10 feet above the ground and 15 feet away horizontally.

   (a) Suppose she shoots the ball at an angle of A degrees above the horizontal ($0 < A < \pi/2$) with an initial speed V. Give the x- and y-coordinates of the position of the basketball at time t. Assume the x-coordinate of the basket is 0 and that the x-coordinate of the shooter is -15. [Hint: There is an acceleration of -32 ft/sec^2 in the y-direction; there is no acceleration in the x-direction. Ignore air resistance.]

(b) Using the parametric equations you obtained in part (a), experiment with different values for V and A, plotting the path of the ball on a graphing calculator or computer to see how close the ball comes to the basket. (The tick marks on the y-axis can be used to locate the basket.) Find some values of V and A for which the shot goes in.

(c) Find the angle A that minimizes the velocity needed for the ball to reach the basket. (This is a lengthy computation. First find an equation in V and A that holds if the path of the ball passes through the point 15 feet from the shooter and 10 feet above the ground. Then minimize V.)

2. Kepler's Second Law

The planets do not orbit in circles with the sun at the center, nor does the moon orbit in a circle with the earth at the center. In fact, the moon's distance from the earth varies from 220,000 to 260,000 miles. In the last half of the 16th century the Danish astronomer Tycho Brahe (1546–1601) made measurements of the positions of the planets. Johann Kepler (1571–1630) studied this data and arrived at three laws now known as *Kepler's Laws*:

I. The orbit of each planet is an ellipse with the sun at one focus. In particular, the orbit lies in a plane containing the sun.

II. As a planet orbits around the sun, the line segment from the sun to the planet sweeps out equal areas in equal times. See Figure 17.59.

III. The ratio p^2/d^3 is the same for every planet orbiting around the sun, where p is the period of the orbit (time to complete one revolution) and d is the mean distance of the orbit (average of the shortest and farthest distances from the sun).

Kepler's Laws, impressive as they are, were purely descriptive; Newton's great achievement was to find an underlying cause for them. In this project, you will derive Kepler's Second Law from Newton's Law of Gravity.

Consider a coordinate system centered at the sun.[2] Let $\vec{r}$ be the position vector of a planet and let $\vec{v}$ and $\vec{a}$ be the planet's velocity and acceleration, respectively. Define $\vec{L} = \vec{r} \times \vec{v}$. (This is a multiple of the planet's angular momentum.)

(a) Show that $\dfrac{d\vec{L}}{dt} = \vec{r} \times \vec{a}$.

(b) Consider the planet moving from $\vec{r}$ to $\vec{r} + \Delta\vec{r}$. Explain why the area ΔA about the origin swept out by the planet is approximately $\frac{1}{2}\|\Delta\vec{r} \times \vec{r}\|$.

(c) Using part (b), explain why $\dfrac{dA}{dt} = \dfrac{1}{2}\|\vec{L}\|$.

(d) Newton's Laws imply that the planet's gravitational acceleration, $\vec{a}$, is directed toward the sun. Using this fact and part (a), explain why $\vec{L}$ is constant.

(e) Use parts (c) and (d) to explain Kepler's Second Law.

(f) Using Kepler's Second Law, determine whether a planet is moving most quickly when it is closest to, or farthest from, the sun.

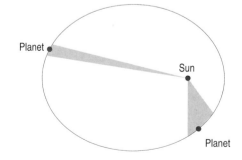

Figure 17.59: The line segment joining a planet to the sun sweeps out equal areas in equal times

---

[2]We are assuming the center of the sun is the same as the center of mass of the planet/sun system. This is only approximately true.

3. Flux Diagrams

A *flux diagram* uses flow lines to represent a vector field. The arrows drawn on a flow line indicate the direction of the vector field. The flow lines are drawn in such a way that their density is proportional to the magnitude of the vector field at each point. (The density is the number of flow lines per unit length along a curve perpendicular to the vector field.)

Figure 17.60 is a flux diagram for the vector field in 2-space $\vec{F} = \vec{r}/\|\vec{r}\|^2$. Since the field points radially away from the origin, the flow lines are straight lines radiating from the origin. The number of flow lines passing through any circle centered at the origin is a constant k. Therefore, the flow lines passing through a small circle are more densely packed than those passing through a large circle, indicating that the magnitude of the vector field decreases as we move away from the origin. In fact,

$$\text{Density of lines} = \frac{\text{Number of lines passing through circle}}{\text{Circumference of circle}} = \frac{k}{2\pi r} = \frac{k}{2\pi}1/r,$$

so that the density is proportional to $1/r$, the magnitude of the field.

Sometimes we have to start new lines to make the density proportional to the magnitude. For example, the flow lines of $\vec{v} = x\vec{i}$ are horizontal straight lines directed away from the y-axis. However, since the magnitude of $\vec{v}$ increases linearly with x, we have to make the density of lines increase linearly with x. We achieve this by starting new lines at regular intervals (see Figure 17.61).

Draw flux diagrams for the following vector fields:

(a) $\vec{v} = \vec{i}$　　**(b)** $\vec{v} = -y\vec{i} + x\vec{j}$　**(c)** $\vec{v} = y\vec{i}$　　**(d)** $\vec{v} = y\vec{j}$

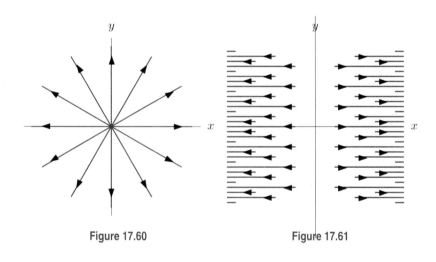

Figure 17.60　　　　　Figure 17.61

Chapter Eighteen

LINE INTEGRALS

When a constant force, $\vec{F}$, acts on an object as it moves through a displacement, $\vec{d}$, the work done by the force is the dot product, $\vec{F} \cdot \vec{d}$. What if the object moves along a curved path through a variable force field? In this chapter we define a *line integral* that calculates the work in this situation. We also define the *circulation* which is used to measure the strength of eddies in a fluid flow.

Line integrals provide us with an analogue of the Fundamental Theorem of Calculus, which tells us how to recover a function from its derivative. The line integral analogue of the Fundamental Theorem shows how a line integral can be used to recover a multivariable function from its gradient field.

In contrast with the one-variable situation, not all vector fields are gradient fields. The line integral can be used to distinguish those that are, using the concept of a *path-independent* (or *conservative*) vector field. We study path-independent fields, which are central in physics, and end with Green's Theorem.

18.1   THE IDEA OF A LINE INTEGRAL

Imagine that you are rowing in a river with a noticeable current. At times you may be working against the current and at other times you may be moving with it. At the end you have a sense of whether, overall, you were helped or hindered by the current. The line integral, defined in this section, measures the extent to which a curve in a vector field is, overall, going with the vector field or against it.

Orientation of a Curve

A curve can be traced out in two directions, as shown in Figure 18.1. We need to choose one direction before we can define a line integral.

> A curve is said to be **oriented** if we have chosen a direction of travel on it.

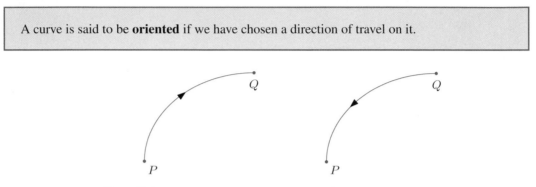

Figure 18.1: A curve with two different orientations represented by arrowheads

Definition of the Line Integral

Consider a vector field $\vec{F}$ and an oriented curve C. We begin by dividing C into n small, almost straight pieces along which $\vec{F}$ is approximately constant. Each piece can be represented by a displacement vector $\Delta \vec{r}_i = \vec{r}_{i+1} - \vec{r}_i$ and the value of $\vec{F}$ at each point of this small piece of C is approximately $\vec{F}(\vec{r}_i)$. See Figures 18.2 and 18.3.

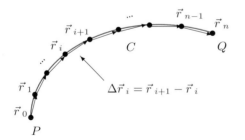

Figure 18.2: The curve C, oriented from P to Q, approximated by straight line segments represented by displacement vectors $\Delta \vec{r}_i = \vec{r}_{i+1} - \vec{r}_i$

Figure 18.3: The vector field $\vec{F}$ evaluated at the points with position vector $\vec{r}_i$ on the curve C oriented from P to Q

For each point with position vector $\vec{r}_i$ on C, we form the dot product $\vec{F}(\vec{r}_i) \cdot \Delta \vec{r}_i$. Summing over all such pieces, we get a Riemann sum:

$$\sum_{i=0}^{n-1} \vec{F}(\vec{r}_i) \cdot \Delta \vec{r}_i.$$

We define the line integral, written $\int_C \vec{F} \cdot d\vec{r}$, by taking the limit as $\|\Delta \vec{r}_i\| \to 0$. Provided the limit exists, we make the following definition:

The **line integral** of a vector field $\vec{F}$ along an oriented curve C is

$$\int_C \vec{F} \cdot d\vec{r} = \lim_{\|\Delta \vec{r}_i\| \to 0} \sum_{i=0}^{n-1} \vec{F}(\vec{r}_i) \cdot \Delta \vec{r}_i.$$

How Does the Limit Defining a Line Integral Work?

The limit in the definition of a line integral exists if $\vec{F}$ is continuous on the curve C and if C is made by joining end to end a finite number of smooth curves. (A vector field $\vec{F} = F_1 \vec{i} + F_2 \vec{j} + F_3 \vec{k}$ is *continuous* if F_1, F_2, and F_3 are continuous, and a *smooth curve* is one that can be parameterized by smooth functions.) We subdivide the curve using a parameterization which goes from one end of the curve to the other, in the forward direction, without retracing any portion of the curve. A subdivision of the parameter interval gives a subdivision of the curve. All the curves we consider in this book are *piecewise smooth* in this sense. Section 18.2 shows how to use a parameterization to compute a line integral.

Example 1   Find the line integral of the constant vector field $\vec{F} = \vec{i} + 2\vec{j}$ along the path from $(1, 1)$ to $(10, 10)$ shown in Figure 18.4.

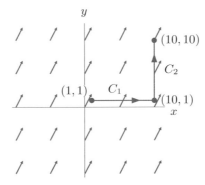

Figure 18.4: The constant vector field $\vec{F} = \vec{i} + 2\vec{j}$ and the path from $(1, 1)$ to $(10, 10)$

Solution   Let C_1 be the horizontal segment of the path going from $(1, 1)$ to $(10, 1)$. When we break this path into pieces, each piece $\Delta \vec{r}$ is horizontal, so $\Delta \vec{r} = \Delta x \vec{i}$ and $\vec{F} \cdot \Delta \vec{r} = (\vec{i} + 2\vec{j}) \cdot \Delta x \vec{i} = \Delta x$. Hence,

$$\int_{C_1} \vec{F} \cdot d\vec{r} = \int_{x=1}^{x=10} dx = 9.$$

Similarly, along the vertical segment C_2, we have $\Delta \vec{r} = \Delta y \vec{j}$ and $\vec{F} \cdot \Delta \vec{r} = (\vec{i} + 2\vec{j}) \cdot \Delta y \vec{j} = 2\Delta y$, so

$$\int_{C_2} \vec{F} \cdot d\vec{r} = \int_{y=1}^{y=10} 2 \, dy = 18.$$

Thus,

$$\int_C \vec{F} \cdot d\vec{r} = \int_{C_1} \vec{F} \cdot d\vec{r} + \int_{C_2} \vec{F} \cdot d\vec{r} = 9 + 18 = 27.$$

What Does the Line Integral Tell Us?

Remember that for any two vectors $\vec{u}$ and $\vec{v}$, the dot product $\vec{u} \cdot \vec{v}$ is positive if $\vec{u}$ and $\vec{v}$ point roughly in the same direction (that is, if the angle between them is less than $\pi/2$). The dot product is zero if $\vec{u}$ is perpendicular to $\vec{v}$ and is negative if they point roughly in opposite directions (that is, if the angle between them is greater than $\pi/2$).

The line integral of adds the dot products of $\vec{F}$ and $\Delta\vec{r}$ along the path. If $||\vec{F}||$ is constant, the line integral gives a positive number if $\vec{F}$ is mostly pointing in the same direction as $\Delta\vec{r}$, a negative number if $\vec{F}$ is mostly pointing in the opposite direction. The line integral is zero if $\vec{F}$ is perpendicular to the path at all points or if the positive and negative contributions cancel out. In general, the line integral of a vector field $\vec{F}$ along a curve C measures the extent to which C is going with $\vec{F}$ or against it.

Example 2   The vector field $\vec{F}$ and the oriented curves C_1, C_2, C_3, C_4 are shown in Figure 18.5. The curves C_1 and C_3 are the same length. Which of the line integrals $\int_{C_i} \vec{F} \cdot d\vec{r}$, for $i = 1, 2, 3, 4$, are positive? Which are negative? Arrange these line integrals in ascending order.

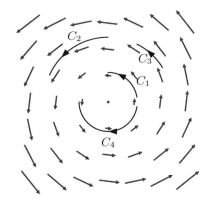

Figure 18.5: Vector field and paths C_1, C_2, C_3, C_4

Solution   The vector field $\vec{F}$ and the line segments $\Delta\vec{r}$ are approximately parallel and in the same direction for the curves C_1, C_2, and C_3. So the contributions of each term $\vec{F} \cdot \Delta\vec{r}$ are positive for these curves. Thus, $\int_{C_1} \vec{F} \cdot d\vec{r}$, $\int_{C_2} \vec{F} \cdot d\vec{r}$, and $\int_{C_3} \vec{F} \cdot d\vec{r}$ are each positive. For the curve C_4, the vector field and the line segments are in opposite directions, so each term $\vec{F} \cdot \Delta\vec{r}$ is negative, and therefore the integral $\int_{C_4} \vec{F} \cdot d\vec{r}$ is negative.

Since the magnitude of the vector field is smaller along C_1 than along C_3, and these two curves are the same length, we have

$$\int_{C_1} \vec{F} \cdot d\vec{r} < \int_{C_3} \vec{F} \cdot d\vec{r}.$$

In addition, the magnitude of the vector field is the same along C_2 and C_3, but the curve C_2 is longer than the curve C_3. Thus,

$$\int_{C_3} \vec{F} \cdot d\vec{r} < \int_{C_2} \vec{F} \cdot d\vec{r}.$$

Putting these results together with the fact that $\int_{C_4} \vec{F} \cdot d\vec{r}$ is negative, we have

$$\int_{C_4} \vec{F} \cdot d\vec{r} < \int_{C_1} \vec{F} \cdot d\vec{r} < \int_{C_3} \vec{F} \cdot d\vec{r} < \int_{C_2} \vec{F} \cdot d\vec{r}.$$

Interpretations of the Line Integral

Work

Recall from Section 13.3 that if a constant force $\vec{F}$ acts on an object while it moves along a straight line through a displacement $\vec{d}$, the work done by the force on the object is

$$\text{Work done} = \vec{F} \cdot \vec{d}.$$

Now suppose we want to find the work done by gravity on an object moving far above the surface of the earth. Since the force of gravity varies with distance from the earth and the path may not be straight, we can't use the formula $\vec{F} \cdot \vec{d}$. We approximate the path by line segments which are small enough that the force is approximately constant on each one. Suppose the force at a point with position vector $\vec{r}$ is $\vec{F}(\vec{r})$, as in Figures 18.2 and 18.3. Then

$$\begin{array}{c}\text{Work done by force } \vec{F}(\vec{r}_i) \\ \text{over small displacement } \Delta\vec{r}_i\end{array} \approx \quad \vec{F}(\vec{r}_i) \cdot \Delta\vec{r}_i,$$

and so,

$$\begin{array}{c}\text{Total work done by force} \\ \text{along oriented curve } C\end{array} \approx \quad \sum_i \vec{F}(\vec{r}_i) \cdot \Delta\vec{r}_i.$$

Taking the limit as $\|\Delta\vec{r}_i\| \to 0$, we get

$$\begin{array}{c}\text{Work done by force } \vec{F}(\vec{r}) \\ \text{along curve } C\end{array} = \lim_{\|\Delta\vec{r}_i\|\to 0} \sum_i \vec{F}(\vec{r}_i) \cdot \Delta\vec{r}_i = \int_C \vec{F} \cdot d\vec{r}.$$

Example 3   A mass lying on a flat table is attached to a spring whose other end is fastened to the wall. (See Figure 18.6.) The spring is extended 20 cm beyond its rest position and released. If the axes are as shown in Figure 18.6, when the spring is extended by a distance of x, the force exerted by the spring on the mass is given by

$$\vec{F}(x) = -kx\vec{i},$$

where k is a positive constant that depends on the strength of the spring.

Suppose the mass moves back to the rest position. How much work is done by the force exerted by the spring?

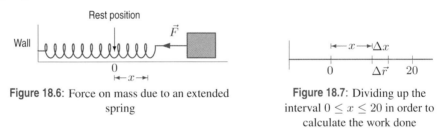

Figure 18.6: Force on mass due to an extended spring

Figure 18.7: Dividing up the interval $0 \le x \le 20$ in order to calculate the work done

Solution   The path from $x = 20$ to $x = 0$ is divided as shown in Figure 18.7, with a typical segment represented by

$$\Delta\vec{r} = \Delta x\vec{i}.$$

Since we are moving from $x = 20$ to $x = 0$, the quantity Δx will be negative. The work done by the force as the mass moves through this segment is approximated by

$$\text{Work done} \approx \vec{F} \cdot \Delta\vec{r} = (-kx\vec{i}) \cdot (\Delta x\vec{i}) = -kx\,\Delta x.$$

Thus, we have

$$\text{Total work done} \approx \sum -kx\,\Delta x.$$

In the limit, as $\|\Delta x\| \to 0$, this sum becomes an ordinary definite integral. Since the path starts at $x = 20$, this is the lower limit of integration; $x = 0$ is the upper limit. Thus, we get

$$\text{Total work done} = \int_{x=20}^{x=0} -kx\,dx = -\frac{kx^2}{2}\bigg|_{20}^{0} = \frac{k(20)^2}{2} = 200k.$$

Note that the work done is positive, since the force acts in the direction of motion.

Example 3 shows how a line integral over a path parallel to the x-axis reduces to a one-variable integral. Section 18.2 shows how to convert *any* line integral into a one-variable integral.

Example 4   A particle with position vector $\vec{r}$ is subject to a force, $\vec{F}$, due to gravity. What is the *sign* of the work done by $\vec{F}$ as the particle moves along the path C_1, a radial line through the center of the earth, starting 8000 km from the center and ending 10,000 km from the center? (See Figure 18.8.)

Solution   We divide the path into small radial segments, $\Delta\vec{r}$, pointing away from the center of the earth and parallel to the gravitational force. The vectors $\vec{F}$ and $\Delta\vec{r}$ point in opposite directions, so each term $\vec{F} \cdot \Delta\vec{r}$ is negative. Adding all these negative quantities and taking the limit results in a negative value for the total work. Thus, the work done by gravity is negative. The negative sign indicates that we would have to do work *against* gravity to move the particle along the path C_1.

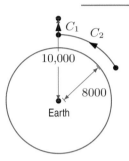

Figure 18.8: The earth

Example 5   Find the sign of the work done by gravity along the curve C_1 in Example 4, but with the opposite orientation.

Solution   Tracing a curve in the opposite direction changes the sign of the line integral because all the segments $\Delta\vec{r}$ change direction, and so every term $\vec{F} \cdot \Delta\vec{r}$ changes sign. Thus, the result will be the negative of the answer found in Example 4. Therefore, the work done by gravity as a particle moves along C_1 toward the center of the earth is positive.

Example 6   Find the work done by gravity as a particle moves along C_2, an arc of a circle 8000 km long at a distance of 8000 km from the center of the earth. (See Figure 18.8.)

Solution   Since C_2 is everywhere perpendicular to the gravitational force, $\vec{F} \cdot \Delta\vec{r} = 0$ for all $\Delta\vec{r}$ along C_2. Thus,

$$\text{Work done} = \int_{C_2} \vec{F} \cdot d\vec{r} = 0,$$

so the work done is zero. This is why satellites can remain in orbit without expending any fuel, once they have attained the correct altitude and velocity.

Circulation

The velocity vector field for the Gulf stream on page 847 shows distinct eddies or regions where the water circulates. We can measure this circulation using a *closed curve*, that is, one that starts and ends at the same point.

> If C is an oriented closed curve, the line integral of a vector field $\vec{F}$ around C is called the **circulation** of $\vec{F}$ around C.

Circulation is a measure of the net tendency of the vector field to point around the curve C. To emphasize that C is closed, the circulation is sometimes denoted $\oint_C \vec{F} \cdot d\vec{r}$, with a small circle on the integral sign.

Example 7   Describe the rotation of the vector fields in Figures 18.9 and 18.10. Find the sign of the circulation of the vector fields around the indicated paths.

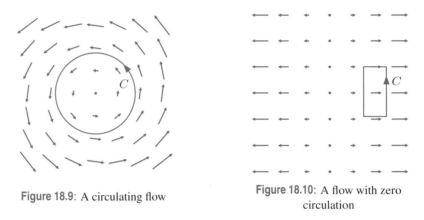

Figure 18.9: A circulating flow

Figure 18.10: A flow with zero circulation

Solution   Consider the vector field in Figure 18.9. If you think of this as representing the velocity of water flowing in a pond, you see that the water is circulating. The line integral around C, measuring the circulation around C, is positive, because the vectors of the field are all pointing in the direction of the path. By way of contrast, look at the vector field in Figure 18.10. Here the line integral around C is zero because the vertical portions of the path are perpendicular to the field and the contributions from the two horizontal portions cancel out. This means that there is no net tendency for the water to circulate around C.

It turns out that the vector field in Figure 18.10 has the property that its circulation around *any* closed path is zero. Water moving according to this vector field has no tendency to circulate around any point and a leaf dropped into the water will not spin. We'll look at such special fields again later when we introduce the notion of the *curl* of a vector field.

Properties of Line Integrals

Line integrals share some basic properties with ordinary one-variable integrals:

> For a scalar constant λ, vector fields $\vec{F}$ and $\vec{G}$, and oriented curves C, C_1, and C_2
>
> 1. $\displaystyle\int_C \lambda\vec{F} \cdot d\vec{r} = \lambda \int_C \vec{F} \cdot d\vec{r}.$ 2. $\displaystyle\int_C (\vec{F} + \vec{G}) \cdot d\vec{r} = \int_C \vec{F} \cdot d\vec{r} + \int_C \vec{G} \cdot d\vec{r}.$
>
> 3. $\displaystyle\int_{-C} \vec{F} \cdot d\vec{r} = -\int_C \vec{F} \cdot d\vec{r}.$ 4. $\displaystyle\int_{C_1+C_2} \vec{F} \cdot d\vec{r} = \int_{C_1} \vec{F} \cdot d\vec{r} + \int_{C_2} \vec{F} \cdot d\vec{r}.$

Properties 3 and 4 are concerned with the curve C over which the line integral is taken. If C is an oriented curve, then $-C$ is the same curve traversed in the opposite direction, that is, with the opposite orientation. (See Figure 18.11 on page 882.) Property 3 holds because if we integrate along $-C$, the vectors $\Delta\vec{r}$ point in the opposite direction and the dot products $\vec{F} \cdot \Delta\vec{r}$ are the negatives of what they were along C.

If C_1 and C_2 are oriented curves with C_1 ending where C_2 begins, we construct a new oriented curve, called $C_1 + C_2$, by joining them together. (See Figure 18.12.) Property 4 is the analogue for line integrals of the property for definite integrals which says that

$$\int_a^b f(x)\,dx = \int_a^c f(x)\,dx + \int_c^b f(x)\,dx.$$

Figure 18.11: A curve, C, and its opposite, $-C$

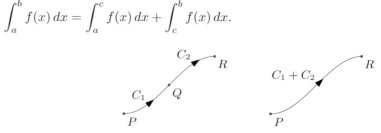

Figure 18.12: Joining two curves, C_1, and C_2, to make a new one, $C_1 + C_2$

Exercises and Problems for Section 18.1

Exercises

In Exercises 1–4, say whether you expect the line integral of the pictured vector field over the given curve to be positive, negative, or zero.

1.

2.

3.

4.

In Exercises 5–13, calculate the line integral of the vector field along the line between the given points.

5. $\vec{F} = x\vec{i}$,   from $(2, 0)$ to $(6, 0)$

6. $\vec{F} = x\vec{j}$,   from $(1, 0)$ to $(3, 0)$

7. $\vec{F} = x\vec{j}$,   from $(2, 0)$ to $(2, 5)$

8. $\vec{F} = 3\vec{i} + 4\vec{j}$,   from $(0, 6)$ to $(0, 13)$

9. $\vec{F} = 6\vec{i} - 7\vec{j}$,   from $(0, 0)$ to $(7, 6)$

10. $\vec{F} = x\vec{i} + y\vec{j}$,   from $(2, 0)$ to $(6, 0)$

11. $\vec{F} = 6\vec{i} + y^2\vec{j}$,   from $(3, 0)$ to $(7, 0)$

12. $\vec{F} = \vec{r}$,   from $(2, 2)$ to $(6, 6)$

13. $\vec{F} = x\vec{i} + 6\vec{j} - \vec{k}$,   from $(0, -2, 0)$ to $(0, -10, 0)$

Problems

14. Given the force field $\vec{F}(x, y) = y\vec{i} + x^2\vec{j}$ and the right-angle curve, C, from the points $(0, -1)$ to $(4, -1)$ to $(4, 3)$ shown in Figure 18.13:

(a) Evaluate $\vec{F}$ at the points $(0, -1)$, $(1, -1)$, $(2, -1)$, $(3, -1)$, $(4, -1)$, $(4, 0)$, $(4, 1)$, $(4, 2)$, $(4, 3)$.
(b) Make a sketch showing the force field along C.
(c) Estimate the work done by the indicated force field on an object traversing the curve C.

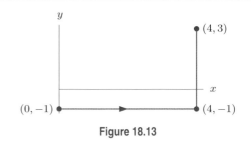

Figure 18.13

15. Consider the vector field $\vec{F}$ shown in Figure 18.14, together with the paths C_1, C_2, and C_3. Arrange the line integrals $\int_{C_1} \vec{F} \cdot d\vec{r}$, $\int_{C_2} \vec{F} \cdot d\vec{r}$ and $\int_{C_3} \vec{F} \cdot d\vec{r}$ in ascending order.

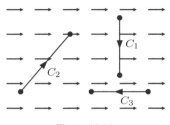

Figure 18.14

16. Let $\vec{F}$ be the constant force field $\vec{j}$ in Figure 18.15. On which of the paths C_1, C_2, C_3 is zero work done by $\vec{F}$? Explain.

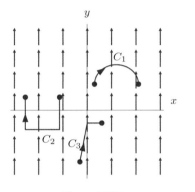

Figure 18.15

In Problems 17–19, let C_1 be the line from $(0,0)$ to $(0,1)$; let C_2 be the line from $(1,0)$ to $(0,1)$; let C_3 be the semi-circle in the upper half plane from $(-1,0)$ to $(1,0)$. Do the line integrals of the vector field along each of the paths C_1, C_2, and C_3 appear to be positive, negative, or zero?

17.   18.

19.

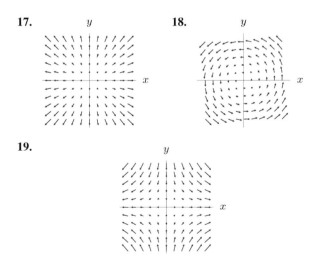

For Problems 20–24, say whether you expect the given vector field to have positive, negative, or zero circulation around

the closed curve $C = C_1 + C_2 + C_3 + C_4$ in Figure 18.16. The segments C_1 and C_3 are circular arcs centered at the origin; C_2 and C_4 are radial line segments. You may find it helpful to sketch the vector field.

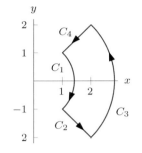

Figure 18.16

20. $\vec{F}(x,y) = x\vec{i} + y\vec{j}$ 21. $\vec{F}(x,y) = -y\vec{i} + x\vec{j}$

22. $\vec{F}(x,y) = y\vec{i} - x\vec{j}$ 23. $\vec{F}(x,y) = x^2\vec{i}$

24. $\vec{F}(x,y) = -\dfrac{y}{x^2+y^2}\vec{i} + \dfrac{x}{x^2+y^2}\vec{j}$

25. (a) For each of the vector fields, $\vec{F}$, shown in Figure 18.17, sketch a curve for which the integral $\int_C \vec{F} \cdot d\vec{r}$ is positive.
    (b) For which of the vector fields is it possible to make your answer to part (a) a closed curve?

(i)   (ii)

(iii)   (iv)

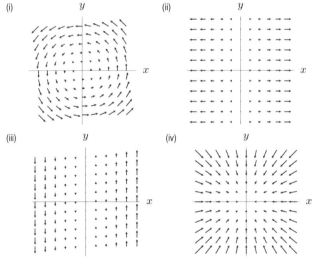

Figure 18.17

26. Along a curve C, a vector field $\vec{F}$ is everywhere tangent to C in the direction of orientation and has constant magnitude $\|\vec{F}\| = m$. Use the definition of the line integral to explain why

$$\int_C \vec{F} \cdot d\vec{r} = m \cdot \text{Length of } C.$$

27. Draw an oriented curve C and a vector field $\vec{F}$ along C that is not always perpendicular to C, but for which $\int_C \vec{F} \cdot d\vec{r} = 0$.

28. Explain why the following statement is true: Whenever the line integral of a vector field around every closed curve is zero, the line integral along a curve with fixed endpoints has a constant value independent of the path taken between the endpoints.

29. Explain why the converse to the statement in Problem 28 is also true: Whenever the line integral of a vector field depends only on endpoints and not on paths, the circulation around every closed curve is zero.

30. A horizontal square has sides of 1000 km running north-south and east-west. A wind blows from the east and decreases in magnitude toward the north at a rate of 6 meter/sec for every 500 km. Compute the circulation of the wind counterclockwise around the square.

In Problems 31–32, use the fact that the force of gravity on a particle of mass m at the point with position vector $\vec{r}$ is

$$\vec{F} = -\frac{GMm\vec{r}}{\|\vec{r}\|^3}$$

where and G is a constant and M is the mass of the earth.

31. Calculate the work done by the force of gravity on a particle of mass m as it moves from 8000 km to 10,000 km from the center of the earth.

32. Calculate the work done by the force of gravity on a particle of mass m as it moves from 8000 km from the center of the earth to infinitely far away.

In Problems 33–36 we explore the notion of the electric potential of an electric field $\vec{E}$. We chose a point P_0 to be the ground, that is, the potential at P_0 is zero. Then the potential $\phi(P)$ at a point P is defined to be the work done moving a particle of charge 1 coulomb from P_0 to P. (The potential does not depend on the path chosen.) Q, E are constants.

33. Write a line integral for $\phi(P)$, using the fact that the electric field acts on a particle with force $\vec{E}$.

34. Let $\vec{E} = \left(\dfrac{Q}{4\pi\epsilon}\right)\dfrac{\vec{r}}{\|\vec{r}\|^3}$, and let P_0 be a point a units from the origin. Describe the set of points with zero potential.

35. An equipotential surface is a surface on which the potential is constant. Describe the equipotential surfaces of $\vec{E} = \dfrac{Q}{4\pi\epsilon}\dfrac{\vec{r}}{\|\vec{r}\|^3}$, where the ground P_0 is chosen to be at distance a from the origin.

36. Let $\vec{E} = \dfrac{Q}{4\pi\epsilon}\dfrac{\vec{r}}{\|\vec{r}\|^3}$ and let the ground P_0 be chosen to be a units from the origin.

    (a) Find a formula for ϕ.
    (b) Engineers often choose the ground point to be "at infinity." Why?

18.2   COMPUTING LINE INTEGRALS OVER PARAMETERIZED CURVES

The goal of this section is to show how to use a parameterization of a curve to convert a line integral into an ordinary one-variable integral.

Using a Parameterization to Evaluate a Line Integral

Recall the definition of the line integral,

$$\int_C \vec{F} \cdot d\vec{r} = \lim_{\|\Delta\vec{r}_i\|\to 0} \sum \vec{F}(\vec{r}_i) \cdot \Delta\vec{r}_i,$$

where the $\vec{r}_i$ are the position vectors of points subdividing the curve into short pieces. Now suppose we have a smooth parameterization, $\vec{r}(t)$, of C for $a \leq t \leq b$, so that $\vec{r}(a)$ is the position vector of the beginning of the curve and $\vec{r}(b)$ is the position vector of the end. Then we can divide C into n pieces by dividing the interval $a \leq t \leq b$ into n pieces, each of size $\Delta t = (b-a)/n$. See Figures 18.18 and 18.19.

At each point $\vec{r}_i = \vec{r}(t_i)$ we want to compute

$$\vec{F}(\vec{r}_i) \cdot \Delta\vec{r}_i.$$

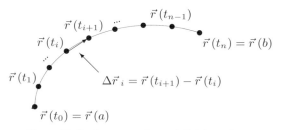

Figure 18.18: Subdivision of the interval $a \leq t \leq b$

Figure 18.19: Corresponding subdivision of the parameterized path C

Since $t_{i+1} = t_i + \Delta t$, the displacement vectors $\Delta \vec{r_i}$ are given by

$$\begin{aligned} \Delta \vec{r_i} &= \vec{r}(t_{i+1}) - \vec{r}(t_i) \\ &= \vec{r}(t_i + \Delta t) - \vec{r}(t_i) \\ &= \frac{\vec{r}(t_i + \Delta t) - \vec{r}(t_i)}{\Delta t} \cdot \Delta t \\ &\approx \vec{r}'(t_i)\Delta t, \end{aligned}$$

where we use the facts that Δt is small and that $\vec{r}(t)$ is differentiable to obtain the last approximation.

Therefore,

$$\int_C \vec{F} \cdot d\vec{r} \approx \sum \vec{F}(\vec{r}_i) \cdot \Delta \vec{r}_i \approx \sum \vec{F}(\vec{r}(t_i)) \cdot \vec{r}'(t_i)\,\Delta t.$$

Notice that $\vec{F}(\vec{r}(t_i)) \cdot \vec{r}'(t_i)$ is the value at t_i of a one-variable function of t, so this last sum is really a one-variable Riemann sum. In the limit as $\Delta t \to 0$, we get a definite integral:

$$\lim_{\Delta t \to 0} \sum \vec{F}(\vec{r}(t_i)) \cdot \vec{r}'(t_i)\,\Delta t = \int_a^b \vec{F}(\vec{r}(t)) \cdot \vec{r}'(t)\,dt.$$

Thus, we have the following result:

If $\vec{r}(t)$, for $a \le t \le b$, is a smooth parameterization of an oriented curve C and $\vec{F}$ is a vector field which is continuous on C, then

$$\int_C \vec{F} \cdot d\vec{r} = \int_a^b \vec{F}(\vec{r}(t)) \cdot \vec{r}'(t)\,dt.$$

In words: To compute the line integral of $\vec{F}$ over C, take the dot product of $\vec{F}$ evaluated on C with the velocity vector, $\vec{r}'(t)$, of the parameterization of C, then integrate along the curve.

Even though we assumed that C is smooth, we can use the same formula to compute line integrals over curves which are only *piecewise smooth*, such as the boundary of a rectangle: If C is piecewise smooth, we apply the formula to each one of the smooth pieces and add the results.

Example 1   Compute $\int_C \vec{F} \cdot d\vec{r}$ where $\vec{F} = (x+y)\vec{i} + y\vec{j}$ and C is the quarter unit circle, oriented counterclockwise as shown in Figure 18.20.

Figure 18.20: The vector field $\vec{F} = (x+y)\vec{i} + y\vec{j}$ and the quarter circle C

Solution   Since all of the vectors in $\vec{F}$ along C point generally in a direction opposite to the orientation of C, we expect our answer to be negative. The first step is to parameterize C by

$$\vec{r}(t) = x(t)\vec{i} + y(t)\vec{j} = \cos t\,\vec{i} + \sin t\,\vec{j}, \quad 0 \le t \le \frac{\pi}{2}.$$

Substituting the parameterization into $\vec{F}$ we get $\vec{F}(x(t), y(t)) = (\cos t + \sin t)\vec{i} + \sin t\vec{j}$. The vector $\vec{r}'(t) = x'(t)\vec{i} + y'(t)\vec{j} = -\sin t\vec{i} + \cos t\vec{j}$. Then

$$\int_C \vec{F} \cdot d\vec{r} = \int_0^{\pi/2} ((\cos t + \sin t)\vec{i} + \sin t\vec{j}) \cdot (-\sin t\vec{i} + \cos t\vec{j})dt$$

$$= \int_0^{\pi/2} (-\cos t \sin t - \sin^2 t + \sin t \cos t)dt$$

$$= \int_0^{\pi/2} -\sin^2 t \, dt = -\frac{\pi}{4} \approx -0.7854.$$

So the answer is negative, as expected.

Example 2   Consider the vector field $\vec{F} = x\vec{i} + y\vec{j}$.

(a) Suppose C_1 is the line segment joining $(1, 0)$ to $(0, 2)$ and C_2 is a part of a parabola with its vertex at $(0, 2)$, joining the same points in the same order. (See Figure 18.21.) Verify that

$$\int_{C_1} \vec{F} \cdot d\vec{r} = \int_{C_2} \vec{F} \cdot d\vec{r}.$$

(b) If C is the triangle shown in Figure 18.22, show that $\int_C \vec{F} \cdot d\vec{r} = 0$.

Figure 18.21   Figure 18.22

Solution   (a) We parameterize C_1 by $\vec{r}(t) = (1 - t)\vec{i} + 2t\vec{j}$ with $0 \leq t \leq 1$. Then $\vec{r}'(t) = -\vec{i} + 2\vec{j}$, so

$$\int_{C_1} \vec{F} \cdot d\vec{r} = \int_0^1 \vec{F}(1 - t, 2t) \cdot (-\vec{i} + 2\vec{j}) \, dt = \int_0^1 ((1 - t)\vec{i} + 2t\vec{j}) \cdot (-\vec{i} + 2\vec{j}) \, dt$$

$$= \int_0^1 (5t - 1) \, dt = \frac{3}{2}.$$

To parameterize C_2, we use the fact that it is part of a parabola with vertex at $(0, 2)$, so its equation is of the form $y = -kx^2 + 2$ for some k. Since the parabola crosses the x-axis at $(1, 0)$, we find that $k = 2$ and $y = -2x^2 + 2$. Therefore, we use the parameterization $\vec{r}(t) = t\vec{i} + (-2t^2 + 2)\vec{j}$ with $0 \leq t \leq 1$, which has $\vec{r}' = \vec{i} - 4t\vec{j}$. This traces out C_2 in reverse, since $t = 0$ gives $(0, 2)$, and $t = 1$ gives $(1, 0)$. Thus, we make $t = 0$ the upper limit of integration and $t = 1$ the lower limit:

$$\int_{C_2} \vec{F} \cdot d\vec{r} = \int_1^0 \vec{F}(t, -2t^2 + 2) \cdot (\vec{i} - 4t\vec{j}) \, dt = -\int_0^1 (t\vec{i} + (-2t^2 + 2)\vec{j}) \cdot (\vec{i} - 4t\vec{j}) \, dt$$

$$= -\int_0^1 (8t^3 - 7t) \, dt = \frac{3}{2}.$$

So the line integrals along C_1 and C_2 have the same value.

(b) We break $\int_C \vec{F} \cdot d\vec{r}$ into three pieces, one of which we have already computed (namely, the piece connecting $(1, 0)$ to $(0, 2)$, where the line integral has value $3/2$). The piece running from $(0, 2)$ to $(0, 0)$ can be parameterized by $\vec{r}(t) = (2 - t)\vec{j}$ with $0 \leq t \leq 2$. The piece running from $(0, 0)$ to $(1, 0)$ can be parameterized by $\vec{r}(t) = t\vec{i}$ with $0 \leq t \leq 1$. Then

$$\int_C \vec{F} \cdot d\vec{r} = \frac{3}{2} + \int_0^2 \vec{F}(0, 2 - t) \cdot (-\vec{j}) \, dt + \int_0^1 \vec{F}(t, 0) \cdot \vec{i} \, dt$$

$$= \frac{3}{2} + \int_0^2 (2 - t)\vec{j} \cdot (-\vec{j}) \, dt + \int_0^1 t\vec{i} \cdot \vec{i} \, dt$$

$$= \frac{3}{2} + \int_0^2 (t - 2) \, dt + \int_0^1 t \, dt = \frac{3}{2} + (-2) + \frac{1}{2} = 0.$$

Example 3   Let C be the closed curve consisting of the upper half-circle of radius 1 and the line forming its diameter along the x-axis, oriented counterclockwise. (See Figure 18.23.) Find $\int_C \vec{F} \cdot d\vec{r}$ where $\vec{F}(x, y) = -y\vec{i} + x\vec{j}$.

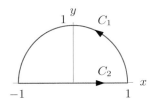

Figure 18.23: The curve $C = C_1 + C_2$ for Example 3

Solution   We write $C = C_1 + C_2$ where C_1 is the half-circle and C_2 is the line, and compute $\int_{C_1} \vec{F} \cdot d\vec{r}$ and $\int_{C_2} \vec{F} \cdot d\vec{r}$ separately. We parameterize C_1 by $\vec{r}(t) = \cos t\vec{i} + \sin t\vec{j}$, with $0 \le t \le \pi$. Then

$$\int_{C_1} \vec{F} \cdot d\vec{r} = \int_0^\pi (-\sin t\vec{i} + \cos t\vec{j}) \cdot (-\sin t\vec{i} + \cos t\vec{j}) \, dt$$

$$= \int_0^\pi (\sin^2 t + \cos^2 t) \, dt = \int_0^\pi 1 \, dt = \pi.$$

For C_2, we have $\int_{C_2} \vec{F} \cdot d\vec{r} = 0$, since the vector field $\vec{F}$ has no $\vec{i}$ component along the x-axis (where $y = 0$) and is therefore perpendicular to C_2 at all points.

Finally, we can write

$$\int_C \vec{F} \cdot d\vec{r} = \int_{C_1} \vec{F} \cdot d\vec{r} + \int_{C_2} \vec{F} \cdot d\vec{r} = \pi + 0 = \pi.$$

It is no accident that the result for $\int_{C_1} \vec{F} \cdot d\vec{r}$ is the same as the length of the curve C_1. See Problem 26 on page 883 and Problem 18 on page 889.

The next example illustrates the computation of a line integral over a path in 3-space.

Example 4   A particle travels along the helix C given by $\vec{r}(t) = \cos t\vec{i} + \sin t\vec{j} + 2t\vec{k}$ and is subject to a force $\vec{F} = x\vec{i} + z\vec{j} - xy\vec{k}$. Find the total work done on the particle by the force for $0 \le t \le 3\pi$.

Solution   The work done is given by a line integral, which we evaluate using the given parameterization:

$$\text{Work done} = \int_C \vec{F} \cdot d\vec{r} = \int_0^{3\pi} \vec{F}(\vec{r}(t)) \cdot \vec{r}'(t) \, dt$$

$$= \int_0^{3\pi} (\cos t\vec{i} + 2t\vec{j} - \cos t \sin t\vec{k}) \cdot (-\sin t\vec{i} + \cos t\vec{j} + 2\vec{k}) \, dt$$

$$= \int_0^{3\pi} (-\cos t \sin t + 2t \cos t - 2\cos t \sin t) \, dt$$

$$= \int_0^{3\pi} (-3\cos t \sin t + 2t \cos t) \, dt = -4.$$

The Notation $\int_C P\,dx + Q\,dy + Rdz$

There is an alternative notation for line integrals that is quite common. Given functions $P(x, y, z)$ and $Q(x, y, z)$ and $R(x, y, z)$, and an oriented curve C, we consider the vector field $\vec{F} = P\vec{i} + Q\vec{j} + R\vec{k}$. We can write

$$\int_C \vec{F} \cdot d\vec{r} = \int_C P(x, y, z)dx + Q(x, y, z)dy + R(x, y, z)dz,$$

The relation between the two notations can be remembered using $d\vec{r} = dx\vec{i} + dy\vec{j} + dz\vec{k}$.

Example 5   Evaluate $\int_C xy\,dx - y^2\,dy$ where C is the line segment from $(0, 0)$ to $(2, 6)$.

Solution   We parameterize C by $\vec{r}(t) = x(t)\vec{i} + y(t)\vec{j} = t\vec{i} + 3t\vec{j}$, for $0 \le t \le 2$. Thus,

$$\int_C xy\,dx - y^2\,dy = \int_C (xy\vec{i} - y^2\vec{j}) \cdot d\vec{r} = \int_0^2 (3t^2\vec{i} - 9t^2\vec{j}) \cdot (\vec{i} + 3\vec{j})\,dt = \int_0^2 (-24t^2)\,dt = -64.$$

Independence of Parameterization

Since there are many different ways of parameterizing a given oriented curve, you may be wondering what happens to the value of a given line integral if you choose another parameterization. The answer is that the choice of parameterization makes no difference. Since we initially defined the line integral without reference to any particular parameterization, this is exactly as we would expect.

Example 6   Consider the oriented path which is a straight line segment L running from $(0, 0)$ to $(1, 1)$. Calculate the line integral of the vector field $\vec{F} = (3x - y)\vec{i} + x\vec{j}$ along L using each of the parameterizations

(a) $A(t) = (t, t), \quad 0 \le t \le 1,$ (b) $D(t) = (e^t - 1, e^t - 1), \quad 0 \le t \le \ln 2.$

Solution   The line L has equation $y = x$. Both $A(t)$ and $D(t)$ give a parameterization of L: each has both coordinates equal and each begins at (0,0) and ends at (1,1). Now let's calculate the line integral of the vector field $\vec{F} = (3x - y)\vec{i} + x\vec{j}$ using each parameterization.

(a) Using $A(t)$, we get

$$\int_L \vec{F} \cdot d\vec{r} = \int_0^1 ((3t - t)\vec{i} + t\vec{j}) \cdot (\vec{i} + \vec{j})\,dt = \int_0^1 3t\,dt = \left.\frac{3t^2}{2}\right|_0^1 = \frac{3}{2}.$$

(b) Using $D(t)$, we get

$$\int_L \vec{F} \cdot d\vec{r} = \int_0^{\ln 2} \left((3(e^t - 1) - (e^t - 1))\,\vec{i} + (e^t - 1)\vec{j} \right) \cdot (e^t\vec{i} + e^t\vec{j})\,dt$$

$$= \int_0^{\ln 2} 3(e^{2t} - e^t)\,dt = 3\left.\left(\frac{e^{2t}}{2} - e^t\right)\right|_0^{\ln 2} = \frac{3}{2}.$$

The fact that both answers are the same illustrates that the value of a line integral is independent of the parameterization of the path. Problems 22–24 at the end of this section give another way of seeing this.

Exercises and Problems for Section 18.2

Exercises

In Exercises 1–16, find $\int_C \vec{F} \cdot d\vec{r}$ for the given $\vec{F}$ and C.

1. $\vec{F} = 2\vec{i} + \vec{j}$; C is the x-axis from $x = 10$ to $x = 7$.

2. $\vec{F} = 3\vec{j} - \vec{i}$; C is the line $y = x$ from $(1,1)$ to $(5,5)$.

3. $\vec{F} = x\vec{i} + y\vec{j}$ and C is the line from $(0,0)$ to $(3,3)$.

4. $\vec{F} = x^2\vec{i} + y^2\vec{j}$ and C is the line from the point $(1,2)$ to the point $(3,4)$.

5. $\vec{F} = y\vec{i} - x\vec{j}$ and C is the right-hand side of the unit circle, starting at $(0,1)$.

6. $\vec{F} = -y\sin x\vec{i} + \cos x\vec{j}$ and C is the parabola $y = x^2$ between $(0,0)$ and $(2,4)$.

7. $\vec{F} = \ln y\vec{i} + \ln x\vec{j}$ and C is the curve given parametrically by $(2t, t^3)$, for $2 \le t \le 4$.

8. $\vec{F} = 2y\vec{i} - (\sin y)\vec{j}$ counterclockwise around the unit circle C starting at the point $(1,0)$.

9. $\vec{F} = y^3\vec{i} + x^2\vec{j}$ and C is the line from $(0,0)$ to $(3,2)$.

10. $\vec{F} = x\vec{i} + 6\vec{j} - \vec{k}$, and C is the line $x = y = z$ from $(0,0,0)$ to $(2,2,2)$.

11. $\vec{F} = e^x\vec{i} + e^y\vec{j}$ and C is the part of the ellipse $x^2 + 4y^2 = 4$ joining the point $(0,1)$ to the point $(2,0)$ in the clockwise direction.

12. $\vec{F} = xy\vec{i} + (x-y)\vec{j}$ and C is the triangle joining $(1,0)$, $(0,1)$ and $(-1,0)$ in the clockwise direction.

13. $\vec{F} = x\vec{i} + 2zy\vec{j} + x\vec{k}$ and C is given by $\vec{r} = t\vec{i} + t^2\vec{j} + t^3\vec{k}$ for $1 \le t \le 2$.

14. $\vec{F} = x^3\vec{i} + y^2\vec{j} + z\vec{k}$ and C is the line from the origin to the point $(2,3,4)$.

15. $\vec{F} = -y\vec{i} + x\vec{j} + 5\vec{k}$ and C is the helix $x = \cos t, y = \sin t, z = t$, for $0 \le t \le 4\pi$.

16. $\vec{F} = e^y\vec{i} + \ln(x^2+1)\vec{j} + \vec{k}$ and C is the circle of radius 2 centered at the origin in the yz-plane in Figure 18.24.

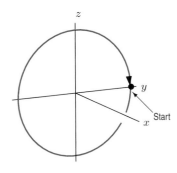

Figure 18.24

Problems

17. In Example 6 on page 888 we integrated $\vec{F} = (3x - y)\vec{i} + x\vec{j}$ over two parameterizations of the line from $(0,0)$ to $(1,1)$, getting $3/2$ each time. Now compute the line integral along two different paths with the same endpoints, and show that the answers are different.

(a) The path (t, t^2), with $0 \le t \le 1$
(b) The path (t^2, t), with $0 \le t \le 1$

18. Let $\vec{F} = -y\vec{i} + x\vec{j}$ and let C be the unit circle oriented counterclockwise.

(a) Show that $\vec{F}$ has a constant magnitude of 1 on C.
(b) Show that $\vec{F}$ is always tangent to the circle C.
(c) Show that $\int_C \vec{F} \cdot d\vec{r} = $ Length of C.

19. A spiral staircase in a building is in the shape of a helix of radius 5 meters. Between two floors of the building, the stairs make one full revolution and climb by 4 meters. A person carries a bag of groceries up two floors. The combined mass of the person and the groceries is 70 kg and the gravitational force is $70g$ downward, where g is the acceleration due to gravity. Calculate the work done by the person against gravity.

20. Suppose C is the line segment from the point $(0,0)$ to the point $(4,12)$ and $\vec{F} = xy\vec{i} + x\vec{j}$.

(a) Is $\int_C \vec{F} \cdot d\vec{r}$ greater than, less than, or equal to zero? Give a geometric explanation.
(b) A parameterization of C is $(x(t), y(t)) = (t, 3t)$ for $0 \le t \le 4$. Use this to compute $\int_C \vec{F} \cdot d\vec{r}$.
(c) Suppose a particle leaves the point $(0,0)$, moves along the line toward the point $(4,12)$, stops before reaching it and backs up, stops again and reverses direction, then completes its journey to the endpoint. All travel takes place along the line segment joining the point $(0,0)$ to the point $(4,12)$. If we call this path C', explain why $\int_{C'} \vec{F} \cdot d\vec{r} = \int_C \vec{F} \cdot d\vec{r}$.
(d) A parameterization for a path like C' is given, for $0 \le t \le 4$, by

$$(x(t), y(t)) = \left(\frac{t^3 - 6t^2 + 11t}{3}, (t^3 - 6t^2 + 11t) \right).$$

Check that this parameterization begins at the point $(0,0)$ and ends at the point $(4,12)$. Check also that all points of C' lie on the line segment connecting the point $(0,0)$ to the point $(4,12)$. What are the values of t at which the particle changes direction?
(e) Find $\int_{C'} \vec{F} \cdot d\vec{r}$ using the parameterization in part (d). Do you get the same answer as in part (b)?

21. Consider the oriented path which is a straight line segment L running from $(0,0)$ to $(1,1)$. Calculate the line integral of the vector field $\vec{F} = (3x - y)\vec{i} + x\vec{j}$ along L using each of the parameterizations

(a) $B(t) = (2t, 2t), \quad 0 \le t \le 1/2,$

(b) $C(t) = \left(\dfrac{t^2 - 1}{3}, \dfrac{t^2 - 1}{3} \right), \quad 1 \le t \le 2,$

In Example 6 on page 888 two parameterizations, $A(t)$, and $D(t)$, are used to convert a line integral into a definite integral. In Problem 21, two other parameterizations, $B(t)$ and $C(t)$, are used on the same line integral. In Problems 22–24 show that two definite integrals corresponding to two of the given parameterizations are equal by finding a substitution which converts one integral to the other. This gives us another way of seeing why changing the parameterization of the curve does not change the value of the line integral.

22. $A(t)$ and $B(t)$ 23. $A(t)$ and $C(t)$

24. $A(t)$ and $D(t)$

18.3 GRADIENT FIELDS AND PATH-INDEPENDENT FIELDS

For a function, f, of one variable, the Fundamental Theorem of Calculus tells us that the definite integral of a rate of change, f', gives the total change in f:

$$\int_a^b f'(t)\, dt = f(b) - f(a).$$

What about functions of two or more variables? The quantity that describes the rate of change is the gradient vector field. If we know the gradient of a function f, can we compute the total change in f between two points? The answer is yes, using a line integral.

Finding the Total Change in f from grad f: The Fundamental Theorem

To find the change in f between two points P and Q, we choose a smooth path C from P to Q, then divide the path into many small pieces. See Figure 18.25.

First we estimate the change in f as we move through a displacement $\Delta \vec{r}_i$ from $\vec{r}_i$ to $\vec{r}_{i+1}$. Suppose $\vec{u}$ is a unit vector in the direction of $\Delta \vec{r}_i$. Then the change in f is given by

$$f(\vec{r}_{i+1}) - f(\vec{r}_i) \approx \text{Rate of change of } f \times \text{Distance moved in direction of } \vec{u}$$
$$= f_{\vec{u}}(\vec{r}_i) \| \Delta \vec{r}_i \|$$
$$= \text{grad } f \cdot \vec{u} \, \| \Delta \vec{r}_i \|$$
$$= \text{grad } f \cdot \Delta \vec{r}_i. \qquad \text{since } \Delta \vec{r}_i = \| \Delta \vec{r}_i \| \vec{u}$$

Therefore, summing over all pieces of the path, the total change in f is given by

$$\text{Total change} = f(Q) - f(P) \approx \sum_{i=0}^{n-1} \text{grad } f(\vec{r}_i) \cdot \Delta \vec{r}_i.$$

In the limit as $\| \Delta \vec{r}_i \|$ approaches zero, this suggests the following result:

Theorem 18.1: The Fundamental Theorem of Calculus for Line Integrals

Suppose C is a piecewise smooth oriented path with starting point P and endpoint Q. If f is a function whose gradient is continuous on the path C, then

$$\int_C \text{grad } f \cdot d\vec{r} = f(Q) - f(P).$$

Notice that there are many different paths from P to Q. (See Figure 18.26.) However, the value of the line integral $\int_C \operatorname{grad} f \cdot d\vec{r}$ depends only on the endpoints of C; it does not depend on where C goes in between. Problem 35 on page 898 shows how the Fundamental Theorem for Line Integrals can be derived from the one-variable Fundamental Theorem of Calculus.

Figure 18.25: Subdivision of the path from P to Q. We estimate the change in f along $\Delta\vec{r}_i$

Figure 18.26: There are many different paths from P to Q: all give the same value of $\int_C \operatorname{grad} f \cdot d\vec{r}$

Example 1 Suppose that $\operatorname{grad} f$ is everywhere perpendicular to the curve joining P and Q shown in Figure 18.27.

(a) Explain why you expect the path joining P and Q to be a contour.
(b) Using a line integral, show that $f(P) = f(Q)$.

Figure 18.27: The gradient vector field of the function f

Solution (a) The gradient of f is everywhere perpendicular to the path from P to Q, as you expect along a contour.

(b) Consider the path from P to Q shown in Figure 18.27 and evaluate the line integral

$$\int_C \operatorname{grad} f \cdot d\vec{r} = f(Q) - f(P).$$

Since $\operatorname{grad} f$ is everywhere perpendicular to the path, the line integral is 0. Thus, $f(Q) = f(P)$.

Example 2 Consider the vector field $\vec{F} = x\vec{i} + y\vec{j}$. In Example 2 on page 886 we calculated $\int_{C_1} \vec{F} \cdot d\vec{r}$ and $\int_{C_2} \vec{F} \cdot d\vec{r}$ over the oriented curves shown in Figure 18.28 and found they were the same. Find a scalar function f with $\operatorname{grad} f = \vec{F}$. Hence, find an easy way to calculate the line integrals, and explain how we could have expected them to be the same.

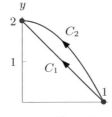

Figure 18.28: Find the line integral of $\vec{F} = x\vec{i} + y\vec{j}$ over the curves C_1 and C_2

Solution   One possibility for f is

$$f(x, y) = \frac{x^2}{2} + \frac{y^2}{2}.$$

You can check that grad $f = x\vec{i} + y\vec{j}$. Now we can use the Fundamental Theorem to compute the line integral. Since $\vec{F}$ = grad f we have

$$\int_{C_1} \vec{F} \cdot d\vec{r} = \int_{C_1} \text{grad } f \cdot d\vec{r} = f(0, 2) - f(1, 0) = \frac{3}{2}.$$

Notice that the calculation looks exactly the same for C_2. Since the value of the integral depends only on the value of f at the endpoints, it is the same no matter what path we choose.

Path-Independent, or Conservative, Vector Fields

In the previous example, the line integral was independent of the path taken between the two (fixed) endpoints. We give vector fields whose line integrals have this property a special name.

> A vector field $\vec{F}$ is said to be **path-independent**, or **conservative**, if for any two points P and Q, the line integral $\int_C \vec{F} \cdot d\vec{r}$ has the same value along any piecewise smooth path C from P to Q lying in the domain of $\vec{F}$.

If, on the other hand, the line integral $\int_C \vec{F} \cdot d\vec{r}$ does depend on the path C joining P to Q, then $\vec{F}$ is said to be a *path-dependent* vector field.

Now suppose that $\vec{F}$ is any continuous gradient field, so $\vec{F}$ = grad f. If C is a path from P to Q, the Fundamental Theorem for Line Integrals tells us that

$$\int_C \vec{F} \cdot d\vec{r} = f(Q) - f(P).$$

Since the right-hand side of this equation does not depend on the path, but only on the endpoints of the path, the vector field $\vec{F}$ is path-independent. Thus, we have the following important result:

> If $\vec{F}$ is a continuous gradient vector field, then $\vec{F}$ is path-independent.

Why Do We Care about Path-Independent, or Conservative, Vector Fields?

Many of the fundamental vector fields of nature are path-independent — for example the gravitational field and the electric field of particles at rest. The fact that the gravitational field is path-independent means that the work done by gravity when an object moves depends only on the starting and ending points and not on the path taken. For example the work done by gravity (computed by the line integral) on a bicycle being carried to a sixth floor apartment is the same whether it is carried up the stairs in a zig-zag path or taken straight up in an elevator.

When a vector field is path-independent we can define the *potential energy* of a body. When the body moves to another position, the potential energy changes by an amount equal to the work done by the vector field, which depends only on the starting and ending positions. If the work done had not been path-independent, the potential energy would depend both on the body's current position *and* on how it got there, making it impossible to define a useful potential energy.

Project 1 on page 913 explains why path-independent force vector fields are also called *conservative* vector fields: When a particle moves under the influence of a conservative vector field, the *total energy* of the particle is *conserved*. It turns out that the force field is obtained from the gradient of the potential energy function.

Path-Independent Fields and Gradient Fields

We have seen that every gradient field is path-independent. What about the converse? That is, given a path-independent vector field $\vec{F}$, can we find a function f such that $\vec{F} = \operatorname{grad} f$? The answer is yes, provided that $\vec{F}$ is continuous.

How to Construct f from $\vec{F}$

First, notice that there are many different choices for f, since we can add a constant to f without changing $\operatorname{grad} f$. If we pick a fixed starting point P, then by adding or subtracting a constant to f we can ensure that $f(P) = 0$. For any other point Q, we define $f(Q)$ by the formula

$$f(Q) = \int_C \vec{F} \cdot d\vec{r}, \quad \text{where } C \text{ is any path from } P \text{ to } Q.$$

Since $\vec{F}$ is path-independent, it does not matter which path we choose from P to Q. On the other hand, if $\vec{F}$ is not path-independent, then different choices might give different values for $f(Q)$, so f would not be a function (a function has to have a single value at each point).

We still have to show that the gradient of the function f really is $\vec{F}$; we do this on page 894. However, by constructing a function f in this manner, we have the following result:

Theorem 18.2: Path-independent Fields are Gradient Fields

If $\vec{F}$ is a continuous path-independent vector field on an open region R, then $\vec{F} = \operatorname{grad} f$ for some f defined on R.

Combining Theorems 18.1 and 18.2, we have

A continuous vector field $\vec{F}$ defined on an open region is path-independent if and only if $\vec{F}$ is a gradient vector field.

The function f is sufficiently important that it is given a special name:

If a vector field $\vec{F}$ is of the form $\vec{F} = \operatorname{grad} f$ for some scalar function f, then f is called a **potential function** for the vector field $\vec{F}$.

Warning

Physicists use the convention that a function ϕ is a potential function for a vector field $\vec{F}$ if $\vec{F} = -\operatorname{grad} \phi$. See Problem 34 on page 898.

Example 3   Show that the vector field $\vec{F}(x, y) = y \cos x \vec{i} + \sin x \vec{j}$ is path-independent.

Solution   If we can find a potential function f, then $\vec{F}$ must be path-independent. We want $\operatorname{grad} f = \vec{F}$. Since

$$\frac{\partial f}{\partial x} = y \cos x,$$

f must be of the form

$$f(x, y) = y \sin x + g(y) \quad \text{where } g(y) \text{ is a function of } y \text{ only.}$$

In addition, since $\operatorname{grad} f = \vec{F}$, we must have

$$\frac{\partial f}{\partial y} = \sin x,$$

and differentiating $f(x, y) = y \sin x + g(y)$ gives

$$\frac{\partial f}{\partial y} = \sin x + g'(y).$$

Thus, we must have $g'(y) = 0$, so $g(y) = C$ where C is some constant. Thus,

$$f(x, y) = y \sin x + C$$

is a potential function for $\vec{F}$. Therefore $\vec{F}$ is path-independent.

Example 4   The gravitational force field, $\vec{F}$, of an object of mass M, is given by

$$\vec{F} = -\frac{GM}{r^3}\vec{r}.$$

Show that $\vec{F}$ is a gradient field by finding a potential function for $\vec{F}$.

Solution   All the force vectors point in toward the origin. If $\vec{F} = \operatorname{grad} f$, the force vectors must be perpendicular to the level surfaces of f, so the level surfaces of f must be spheres. Also, if $\operatorname{grad} f = \vec{F}$, then $\| \operatorname{grad} f \| = \| \vec{F} \| = GM/r^2$ is the rate of change of f in the direction toward the origin. Now, differentiating with respect to r gives the rate of change in a radially outward direction. Thus, if $w = f(x, y, z)$ we have

$$\frac{dw}{dr} = -\frac{GM}{r^2} = GM\left(-\frac{1}{r^2}\right) = GM\frac{d}{dr}\left(\frac{1}{r}\right).$$

So let's try

$$w = \frac{GM}{r} \quad \text{or} \quad f(x, y, z) = \frac{GM}{\sqrt{x^2 + y^2 + z^2}}.$$

We calculate

$$f_x = \frac{\partial}{\partial x}\frac{GM}{\sqrt{x^2 + y^2 + z^2}} = \frac{-GMx}{(x^2 + y^2 + z^2)^{3/2}},$$

$$f_y = \frac{\partial}{\partial y}\frac{GM}{\sqrt{x^2 + y^2 + z^2}} = \frac{-GMy}{(x^2 + y^2 + z^2)^{3/2}},$$

$$f_z = \frac{\partial}{\partial z}\frac{GM}{\sqrt{x^2 + y^2 + z^2}} = \frac{-GMz}{(x^2 + y^2 + z^2)^{3/2}}.$$

So

$$\operatorname{grad} f = f_x\vec{i} + f_y\vec{j} + f_z\vec{k} = \frac{-GM}{(x^2 + y^2 + z^2)^{3/2}}(x\vec{i} + y\vec{j} + z\vec{k}) = \frac{-GM}{r^3}\vec{r} = \vec{F}.$$

Our computations show that $\vec{F}$ is a gradient field and that $f = GM/r$ is a potential function for $\vec{F}$.

Why Path-Independent Vector Fields are Gradient Fields: Showing grad $f = \vec{F}$

Suppose $\vec{F}$ is a path-independent vector field. On page 893 we defined the function f, which we hope will satisfy $\operatorname{grad} f = \vec{F}$, as follows:

$$f(x_0, y_0) = \int_C \vec{F} \cdot d\vec{r},$$

where C is a path from a fixed starting point P to a point $Q = (x_0, y_0)$. This integral has the same value for any path from P to Q because $\vec{F}$ is path-independent. Now we show why $\operatorname{grad} f = \vec{F}$. We consider vector fields in 2-space; the argument in 3-space is essentially the same.

First, we write the line integral in terms of the components $\vec{F}(x, y) = F_1(x, y)\vec{i} + F_2(x, y)\vec{j}$ and the components $d\vec{r} = dx\vec{i} + dy\vec{j}$:

$$f(x_0, y_0) = \int_C F_1(x, y)dx + F_2(x, y)dy.$$

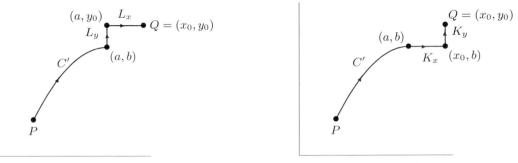

Figure 18.29: The path $C' + L_y + L_x$ is used to show $f_x = F_1$

Figure 18.30: The path $C' + K_x + K_y$ is used to show $f_y = F_2$

We want to compute the partial derivatives of f, that is, the rate of change of f at (x_0, y_0) parallel to the axes. To do this easily, we choose a path which reaches the point (x_0, y_0) on a horizontal or vertical line segment. Let C' be a path from P which stops short of Q at a fixed point (a, b) and let L_x and L_y be the paths shown in Figure 18.29. Then we can split the line integral into three pieces. Since $d\vec{r} = \vec{j}\, dy$ on L_y and $d\vec{r} = \vec{i}\, dx$ on L_x, we have:

$$f(x_0, y_0) = \int_{C'} \vec{F} \cdot d\vec{r} + \int_{L_y} \vec{F} \cdot d\vec{r} + \int_{L_x} \vec{F} \cdot d\vec{r} = \int_{C'} \vec{F} \cdot d\vec{r} + \int_b^{y_0} F_2(a, y)dy + \int_a^{x_0} F_1(x, y_0)dx.$$

The first two integrals do not involve x_0. Thinking of x_0 as a variable and differentiating with respect to it gives

$$f_{x_0}(x_0, y_0) = \frac{\partial}{\partial x_0} \int_{C'} \vec{F} \cdot d\vec{r} + \frac{\partial}{\partial x_0} \int_b^{y_0} F_2(a, y)dy + \frac{\partial}{\partial x_0} \int_a^{x_0} F_1(x, y_0)dx$$
$$= 0 + 0 + F_1(x_0, y_0) = F_1(x_0, y_0),$$

and thus

$$f_x(x, y) = F_1(x, y).$$

A similar calculation for y using the path from P to Q shown in Figure 18.30 gives

$$f_{y_0}(x_0, y_0) = F_2(x_0, y_0).$$

Therefore, as we claimed

$$\text{grad } f = f_x\vec{i} + f_y\vec{j} = F_1\vec{i} + F_2\vec{j} = \vec{F}.$$

Summary

We have studied two apparently different types of vector field: path-independent vector fields and gradient vector fields. It turns out that these are the same. Here is a summary of the definitions and properties of these vector fields:

- **Path-independent vector fields** have the property that for any two points P and Q, the line integral along a path from P to Q is the same no matter what path we choose.

- **Gradient vector fields** are of the form grad f for some scalar function f, called the potential function of the vector field.

- **Gradient vector fields are path-independent** by the Fundamental Theorem of Line Integrals,

$$\int_C \text{grad } f \cdot d\vec{r} = f(Q) - f(P).$$

- **Path-independent vector fields are gradient fields** because we can use a line integral and path independence to construct a potential function.

Exercises and Problems for Section 18.3

Exercises

1. The vector field $\vec{F}(x, y) = x\vec{i} + y\vec{j}$ is path-independent. Compute geometrically the line integrals over the three paths A, B, and C shown in Figure 18.31 from $(1, 0)$ to $(0, 1)$ and check that they are equal. Here A is a portion of a circle, B is a line, and C consists of two line segments meeting at a right angle.

Figure 18.31

2. The vector field $\vec{F}(x, y) = x\vec{i} + y\vec{j}$ is path-independent. Compute algebraically the line integrals over the three paths A, B, and C shown in Figure 18.32 from $(0, 0)$ to $(1, 1)$ and check that they are equal. Here A is a line segment, B is part of the graph of $f(x) = x^2$, and C consists of two line segments meeting at a right angle.

Figure 18.32

In Exercises 3–6, decide whether the vector field could be a gradient vector field. Justify your answer.

3. $\vec{F}(x, y) = x\vec{i}$

4. $\vec{G}(x, y) = (x^2 - y^2)\vec{i} - 2xy\vec{j}$

5. $\vec{F}(x, y, z) = \dfrac{-z}{\sqrt{x^2 + z^2}}\vec{i} + \dfrac{y}{\sqrt{x^2 + z^2}}\vec{j} + \dfrac{x}{\sqrt{x^2 + z^2}}\vec{k}$

6. $\vec{F}(\vec{r}) = \vec{r}/||\vec{r}||^3$,   where $\vec{r} = x\vec{i} + y\vec{j} + z\vec{k}$

Do the vector fields in Exercises 7–10 appear to be path-independent (conservative)?

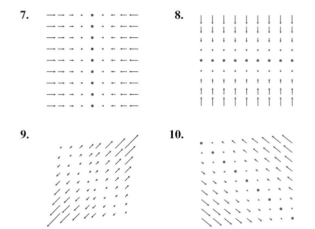

7.

8.

9.

10.

In Exercises 11–16, use the Fundamental Theorem of Line Integrals to calculate $\int_C \vec{F} \cdot d\vec{r}$ exactly.

11. $\vec{F} = (x+2)\vec{i} + (2y+3)\vec{j}$ and C is the line from $(1, 0)$ to $(3, 1)$.

12. $\vec{F} = 2x\vec{i} - 4y\vec{j} + (2z - 3)\vec{k}$ and C is the line from $(1, 1, 1)$ to $(2, 3, -1)$.

13. $\vec{F} = y\sin(xy)\vec{i} + x\sin(xy)\vec{j}$ and C is the parabola $y = 2x^2$ from $(1, 2)$ to $(3, 18)$.

14. $\vec{F} = x^{2/3}\vec{i} + e^{7y}\vec{j}$, and C is the unit circle oriented clockwise.

15. $\vec{F} = x^{2/3}\vec{i} + e^{7y}\vec{j}$, and C is the quarter of the unit circle in the first quadrant, traced counterclockwise from $(1, 0)$ to $(0, 1)$.

16. $\vec{F} = 2xy^2ze^{x^2y^2z}\vec{i} + 2x^2yze^{x^2y^2z}\vec{j} + x^2y^2e^{x^2y^2z}\vec{k}$ and C is the circle of radius 1 in the plane $z = 1$, centered on the z-axis, starting at $(1, 0, 1)$ and oriented counterclockwise viewed from above.

Problems

17. Let $\operatorname{grad} f = 2xe^{x^2}\sin y\vec{i} + e^{x^2}\cos y\vec{j}$. Find the change in f between $(0, 0)$ and $(1, \pi/2)$: **(a)** By computing a line integral   **(b)** By computing f.

18. The line integral of $\vec{F} = (x + y)\vec{i} + x\vec{j}$ along each of the following paths is $3/2$:

(i) The path (t, t^2), with $0 \le t \le 1$
(ii) The path (t^2, t), with $0 \le t \le 1$

(iii) The path (t, t^n), with $n > 0$ and $0 \le t \le 1$

Show this

(a) Using the given parameterization to compute the line integral.
(b) Using the Fundamental Theorem of Calculus for Line Integrals.

Problems 19–20, refer to the oriented curves from $(0,0)$ to (a,b) in Figure 18.33.

Figure 18.33

19. Evaluate $f(a,b) = \int_C \vec{F} \cdot d\vec{r}$ where
$\vec{F}(x,y) = (x+y)\vec{i} + (2x+y)\vec{j}$ and

(a) $C = C_1$ (b) $C = C_2$ (c) $C = C_3$

20. Evaluate $f(a,b) = \int_C \vec{F} \cdot d\vec{r}$ where
$\vec{F}(x,y) = (x+y)\vec{i} + (x+2y)\vec{j}$ and

(a) $C = C_1$ (b) $C = C_2$ (c) $C = C_3$

For the vector fields in Problems 21–24, find the line integral along the curve C from the origin along the x-axis to the point $(3,0)$ and then counterclockwise around the circumference of the circle $x^2 + y^2 = 9$ to the point $(3/\sqrt{2}, 3/\sqrt{2})$.

21. $\vec{F} = x\vec{i} + y\vec{j}$

22. $\vec{H} = -y\vec{i} + x\vec{j}$

23. $\vec{F} = y(x+1)^{-1}\vec{i} + \ln(x+1)\vec{j}$

24. $\vec{G} = (ye^{xy} + \cos(x+y))\vec{i} + (xe^{xy} + \cos(x+y))\vec{j}$

25. Let $f(x,y,z) = x^2 + 2y^3 + 3z^4$ and $\vec{F} = \text{grad } f$. Find $\int_C \vec{F} \cdot d\vec{r}$ where C consists of four line segments from $(4,0,0)$ to $(4,3,0)$ to $(0,3,0)$ to $(0,3,5)$ to $(0,0,5)$.

26. Let C be the helix $x = \cos t$, $y = \sin t$, $z = t$ for $0 \le t \le 1.25\pi$. Find $\int_C \vec{F} \cdot d\vec{r}$ exactly for

$$\vec{F} = yz^2 e^{xyz^2}\vec{i} + xz^2 e^{xyz^2}\vec{j} + 2xyze^{xyz^2}\vec{k}.$$

27. Calculate the line integral $\int_C \vec{F} \cdot d\vec{r}$ exactly, where C is the curve from P to Q in Figure 18.34 and

$$\vec{F} = \sin\left(\frac{x}{2}\right)\sin\left(\frac{y}{2}\right)\vec{i} - \cos\left(\frac{x}{2}\right)\cos\left(\frac{y}{2}\right)\vec{j}.$$

The curves PR, RS and SQ are trigonometric functions of period 2π and amplitude 1.

Figure 18.34

28. Figure 18.35 shows level curves of the function $f(x,y)$.

(a) Sketch ∇f at P.
(b) Is the vector ∇f at P longer than, shorter than, or the same length as, ∇f at Q?
(c) If C is a curve from P to Q, evaluate $\int_C \nabla f \cdot d\vec{r}$.

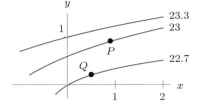

Figure 18.35

29. Figure 18.36 shows the vector field $\vec{F}(x,y) = x\vec{j}$.

(a) Find paths C_1, C_2, and C_3 from P to Q such that

$$\int_{C_1} \vec{F} \cdot d\vec{r} = 0, \quad \int_{C_2} \vec{F} \cdot d\vec{r} > 0, \quad \int_{C_3} \vec{F} \cdot d\vec{r} < 0.$$

(b) Is $\vec{F}$ a gradient field? Explain.

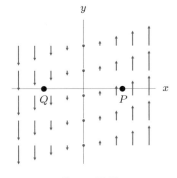

Figure 18.36

30. Consider the vector field $\vec{F}$ shown in Figure 18.37.

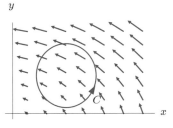

Figure 18.37

(a) Is $\int_C \vec{F} \cdot d\vec{r}$ positive, negative, or zero?
(b) From your answer to part (a), can you determine whether or not $\vec{F} = \text{grad } f$ for some function f?
(c) Which of the following formulas best fits $\vec{F}$?

$$\vec{F_1} = \frac{x}{x^2 + y^2}\vec{i} + \frac{y}{x^2 + y^2}\vec{j},$$
$$\vec{F_2} = -y\vec{i} + x\vec{j},$$
$$\vec{F_3} = \frac{-y}{(x^2 + y^2)^2}\vec{i} + \frac{x}{(x^2 + y^2)^2}\vec{j}.$$

31. The force exerted by gravity on a refrigerator of mass m is $\vec{F} = -mg\vec{k}$.

  (a) Find the work done against this force in moving from the point $(1, 0, 0)$ to the point $(1, 0, 2\pi)$ along the curve $x = \cos t, y = \sin t, z = t$ by calculating a line integral.

  (b) Is $\vec{F}$ conservative (that is, path independent)? Give a reason for your answer.

32. A particle subject to a force $\vec{F}(x, y) = y\vec{i} - x\vec{j}$ moves clockwise along the arc of the unit circle, centered at the origin, that begins at $(-1, 0)$ and ends at $(0, 1)$.

  (a) Find the work done by $\vec{F}$. Explain the sign of your answer.

  (b) Is $\vec{F}$ path-independent? Explain.

33. Let $\vec{r} = x\vec{i} + y\vec{j} + z\vec{k}$ and $\vec{a} = a_1\vec{i} + a_2\vec{j} + a_3\vec{k}$, a constant vector.

  (a) Find $\mathrm{grad}(\vec{r} \cdot \vec{a})$

  (b) Let C be a path from the origin to the point with position vector $\vec{r_0}$. Find $\int_C \mathrm{grad}(\vec{r} \cdot \vec{a}) \cdot d\vec{r}$.

  (c) If $\|\vec{r_0}\| = 10$, what is the maximum possible value of $\int_C \mathrm{grad}(\vec{r} \cdot \vec{a}) \cdot d\vec{r}$? Explain.

34. Let $\vec{F}$ be a path-independent vector field. In physics, the potential function ϕ is usually required to satisfy the equation $\vec{F} = -\nabla\phi$. This problem illustrates the significance of the negative sign.[1]

  (a) Let the xy-plane represent part of the earth's surface with the z-axis pointing upward. (The scale is small enough that a flat plane is a good approximation to the earth's surface.) Let $\vec{r} = x\vec{i} + y\vec{j} + z\vec{k}$, with $z \geq 0$, and x, y, z in meters, be the position vector of a rock of unit mass. The gravitational potential energy function for the rock is $\phi(x, y, z) = gz$, where $g \approx 9.8 \text{ m/sec}^2$. Describe in words the level surfaces of ϕ. Does the potential energy increase or decrease with height above the earth?

  (b) What is the relation between the gravitational vector, $\vec{F}$, and the vector $\nabla\phi$? Explain the significance of the negative sign in the equation $\vec{F} = -\nabla\phi$.

35. In this problem, we see how the Fundamental Theorem for Line Integrals can be derived from the Fundamental Theorem for ordinary definite integrals. Suppose that $(x(t), y(t))$, for $a \leq t \leq b$, is a parameterization of C, with endpoints $P = (x(a), y(a))$ and $Q = (x(b), y(b))$. The values of f along C are given by the single variable function $h(t) = f(x(t), y(t))$.

  (a) Use the chain rule to show that
  $$h'(t) = f_x(x(t), y(t))x'(t) + f_y(x(t), y(t))y'(t).$$

  (b) Use the Fundamental Theorem of Calculus applied to $h(t)$ to show

  $$\int_C \mathrm{grad}\, f \cdot d\vec{r} = f(Q) - f(P).$$

18.4   PATH-DEPENDENT VECTOR FIELDS AND GREEN'S THEOREM

Suppose we are given a vector field but are not told whether it is path-independent. How can we tell if it has a potential function, that is, if it is a gradient field?

How to Tell if a Vector Field is Path-Dependent Using Line Integrals

Example 1   Is the vector field, $\vec{F}$, shown in Figure 18.38 path-independent? At any point $\vec{F}$ has magnitude equal to the distance from the origin and direction perpendicular to the line joining the point to the origin.

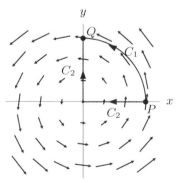

Figure 18.38: Is this vector field path-independent?

[1] Adapted from V.I. Arnold, *Mathematical Methods of Classical Mechanics*, 2nd Edition, Graduate Texts in Mathematics, Springer. Potential energy is also discussed in Project 1 on page 913.

Solution   We choose $P = (1, 0)$ and $Q = (0, 1)$ and two paths between them: C_1, a quarter circle of radius 1, and C_2, formed by parts of the x- and y-axes. (See Figure 18.38.) Along C_1, the line integral $\int_{C_1} \vec{F} \cdot d\vec{r} > 0$, since $\vec{F}$ points in the direction of the curve. Along C_2, however, we have $\int_{C_2} \vec{F} \cdot d\vec{r} = 0$, since $\vec{F}$ is perpendicular to C_2 everywhere. Thus, $\vec{F}$ is not path-independent.

Path-Dependent Fields and Circulation

Notice that the vector field in the previous example has nonzero circulation around the origin. What can we say about the circulation of a general path-independent vector field around a closed curve, C? Suppose C is a *simple* closed curve, that is, a curve which does not cross itself. If P and Q are any two points on the path, then we can think of C (oriented as shown in Figure 18.39) as made up of the path C_1 followed by $-C_2$. Since $\vec{F}$ is path-independent, we know that

$$\int_{C_1} \vec{F} \cdot d\vec{r} = \int_{C_2} \vec{F} \cdot d\vec{r}.$$

Thus, we see that the circulation around C is zero:

$$\int_C \vec{F} \cdot d\vec{r} = \int_{C_1} \vec{F} \cdot d\vec{r} + \int_{-C_2} \vec{F} \cdot d\vec{r} = \int_{C_1} \vec{F} \cdot d\vec{r} - \int_{C_2} \vec{F} \cdot d\vec{r} = 0.$$

If the curve C does cross itself, we break it into simple closed curves as shown in Figure 18.40 and apply the same argument to each one.

Now suppose we know that the line integral around any closed curve is zero. For any two points, P and Q, with two paths, C_1 and C_2, between them, create a closed curve, C, as in Figure 18.39. Since the circulation around this closed curve, C, is zero, the line integrals along the two paths, C_1 and C_2, are equal. Thus, $\vec{F}$ is path-independent.

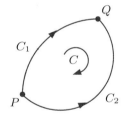

Figure 18.39: A simple closed curve C broken into two pieces, C_1 and C_2

Figure 18.40: A curve C which crosses itself can be broken into simple closed curves

Thus, we have the following result:

A vector field is path-independent if and only if $\int_C \vec{F} \cdot d\vec{r} = 0$ for every closed curve C.

Hence, to see if a field is *path-dependent*, we look for a closed path with nonzero circulation. For instance, the vector field in Example 1 has nonzero circulation around a circle around the origin, showing it is path-dependent.

How to Tell If a Vector Field is Path-Dependent Algebraically: The Curl

Example 2   Does the vector field $\vec{F} = 2xy\vec{i} + xy\vec{j}$ have a potential function? If so, find it.

Solution   Let's suppose $\vec{F}$ does have a potential function, f, so $\vec{F} = \operatorname{grad} f$. This means that

$$\frac{\partial f}{\partial x} = 2xy \quad \text{and} \quad \frac{\partial f}{\partial y} = xy.$$

Integrating the expression for $\partial f/\partial x$ shows that we must have

$$f(x, y) = x^2 y + C(y) \qquad \text{where } C(y) \text{ is a function of } y.$$

Differentiating this expression for $f(x, y)$ with respect to y and using the fact that $\partial f/\partial y = xy$, we get

$$\frac{\partial f}{\partial y} = x^2 + C'(y) = xy.$$

Thus, we must have

$$C'(y) = xy - x^2.$$

But this expression for $C'(y)$ is impossible because $C'(y)$ is a function of y alone. This argument shows that there is no potential function for the vector field $\vec{F}$.

Is there an easier way to see that a vector field has no potential function, other than by trying to find the potential function and failing? The answer is yes. First we look at a 2-dimensional vector field $\vec{F} = F_1 \vec{i} + F_2 \vec{j}$. If $\vec{F}$ is a gradient field, then there is a potential function f such that

$$\vec{F} = F_1 \vec{i} + F_2 \vec{j} = \frac{\partial f}{\partial x} \vec{i} + \frac{\partial f}{\partial y} \vec{j}.$$

Thus,

$$F_1 = \frac{\partial f}{\partial x} \quad \text{and} \quad F_2 = \frac{\partial f}{\partial y}.$$

Let us assume that f has continuous second partial derivatives. Then, by the equality of mixed partial derivatives

$$\frac{\partial F_1}{\partial y} = \frac{\partial^2 f}{\partial y \partial x} = \frac{\partial^2 f}{\partial x \partial y} = \frac{\partial F_2}{\partial x}$$

Thus we have the following result:

If $\vec{F}(x, y) = F_1 \vec{i} + F_2 \vec{j}$ is a gradient vector field with continuous partial derivatives, then

$$\frac{\partial F_2}{\partial x} - \frac{\partial F_1}{\partial y} = 0.$$

We call $\dfrac{\partial F_2}{\partial x} - \dfrac{\partial F_1}{\partial y}$ the 2-dimensional or scalar **curl** of the vector field $\vec{F}$.

Notice that we now know that if $\vec{F}$ is a gradient field, then its curl is 0. We do not (yet) know whether the converse is true. (That is: If the curl is 0, does $\vec{F}$ have to be a gradient field?) However, the curl already enables us to show that a vector field is *not* a gradient field.

Example 3   Show that $\vec{F} = 2xy\vec{i} + xy\vec{j}$ cannot be a gradient vector field.

Solution   We have $F_1 = 2xy$ and $F_2 = xy$. Since $\partial F_1/\partial y = 2x$ and $\partial F_2/\partial x = y$, in this case

$$\partial F_2/\partial x - \partial F_1/\partial y \neq 0$$

so $\vec{F}$ cannot be a gradient field.

We now have two ways of seeing that a vector field $\vec{F}$ in the plane is path-dependent. We can evaluate $\int_C \vec{F} \cdot d\vec{r}$ for some closed curve and find it is not zero, or we can show that $\partial F_2/\partial x - \partial F_1/\partial y \neq 0$. It's natural to think that

$$\int_C \vec{F} \cdot d\vec{r} \quad \text{and} \quad \frac{\partial F_2}{\partial x} - \frac{\partial F_1}{\partial y}$$

might be related. The relation is called Green's Theorem.

Green's Theorem

Figure 18.41: Region R bounded by a closed curve C and split into many small regions, ΔR

Figure 18.42: Two adjacent small closed curves

Figure 18.43: A small closed curve ΔC broken into C_1, C_2, C_3, C_4

We obtain the statement of Green's Theorem by splitting a region up into small pieces and looking at the relation between the line integral and the curl on each one.

In 2-space, a simple closed curve C separates the plane into points inside and points outside C. We consider $\int_C \vec{F} \cdot d\vec{r}$, where C is oriented as shown in Figure 18.41. We divide the region R inside C into small pieces, each bounded by a closed curve with the orientation shown. Figure 18.42 shows that if we add the circulation around all of these small closed curves, each common edge of a pair of adjacent curves is counted twice, once in each direction. Hence the integrals along these edges cancel. Thus, the line integrals along all the edges inside the region R cancel, giving

$$\begin{matrix} \text{Circulation of } \vec{F} \\ \text{around } C \end{matrix} = \sum_{\Delta C} \begin{matrix} \text{Circulation of } \vec{F} \\ \text{around small curve, } \Delta C \end{matrix}$$

Now we estimate the line integral around one of these small closed curves, ΔC. We break ΔC into C_1, C_2, C_3, and C_4, as shown in Figure 18.43. Then we calculate the line integrals along C_1 and C_3, where $\Delta \vec{r}$ is parallel to the x-axis, so

$$\Delta \vec{r} = \Delta x \vec{i}.$$

Thus, on C_1 and C_3,

$$\vec{F} \cdot \Delta \vec{r} = (F_1 \vec{i} + F_2 \vec{j}) \cdot \Delta x \vec{i} = F_1 \Delta x.$$

However, the function F_1 is evaluated at (x, y_0) on C_1 and at $(x, y_0 + \Delta y)$ on C_3, so

$$\int_{C_1} \vec{F} \cdot d\vec{r} + \int_{C_3} \vec{F} \cdot d\vec{r} = = \int_{C_1} F_1(x, y_0)\, dx + \int_{C_3} F_1(x, y_0 + \Delta y)\, dx$$

$$= \int_{x=x_0}^{x_0+\Delta x} F_1(x, y_0)\, dx + \int_{x_0+\Delta x}^{x=x_0} F_1(x, y_0 + \Delta y)\, dx$$

$$= \int_{x=x_0}^{x_0+\Delta x} F_1(x, y_0)\, dx - \int_{x=x_0}^{x_0+\Delta x} F_1(x, y_0 + \Delta y)\, dx$$

$$= \int_{x=x_0}^{x_0+\Delta x} \left(F_1(x, y_0) - F_1(x, y_0 + \Delta y) \right) dx.$$

Since F_1 is differentiable and Δy is small,

$$F_1(x, y_0) - F_1(x, y_0 + \Delta y) = -(F_1(x, y_0 + \Delta y) - F_1(x, y_0)) \approx -\frac{\partial F_1}{\partial y}(x, y_0)\, \Delta y,$$

so we have

$$\int_{x=x_0}^{x_0+\Delta x} \left(F_1(x, y_0) - F_1(x, y_0 + \Delta y) \right) dx \approx - \left(\int_{x=x_0}^{x_0+\Delta x} \frac{\partial F_1}{\partial y}\, dx \right) \Delta y.$$

Assuming $\dfrac{\partial F_1}{\partial y}(x, y_0)$ is approximately constant over the interval $x_0 \leq x \leq x_0 + \Delta x$, we get

$$-\left(\int_{x=x_0}^{x_0+\Delta x} \frac{\partial F_1}{\partial y}\, dx \right) \Delta y \approx -\frac{\partial F_1}{\partial y}(x_0, y_0) \left(\int_{x=x_0}^{x_0+\Delta x} dx \right) \Delta y = -\frac{\partial F_1}{\partial y}(x_0, y_0)\, \Delta x\, \Delta y.$$

By a similar argument on C_2 and C_4, we have

$$\int_{C_2} \vec{F} \cdot d\vec{r} + \int_{C_4} \vec{F} \cdot d\vec{r} \approx \frac{\partial F_2}{\partial x}\, \Delta x\, \Delta y.$$

Combining these results for $C_1, C_2, C_3,$ and C_4, we get

$$\int_{\Delta C} \vec{F} \cdot d\vec{r} = \int_{C_1} \vec{F} \cdot d\vec{r} + \int_{C_2} \vec{F} \cdot d\vec{r} + \int_{C_3} \vec{F} \cdot d\vec{r} + \int_{C_4} \vec{F} \cdot d\vec{r} \approx \frac{\partial F_2}{\partial x}\, \Delta x\, \Delta y - \frac{\partial F_1}{\partial y}\, \Delta x\, \Delta y.$$

Summing over all small regions ΔC gives

$$\int_C \vec{F} \cdot d\vec{r} \approx \sum_{\Delta C} \int_{\Delta C} \vec{F} \cdot d\vec{r} \approx \sum_{\Delta R} \left(\frac{\partial F_2}{\partial x} - \frac{\partial F_1}{\partial y} \right) \Delta x\, \Delta y.$$

The last sum is a Riemann sum approximating a double integral; taking the limit as $\Delta x,\ \Delta y$ tend to zero, we get

---

Theorem 18.3: Green's Theorem

Suppose C is a piecewise smooth simple closed curve that is the boundary of an open region R in the plane and oriented so that the region is on the left as we move around the curve. Suppose $\vec{F} = F_1 \vec{i} + F_2 \vec{j}$ is a smooth vector field on an open region containing R and C. Then

$$\int_C \vec{F} \cdot d\vec{r} = \int_R \left(\frac{\partial F_2}{\partial x} - \frac{\partial F_1}{\partial y} \right) dx\, dy.$$

The online supplement at www.wiley.com/college/hugheshallett contains a proof of Green's Theorem with different, but equivalent, conditions on the region R.

The Curl Test for Vector Fields in the Plane

We already know that if $\vec{F} = F_1 \vec{i} + F_2 \vec{j}$ is a gradient field with continuous partial derivatives, then

$$\frac{\partial F_2}{\partial x} - \frac{\partial F_1}{\partial y} = 0.$$

Now we show that the converse is true if the domain of $\vec{F}$ has no holes in it. This means that we assume that

$$\frac{\partial F_2}{\partial x} - \frac{\partial F_1}{\partial y} = 0$$

and show that $\vec{F}$ is path-independent. If C is any oriented closed curve in the domain of $\vec{F}$ and R is the region inside C, then

$$\int_R \left(\frac{\partial F_2}{\partial x} - \frac{\partial F_1}{\partial y} \right) dx\, dy = 0$$

since the integrand is identically 0. Therefore, by Green's Theorem

$$\int_C \vec{F} \cdot d\vec{r} = \int_R \left(\frac{\partial F_2}{\partial x} - \frac{\partial F_1}{\partial y} \right) dx dy = 0.$$

Thus, $\vec{F}$ is path-independent and therefore a gradient field. This argument is valid for every closed curve, C, provided the region R is entirely in the domain of $\vec{F}$. Thus we have the following result:

The Curl Test for Vector Fields in 2-Space

Suppose $\vec{F} = F_1\vec{i} + F_2\vec{j}$ is a vector field with continuous partial derivatives, such that
- The domain of $\vec{F}$ has the property that every closed curve in it encircles a region that lies entirely within the domain. In particular, the domain of $\vec{F}$ has no holes.
- $\dfrac{\partial F_2}{\partial x} - \dfrac{\partial F_1}{\partial y} = 0.$

Then $\vec{F}$ is path-independent, so $\vec{F}$ is a gradient field and has a potential function.

Why Are Holes in the Domain of the Vector Field Important?

The reason for assuming that the domain of the vector field $\vec{F}$ has no holes is to ensure that the region R inside C is actually contained in the domain of $\vec{F}$. Otherwise, we cannot apply Green's Theorem. The next two examples show that if $\partial F_2/\partial x - \partial F_1/\partial y = 0$ but the domain of $\vec{F}$ contains a hole, then $\vec{F}$ can either be path-independent or path-dependent.

Example 4 Let $\vec{F}$ be the vector field given by $\vec{F}(x, y) = \dfrac{-y\vec{i} + x\vec{j}}{x^2 + y^2}$.

(a) Calculate $\dfrac{\partial F_2}{\partial x} - \dfrac{\partial F_1}{\partial y}$. Does the curl test imply that $\vec{F}$ is path-independent?

(b) Calculate $\displaystyle\int_C \vec{F} \cdot d\vec{r}$, where C is the unit circle centered at the origin and oriented counterclockwise. Is $\vec{F}$ a path-independent vector field?

(c) Explain why the answers to parts (a) and (b) do not contradict Green's Theorem.

Solution (a) Taking partial derivatives, we have

$$\frac{\partial F_2}{\partial x} = \frac{\partial}{\partial x}\left(\frac{x}{x^2 + y^2}\right) = \frac{1}{x^2 + y^2} - \frac{x \cdot 2x}{(x^2 + y^2)^2} = \frac{y^2 - x^2}{(x^2 + y^2)^2}.$$

Similarly,

$$\frac{\partial F_1}{\partial y} = \frac{\partial}{\partial y}\left(\frac{-y}{x^2 + y^2}\right) = \frac{-1}{x^2 + y^2} + \frac{y \cdot 2y}{(x^2 + y^2)^2} = \frac{y^2 - x^2}{(x^2 + y^2)^2}.$$

Thus,

$$\frac{\partial F_2}{\partial x} - \frac{\partial F_1}{\partial y} = 0.$$

Since $\vec{F}$ is undefined at the origin, the domain of $\vec{F}$ contains a hole. Therefore, the curl test does not apply.

(b) On the unit circle, $\vec{F}$ is tangent to the circle and $||\vec{F}|| = 1$. Thus,

$$\int_C \vec{F} \cdot d\vec{r} = ||\vec{F}|| \cdot \text{Length of curve} = 1 \cdot 2\pi = 2\pi.$$

Since the line integral around the closed curve C is nonzero, $\vec{F}$ is not path-independent.

(c) The domain of $\vec{F}$ is the "punctured plane," as shown in Figure 18.44. Since $\vec{F}$ is not defined at the origin, which is inside C, Green's Theorem does not apply. In this case

$$2\pi = \int_C \vec{F} \cdot d\vec{r} \neq \int_R \left(\frac{\partial F_2}{\partial x} - \frac{\partial F_1}{\partial y}\right) dx\, dy = 0.$$

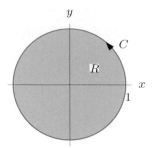

Figure 18.44: The domain of $\vec{F}(x, y) = \frac{-y\vec{i} + x\vec{j}}{x^2 + y^2}$ is the plane minus the origin

Figure 18.45: The region R is *not* contained in the domain of $\vec{F}(x, y) = \frac{-y\vec{i} + x\vec{j}}{x^2 + y^2}$

Although the vector field $\vec{F}$ in the last example was not defined at the origin, this by itself does not prevent the vector field from being path-independent as we see in the following example.

Example 5   Consider the vector field $\vec{F}$ given by $\vec{F}(x, y) = \dfrac{x\vec{i} + y\vec{j}}{x^2 + y^2}$.

(a) Calculate $\dfrac{\partial F_2}{\partial x} - \dfrac{\partial F_1}{\partial y}$. Does the curl test imply that $\vec{F}$ is path-independent?

(b) Explain how we know that $\displaystyle\int_C \vec{F} \cdot d\vec{r} = 0$, where C is the unit circle centered at the origin and oriented counterclockwise. Does this imply that $\vec{F}$ is path-independent?

(c) Check that $f(x, y) = \frac{1}{2}\ln(x^2 + y^2)$ is a potential function for $\vec{F}$. Does this imply that $\vec{F}$ is path-independent?

Solution   (a) Taking partial derivatives, we have

$$\frac{\partial F_2}{\partial x} = \frac{\partial}{\partial x}\left(\frac{y}{x^2 + y^2}\right) = \frac{-2xy}{(x^2 + y^2)^2}, \qquad \text{and} \qquad \frac{\partial F_1}{\partial y} = \frac{\partial}{\partial y}\left(\frac{x}{x^2 + y^2}\right) = \frac{-2xy}{(x^2 + y^2)^2}.$$

Therefore,

$$\frac{\partial F_2}{\partial x} - \frac{\partial F_1}{\partial y} = 0.$$

This does *not* imply that $\vec{F}$ is path-independent: The domain of $\vec{F}$ contains a hole since $\vec{F}$ is undefined at the origin. Thus, the curl test does not apply.

(b) Since $\vec{F}(x, y) = x\vec{i} + y\vec{j} = \vec{r}$ on the unit circle C, the field $\vec{F}$ is everywhere perpendicular to C. Thus

$$\int_C \vec{F} \cdot d\vec{r} = 0.$$

The fact that $\int_C \vec{F} \cdot d\vec{r} = 0$ when C is the unit circle does *not* imply that $\vec{F}$ is path-independent. To be sure that $\vec{F}$ is path-independent, we would have to show that $\int_C \vec{F} \cdot d\vec{r} = 0$ for *every* closed curve C in the domain of $\vec{F}$, not just the unit circle.

(c) To check that $\text{grad } f = \vec{F}$, we differentiate f:

$$f_x = \frac{1}{2}\frac{\partial}{\partial x}\ln(x^2 + y^2) = \frac{1}{2}\frac{2x}{x^2 + y^2} = \frac{x}{x^2 + y^2},$$

and

$$f_y = \frac{1}{2}\frac{\partial}{\partial y}\ln(x^2 + y^2) = \frac{1}{2}\frac{2y}{x^2 + y^2} = \frac{y}{x^2 + y^2},$$

so that

$$\text{grad } f = \frac{x\vec{i} + y\vec{j}}{x^2 + y^2} = \vec{F}.$$

Thus, $\vec{F}$ is a gradient field and therefore is path-independent — even though $\vec{F}$ is undefined at the origin.

The Curl Test for Vector Fields in 3-Space

The curl test is a convenient way of deciding whether a 2-dimensional vector field is path-independent. Fortunately, there is an analogous test for 3-dimensional vector fields, although we cannot justify it until Chapter 20.

If $\vec{F}(x, y, z) = F_1\vec{i} + F_2\vec{i} + F_3\vec{k}$ is a vector field on 3-space we define a new vector field, curl $\vec{F}$, on 3-space by

$$\text{curl } \vec{F} = \left(\frac{\partial F_3}{\partial y} - \frac{\partial F_2}{\partial z}\right)\vec{i} + \left(\frac{\partial F_1}{\partial z} - \frac{\partial F_3}{\partial x}\right)\vec{j} + \left(\frac{\partial F_2}{\partial x} - \frac{\partial F_1}{\partial y}\right)\vec{k}.$$

The vector field curl $\vec{F}$ can be used to determine whether the vector field $\vec{F}$ is path-independent.

The Curl Test for Vector Fields in 3-Space

Suppose $\vec{F}$ is a vector field on 3-space with continuous partial derivatives, such that
- The domain of $\vec{F}$ has the property that every closed curve in it can be contracted to a point in a smooth way, staying at all times within the domain.
- curl $\vec{F} = \vec{0}$.

Then $\vec{F}$ is path-independent, so $\vec{F}$ is a gradient field and has a potential function.

For the 2-dimensional curl test, the domain of $\vec{F}$ must have no holes. This meant that if $\vec{F}$ was defined on a closed curve C, then it was also defined at all points inside C. One way to test for holes is to try to "lasso" them with a closed curve. If every closed curve in the domain can be pulled to a point without hitting a hole, that is, without straying outside the domain, then the domain has no holes. In 3-space, we need the same condition to be satisfied; we must be able to pull every closed curve to a point, like a lasso, without straying outside the domain.

Example 6   Decide if the following vector fields are path-independent and whether or not the curl test applies.

(a) $\vec{F} = \dfrac{x\vec{i} + y\vec{j} + z\vec{k}}{(x^2 + y^2 + z^2)^{3/2}}$ (b) $\vec{G} = \dfrac{-y\vec{i} + x\vec{j}}{x^2 + y^2}$

Solution   (a) Suppose $f = -(x^2 + y^2 + z^2)^{-1/2}$. Then $f_x = x(x^2 + y^2 + z^2)^{-3/2}$ and similarly grad $f = \vec{F}$. Thus, $\vec{F}$ is a gradient field and therefore path-independent. Calculations show curl $\vec{F} = \vec{0}$. The domain of $\vec{F}$ is all of 3-space minus the origin, and any closed curve in the domain can be pulled to a point without leaving the domain. Thus, the curl test applies.

(b) Let C be the circle $x^2 + y^2 = 1, z = 0$ traversed counterclockwise when viewed from the positive z-axis. The vector field is everywhere tangent to this curve and of magnitude 1,[2] so

$$\int_C \vec{G} \cdot d\vec{r} = \|\vec{G}\| \cdot \text{Length of curve} = 1 \cdot 2\pi = 2\pi.$$

Since the line integral around this closed curve is nonzero, $\vec{G}$ is path-dependent. Computations show $\text{curl}\, \vec{G} = \vec{0}$. However, the domain of $\vec{G}$ is all of 3-space minus the z-axis, and it does not satisfy the curl test domain criterion. For example, the circle, C, is lassoed around the z-axis, and cannot be pulled to a point without hitting the z-axis. Thus the curl test does not apply.

Exercises and Problems for Section 18.4

Exercises

1. Find f if $\text{grad}\, f = 2xy\vec{i} + (x^2 + 8y^3)\vec{j}$

2. Find f if $\text{grad}\, f = (yze^{xyz} + z^2 \cos(xz^2))\vec{i} + xze^{xyz}\vec{j} + (xye^{xyz} + 2xz \cos(xz^2))\vec{k}$

7. $\dfrac{\vec{i}}{x} + \dfrac{\vec{j}}{y} + \dfrac{\vec{k}}{xy}$

8. $2x \cos(x^2 + z^2)\vec{i} + \sin(x^2 + z^2)\vec{j} + 2z \cos(x^2 + z^2)\vec{k}$

In Exercises 3–8, decide if the given vector field is the gradient of a function f. If so, find f. If not, explain why not.

3. $y\vec{i} + y\vec{j}$

4. $(x^2 + y^2)\vec{i} + 2xy\vec{j}$

5. $(2xy^3 + y)\vec{i} + (3x^2y^2 + x)\vec{j}$

6. $\dfrac{\vec{i}}{x} + \dfrac{\vec{j}}{y} + \dfrac{\vec{k}}{z}$

In Exercises 9–12, use Green's Theorem to calculate the circulation of $\vec{F}$ around the curve, oriented counterclockwise.

9. $\vec{F} = y\vec{i} - x\vec{j}$ around the unit circle.

10. $\vec{F} = xy\vec{j}$ around the square $0 \le x \le 1, 0 \le y \le 1$.

11. $\vec{F} = (2x^2 + 3y)\vec{i} + (2x + 3y^2)\vec{j}$ around the triangle with vertices $(2, 0)$, $(0, 3)$, $(-2, 0)$.

12. $\vec{F} = 3y\vec{i} + xy\vec{j}$ around the unit circle.

Problems

13. Example 1 on page 898 showed that the vector field in Figure 18.46 could not be a gradient field by showing that it is not path-independent. Here is another way to see the same thing. Suppose that the vector field were the gradient of a function f. Draw and label a diagram showing what the contours of f would have to look like, and explain why it would not be possible for f to have a single value at any given point.

14. Repeat Problem 13 for the vector field in Problem 29 on page 897.

15. (a) Sketch $\vec{F} = y\vec{i}$ and hence decide the sign of the circulation of $\vec{F}$ around the unit circle centered at the origin and traversed counterclockwise.

(b) Use Green's Theorem to compute the circulation in part (a) exactly.

16. Let $\vec{F} = (\sin x)\vec{i} + (x+y)\vec{j}$. Find the line integral of $\vec{F}$ around the perimeter of the rectangle with corners $(3, 0)$, $(3, 5)$, $(-1, 5)$, $(-1, 0)$, traversed in that order.

17. Find the line integral of $\vec{F} = (x - y)\vec{i} + x\vec{j}$ around the closed curve in Figure 18.47. (The arc is part of a circle.)

Figure 18.46

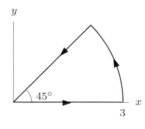

Figure 18.47

[2] See Problem 26 on page 883.

18. Find the line integral of $\vec{F} = (x+y)\vec{i} + \sin y\vec{j}$ around the closed curve in Figure 18.47. (The arc is part of a circle.)

19. Let $\vec{F} = 2xe^y\vec{i} + x^2e^y\vec{j}$ and $\vec{G} = (x-y)\vec{i} + (x+y)\vec{j}$. Let C be the path consisting of lines from $(0,0)$ to $(3,0)$ to $(3,8)$ to $(0,0)$. Find exactly:

(a) $\int_C \vec{F} \cdot d\vec{r}$ (b) $\int_C \vec{G} \cdot d\vec{r}$

20. (a) Sketch the curves C_1 and C_2:

C_1 is $(x(t), y(t)) = (0, t)$ for $-1 \leq t \leq 1$

C_2 is $(x(t), y(t)) = (\cos t, \sin t)$ for $\dfrac{\pi}{2} \leq t \leq \dfrac{3\pi}{2}$.

(b) Find $\int_C \vec{F} \cdot d\vec{r}$, where $\vec{F} = (x+3y)\vec{i} + y\vec{j}$ and $C = C_1 + C_2$.

21. Show that the line integral of $\vec{F} = x\vec{j}$ around a closed curve in the xy-plane, oriented as in Green's Theorem, measures the area of the region enclosed by the curve.

Use the result of Problem 21 to calculate the area of the region within the parameterized curves in Problems 22–24. In each case, sketch the curve.

22. The ellipse $x^2/a^2 + y^2/b^2 = 1$ parameterized by $x = a\cos t$, $y = b\sin t$, for $0 \leq t \leq 2\pi$.

23. The hypocycloid $x^{2/3} + y^{2/3} = a^{2/3}$ parameterized by $x = a\cos^3 t$, $y = a\sin^3 t$, $0 \leq t \leq 2\pi$.

24. The folium of Descartes, $x^3 + y^3 = 3xy$, parameterized by $x = \dfrac{3t}{1+t^3}$, $y = \dfrac{3t^2}{1+t^3}$, for $0 \leq t < \infty$.

25. Calculate $\int_C \left((x^2 - y)\vec{i} + (y^2 + x)\vec{j}\right) \cdot d\vec{r}$ if:

(a) C is the circle $(x-5)^2 + (y-4)^2 = 9$ oriented counterclockwise.

(b) C is the circle $(x-a)^2 + (y-b)^2 = R^2$ in the xy-plane oriented counterclockwise.

26. The following vector field $\vec{F}$ is defined on the disk D of radius 5 centered at the origin in the plane:

$\vec{F} = (-y^3 + y\sin(xy))\vec{i} + (4x(1-y^2) + x\sin(xy))\vec{j}$

Consider the line integral $\int_C \vec{F} \cdot d\vec{r}$, where C is some closed curve contained in D. For which C is the value of this integral the largest? [Hint: Assume C is a closed curve, made up of smooth pieces and never crossing itself, and oriented counterclockwise.]

27. At all points in the xy-plane, let $\vec{F} = F_1(x,y)\vec{i} + F_2(x,y)\vec{j}$ satisfy

$$\frac{\partial F_2}{\partial x} - \frac{\partial F_1}{\partial y} = 3$$

(a) Calculate $\int_{C_1} \vec{F} \cdot d\vec{r}$ where C_1 is the unit circle in the xy-plane centered at the origin, oriented counterclockwise.

(b) Calculate $\int_{C_2} \vec{F} \cdot d\vec{r}$ where C_2 is the boundary of the rectangle of $4 \leq x \leq 7$, $5 \leq y \leq 7$, oriented counterclockwise.

(c) Let C_3 be the circle of radius 7 centered at the point $(10, 2)$; let C_4 be the circle of radius 8 centered at the origin; let C_5 be the square of side 14 centered at $(7, 7)$ with sides parallel to the axes; C_3, C_4, C_5 are all oriented counterclockwise. Arrange the integrals $\int_{C_3} \vec{F} \cdot d\vec{r}$, $\int_{C_4} \vec{F} \cdot d\vec{r}$, $\int_{C_5} \vec{F} \cdot d\vec{r}$ in increasing order.

28. (a) By finding potential functions, show that each of the vector fields $\vec{F}$, $\vec{G}$, $\vec{H}$ is a gradient vector field.

(b) Find the line integrals of $\vec{F}$, $\vec{G}$, $\vec{H}$ around the unit circle in the xy-plane, centered at the origin, and traversed counterclockwise.

(c) For which of the three vector fields can Green's Theorem be used to calculate the line integral in part (b)? Why?

$$\vec{F} = y\vec{i} + x\vec{j}, \quad \vec{G} = \frac{y\vec{i} - x\vec{j}}{x^2 + y^2}, \quad \vec{H} = \frac{x\vec{i} + y\vec{j}}{(x^2 + y^2)^{1/2}}$$

CHAPTER SUMMARY (see also Ready Reference at the end of the book)

- **Line Integrals**
  Oriented curves, definition as a limit of a Riemann sum, work interpretation, circulation, algebraic properties, computing line integrals over parameterized curves, independence of parameterization.

- **Gradient Fields**

Fundamental Theorem for line integrals, path-independent (conservative) fields and their relation to gradient fields, potential functions.

- **Green's Theorem**
  Statement of the theorem, curl test for path-independence.

REVIEW EXERCISES AND PROBLEMS FOR CHAPTER EIGHTEEN

Exercises

1. Figure 18.48 shows a vector field $\vec{F}$ and a curve C. Decide if $\int_C \vec{F} \cdot d\vec{r}$ is positive, zero, or negative.

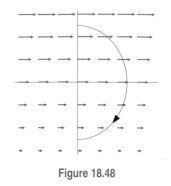

Figure 18.48

For the vector fields in Exercises 2–3, is the line integral positive, negative, or zero along

(a) A?   **(b)** C_1, C_2, C_3, C_4?
(c) C, the closed curve consisting of all the C's together?

2.

3.

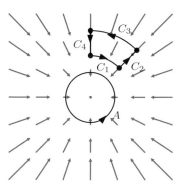

In Exercises 4–9, calculate the line integral of $\vec{F} = 5x\vec{i} + 3x\vec{j}$ along the path C.

4. C is the line from $(2, 3)$ to $(2, 8)$.

5. C is the line from $(2, 3)$ to $(12, 3)$.

6. C is the curve $y = x^2$ from $(1, 1)$ to $(2, 4)$.

7. C is the semicircle of radius 3 from $(3, 0)$ to $(-3, 0)$ in the upper half plane.

8. C is the path in Figure 18.49.

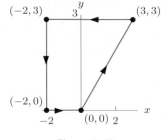

Figure 18.49

9. C is the path in Figure 18.50.

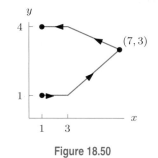

Figure 18.50

In Exercises 10–18, find $\int_C \vec{F} \cdot d\vec{r}$ for the given $\vec{F}$ and C.

10. $\vec{F} = 6\vec{i} - 7\vec{j}$, and C is an oriented curve from $(2, -6)$ to $(4, 4)$.

11. $\vec{F} = 5\vec{i} + 7\vec{j}$, and C is the x-axis from $(-1, 0)$ to $(-9, 0)$

12. $\vec{F} = x\vec{i} + y\vec{j}$ and C is the unit circle in xy-plane oriented counterclockwise.

13. $\vec{F} = x\vec{i} + y\vec{j}$ and C is the y-axis from the origin to $(0, 10)$.

14. $\vec{F} = 6x\vec{i} + (x + y^2)\vec{j}$; C is the y-axis from $(0, 3)$ to $(0, 5)$.

15. $\vec{F} = x^2\vec{i} + y^2\vec{j}$, and C is the x-axis from $(2, 0)$ to $(3, 0)$.

16. $\vec{F} = (x^2 - y)\vec{i} + (y^2 + x)\vec{j}$ and C is the parabola $y = x^2 + 1$ traversed from $(0, 1)$ to $(1, 2)$.

17. $\vec{F} = x\vec{i} + y\vec{j} + z\vec{k}$ and C is path consisting of a line from $(2, 3, 0)$ to $(4, 5, 0)$, followed by a line from $(4, 5, 0)$ to $(0, 0, 7)$.

18. $\vec{F} = (2x - y + 4)\vec{i} + (5y + 3x - 6)\vec{j}$ and C is the triangle with vertices $(0, 0), (3, 0), (3, 2)$ traversed counterclockwise.

Which of the vector fields in Exercises 19–23 are path independent on all of 3-space?

19. $y\vec{i}$ **20.** $y\vec{j}$

21. $y\vec{i} + x\vec{j}$

22. $(x+y)\vec{i}$

23. $yz\vec{i} + zx\vec{j} + xy\vec{k}$

Problems

24. Which two of the vector fields (i)-(iv) could represent gradient vector fields on the whole plane? Give reasons for your answer.

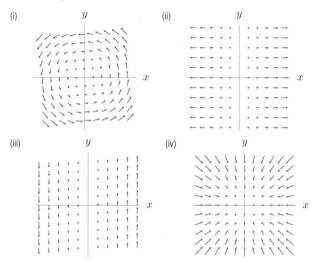

25. Let $\vec{F}(x,y)$ be the path-independent vector field in Figure 18.51. The vector field $\vec{F}$ associates with each point a unit vector pointing radially outward. The curves $C_1, C_2, \ldots, C_7$ have the directions shown. Consider the line integrals $\int_{C_i} \vec{F} \cdot d\vec{r}$, $i = 1, \ldots, 7$. Without computing any integrals

(a) List all the line integrals which you expect to be zero.

(b) List all the line integrals which you expect to be negative.

(c) Arrange the positive line integrals in ascending order.

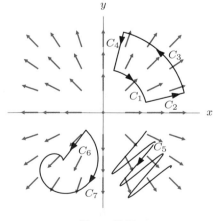

Figure 18.51

26. If $\vec{F}(x,y,z) = 2xe^{x^2+yz}\vec{i} + ze^{x^2+yz}\vec{j} + ye^{x^2+yz}\vec{k}$, find exactly the line integral of $\vec{F}$ along the curve consisting of the two half circles in the plane $z = 0$ in Figure 18.52.

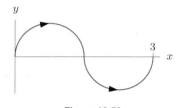

Figure 18.52

27. Let $\vec{F} = (x^2 + 3x^2y^4)\vec{i} + 4x^3y^3\vec{j}$ and $\vec{G} = (x^4 + x^3y^2)\vec{i} + x^2y^3\vec{j}$. Let C_1 be the path along the x-axis from $(2,0)$ to $(-2,0)$; let C_2 be the semi-circle in the upper half plane from $(2,0)$ to $(-2,0)$. Find exactly:

(a) $\int_{C_1} \vec{F} \cdot d\vec{r}$ **(b)** $\int_{C_2} \vec{F} \cdot d\vec{r}$

(c) $\int_{C_1} \vec{G} \cdot d\vec{r}$ **(d)** $\int_{C_2} \vec{G} \cdot d\vec{r}$

28. Let $\vec{F} = 2xe^y\vec{i} + x^2e^y\vec{j}$ and $\vec{G} = (x-y)\vec{i} + (x+y)\vec{j}$. Let C be the line from $(0,0)$ to $(2,4)$. Find exactly:

(a) $\int_C \vec{F} \cdot d\vec{r}$ **(b)** $\int_C \vec{G} \cdot d\vec{r}$

29. Let $\vec{F} = x\vec{i} + y\vec{j}$. Find the line integral of $\vec{F}$:

(a) Along the x-axis from the origin to the point $(3,0)$.

(b) Around the path from A to B to O in Figure 18.53. (The curve is part of a circle centered at the origin.)

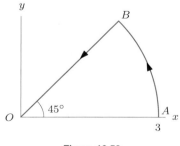

Figure 18.53

30. Calculate the line integral of $\vec{F} = -y\vec{i} + x\vec{j}$ along the following paths in the xy-plane.

   (a) Straight line from the origin to the point $(2, 3)$.
   (b) Straight line from $(2, 3)$ to $(0, 3)$.
   (c) Counterclockwise around a circle of radius 5 centered at the origin, starting from $(5, 0)$ and ending at $(0, -5)$.
   (d) Counterclockwise around the perimeter of a triangle of area 7.

31. Let C_1 and C_2 be the curves in Figure 18.54. Let $\vec{F} = (6x + y^2)\vec{i} + 2xy\vec{j}$ and $\vec{G} = (x - y)\vec{i} + (x + y)\vec{j}$. [Note C_1 is made up of line segments and C_2 is part of a circle.] Compute the following line integrals.

   (a) $\displaystyle\int_{C_1} \vec{F} \cdot d\vec{r}$ **(b)** $\displaystyle\int_{C_1} \vec{G} \cdot d\vec{r}$

   (c) $\displaystyle\int_{C_2} \vec{F} \cdot d\vec{r}$ **(d)** $\displaystyle\int_{C_2} \vec{G} \cdot d\vec{r}$

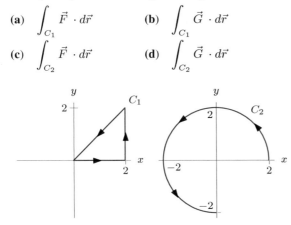

Figure 18.54

32. Let $\vec{F} = x\vec{i} + y\vec{j}$ and let C_1 be the line joining $(1, 0)$ to $(0, 2)$ and let C_2 be the line joining $(0, 2)$ to $(-1, 0)$. Is $\int_{C_1} \vec{F} \cdot d\vec{r} = -\int_{C_2} \vec{F} \cdot d\vec{r}$? Explain.

33. (a) Figure 18.55 shows level curves of $f(x, y)$. Sketch a vector at P in the direction of grad f.
   (b) Is the length of grad f at P longer, shorter, or the same length as the length of grad f at Q?
   (c) If C is a curve going from P to Q, evaluate $\int_C$ grad $f \cdot d\vec{r}$.

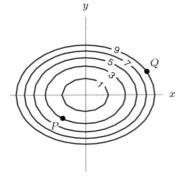

Figure 18.55

34. Consider the line integrals, $\int_{C_i} \vec{F} \cdot d\vec{r}$, for $i = 1, 2, 3, 4$, where C_i is the path from P_i to Q_i shown in Figure 18.56 and $\vec{F} = $ grad f. Level curves of f are also shown in Figure 18.56.

   (a) Which of the line integrals is (are) zero?
   (b) Arrange the four line integrals in ascending order (from least to greatest).
   (c) Two of the nonzero line integrals have equal and opposite values. Which are they? Which is negative and which is positive?

Figure 18.56

35. Suppose P and Q both lie on the same contour of f. What can you say about the total change in f from P to Q? Explain your answer in terms of $\int_C$ grad $f \cdot d\vec{r}$ where C is a part of the contour that goes from P to Q.

36. The vector field $\vec{F}$ has $||\vec{F}|| \leq 7$ everywhere and C is the circle of radius 1 centered at the origin. What is the largest possible value of $\int_C \vec{F} \cdot d\vec{r}$? The smallest possible value? What conditions lead to these values?

37. The fact that an electric current gives rise to a magnetic field is the basis for some electric motors. Ampère's Law relates the magnetic field $\vec{B}$ to a steady current I. It says

$$\int_C \vec{B} \cdot d\vec{r} = kI$$

where I is the current[3] flowing through a closed curve C and k is a constant. Figure 18.57 shows a rod carrying a current and the magnetic field induced around the rod. If the rod is very long and thin, experiments show that the magnetic field $\vec{B}$ is tangent to every circle that is perpendicular to the rod and has center on the axis of the rod (like C in Figure 18.57). The magnitude of $\vec{B}$ is constant along every such circle. Use Ampère's Law to show that around a circle of radius r, the magnetic field due to a current I has magnitude given by

$$||\vec{B}|| = \frac{kI}{2\pi r}.$$

---

[3] More precisely, I is the net current through any surface that has C as its boundary.

(In other words, the strength of the field is inversely proportional to the radial distance from the rod.)

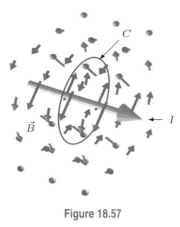

Figure 18.57

38. Figure 18.58 shows the tangential velocity as a function of radius for the tornado that hit Dallas on April 2, 1957. Use it and Problem 40 to estimate K and ω for the Rankine model of this tornado.[4]

Figure 18.58

39. A *central vector field* is a vector field whose direction is always toward (or away from) a fixed point O (the center) and whose magnitude at a point P is a function only of the distance from P to O. In two dimensions this means that the vector field has constant magnitude on circles centered at O. The gravitational and electrical fields of spherically symmetric sources are both central fields.

(a) Sketch an example of a central vector field.
(b) Suppose that the central field $\vec{F}$ is a gradient field, that is, $\vec{F} = \operatorname{grad} f$. What must be the shape of the contours of f? Sketch some contours for this case.

(c) Is every gradient field a central vector field? Explain.
(d) In Figure 18.59, two paths are shown between the points Q and P. Assuming that the three circles C_1, C_2, and C_3 are centered at O, explain why the work done by a central vector field $\vec{F}$ is the same for either path.
(e) It is in fact true that every central vector field is a gradient field. Use an argument suggested by Figure 18.59 to explain why any central vector field must be path-independent.

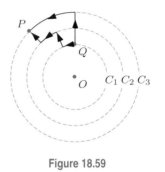

Figure 18.59

40. A *free vortex* circulating about the origin in the xy-plane (or about the z-axis in 3-space) has vector field $\vec{v} = K(x^2 + y^2)^{-1}(-y\vec{i} + x\vec{j})$ where K is a constant. The Rankine model of a tornado hypothesizes an inner core that rotates at constant angular velocity, surrounded by a free vortex. Suppose that the inner core has radius 100 meters and that $\|\vec{v}\| = 3 \cdot 10^5$ meters/hr at a distance of 100 meters from the center.

(a) Assuming that the tornado rotates counterclockwise (viewed from above the xy-plane) and that $\vec{v}$ is continuous, determine ω and K such that

$$\vec{v} = \begin{cases} \omega(-y\vec{i} + x\vec{j}) & \text{if } \sqrt{x^2 + y^2} < 100 \\ K(x^2 + y^2)^{-1}(-y\vec{i} + x\vec{j}) \\ \qquad \text{if } \sqrt{x^2 + y^2} \geq 100. \end{cases}$$

(b) Sketch the vector field $\vec{v}$.
(c) Find the circulation of $\vec{v}$ around the circle of radius r centered at the origin, traversed counterclockwise.

CAS Challenge Problems

41. Let C_a be the circle of radius a, centered at the origin, oriented in the counterclockwise direction, and let

$$\vec{F} = (-y + \tfrac{2}{3}y^3)\vec{i} + (2x - \tfrac{x^3}{3} + xy^2)\vec{j}.$$

(a) Evaluate $\int_{C_a} \vec{F} \cdot d\vec{r}$. For what positive value of a does the integral take its maximum value?

(b) Use Green's theorem to convert the integral to a double integral. Without evaluating the double integral, give a geometric explanation of the value of a you found in part (a).

----
[4]Adapted from *Encyclopedia Britannica, Macropedia*, Vol. 16, page 477, "Climate and the Weather", Tornados and Waterspouts, 1991.

42. If f is a potential function for the two-dimensional vector field $\vec{F}$ then the Fundamental Theorem of Calculus for Line Integrals says that $\int_C \vec{F} \cdot d\vec{r} = f(x, y) - f(0, 0)$ where C is any path from $(0, 0)$ to (x, y). Using this fact and choosing C to be a straight line, find potential functions for the following conservative fields (where a, b, c are constants):

(a) $\vec{F} = ay\vec{i} + ax\vec{j}$
(b) $\vec{F} = abye^{bxy}\vec{i} + (c + abxe^{bxy})\vec{j}$

43. Let $\vec{F} = (ax + by)\vec{i} + (cx + dy)\vec{j}$. Evaluate the line integral of $\vec{F}$ along the paths

$$C_1 : \vec{r}(t) = 2t\vec{i} + t^2\vec{j}, \quad 0 \le t \le 3$$
$$C_2 : \vec{r}(t) = 2(3 - t)\vec{i} + (3 - t)^2\vec{j}, \quad 0 \le t \le 3$$

Describe and explain the relationship between the two integrals.

CHECK YOUR UNDERSTANDING

In Problems 1–4, each of the statements is *false*. Explain why or give a counterexample.

1. If $\int_C \vec{F} \cdot d\vec{r} = 0$ for one particular closed path C, then $\vec{F}$ is path-independent.

2. $\int_C \vec{F} \cdot d\vec{r}$ is the total change in $\vec{F}$ along C.

3. If the vector fields $\vec{F}$ and $\vec{G}$ have $\int_C \vec{F} \cdot d\vec{r} = \int_C \vec{G} \cdot d\vec{r}$ for a particular path C, then $\vec{F} = \vec{G}$.

4. If the total change of a function f along a curve C is zero, then C must be a contour of f.

Are the statements in Problems 5–8 true or false? Explain why or give a counterexample.

5. $\int_C \vec{F} \cdot d\vec{r}$ is a vector.

6. $\int_C \vec{F} \cdot d\vec{r} = \vec{F}(Q) - \vec{F}(P)$ when P and Q are the endpoints of C.

7. The fact that the line integral of a vector field $\vec{F}$ is zero around the unit circle $x^2 + y^2 = 1$ means that $\vec{F}$ must be a gradient vector field.

8. Suppose C_1 is the unit square joining the points $(0, 0)$, $(1, 0)$, $(1, 1)$, $(0, 1)$ oriented clockwise and C_2 is the same square but traversed twice in the opposite direction. If $\int_{C_1} \vec{F} \cdot d\vec{r} = 3$, then $\int_{C_2} \vec{F} \cdot d\vec{r} = -6$.

Are the statements in Problems 9–41 true or false? Give reasons for your answer.

9. The line integral $\int_C \vec{F} \cdot d\vec{r}$ is a scalar.

10. If C_1 and C_2 are oriented curves, and the length of C_1 is greater than the length of C_2, then $\int_{C_1} \vec{F} \cdot d\vec{r} > \int_{C_2} \vec{F} \cdot d\vec{r}$.

11. If C is an oriented curve and $\int_C \vec{F} \cdot d\vec{r} = 0$, then $\vec{F} = \vec{0}$.

12. If $\vec{F} = \vec{i}$ is a vector field in 2-space, then $\int_C \vec{F} \cdot d\vec{r} > 0$, where C is the oriented line from $(0, 0)$ to $(1, 0)$.

13. If $\vec{F} = \vec{i}$ is a vector field in 2-space, then $\int_C \vec{F} \cdot d\vec{r} > 0$, where C is the oriented line from $(0, 0)$ to $(0, 1)$.

14. If C_1 is the upper semicircle $x^2 + y^2 = 1, y \ge 0$ and C_2 is the lower semicircle $x^2 + y^2 = 1, y \le 0$, both oriented counterclockwise, then for any vector field $\vec{F}$, we have $\int_{C_1} \vec{F} \cdot d\vec{r} = -\int_{C_2} \vec{F} \cdot d\vec{r}$.

15. If C_1 and C_2 are oriented curves with C_2 beginning where C_1 ends, then $\int_{C_1 + C_2} \vec{F} \cdot d\vec{r} > \int_{C_1} \vec{F} \cdot d\vec{r}$.

16. The work done by the force $\vec{F} = -y\vec{i} + x\vec{j}$ on a particle moving clockwise around the boundary of the square $-1 \le x \le 1, -1 \le y \le 1$ is positive.

17. The circulation of any vector field $\vec{F}$ around any closed curve C is zero.

18. The line integral $\int_C 4\vec{i} \cdot d\vec{r}$ over the curve C parameterized by $\vec{r}(t) = t\vec{i} + t^2\vec{j}$, for $0 \le t \le 2$, is positive.

19. If C_1 is the curve parameterized by $\vec{r}_1(t) = \cos t\vec{i} + \sin t\vec{j}$, with $0 \le t \le \pi$, and C_2 is the curve parameterized by $\vec{r}_2(t) = \cos t\vec{i} - \sin t\vec{j}, 0 \le t \le \pi$, then for any vector field $\vec{F}$ we have $\int_{C_1} \vec{F} \cdot d\vec{r} = \int_{C_2} \vec{F} \cdot d\vec{r}$.

20. If C_1 is the curve parameterized by $\vec{r}_1(t) = \cos t\vec{i} + \sin t\vec{j}$, with $0 \le t \le \pi$, and C_2 is the curve parameterized by $\vec{r}_2(t) = \cos(2t)\vec{i} + \sin(2t)\vec{j}, 0 \le t \le \frac{\pi}{2}$, then for any vector field $\vec{F}$ we have $\int_{C_1} \vec{F} \cdot d\vec{r} = \int_{C_2} \vec{F} \cdot d\vec{r}$.

21. If C is the curve parameterized by $\vec{r}(t)$, for $a \le t \le b$ with $\vec{r}(a) = \vec{r}(b)$, then $\int_C \vec{F} \cdot d\vec{r} = 0$ for any vector field $\vec{F}$. (Note that C starts and ends at the same place.)

22. If C_1 is the line segment from $(0, 0)$ to $(1, 0)$ and C_2 is the line segment from $(0, 0)$ to $(2, 0)$, then for any vector field $\vec{F}$, we have $\int_{C_2} \vec{F} \cdot d\vec{r} = 2 \int_{C_1} \vec{F} \cdot d\vec{r}$.

23. If C is a circle of radius a, centered at the origin and oriented counterclockwise, then $\int_C 2x\vec{i} + y\vec{j} \cdot d\vec{r} = 0$.

24. If C is a circle of radius a, centered at the origin and oriented counterclockwise, then $\int_C 2y\vec{i} + x\vec{j} \cdot d\vec{r} = 0$.

25. If C_1 is the curve parameterized by $\vec{r}_1(t) = t\vec{i} + t^2\vec{j}$, with $0 \le t \le 2$, and C_2 is the curve parameterized by $\vec{r}_2(t) = (2 - t)\vec{i} + (2 - t)^2\vec{j}, 0 \le t \le 2$, then for any vector field $\vec{F}$ we have $\int_{C_1} \vec{F} \cdot d\vec{r} = -\int_{C_2} \vec{F} \cdot d\vec{r}$.

26. If $\vec{F} = \text{grad } f$, then $\vec{F}$ is path-independent.

27. If $\vec{F}$ is path-independent, then $\int_{C_1} \vec{F} \cdot d\vec{r} = \int_{C_2} \vec{F} \cdot d\vec{r}$, where C_1 and C_2 are any paths.

28. The line integral $\int_C \vec{F} \cdot d\vec{r}$ is the total change of $\vec{F}$ along C.

29. If $\vec{F}$ is path-independent, then there is a potential function for $\vec{F}$.

30. If $f(x,y) = e^{\cos(xy)}$, and C_1 is the upper semicircle $x^2 + y^2 = 1$ from $(-1,0)$ to $(1,0)$, and C_2 is the line from $(-1,0)$ to $(1,0)$, then $\int_{C_1} \operatorname{grad} f \cdot d\vec{r} = \int_{C_2} \operatorname{grad} f \cdot d\vec{r}$.

31. If $\vec{F}$ is path-independent, and C is any closed curve, then $\int_C \vec{F} \cdot d\vec{r} = 0$.

32. The vector field $\vec{F}(x,y) = y^2\vec{i} + k\vec{j}$, where k is constant, is a gradient field.

33. If $\int_C \vec{F} \cdot d\vec{r} = 0$, where C is any circle of the form $x^2 + y^2 = a^2$, then $\vec{F}$ is path-independent.

34. If $f(x)$ and $g(y)$ are continuous one-variable functions, then the vector field $\vec{F} = f(x)\vec{i} + g(y)\vec{j}$ is path-independent.

35. If $\vec{F} = \operatorname{grad} f$, and C is the perimeter of a square of side length a oriented counterclockwise and surrounding the region R, then

$$\int_C \vec{F} \cdot d\vec{r} = \int_R f\, dA.$$

36. If $\vec{F}$ and $\vec{G}$ are both path-independent vector fields, then $\vec{F} + \vec{G}$ is path-independent.

37. If $\vec{F}$ and $\vec{G}$ are both path-dependent vector fields, then $\vec{F} + \vec{G}$ is path-dependent.

38. The vector field $\vec{F}(\vec{r}) = \vec{r}$ in 3-space is path-independent.

39. A constant vector field $\vec{F} = a\vec{i} + b\vec{j}$ is path-independent.

40. If $\vec{F}$ is path-independent and k is a constant, then the vector field $k\vec{F}$ is path-independent.

41. If $\vec{F}$ is path-independent and $h(x,y)$ is a scalar function, then the vector field $h(x,y)\vec{F}$ is path-independent.

PROJECTS FOR CHAPTER EIGHTEEN

1. Conservation of Energy

(a) A particle moves with position vector $\vec{r}(t) = x(t)\vec{i} + y(t)\vec{j} + z(t)\vec{k}$. Let $\vec{v}(t)$ and $\vec{a}(t)$ be its velocity and acceleration vectors. Show that

$$\frac{1}{2}\frac{d}{dt}\|\vec{v}(t)\|^2 = \vec{a}(t) \cdot \vec{v}(t).$$

(b) We now derive the principle of Conservation of Energy. The kinetic energy of a particle of mass m moving with speed v is $(1/2)mv^2$. Suppose the particle has potential energy $f(\vec{r})$ at the position $\vec{r}$ due to a force field $\vec{F} = -\nabla f$. If the particle moves with position vector $\vec{r}(t)$ and velocity $\vec{v}(t)$, then the Conservation of Energy principle says that

Total energy = Kinetic energy + Potential energy = $\dfrac{1}{2}m\|\vec{v}(t)\|^2 + f(\vec{r}(t))$ = Constant.

Let P and Q be two points in space and let C be a path from P to Q parameterized by $\vec{r}(t)$ for $t_0 \le t \le t_1$, where $\vec{r}(t_0) = P$ and $\vec{r}(t_1) = Q$.

(i) Using part (a) and Newton's law $\vec{F} = m\vec{a}$, show

Work done by $\vec{F}$ as particle moves along C = Kinetic energy at Q − Kinetic energy at P.

(ii) Use the Fundamental Theorem of Calculus for Line Integrals to show that

Work done by $\vec{F}$ as particle moves along C = Potential energy at P − Potential energy at Q.

(iii) Use parts (a) and (b) to show that the total energy at P is the same as at Q.

This problem explains why force vector fields which are *path-independent* are usually called *conservative* (force) vector fields.

2. **Ampère's Law**

Ampère's Law, introduced in Problem 37 on page 911 , relates the net current, I, flowing through a surface to the magnetic field around the boundary, C, of the surface. The law says that

$$\int_C \vec{B} \cdot d\vec{r} = kI, \quad \text{for some constant } k.$$

The orientation of C is given by the right hand rule; if the fingers of your right hand lie against the surface and curl in the direction of the oriented curve C, then your thumb points in the direction of positive current.

(a) We consider an infinitely long cylindrical wire having a radius r_0, where $r_0 > 0$. Suppose the wire is centered on the z-axis and carries a constant current, I, uniformly distributed across a cross-section of the wire. The magnitude of $\vec{B}$ is constant along any circle which is centered on and perpendicular to the z-axis. The direction of $\vec{B}$ is tangent to such circles. Show that the magnitude of $\vec{B}$, at a distance r from the z-axis, is given by

$$\|\vec{B}\| = \begin{cases} \dfrac{aI}{2\pi r} & \text{for } r \geq r_0 \\ \dfrac{aIr}{2\pi r_0^2} & \text{for } r < r_0. \end{cases}$$

(b) A *torus* is a doughnut-shaped surface obtained by rotating around the z-axis the circle $(x-\beta)^2 + z^2 = \alpha^2$ of radius α and center $(\beta, 0, 0)$, where $0 < \alpha < \beta$. A *toroidal solenoid* is constructed by wrapping a thin wire a large number, say N, times around the torus. Experiments show that if a constant current I flows though the wire, then the magnitude of the magnetic field is constant on circles inside the torus which are centered on and perpendicular to the z-axis. The direction of $\vec{B}$ is tangent to such circles. Explain why, at all points inside the torus,

$$\|\vec{B}\| = \frac{aNI}{2\pi r},$$

and $\|\vec{B}\| = 0$ otherwise. [Hint: Apply Ampère's Law to a suitably chosen surface S with boundary curve C. What is the net current passing through S?]

Chapter Nineteen

FLUX INTEGRALS

In the previous chapter we saw how to integrate vector fields along curves. In this chapter we shall define a new sort of integral, the flux integral, which goes over a surface rather than along a curve. If we view a vector field as representing the velocity of a fluid flow, the flux integral tells us about the rate at which fluid is flowing through the surface. In addition, the flux integral appears in the theory of electricity and magnetism.

19.1 THE IDEA OF A FLUX INTEGRAL

Flow Through a Surface

Imagine water flowing through a fishing net stretched across a stream. Suppose we want to measure the flow rate of water through the net, that is, the volume of fluid that passes through the surface per unit time. (See Figure 19.1.) This flow rate is called the *flux* of the fluid through the surface. We can also compute the flux of vector fields, such as electric and magnetic fields, where no flow is actually taking place.

Figure 19.1: Flux measures rate of flow
through a surface

Orientation of a Surface

Before computing the flux of a vector field through a surface, we need to decide which direction of flow through the surface is the positive direction; this is described as choosing an orientation. [1]

> At each point on a smooth surface there are two unit normals, one in each direction. **Choosing an orientation** means picking one of these normals at every point of the surface in a continuous way. The normal vector in the direction of the orientation is denoted by $\vec{n}$. For a closed surface, we usually choose the outward orientation.

We say the flux through a piece of surface is positive if the flow is in the direction of the orientation and negative if it is in the opposite direction. (See Figure 19.2.)

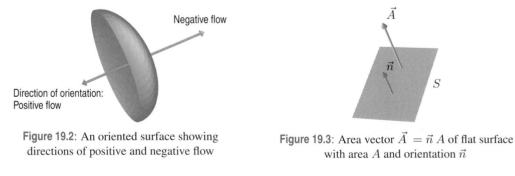

Figure 19.2: An oriented surface showing
directions of positive and negative flow

Figure 19.3: Area vector $\vec{A} = \vec{n}\,A$ of flat surface
with area A and orientation $\vec{n}$

The Area Vector

The flux through a flat surface depends both on the area of the surface and its orientation. Thus, it is useful to represent its area by a vector as shown in Figure 19.3.

---

[1] Although we will not study them, there are a few surfaces for which this cannot be done. See page 921.

The **area vector** of a flat, oriented surface is a vector $\vec{A}$ such that
- The magnitude of $\vec{A}$ is the area of the surface.
- The direction of $\vec{A}$ is the direction of the orientation vector $\vec{n}$.

The Flux of a Constant Vector Field Through a Flat Surface

Suppose the velocity vector field, $\vec{v}$, of a fluid is constant and $\vec{A}$ is the area vector of a flat surface. The flux through this surface is the volume of fluid that flows through in one unit of time. The volume of the skewed box in Figure 19.4 has cross-sectional area $\|\vec{A}\|$ and height $\|\vec{v}\|\cos\theta$, so its volume is $(\|\vec{v}\|\cos\theta)\|\vec{A}\| = \vec{v}\cdot\vec{A}$. Thus we have the following result:

If $\vec{v}$ is constant and $\vec{A}$ is the area vector of a flat surface, then

$$\text{Flux through surface} = \vec{v}\cdot\vec{A}.$$

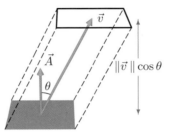

Figure 19.4: Flux of $\vec{v}$ through a surface with area vector $\vec{A}$ is the volume of this skewed box

Example 1   Water is flowing down a cylindrical pipe 2 cm in radius with a velocity of 3 cm/sec. Find the flux of the velocity vector field through the ellipse-shaped region shown in Figure 19.5. The normal to the ellipse makes an angle of θ with the direction of flow and the area of the ellipse is $4\pi/(\cos\theta)$ cm^2.

Figure 19.5: Flux through ellipse-shaped region across a cylindrical pipe

Solution   There are two ways to approach this problem. One is to use the formula we just derived which gives

$$\text{Flux through ellipse} = \vec{v}\cdot\vec{A} = \|\vec{v}\|\|\vec{A}\|\cos\theta = 3(\text{Area of ellipse})\cos\theta$$

$$= 3\left(\frac{4\pi}{\cos\theta}\right)\cos\theta = 12\pi \text{ cm}^3/\text{sec}.$$

The second way is to notice that the flux through the ellipse is equal to the flux through the circle perpendicular to the pipe in Figure 19.5. Since the flux is the rate at which water is flowing down the pipe, we have

$$\text{Flux through circle} = \frac{\text{Velocity}}{\text{of water}} \times \frac{\text{Area of}}{\text{circle}} = \left(3\,\frac{\text{cm}}{\text{sec}}\right)(\pi 2^2 \text{ cm}^2) = 12\pi \text{ cm}^3/\text{sec}.$$

When the vector field is not constant or the surface is not flat, we divide the surface up into small, almost flat pieces such that the vector field is approximately constant on each one, as follows.

The Flux Integral

To calculate the flux of a vector field $\vec{F}$ which is not necessarily constant through a curved, oriented surface S, we divide the surface into a patchwork of small, approximately flat pieces (like a wireframe representation of the surface) as shown in Figure 19.6. Suppose a particular patch has area ΔA. We pick an orientation vector $\vec{n}$ at a point on the patch and define the area vector of the patch, $\Delta \vec{A}$, as

$$\Delta \vec{A} = \vec{n}\, \Delta A.$$

(See Figure 19.6.) If the patches are small enough, we can assume that $\vec{F}$ is approximately constant on each piece. Then we know that

$$\text{Flux through patch} \approx \vec{F} \cdot \Delta \vec{A}$$

and so

$$\text{Flux through whole surface} \approx \sum \vec{F} \cdot \Delta \vec{A},$$

where the sum adds the fluxes through all the small pieces. As each patch becomes smaller and $\|\Delta \vec{A}\| \to 0$, the approximation gets better and we get

$$\text{Flux through } S = \lim_{\|\Delta \vec{A}\| \to 0} \sum \vec{F} \cdot \Delta \vec{A}.$$

Thus, provided the limit exists, we make the following definition:

> The **flux integral** of the vector field $\vec{F}$ through the oriented surface S is
>
> $$\int_S \vec{F} \cdot d\vec{A} = \lim_{\|\Delta \vec{A}\| \to 0} \sum \vec{F} \cdot \Delta \vec{A}.$$

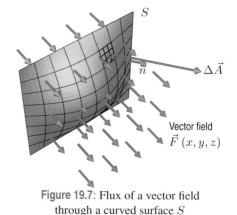

Figure 19.6: Surface S divided into small, almost flat pieces, showing a typical orientation vector $\vec{n}$ and area vector $\Delta \vec{A}$

Figure 19.7: Flux of a vector field through a curved surface S

In computing a flux integral, we have to divide the surface up in a reasonable way, or the limit might not exist. In practice this problem seldom arises; however, one way to avoid it is to define flux integrals by the method used to compute them shown in Section 19.3.

Flux and Fluid Flow

If $\vec{v}$ is the velocity vector field of a fluid, we have

$$\begin{array}{ccccc} \text{Rate fluid flows} \\ \text{through surface } S \end{array} = \begin{array}{c} \text{Flux of } \vec{v} \\ \text{through } S \end{array} = \int_S \vec{v} \cdot d\vec{A}$$

The rate of fluid flow is measured in units of volume per unit time.

Example 2 Find the flux of the vector field $\vec{B}(x, y, z)$ shown in Figure 19.8 through the square S of side 2 shown in Figure 19.9, oriented in the $\vec{j}$ direction, where

$$\vec{B}(x, y, z) = \frac{-y\vec{i} + x\vec{j}}{x^2 + y^2}.$$

Figure 19.8: The vector field $\vec{B}(x, y, z) = \frac{-y\vec{i} + x\vec{j}}{x^2 + y^2}$

Figure 19.9: Flux of $\vec{B}$ through the square S of side 2 in xy-plane and oriented in $\vec{j}$ direction

Figure 19.10: A small patch of surface with area $\|\Delta\vec{A}\| = \Delta x \Delta z$

Solution Consider a small rectangular patch with area vector $\Delta\vec{A}$ in S, with sides Δx and Δz so that $\|\Delta\vec{A}\| = \Delta x \Delta z$. Since $\Delta\vec{A}$ points in the $\vec{j}$ direction we have $\Delta\vec{A} = \vec{j} \Delta x \Delta z$. (See Figure 19.10.)

At the point $(x, 0, z)$ in S, substituting $y = 0$ into $\vec{B}$ gives $\vec{B}(x, 0, z) = (1/x)\vec{j}$. Thus, we have

$$\text{Flux through small patch} \approx \vec{B} \cdot \Delta\vec{A} = \left(\frac{1}{x}\vec{j}\right) \cdot (\vec{j} \Delta x \Delta z) = \frac{1}{x} \Delta x \Delta z.$$

Therefore,

$$\text{Flux through surface} = \int_S \vec{B} \cdot d\vec{A} = \lim_{\|\Delta\vec{A}\| \to 0} \sum \vec{B} \cdot \Delta\vec{A} = \lim_{\substack{\Delta x \to 0 \\ \Delta z \to 0}} \sum \frac{1}{x} \Delta x \Delta z.$$

This last expression is a Riemann sum for the double integral $\int_R \frac{1}{x} dA$, where R is the square $1 \leq x \leq 3$, $0 \leq z \leq 2$. Thus,

$$\text{Flux through surface} = \int_S \vec{B} \cdot d\vec{A} = \int_R \frac{1}{x} dA = \int_0^2 \int_1^3 \frac{1}{x} dx\, dz = 2\ln 3.$$

The result is positive since the vector field is passing through the surface in the positive direction.

Example 3 Each of the vector fields in Figure 19.11 consists entirely of vectors parallel to the xy-plane, and is constant in the z direction (that is, the vector field looks the same in any plane parallel to the xy-plane). For each one, say whether you expect the flux through a closed surface surrounding the origin to be positive, negative, or zero. In part (a) the surface is a closed cube with faces parallel to the axes; in parts (b) and (c) the surface is a closed cylinder. In each case we choose the outward orientation. (See Figure 19.12 on page 920.)

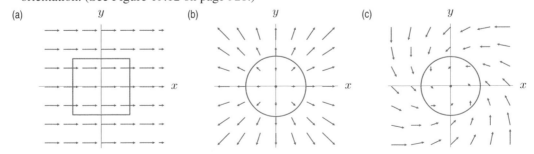

Figure 19.11: Flux of a vector field through the closed surfaces whose cross-sections are shown in the xy-plane

Figure 19.12: The closed cube and closed cylinder, both oriented outward

Solution

(a) Since the vector field appears to be parallel to the faces of the cube which are perpendicular to the y- and z-axes, we expect the flux through these faces to be zero. The fluxes through the two faces perpendicular to the x-axis appear to be equal in magnitude and opposite in sign, so we expect the net flux to be zero.

(b) Since the top and bottom of the cylinder are parallel to the flow, the flux through them is zero. On the round surface of the cylinder, $\vec{v}$ and $\Delta\vec{A}$ appear to be everywhere parallel and in the same direction, so we expect each term $\vec{v} \cdot \Delta\vec{A}$ to be positive, and therefore the flux integral $\int_S \vec{v} \cdot d\vec{A}$ to be positive.

(c) As in part (b), the flux through the top and bottom of the cylinder is zero. In this case $\vec{v}$ and $\Delta\vec{A}$ are not parallel on the round surface of the cylinder, but since the fluid appears to be flowing inward as well as swirling, we expect each term $\vec{v} \cdot \Delta\vec{A}$ to be negative, and therefore the flux integral to be negative.

Calculating Flux Integrals Using $d\vec{A} = \vec{n}\,dA$

For a small patch of surface ΔS with normal $\vec{n}$ and area ΔA, the area vector is $\Delta\vec{A} = \vec{n}\,\Delta A$. The next example shows how we can use this relationship to compute a flux integral.

Example 4

An electric charge q is placed at the origin in 3-space. The resulting electric field $\vec{E}(\vec{r})$ at the point with position vector $\vec{r}$ is given by

$$\vec{E}(\vec{r}) = q\frac{\vec{r}}{\|\vec{r}\|^3}, \qquad \vec{r} \neq \vec{0}.$$

Find the flux of $\vec{E}$ out of the sphere of radius R centered at the origin. (See Figure 19.13.)

Figure 19.13: Flux of $\vec{E} = q\vec{r}/\|\vec{r}\|^3$ through the surface of a sphere of radius R centered at the origin

Solution

This vector field points radially outward from the origin in the same direction as $\vec{n}$. Thus, since $\vec{n}$ is a unit vector,

$$\vec{E} \cdot \Delta\vec{A} = \vec{E} \cdot \vec{n}\,\Delta A = \|\vec{E}\|\,\Delta A.$$

On the sphere, $\|\vec{E}\| = q/R^2$, so

$$\int_S \vec{E} \cdot d\vec{A} = \lim_{\|\Delta\vec{A}\| \to 0} \sum \vec{E} \cdot \Delta\vec{A} = \lim_{\Delta A \to 0} \sum \frac{q}{R^2}\,\Delta A = \frac{q}{R^2} \lim_{\Delta A \to 0} \sum \Delta A.$$

The last sum approximates the surface area of the sphere. In the limit as the subdivisions get finer we have

$$\lim_{\Delta A \to 0} \sum \Delta A = \text{Surface area of sphere.}$$

Thus, the flux is given by

$$\int_S \vec{E} \cdot d\vec{A} = \frac{q}{R^2} \lim_{\Delta A \to 0} \sum \Delta A = \frac{q}{R^2} \cdot \text{Surface area of sphere} = \frac{q}{R^2}(4\pi R^2) = 4\pi q.$$

This result is known as Gauss's law.

Instead of using Riemann sums, we often write $d\vec{A} = \vec{n}\, dA$, as in the next example.

Example 5   Suppose S is the surface of the cube bounded by the six planes $x = \pm 1$, $y = \pm 1$, and $z = \pm 1$. Compute the flux of the electric field $\vec{E}$ of the previous example outward through S.

Solution   It is enough to compute the flux of $\vec{E}$ through a single face, say the top face S_1 defined by $z = 1$, where $-1 \le x \le 1$ and $-1 \le y \le 1$. By symmetry, the flux of $\vec{E}$ through the other five faces of S must be the same.

On the top face, S_1, we have $d\vec{A} = \vec{n}\, dA = \vec{k}\, dx\, dy$ and

$$\vec{E}(x, y, 1) = q\frac{x\vec{i} + y\vec{j} + \vec{k}}{(x^2 + y^2 + 1)^{3/2}}.$$

The corresponding flux integral is given by

$$\int_{S_1} \vec{E} \cdot d\vec{A} = q\int_{-1}^{1}\int_{-1}^{1} \frac{x\vec{i} + y\vec{j} + \vec{k}}{(x^2 + y^2 + 1)^{3/2}} \cdot \vec{k}\, dx\, dy = q\int_{-1}^{1}\int_{-1}^{1} \frac{1}{(x^2 + y^2 + 1)^{3/2}}\, dx\, dy.$$

Computing this integral numerically shows that

$$\text{Flux through top face} = \int_{S_1} \vec{E} \cdot d\vec{A} \approx 2.0944q.$$

Thus,

$$\text{Total flux of } \vec{E} \text{ out of cube} = \int_S \vec{E} \cdot d\vec{A} \approx 6(2.0944q) = 12.5664q.$$

Example 4 on page 920 showed that the flux of $\vec{E}$ through a sphere of radius R centered at the origin is $4\pi q$. Since $4\pi \approx 12.5664$, Example 5 suggests that

$$\text{Total flux of } \vec{E} \text{ out of cube} = 4\pi q.$$

By computing the flux integral in Example 5 exactly, it is possible to verify that the flux of $\vec{E}$ through the cube and the sphere are exactly equal. When we encounter the Divergence Theorem in Chapter 20 we will see why this is so.

Notes on Orientation

Two difficulties can occur in choosing an orientation. The first is that if the surface is not smooth, it may not have a normal vector at every point. For example, a cube does not have a normal vector along its edges. When we have a surface, such as a cube, which is made of a finite number of smooth pieces, we choose an orientation for each piece separately. The best way to do this is usually clear. For example, on the cube we choose the outward orientation on each face. (See Figure 19.14 on page 922.)

Figure 19.14: The orientation vector field $\vec{n}$ on the cube surface determined by the choice of unit normal vector at the point P

Figure 19.15: The Möbius strip is an example of a non-orientable surface

The second difficulty is that there are some surfaces which cannot be oriented at all, such as the *Möbius strip* in Figure 19.15.

Exercises and Problems for Section 19.1

Exercises

1. Let $\vec{F}(x, y, z) = z\vec{i}$. For each of the surfaces in (a)–(e), say whether the flux of $\vec{F}$ through the surface is positive, negative, or zero. In each case, the orientation of the surface is indicated by the given normal vector.

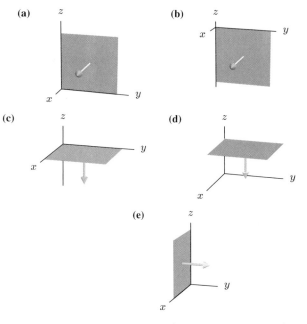

(a)

(b)

(c)

(d)

(e)

2. Repeat Exercise 1 with $\vec{F}(x, y, z) = -z\vec{i} + x\vec{k}$.

3. Repeat Exercise 1 with the vector field $\vec{F}(\vec{r}) = \vec{r}$.

Compute the flux of the vector field $\vec{v} = 2\vec{i} + 3\vec{j} + 5\vec{k}$ through each of the rectangular regions in Exercises 4–7, assuming each is oriented as shown.

4.

5.

6.

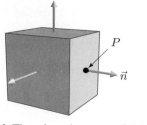

7.

In Exercises 8–24, calculate the flux of the vector field through the surface.

8. $\vec{F} = x\vec{i}$ through a square of side 3 in the plane $x = -5$. The square is centered on the x-axis, has sides parallel to the axes, and is oriented in the positive x-direction.

9. $\vec{F} = y\vec{j}$ through the square of side 4 in the plane $y = 5$. The square is centered on the y-axis, has sides parallel to the axes, and is oriented in the positive y-direction.

10. $\vec{F} = (y+3)\vec{j}$ through a square of side 2 in the xz-plane, oriented in the negative y-direction.

11. $\vec{F} = 2\vec{i} + 3\vec{j}$ through a disk of radius 5 in the plane $y = 2$ oriented in the direction of increasing y.

12. $\vec{F} = \vec{i} + 2\vec{j}$ through a sphere of radius 3 at the origin.

13. $\vec{F} = x\vec{k}$ through the square $0 \leq x \leq 3, 0 \leq y \leq 3$ in the xy-plane, with sides parallel to the axes, and oriented upward.

14. $\vec{F} = 6\vec{i} + x^2\vec{j} - \vec{k}$, through a square of side 2 in the plane $z = 3$, oriented upward.

15. $\vec{F} = 6\vec{i} + x^2\vec{j} - \vec{k}$, through the square of side 4 in the plane $y = 3$, centered on the y-axis, with sides parallel to the x and z axes, and oriented in the positive y-direction.

16. $\vec{F} = \vec{i} - \vec{j} - \vec{k}$ through a cube of side 2 with sides parallel to the axes.

17. $\vec{F} = 7\vec{r}$ through the sphere of radius 3 centered at the origin.

18. $\vec{F} = (x + 3)\vec{i} + (y + 5)\vec{j} + (z + 7)\vec{k}$ through is the rectangle $x = 4, 0 \leq y \leq 2, 0 \leq z \leq 3$, oriented in the positive x-direction.

19. $\vec{F} = -3\vec{r}$ through the sphere of radius 2 centered at the origin.

20. $\vec{F} = 2z\vec{i} + x\vec{j} + x\vec{k}$ through the rectangle $x = 4$, $0 \le y \le 2$, $0 \le z \le 3$, oriented in the positive x-direction.

21. $\vec{F} = \vec{i} + 2\vec{j}$ through a square of side 2 lying in the plane $x + y + z = 1$, oriented away from the origin.

22. $\vec{F} = (x^2 + y^2)\vec{k}$ through the disk of radius 3 in the xy-plane and centered at the origin and oriented upward.

23. $\vec{F} = e^{y^2+z^2}\vec{i}$ through the disk of radius 2 in the yz-plane, centered at the origin and oriented in the positive x-direction.

24. $\vec{F} = \cos(x^2 + y^2)\vec{k}$ through the disk $x^2 + y^2 \le 9$ oriented upward in the plane $z = 1$.

25. Find the flux of $\vec{F} = 5\vec{i} + 7\vec{j} + z\vec{k}$ out of a closed cylinder of radius 3 centered on the z-axis, with $-2 \le z \le 2$.

26. Find the flux of $\vec{F} = x\vec{i} + y\vec{j} + z\vec{k}$ out of a closed cylinder of radius 2 centered on the y-axis, with $-3 \le y \le 3$.

Problems

27. Let S be the cube with side length 2, faces parallel to the coordinate planes, and centered at the origin.

    (a) Calculate the total flux of the constant vector field $\vec{v} = -\vec{i} + 2\vec{j} + \vec{k}$ out of S by computing the flux through each face separately.

    (b) Calculate the flux out of S for any constant vector field $\vec{v} = a\vec{i} + b\vec{j} + c\vec{k}$.

    (c) Explain why the answers to parts (a) and (b) make sense.

28. Let S be the tetrahedron with vertices at the origin and at $(1, 0, 0)$, $(0, 1, 0)$ and $(0, 0, 1)$.

    (a) Calculate the total flux of the constant vector field $\vec{v} = -\vec{i} + 2\vec{j} + \vec{k}$ out of S by computing the flux through each face separately.

    (b) Calculate the flux out of S in part (a) for any constant vector field $\vec{v}$.

    (c) Explain why the answers to parts (a) and (b) make sense.

29. The flux of the constant vector field $a\vec{i} + b\vec{j} + c\vec{k}$ through the square of side 2 in the plane $x = 5$, oriented in the positive x-direction, is 24. Which of the constants a, b, c can be determined from the information given? Give the value(s).

30. **(a)** What do you think will be the electric flux through the cylindrical surface that is placed as shown in the constant electric field in Figure 19.16? Why?

    (b) What if the cylinder is placed upright, as shown in Figure 19.17? Explain.

Figure 19.17

31. **(a)** Let $\vec{F} = (x^2 + 4)\vec{i} + y\vec{j}$. Which of the following flux integrals is the largest? Explain.

        $\int_{S_1} \vec{F} \cdot d\vec{A}$, where S_1 is the disk of radius 1 in the plane $x = 2$ centered on the x-axis and oriented away from the origin.

        $d\int_{S_2} \vec{F} \cdot d\vec{A}$, where S_2 is the disk of radius 1 in the plane $y = 4$ centered on the y-axis and oriented away from the origin.

        $\int_{S_3} \vec{F} \cdot d\vec{A}$, where S_3 is the disk of radius 1 in the plane $x = -3$ centered on the x-axis and oriented toward the origin.

    (b) Calculate the integral you chose in part (a).

32. Figure 19.18 shows a cross-section of the earth's magnetic field. Assume that the earth's magnetic and geographic poles coincide. Is the magnetic flux through a horizontal plate, oriented skyward, positive, negative, or zero if the plate is

    (a) At the north pole?   **(b)** At the south pole?

    (c) On the equator?

Figure 19.16

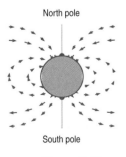

North pole

South pole

Figure 19.18

33. Explain why if $\vec{F}$ has constant magnitude on S and is everywhere normal to S and in the direction of orientation, then

$$\int_S \vec{F} \cdot d\vec{A} = \|\vec{F}\| \cdot \text{Area of } S.$$

34. The z-axis carries a constant electric charge density of λ units of charge per unit length, with $\lambda > 0$, and that $\vec{E}$ is the resulting electric field.

(a) Sketch the electric field, $\vec{E}$, in the xy-plane, given

$$\vec{E}(x, y, z) = 2\lambda \frac{x\vec{i} + y\vec{j}}{x^2 + y^2}.$$

(b) Compute the flux of $\vec{E}$ outward through the cylinder $x^2 + y^2 = R^2$, for $0 \le z \le h$.

35. Let $P(x, y, z)$ be the pressure at the point (x, y, z) in a fluid. Let $\vec{F}(x, y, z) = P(x, y, z)\vec{k}$. Let S be the surface of a body submerged in the fluid. If S is oriented inward, show that $\int_S \vec{F} \cdot d\vec{A}$ is the buoyant force on the body, that is, the force upward on the body due to the pressure of the fluid surrounding it. [Hint: $\vec{F} \cdot d\vec{A} = P(x, y, z)\vec{k} \cdot d\vec{A} = (P(x, y, z) \, d\vec{A}) \cdot \vec{k}$.]

36. A closed surface S encloses a volume W. The function $\rho(x, y, z)$ gives the electrical charge density at points in space. The vector field $\vec{J}(x, y, z)$ gives the electric current density at any point in space and is defined so that the current through a small area $d\vec{A}$ is given by

$$\text{Current through small area} \approx \vec{J} \cdot d\vec{A}.$$

(a) What do the following integrals represent, in terms of electricity?

(i) $\displaystyle\int_W \rho \, dV$ (ii) $\displaystyle\int_S \vec{J} \cdot d\vec{A}$

(b) Using the fact that an electric current through a surface is the rate at which electric charge passes through the surface per unit time, explain why

$$\int_S \vec{J} \cdot d\vec{A} = -\frac{\partial}{\partial t}\left(\int_W \rho \, dV\right).$$

37. A fluid is flowing along a cylindrical pipe of radius a in the $\vec{i}$ direction. The velocity of the fluid at a radial distance r from the center of the pipe is $\vec{v} = u(1 - r^2/a^2)\vec{i}$.

(a) What is the significance of the constant u?
(b) What is the velocity of the fluid at the wall of the pipe?
(c) Find the flux through a circular cross-section of the pipe.

38. A region of 3-space has a temperature which varies from point to point. Let $T(x, y, z)$ be the temperature at a point (x, y, z). Newton's law of cooling says that $\operatorname{grad} T$ is proportional to the heat flow vector field, $\vec{F}$, where $\vec{F}$ points in the direction in which heat is flowing and has magnitude equal to the rate of flow of heat.

(a) Suppose $\vec{F} = k \operatorname{grad} T$ for some constant k. What is the sign of k?
(b) Explain why this form of Newton's law of cooling makes sense.
(c) Let W be a region of space bounded by the surface S. Explain why

$$\begin{array}{l}\text{Rate of heat} \\ \text{loss from } W\end{array} = k \int_S (\operatorname{grad} T) \cdot d\vec{A}.$$

19.2   FLUX INTEGRALS FOR GRAPHS, CYLINDERS, AND SPHERES

In Section 19.1 we computed flux integrals in certain simple cases. In this section we see how to compute flux through surfaces that are graphs of functions, through cylinders, and through spheres.

Flux of a Vector Field through the Graph of $z = f(x, y)$

Suppose S is the graph of the differentiable function $z = f(x, y)$, oriented upward, and that $\vec{F}$ is a smooth vector field. In Section 19.1 we subdivided the surface into small pieces with area vector $\Delta\vec{A}$ and defined the flux of $\vec{F}$ through S as follows:

$$\int_S \vec{F} \cdot d\vec{A} = \lim_{\|\Delta\vec{A}\| \to 0} \sum \vec{F} \cdot \Delta\vec{A}.$$

How do we divide S into small pieces? One way is to use the cross-sections of f with x or y constant and take the patches in a wire frame representation of the surface. So we must calculate the area vector of one of these patches, which is approximately a parallelogram.

The Area Vector of a Parallelogram-Shaped Patch

According to the geometric definition of the cross product on page 674, the vector $\vec{v} \times \vec{w}$ has magnitude equal to the area of the parallelogram formed by $\vec{v}$ and $\vec{w}$ and direction perpendicular to this parallelogram and determined by the right-hand rule. Thus, we have

$$\text{Area vector of parallelogram} = \vec{A} = \vec{v} \times \vec{w}.$$

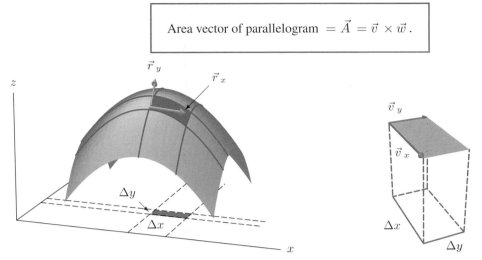

Figure 19.19: Surface showing parameter rectangle and tangent vectors $\vec{r}_x$ and $\vec{r}_y$

Figure 19.20: Parallelogram-shaped patch in the tangent plane to the surface

Consider the patch of surface above the rectangular region with sides Δx and Δy in the xy-plane shown in Figure 19.19. We approximate the area vector, $\Delta \vec{A}$, of this patch by the area vector of the corresponding patch on the tangent plane to the surface. See Figure 19.20 This patch is the parallelogram determined by the vectors $\vec{v}_x$ and $\vec{v}_y$, so its area vector is given by

$$\Delta \vec{A} \approx \vec{v}_x \times \vec{v}_y.$$

To find $\vec{v}_x$ and $\vec{v}_y$, notice that a point on the surface has position vector $\vec{r} = x\vec{i} + y\vec{j} + f(x,y)\vec{k}$. Thus, a cross-section of S with y constant has tangent vector

$$\vec{r}_x = \frac{\partial \vec{r}}{\partial x} = \vec{i} + f_x\vec{k},$$

and a cross-section with x constant has tangent vector

$$\vec{r}_y = \frac{\partial \vec{r}}{\partial y} = \vec{j} + f_y\vec{k}.$$

The vectors $\vec{r}_x$ and $\vec{v}_x$ are parallel because they are both tangent to the surface and in the xz-plane. Since the x-component of $\vec{r}_x$ is $\vec{i}$ and the x-component of $\vec{v}_x$ is $(\Delta x)\vec{i}$, we have $\vec{v}_x = (\Delta x)\vec{r}_x$. Similarly, we have $\vec{v}_y = (\Delta y)\vec{r}_y$. So the upward pointing area vector of the parallelogram is

$$\Delta \vec{A} \approx \vec{v}_x \times \vec{v}_y = (\vec{r}_x \times \vec{r}_y)\,\Delta x\,\Delta y = \left(-f_x\vec{i} - f_y\vec{j} + \vec{k}\right)\Delta x\,\Delta y.$$

This is our approximation for the area vector $\Delta \vec{A}$ on the surface. Replacing $\Delta \vec{A}$, Δx, and Δy by $d\vec{A}$, dx and dy, we write

$$d\vec{A} = \left(-f_x\vec{i} - f_y\vec{j} + \vec{k}\right)dx\,dy.$$

The Flux of $\vec{F}$ through a Surface given by a Graph of $z = f(x, y)$

Suppose the surface S is the part of the graph of $z = f(x, y)$ above a region R in the xy-plane, and suppose S is oriented upward. The flux of $\vec{F}$ through S is

$$\int_S \vec{F} \cdot d\vec{A} = \int_R \vec{F}(x, y, f(x, y)) \cdot \left(-f_x\vec{i} - f_y\vec{j} + \vec{k}\right)dx\,dy.$$

Example 1   Compute $\int_S \vec{F} \cdot d\vec{A}$ where $\vec{F}(x, y, z) = z\vec{k}$ and S is the rectangular plate with corners $(0, 0, 0)$, $(1, 0, 0)$, $(0, 1, 3)$, $(1, 1, 3)$, oriented upward. See Figure 19.21 on page 926.

Figure 19.21: The vector field $\vec{F} = z\vec{k}$ on the rectangular surface S

Solution   We find the equation for the plane S in the form $z = f(x, y)$. Since f is linear, with x-slope equal to 0 and y-slope equal to 3, and $f(0,0) = 0$, we have

$$z = f(x, y) = 0 + 0x + 3y = 3y.$$

Thus, we have

$$d\vec{A} = (-f_x\vec{i} - f_y\vec{j} + \vec{k}) \, dx \, dy = (0\vec{i} - 3\vec{j} + \vec{k}) \, dx \, dy = (-3\vec{j} + \vec{k}) \, dx \, dy.$$

The flux integral is therefore

$$\int_S \vec{F} \cdot d\vec{A} = \int_0^1 \int_0^1 3y\vec{k} \cdot (-3\vec{j} + \vec{k}) \, dx \, dy = \int_0^1 \int_0^1 3y \, dx \, dy = 1.5.$$

Flux of a Vector Field through a Cylindrical Surface

Consider the cylinder of radius R centered on the z-axis illustrated in Figure 19.22 and oriented away from the z-axis. The small patch of surface, or parameter rectangle, in Figure 19.23 has surface area given by

$$\Delta A \approx R \, \Delta\theta \, \Delta z.$$

The outward unit normal $\vec{n}$ points in the direction of $x\vec{i} + y\vec{j}$, so

$$\vec{n} = \frac{x\vec{i} + y\vec{j}}{\|x\vec{i} + y\vec{j}\|} = \frac{R\cos\theta\vec{i} + R\sin\theta\vec{j}}{R} = \cos\theta\vec{i} + \sin\theta\vec{j}.$$

Figure 19.22: Outward oriented cylinder

Figure 19.23: Small patch with area $\Delta\vec{A}$ on surface of a cylinder

Therefore, the area vector of the parameter rectangle is approximated by

$$\Delta \vec{A} = \vec{n} \; \Delta A \approx \left(\cos \theta \vec{i} + \sin \theta \vec{j} \right) R \, \Delta z \, \Delta \theta.$$

Replacing $\Delta \vec{A}$, Δz, and $\Delta \theta$ by $d\vec{A}$, dz, and $d\theta$, we write

$$d\vec{A} = \left(\cos \theta \vec{i} + \sin \theta \vec{j} \right) R \, dz \, d\theta.$$

This gives the following result:

The Flux of a Vector Field through a Cylinder

The flux of $\vec{F}$ through the cylindrical surface S, of radius R and oriented away from the z-axis, is given by

$$\int_S \vec{F} \cdot d\vec{A} = \int_T \vec{F}(R, \theta, z) \cdot \left(\cos \theta \vec{i} + \sin \theta \vec{j} \right) R \, dz \, d\theta,$$

where T is the θz-region corresponding to S.

Example 2   Compute $\int_S \vec{F} \cdot d\vec{A}$ where $\vec{F}(x, y, z) = y\vec{j}$ and S is the part of the cylinder of radius 2 centered on the z-axis with $x \geq 0, y \geq 0$, and $0 \leq z \leq 3$. The surface is oriented toward the z-axis.

Figure 19.24: The vector field $\vec{F} = y\vec{j}$ on the surface S

Solution   In cylindrical coordinates, we have $R = 2$ and $\vec{F} = y\vec{j} = 2 \sin \theta \vec{j}$. Since the orientation of S is toward the z-axis, the flux across S is given by

$$\int_S \vec{F} \cdot d\vec{A} = -\int_T 2 \sin \theta \vec{j} \cdot (\cos \theta \vec{i} + \sin \theta \vec{j}) 2 \, dz \, d\theta = -4 \int_0^{\pi/2} \int_0^3 \sin^2 \theta \, dz \, d\theta = -3\pi.$$

Flux of a Vector Field through a Spherical Surface

Consider the piece of the sphere of radius R centered at the origin, oriented outward illustrated in Figure 19.25 on page 928. The small parameter rectangle in Figure 19.25 has surface area given by

$$\Delta A \approx R^2 \sin \phi \, \Delta \phi \, \Delta \theta.$$

The outward unit normal $\vec{n}$ points in the direction of $\vec{r} = x\vec{i} + y\vec{j} + z\vec{k}$, so

$$\vec{n} = \frac{\vec{r}}{\|\vec{r}\|} = \sin \phi \cos \theta \vec{i} + \sin \phi \sin \theta \vec{j} + \cos \phi \vec{k}.$$

Therefore, the area vector of the parameter rectangle is approximated by

$$\Delta \vec{A} \approx \vec{n}\ \Delta A = \frac{\vec{r}}{\|\vec{r}\|}\ \Delta A = \left(\sin\phi\cos\theta\vec{i} + \sin\phi\sin\theta\vec{j} + \cos\phi\vec{k} \right) R^2 \sin\phi\ \Delta\phi\ \Delta\theta.$$

Replacing $\Delta\vec{A}$, $\Delta\phi$, and $\Delta\theta$ by $d\vec{A}$, $d\phi$, and $d\theta$, we write

$$d\vec{A} = \frac{\vec{r}}{\|\vec{r}\|}\ dA = \left(\sin\phi\cos\theta\vec{i} + \sin\phi\sin\theta\vec{j} + \cos\phi\vec{k} \right) R^2 \sin\phi\ d\phi\ d\theta.$$

Thus, we obtain the following result:

The Flux of a Vector Field through a Sphere

The flux of $\vec{F}$ through the spherical surface S, with radius R and oriented away from the origin, is given by

$$\int_S \vec{F}\cdot d\vec{A} = \int_S \vec{F}\cdot \frac{\vec{r}}{\|\vec{r}\|}\ dA$$

$$= \int_T \vec{F}(R,\theta,\phi)\cdot \left(\sin\phi\cos\theta\vec{i} + \sin\phi\sin\theta\vec{j} + \cos\phi\vec{k} \right) R^2 \sin\phi\ d\phi\ d\theta,$$

where T is the $\theta\phi$-region corresponding to S.

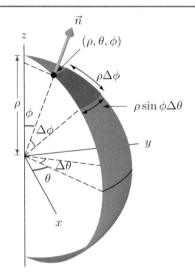

Figure 19.25: Small patch with area $\Delta\vec{A}$ on surface of a sphere

---

Example 3   Find the flux of $\vec{F} = z\vec{k}$ through S, the upper hemisphere of radius 2 centered at the origin, oriented outward.

Solution   The hemisphere S is parameterized by spherical coordinates θ and ϕ, with $0 \le \theta \le 2\pi$ and $0 \le \phi \le \pi/2$. Since $R = 2$ and $\vec{F} = z\vec{k} = 2\cos\phi\vec{k}$, the flux is

$$\int_S \vec{F}\cdot d\vec{A} = \int_S 2\cos\phi\vec{k}\cdot(\sin\phi\cos\theta\vec{i} + \sin\phi\sin\theta\vec{j} + \cos\phi\vec{k})4\sin\phi\ d\phi\ d\theta$$

$$= \int_0^{2\pi}\int_0^{\pi/2} 8\sin\phi\cos^2\phi\ d\phi\ d\theta = 2\pi\left(8\left(\frac{-\cos^3\phi}{3} \right)\bigg|_{\phi=0}^{\pi/2} \right) = \frac{16\pi}{3}.$$

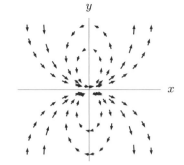

Figure 19.26: The magnetic field of a dipole, $\vec{i}$, at the origin:
$$\vec{B} = \frac{-\vec{i}}{\|\vec{r}\|^3} + \frac{3(\vec{i} \cdot \vec{r})\vec{r}}{\|\vec{r}\|^5}$$

Example 4   The magnetic field $\vec{B}$ due to an *ideal magnetic dipole*, $\vec{\mu}$, located at the origin is defined to be

$$\vec{B}(\vec{r}) = -\frac{\vec{\mu}}{\|\vec{r}\|^3} + \frac{3(\vec{\mu} \cdot \vec{r})\vec{r}}{\|\vec{r}\|^5}.$$

Figure 19.26 shows a sketch of $\vec{B}$ in the plane $z = 0$ for the dipole $\vec{\mu} = \vec{i}$. Notice that $\vec{B}$ is similar to the magnetic field of a bar magnet with its north pole at the tip of the vector $\vec{i}$ and its south pole at the tail of the vector $\vec{i}$.

Compute the flux of $\vec{B}$ outward through the sphere S with center at the origin and radius R.

Solution   Since $\vec{i} \cdot \vec{r} = x$ and $\|\vec{r}\| = R$ on the sphere of radius R, we have

$$\int_S \vec{B} \cdot d\vec{A} = \int_S \left(-\frac{\vec{i}}{\|\vec{r}\|^3} + \frac{3(\vec{i} \cdot \vec{r})\vec{r}}{\|\vec{r}\|^5} \right) \cdot \frac{\vec{r}}{\|\vec{r}\|} \, dA = \int_S \left(-\frac{\vec{i} \cdot \vec{r}}{\|\vec{r}\|^4} + \frac{3(\vec{i} \cdot \vec{r})\|\vec{r}\|^2}{\|\vec{r}\|^6} \right) dA$$

$$= \int_S \frac{2\vec{i} \cdot \vec{r}}{\|\vec{r}\|^4} \, dA = \int_S \frac{2x}{\|\vec{r}\|^4} \, dA = \frac{2}{R^4} \int_S x \, dA,$$

But the sphere S is centered at the origin. Thus, the contribution to the integral from each positive x- value is canceled by the contribution from the corresponding negative x-value; so $\int_S x \, dA = 0$. Therefore,

$$\int_S \vec{B} \cdot d\vec{A} = \frac{2}{R^4} \int_S x \, dA = 0.$$

Exercises and Problems for Section 19.2

Exercises

In Exercises 1–12 compute the flux of the vector field, $\vec{F}$, through the surface, S.

1. $\vec{F} = (x-y)\vec{i} + z\vec{j} + 3x\vec{k}$ and S is the part of the plane $z = x + y$ above the rectangle $0 \leq x \leq 2, 0 \leq y \leq 3$, oriented upward.

2. $\vec{F} = \vec{r}$ and S is the part of the plane $x + y + z = 1$ above the rectangle $0 \leq x \leq 2, 0 \leq y \leq 3$, oriented downward.

3. $\vec{F} = \vec{r}$ and S is the part of the surface $z = x^2 + y^2$ above the disk $x^2 + y^2 \leq 1$, oriented downward.

4. $\vec{F}(x, y, z) = 2x\vec{j} + y\vec{k}$ and S is the part of the surface $z = -y + 1$ above the square $0 \leq x \leq 1, 0 \leq y \leq 1$, oriented upward.

5. $\vec{F} = 3x\vec{i} + y\vec{j} + z\vec{k}$ and S is the part of the surface $z = -2x - 4y + 1$, oriented upward, with (x, y) in the triangle R with vertices $(0, 0), (0, 2), (1, 0)$.

6. $\vec{F} = x\vec{i} + y\vec{j}$ and S is the part of the surface $z = 25 - (x^2 + y^2)$ above the disk of radius 5 centered at the origin, oriented upward.

7. $\vec{F} = \cos(x^2 + y^2)\vec{k}$ and S is as in Exercise 6.

8. $\vec{F} = -y\vec{j} + z\vec{k}$ and S is the part of the surface $z = y^2 + 5$ over the rectangle $-2 \leq x \leq 1, 0 \leq y \leq 1$, oriented upward.

9. $\vec{F}(x, y, z) = -xz\vec{i} - yz\vec{j} + z^2\vec{k}$ and S is the cone $z = \sqrt{x^2 + y^2}$ for $0 \leq z \leq 6$, oriented upward.

10. $\vec{F} = y\vec{i} + \vec{j} - xz\vec{k}$ and S is the surface $y = x^2 + z^2$, with $x^2 + z^2 \leq 1$, oriented in the positive y-direction.

11. $\vec{F} = xz\vec{i} + y\vec{k}$ and S is the hemisphere $x^2 + y^2 + z^2 = 9, z \geq 0$, oriented upward.

12. $\vec{F} = x^2\vec{i} + y^2\vec{j} + z^2\vec{k}$ and S is the oriented triangular surface shown in Figure 19.27.

Figure 19.27

Figure 19.28:

In Exercises 13–14, compute the flux of the vector field, $\vec{F}$, through the cylindrical surface shown in Figure 19.28, oriented away from the z-axis.

13. $\vec{F} = x\vec{i} + y\vec{j}$

14. $\vec{F} = xz\vec{i} + yz\vec{j} + z^3\vec{k}$

In Exercises 15–16 compute the flux of the vector field, $\vec{F}$, through the given spherical surface, S.

15. $\vec{F} = z^2\vec{k}$ and S is the upper hemisphere of the sphere $x^2 + y^2 + z^2 = 25$, oriented away from the origin.

16. $\vec{F} = x\vec{i} + y\vec{j} + z\vec{k}$ and S is the surface of the sphere $x^2 + y^2 + z^2 = a^2$, oriented outward.

Problems

17. Let $\vec{F} = (xze^{yz})\vec{i} + xz\vec{j} + (5 + x^2 + y^2)\vec{k}$. Calculate the flux of $\vec{F}$ through the disk $x^2 + y^2 \leq 1$ in the xy-plane, oriented upward.

18. Let $\vec{H} = (e^{xy} + 3z + 5)\vec{i} + (e^{xy} + 5z + 3)\vec{j} + (3z + e^{xy})\vec{k}$. Calculate the flux of $\vec{H}$ through the square of side 2 with one vertex at the origin, one edge along the positive y-axis, one edge in the xz-plane with $x > 0$, $z > 0$, and the normal $\vec{n} = \vec{i} - \vec{k}$.

19. Electric charge is distributed in space with density (in coulomb/m^3) given in spherical coordinates by

$$\delta(\rho, \phi, \theta) = \begin{cases} \delta_0 \text{ (a constant)} & \rho \leq a \\ 0 & \rho > a. \end{cases}$$

   (a) Describe the charge distribution in words.
   (b) Find the electric field $\vec{E}$ due to δ. Assume that $\vec{E}$ can be written in spherical coordinates as $\vec{E} = E(\rho)\vec{e}_\rho$, where $\vec{e}_\rho$ is the unit outward normal to the sphere of radius ρ. In addition, $\vec{E}$ satisfies Gauss's Law for any simple closed surface S enclosing a volume W:

$$\int_S \vec{E} \cdot d\vec{A} = k \int_W \delta \, dV, \quad k \text{ a constant.}$$

20. Electric charge is distributed in space with density (in coulomb/m^3) given in cylindrical coordinates by

$$\delta(r, \theta, z) = \begin{cases} \delta_0 \text{ (a constant)} & \text{if } r \leq a \\ 0 & \text{if } r > a \end{cases}$$

   (a) Describe the charge distribution in words.
   (b) Find the electric field $\vec{E}$ due to δ. Assume that $\vec{E}$ can be written in cylindrical coordinates as $\vec{E} = E(r)\vec{e}_r$, where $\vec{e}_r$ is the unit outward vector to the cylinder of radius r, and that $\vec{E}$ satisfies Gauss's Law (see Problem 19).

19.3 FLUX INTEGRALS OVER PARAMETERIZED SURFACES

Most of the flux integrals we are likely to encounter can be computed using the methods of Sections 19.1 and 19.2. In this section we briefly consider the general case: how to compute the flux of a smooth vector field $\vec{F}$ through a smooth oriented surface, S, parameterized by

$$\vec{r} = \vec{r}(s, t),$$

for (s, t) in some region R of the parameter space. The method is similar to the one used for graphs in Section 19.2. We consider a parameter rectangle on the surface S corresponding to a rectangular region with sides Δs and Δt in the parameter space. (See Figure 19.29.)

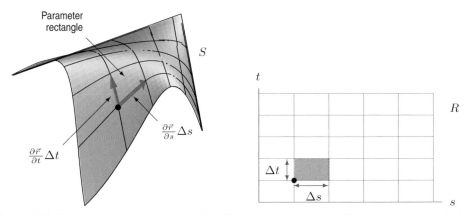

Figure 19.29: Parameter rectangle on the surface S corresponding to a small rectangular region in the parameter space, R

If Δs and Δt are small, the area vector, $\Delta \vec{A}$, of the patch is approximately the area vector of the parallelogram defined by the vectors

$$\vec{r}(s + \Delta s, t) - \vec{r}(s, t) \approx \frac{\partial \vec{r}}{\partial s} \Delta s, \qquad \text{and} \qquad \vec{r}(s, t + \Delta t) - \vec{r}(s, t) \approx \frac{\partial \vec{r}}{\partial t} \Delta t.$$

Thus

$$\Delta \vec{A} \approx \frac{\partial \vec{r}}{\partial s} \times \frac{\partial \vec{r}}{\partial t} \Delta s \, \Delta t.$$

We assume that the vector $\partial \vec{r} / \partial s \times \partial \vec{r} / \partial t$ is never zero and points in the direction of the unit normal orientation vector $\vec{n}$. If the vector $\partial \vec{r} / \partial s \times \partial \vec{r} / \partial t$ points in the opposite direction to $\vec{n}$, we reverse the order of the cross-product. Replacing $\Delta \vec{A}$, Δs, and Δt by $d\vec{A}$, ds, and dt, we write

$$d\vec{A} = \left(\frac{\partial \vec{r}}{\partial s} \times \frac{\partial \vec{r}}{\partial t} \right) ds \, dt.$$

The Flux of a Vector Field through a Parameterized Surface

The flux of a smooth vector field $\vec{F}$ through a smooth oriented surface S parameterized by $\vec{r} = \vec{r}(s, t)$, where (s, t) varies in a parameter region R, is given by

$$\int_S \vec{F} \cdot d\vec{A} = \int_R \vec{F}(\vec{r}(s, t)) \cdot \left(\frac{\partial \vec{r}}{\partial s} \times \frac{\partial \vec{r}}{\partial t} \right) ds \, dt.$$

We choose the parameterization so that $\partial \vec{r} / \partial s \times \partial \vec{r} / \partial t$ is never zero and points in the direction of $\vec{n}$ everywhere.

Example 1   Find the flux of the vector field $\vec{F} = x\vec{i} + y\vec{j}$ through the surface S, oriented downward and given by

$$x = 2s, \quad y = s + t, \quad z = 1 + s - t, \qquad \text{where } 0 \leq s \leq 1, \quad 0 \leq t \leq 1.$$

Solution   Since S is parameterized by

$$\vec{r}(s,t) = 2s\vec{i} + (s+t)\vec{j} + (1+s-t)\vec{k},$$

we have

$$\frac{\partial \vec{r}}{\partial s} = 2\vec{i} + \vec{j} + \vec{k} \quad \text{and} \quad \frac{\partial r}{\partial t} = \vec{j} - \vec{k},$$

so

$$\frac{\partial \vec{r}}{\partial s} \times \frac{\partial \vec{r}}{\partial t} = \begin{vmatrix} \vec{i} & \vec{j} & \vec{k} \\ 2 & 1 & 1 \\ 0 & 1 & -1 \end{vmatrix} = -2\vec{i} + 2\vec{j} + 2\vec{k}.$$

Since the vector $-2\vec{i} + 2\vec{j} + 2\vec{k}$ points upward, we use $2\vec{i} - 2\vec{j} - 2\vec{k}$ for downward orientation. Thus, the flux integral is given by

$$\int_S \vec{F} \cdot d\vec{A} = \int_0^1 \int_0^1 (2s\vec{i} + (s+t)\vec{j}) \cdot (2\vec{i} - 2\vec{j} - 2\vec{k}) \, ds \, dt$$

$$= \int_0^1 \int_0^1 (4s - 2s - 2t) \, ds dt = \int_0^1 \int_0^1 (2s - 2t) \, ds \, dt$$

$$= \int_0^1 \left(s^2 - 2st \Big|_{s=0}^{s=1} \right) dt = \int_0^1 (1 - 2t) \, dt = t - t^2 \Big|_0^1 = 0.$$

Area of a Parameterized Surface

The area ΔA of a small parameter rectangle is the magnitude of its area vector $\Delta \vec{A}$. Therefore,

$$\text{Area of } S = \sum \Delta A = \sum \|\Delta \vec{A}\| \approx \sum \left\| \frac{\partial \vec{r}}{\partial s} \times \frac{\partial \vec{r}}{\partial t} \right\| \Delta s \, \Delta t.$$

Taking the limit as the area of the parameter rectangles tends to zero, we are led to the following expression for the area of S.

The Area of a Parameterized Surface

The area of a surface S which is parameterized by $\vec{r} = \vec{r}(s,t)$, where (s,t) varies in a parameter region R, is given by

$$\int_S dA = \int_R \left\| \frac{\partial \vec{r}}{\partial s} \times \frac{\partial \vec{r}}{\partial t} \right\| ds \, dt.$$

Example 2   Compute the surface area of a sphere of radius a.

Solution   We take the sphere S of radius a centered at the origin and parameterize it with the spherical coordinates ϕ and θ. The parameterization is

$$x = a \sin \phi \cos \theta, \quad y = a \sin \phi \sin \theta, \quad z = a \cos \phi, \quad \text{for} \quad 0 \le \theta \le 2\pi, \quad 0 \le \phi \le \pi.$$

We compute

$$\frac{\partial \vec{r}}{\partial \phi} \times \frac{\partial \vec{r}}{\partial \theta} = (a\cos\phi\cos\theta\vec{i} + a\cos\phi\sin\theta\vec{j} - a\sin\phi\vec{k}) \times (-a\sin\phi\sin\theta\vec{i} + a\sin\phi\cos\theta\vec{j})$$

$$= a^2(\sin^2\phi\cos\theta\vec{i} + \sin^2\phi\sin\theta\vec{j} + \sin\phi\cos\phi\vec{k})$$

and so

$$\left\|\frac{\partial \vec{r}}{\partial \phi} \times \frac{\partial \vec{r}}{\partial \theta}\right\| = a^2\sin\phi.$$

Thus, we see that the surface area of the sphere S is given by

$$\text{Surface area} = \int_S dA = \int_R \left\|\frac{\partial \vec{r}}{\partial \phi} \times \frac{\partial \vec{r}}{\partial \theta}\right\| d\phi d\theta = \int_{\phi=0}^{\pi}\int_{\theta=0}^{2\pi} a^2\sin\phi \, d\theta \, d\phi = 4\pi a^2.$$

Exercises and Problems for Section 19.3

Exercises

In Exercises 1–5 compute the flux of the vector field, $\vec{F}$, through the parameterized surface, S.

1. $\vec{F} = z\vec{k}$ and S is oriented toward the z-axis and given, for $0 \le s \le 1$, $0 \le t \le 1$, by

$$x = s + t, \quad y = s - t, \quad z = s^2 + t^2.$$

2. $\vec{F} = y\vec{i} + x\vec{j}$ and S is oriented away from the z-axis and given, for $0 \le s \le \pi$, $0 \le t \le 1$, by

$$x = 3\sin s, \quad y = 3\cos s, \quad z = t + 1.$$

3. $\vec{F} = z\vec{i} + x\vec{j}$ and S is oriented upward and given, for $0 \le s \le 1$, $1 \le t \le 3$ by

$$x = s^2, \quad y = 2s + t^2, \quad z = 5t.$$

4. $\vec{F} = -\frac{2}{x}\vec{i} + \frac{2}{y}\vec{j}$ and S is oriented upward and parameterized by a and θ, where, for $1 \le a \le 3$, $0 \le \theta \le \pi$,

$$x = a\cos\theta, \quad y = a\sin\theta, \quad z = \sin a^2.$$

5. $\vec{F} = x^2 y^2 z \vec{k}$ and S is the cone $\sqrt{x^2 + y^2} = z$, with $0 \le z \le R$, oriented downward. Parameterize the cone using cylindrical coordinates. (See Figure 19.30.)

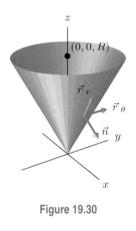

Figure 19.30

Problems

6. Find the area of the ellipse S on the plane $2x + y + z = 2$ cut out by the circular cylinder $x^2 + y^2 = 2x$. (See Figure 19.31.)

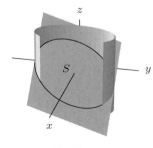

Figure 19.31

7. Evaluate $\int_S \vec{F} \cdot d\vec{A}$, where $\vec{F} = (bx/a)\vec{i} + (ay/b)\vec{j}$ and S is the elliptic cylinder oriented away from the z-axis, and given by $x^2/a^2 + y^2/b^2 = 1$, $|z| \le c$, where a, b, c are positive constants.

8. Consider the surface S formed by rotating the graph of $y = f(x)$ around the x-axis between $x = a$ and $x = b$. Assume that $f(x) \ge 0$ for $a \le x \le b$. Show that the surface area of S is $2\pi \int_a^b f(x)\sqrt{1 + f'(x)^2} \, dx$.

As we remarked in Section 19.1, the limit defining a flux integral might not exist if we subdivide the surface in the wrong way. One way to get around this is to take the formula for a flux integral over a parameterized surface that we have developed in this section and to use it as the *definition* of the flux integral. In Problems 9–12 we explore how this works.

9. Use a parameterization to verify the formula for a flux integral over a surface graph on page 925.

10. Use a parameterization to verify the formula for a flux integral over a cylindrical surface on page 927.

11. Use a parameterization to verify the formula for a flux integral over a spherical surface on page 928.

12. One problem with defining the flux integral using a parameterization is that the integral appears to depend

on the choice of parameterization. However, the flux through a surface ought not to depend on how the surface is parameterized. Suppose that the surface S has two parameterizations, $\vec{r} = \vec{r}(s, t)$ for (s, t) in the region R of st-space, and also $\vec{r} = r(u, v)$ for (u, v) in the region T in uv-space, and suppose that the two parameterizations are related by the change of variables

$$u = u(s, t) \quad v = v(s, t).$$

Suppose that the Jacobian determinant $\partial(u, v)/\partial(s, t)$ is positive at every point (s, t) in R. Use the change of variables formula for double integrals on page 823 to show that computing the flux integral using either parameterization gives the same result.

CHAPTER SUMMARY (see also Ready Reference at the end of the book)

- **Flux Integrals**
  Oriented surfaces, definition of flux through a surface, definition of flux integral as a limit of a Riemann sum

- **Calculating flux integrals** over surfaces
  Graphs: $d\vec{A} = (-f_x\vec{i} - f_y\vec{j} + \vec{k})\,dx\,dy$
  Cylinders: $d\vec{A} = (\cos\theta\vec{i} + \sin\theta\vec{j})R\,dz\,d\theta$

  Spheres: $d\vec{A} = (\sin\phi\cos\theta\vec{i} + \sin\phi\sin\theta\vec{j} + \cos\phi\vec{k})R^2\,d\phi\,d\theta$
  Surfaces parameterized by $\vec{r}(s, t)$: $d\vec{A} = (\partial\vec{r}/\partial s) \times (\partial\vec{r}/\partial t)\,ds\,dt$

- **Area of parameterized surface**
  $dA = \|(\partial\vec{r}/\partial s) \times (\partial\vec{r}/\partial t)\|\,ds\,dt$

REVIEW EXERCISES AND PROBLEMS FOR CHAPTER NINETEEN

Exercises

Find the flux of the vector fields in Exercises 1–3 out of the closed box $0 \le x \le 1, 0 \le y \le 2, 0 \le z \le 3$.

1. $\vec{F} = 3\vec{i} + 2\vec{j} + \vec{k}$

2. $\vec{G} = x\vec{i}$

3. $\vec{H} = zx\vec{k}$

In Exercises 4–6, find the flux $\vec{H} = 2\vec{i} + 3\vec{k} - 5\vec{k}$ through the surface.

4. S is the cylinder $x^2 + y^2 = 1$, closed at the ends by the planes $z = 0$ and $z = 1$ and oriented outward.

5. S is the disk of radius 1 in the plane $x = 2$ oriented in the positive x-direction.

6. S is the disk of radius 1 in the plane $x + y + z = 1$ oriented in upward.

Find the flux of $\vec{F} = \vec{r} = x\vec{i} + y\vec{j} + z\vec{k}$ through the surfaces in Exercises 7–9.

7. Through a sphere of radius π centered at the origin.

8. Through a square, oriented upward, with corners $(0, 0, 2), (3, 0, 2), (3, 3, 2), (0, 3, 2)$.

9. Through a circular disk of radius 3 in the plane $x = 5$, centered at $(5, 2, 4)$ and oriented in the positive x-direction.

Find the flux of $\vec{G} = 2\vec{j} + 3\vec{k}$ through the surface in Exercises 10–13.

10. The unit sphere oriented outward.

11. A square of side 7 perpendicular to the y-axis, oriented in the negative y-direction.

12. A circular disk of radius 5 in the plane $z = 2$, centered at $(1, 3, 2)$, and oriented upward.

13. The part of the plane $z = y$ above the square $0 \le x \le 1$, $0 \le y \le 1$, oriented upward.

Find the flux of $\vec{F} = 2\vec{i} + 3\vec{j} + z\vec{k}$ through the surfaces in Exercises 14–16.

14. A disk of radius 5 in the plane $x = 4$, oriented away from the origin.

15. A square of side 5 in the plane $z = 4$, oriented upward.

16. A disk of radius 5 in the plane $x + y = 1$, oriented away from the origin.

In Exercises 17–28, calculate the flux of the vector field through the given surface.

17. $\vec{F} = 7\vec{i} + 6\vec{j} + 5\vec{k}$ and S is a disk of radius 2 in the yz-plane, centered at the origin and oriented in the positive x-direction.

18. $\vec{F} = x\vec{i} + 2y\vec{j} + 3z\vec{k}$ and S is a square of side 2 in the plane $y = 3$, oriented in the positive y-direction.

19. $\vec{F} = 7\vec{i} + 6\vec{j} + 5\vec{k}$ and S is a sphere of radius π centered at the origin.

20. $\vec{F} = -5\vec{r}$ through the sphere of radius 2 centered at the origin.

21. $\vec{F} = x\vec{i} + y\vec{j} + (z^2 + 3)\vec{k}$ and S is the rectangle $z = 4$, $0 \le x \le 2$, $0 \le y \le 3$, oriented in the positive z-direction.

22. $\vec{F} = z\vec{i} + y\vec{j} + 2x\vec{k}$ and S is the rectangle $z = 4$, $0 \le x \le 2$, $0 \le y \le 3$, oriented in the positive z-direction.

23. $\vec{F} = (x + \cos z)\vec{i} + y\vec{j} + 2x\vec{k}$ and S is the rectangle $x = 2$, $0 \le y \le 3$, $0 \le z \le 4$, oriented in the positive x-direction.

24. $\vec{F} = x^2\vec{i} + (x + e^y)\vec{j} - \vec{k}$, and S is the rectangle $y = -1$, $0 \le x \le 2$, $0 \le z \le 4$, oriented in the negative y-direction.

25. $\vec{F} = (5 + xy)\vec{i} + z\vec{j} + yz\vec{k}$ and S is the 2×2 square plate in the yz-plane centered at the origin, oriented in the positive x-direction.

26. $\vec{F} = x\vec{i} + y\vec{j}$ and S is the surface of a closed cylinder of radius 2 and height 3 centered on the z-axis with its base in the xy-plane.

27. $\vec{F} = -y\vec{i} + x\vec{j} + z\vec{k}$ and S is the surface of a closed cylinder of radius 1 centered on the z-axis with base in the plane $z = -1$ and top in the plane $z = 1$.

28. $\vec{F} = x^2\vec{i} + y^2\vec{j} + z\vec{k}$ and S is the cone $z = \sqrt{x^2 + y^2}$, oriented upward with $x^2 + y^2 \le 1$, $x \ge 0$, $y \ge 0$.

Problems

29. Arrange the following flux integrals, $\int_{S_i} \vec{F} \cdot d\vec{A}$, with $i = 1, 2, 3, 4$, in ascending order if $\vec{F} = -\vec{i} - \vec{j} + \vec{k}$ and if the S_i are the following surfaces:

- S_1 is a horizontal square of side 1 with one corner at $(0, 0, 2)$, above the first quadrant of the xy-plane, oriented upward.
- S_2 is a horizontal square of side 1 with one corner at $(0, 0, 3)$, above the third quadrant of the xy-plane, oriented upward.
- S_3 is a square of side $\sqrt{2}$ in the xz-plane with one corner at the origin, one edge along the positive x-axis, one along the negative z-axis, oriented in the negative y-direction.
- S_4 is a square of side $\sqrt{2}$ with one corner at the origin, one edge along the positive y-axis, one corner at $(1, 0, 1)$, oriented upward.

30. (a) Describe in words the vector field $\vec{F} = c\vec{r}$, where $\vec{r} = x\vec{i} + y\vec{j} + z\vec{k}$ and $c > 0$ is a constant.
(b) Explain how you can tell the sign of the flux of $\vec{F}$ out of the sphere of radius 3 centered at the origin without calculating it.
(c) Calculate the flux described in part (b).

In Problems 31–33, find the flux of the vector field through the surface.

31. $\vec{D}(\vec{r}) = \dfrac{Q}{4\pi\|\vec{r}\|^3}\vec{r}$, where Q is a constant, and S is a sphere of radius R centered at the origin.

32. $\vec{D}(\vec{r}) = \dfrac{\rho_0}{3}\vec{r}$, where ρ_0 is a constant, and S is a sphere of radius R centered at the origin.

33. $\vec{E} = \dfrac{\rho}{2\epsilon_0}(x\vec{i} + y\vec{j})$, where ρ and ϵ_0 are constants, and S is a cylinder (without ends) of height h and radius R

centered on the z-axis and oriented outward. Would the answer change if the cylinder included the ends?

34. Water is flowing down a cylindrical pipe of radius 2 cm and that the speed is $(3 - (3/4)r^2)$ cm/sec at a distance r cm from the center of the pipe. Find the flux through the circular cross-section of the pipe, oriented so that the flow is positive.

35. Suppose that $\vec{E}$ is a uniform electric field on 3-space, so $\vec{E}(x, y, z) = a\vec{i} + b\vec{j} + c\vec{k}$, for all points (x, y, z), where a, b, c are constants. Show, with the aid of symmetry, that the flux of $\vec{E}$ through each of the following closed surfaces S is zero:

(a) S is the cube bounded by the planes $x = \pm 1$, $y = \pm 1$, and $z = \pm 1$
(b) S is the sphere $x^2 + y^2 + z^2 = 1$
(c) S is the cylinder bounded by $x^2 + y^2 = 1$, $z = 0$, and $z = 2$

36. According to Coulomb's Law the electrostatic field $\vec{E}$, at the point with position vector $\vec{r}$ in 3-space, due to a charge q at the origin is given by

$$\vec{E}(\vec{r}) = q\frac{\vec{r}}{\|\vec{r}\|^3}.$$

Let S_a be the outward oriented sphere in 3-space of radius $a > 0$ and center at the origin. Show that the flux of the resulting electric field $\vec{E}$ through the surface S_a is equal to $4\pi q$, for any radius a. This is Gauss's Law for a single point charge.

37. An infinitely long straight wire lying along the z-axis carries an electric current I flowing in the $\vec{k}$ direction. Ampère's Law in magnetostatics says that the current gives rise to a magnetic field $\vec{B}$ given by

$$\vec{B}(x,y,z) = \frac{I}{2\pi}\frac{-y\vec{i} + x\vec{j}}{x^2 + y^2}.$$

(a) Sketch the field $\vec{B}$ in the xy-plane.

(b) Suppose S_1 is a disk with center at $(0,0,h)$, radius a, and parallel to the xy-plane, oriented in the $\vec{k}$ direction. What is the flux of $\vec{B}$ through S_1? Is your answer reasonable?

(c) Suppose S_2 is the rectangle given by $x = 0$, $a \le y \le b$, $0 \le z \le h$, and oriented in the $-\vec{i}$ direction. What is the flux of $\vec{B}$ through S_2? Does your answer seem reasonable?

38. An ideal electric dipole in electrostatics is characterized by its position in 3-space and its dipole moment vector $\vec{p}$. The electric field $\vec{D}$, at the point with position vector $\vec{r}$, of an ideal electric dipole located at the origin with dipole moment $\vec{p}$ is given by

$$\vec{D}(\vec{r}) = 3\frac{(\vec{r}\cdot\vec{p})\vec{r}}{\|\vec{r}\|^5} - \frac{\vec{p}}{\|\vec{r}\|^3}.$$

Assume $\vec{p} = p\vec{k}$, so the dipole points in the $\vec{k}$ direction and has magnitude p.

(a) What is the flux of $\vec{D}$ through a sphere S with center at the origin and radius $a > 0$?

(b) The field $\vec{D}$ is a useful approximation to the electric field $\vec{E}$ produced by two "equal and opposite" charges, q at $\vec{r}_2$ and $-q$ at $\vec{r}_1$, where the distance $\|\vec{r}_2 - \vec{r}_1\|$ is small. The dipole moment of this configuration of charges is defined to be $q(\vec{r}_2 - \vec{r}_1)$. Gauss's Law in electrostatics says that the flux of $\vec{E}$ through S is equal to 4π times the total charge enclosed by S. What is the flux of $\vec{E}$ through S if the charges at $\vec{r}_1$ and $\vec{r}_2$ are enclosed by S? How does this compare with your answer for the flux of $\vec{D}$ through S if $\vec{p} = q(\vec{r}_2 - \vec{r}_1)$?

CAS Challenge Problems

39. Let S be the part of the ellipsoid $x^2 + y^2 + 2z^2 = 1$ lying above the rectangle $-1/2 \le x \le 1/2$, $-1/2 \le y \le 1/2$, oriented upward. For each vector field (a)–(c), say whether you expect $\int_S \vec{F} \cdot d\vec{A}$ to be positive, negative, or zero. Then evaluate the integral exactly using a computer algebra system and find numerical approximations for your answers. Describe and explain what you notice.

(a) $\vec{F} = x\vec{i}$

(b) $\vec{F} = (x+1)\vec{i}$

(c) $\vec{F} = y\vec{j}$

40. Let $\vec{F} = (z+4)\vec{k}$, and let S be the surface with normal pointing in the direction of the negative y-axis parameterized, for $0 \le s \le 2$, $0 \le t \le 2\pi$, by

$$\vec{r}(s,t) = s^2\cos t\,\vec{i} + s\vec{j} + s^2\sin t\,\vec{k}.$$

(a) Sketch S and, without evaluating the integral, say whether $\int_S \vec{F} \cdot d\vec{A}$ is positive, negative, or zero. Give a geometric explanation for your answer.

(b) Evaluate the flux integral. Does it agree with your answer to part (a)?

CHECK YOUR UNDERSTANDING

Are the statements in Problems 1–16 true or false? Give reasons for your answer.

1. The value of a flux integral is a scalar.

2. The area vector $\vec{A}$ of a flat, oriented surface is parallel to the surface.

3. If S is the unit sphere centered at the origin, oriented outward and the flux integral $\int_S \vec{F} \cdot d\vec{A}$ is zero, then $\vec{F} = \vec{0}$.

4. The flux of the vector field $\vec{F} = \vec{i}$ through the plane $x = 0$, with $0 \le y \le 1$, $0 \le z \le 1$, oriented in the $\vec{i}$ direction is positive.

5. If S is the unit sphere centered at the origin, oriented outward and $\vec{F} = x\vec{i} + y\vec{j} + z\vec{k} = \vec{r}$, then the flux integral $\int_S \vec{F} \cdot d\vec{A}$ is positive.

6. If S is the cube bounded by the six planes $x = \pm 1$, $y = \pm 1$, $z = \pm 1$, oriented outward, and $\vec{F} = \vec{k}$, then

$$\int_S \vec{F} \cdot d\vec{A} = 0.$$

7. If S is an oriented surface in 3-space, and $-S$ is the same surface, but with the opposite orientation, then $\int_S \vec{F} \cdot d\vec{A} = -\int_{-S} \vec{F} \cdot d\vec{A}$.

8. If S_1 is a rectangle with area 1 and S_2 is a rectangle with area 2, then $2\int_{S_1} \vec{F} \cdot d\vec{A} = \int_{S_2} \vec{F} \cdot d\vec{A}$.

9. If $\vec{F} = 2\vec{G}$, then $\int_S \vec{F} \cdot d\vec{A} = 2\int_S \vec{G} \cdot d\vec{A}$.

10. If $\int_S \vec{F} \cdot d\vec{A} > \int_S \vec{G} \cdot d\vec{A}$ then $\|\vec{F}\| > \|\vec{G}\|$ at all points on the surface S.

11. If S is parameterized by $\vec{r} = \vec{r}(s,t)$ and $\vec{r} \cdot \vec{F} = 0$ at each point of S, then $\int_S \vec{F} \cdot d\vec{A} = 0$.

12. If S is the part of the graph of f lying above $a \le x \le b$, $c \le y \le d$, then S has surface area $\int_a^b \int_c^d \sqrt{f_x^2 + f_y^2 + 1}\,dx\,dy$.

13. If $\vec{A}(x, y)$ is the area vector for $z = f(x, y)$ oriented upward and $\vec{B}(x, y)$ is the area vector for $z = -f(x, y)$ oriented upward, then $\vec{A}(x, y) = -\vec{B}(x, y)$.

14. If S is the sphere $x^2 + y^2 + z^2 = 1$ oriented outward and $\int_S \vec{F} \cdot d\vec{A} = 0$, then $\vec{F}(x, y, z)$ is perpendicular to $x\vec{i} + y\vec{j} + z\vec{k}$ at every point of S.

15. For any parameterization of the surface $x^2 - y^2 + z^2 = 6$, $d\vec{A}$ at $(1, 2, 3)$ is a multiple of $(2\vec{i} - 4\vec{j} + 6\vec{k})dx\,dy$.

16. If you parameterize the plane $3x + 4y + 5z = 7$, then there is a constant c such that, at any point (x, y, z), $d\vec{A} = c(3\vec{i} + 4\vec{j} + 5\vec{k})dx\,dy$.

PROJECTS FOR CHAPTER NINETEEN

1. **Gauss' Law Applied to a Charged Wire and a Charged Sheet**

   Gauss' Law states that the flux of an electric field through a closed surface, S, is proportional to the quantity of charge, q, enclosed within S. That is,

   $$\int_S \vec{E} \cdot d\vec{A} = kq.$$

   In this project we use Gauss' Law to calculate the electric field of a uniformly charged wire in part (a) and a flat sheet of charge in part (b).

   (a) Consider the electric field due to an infinitely long, straight, uniformly charged wire. (There is no current running through the wire—all charges are fixed.) Assuming that the wire is infinitely long means that we can assume that the electric field is perpendicular to any cylinder that has the wire as an axis and that the magnitude of the field is constant on any such cylinder. Denote by E_r the magnitude of the electric field due to the wire on a cylinder of radius r. (See Figure 19.32.)

   Imagine a closed surface S made up of two cylinders, one of radius a and one of larger radius b, both coaxial with the wire, and the two washers that cap the ends. (See Figure 19.33.) The outward orientation of S means that a normal on the outer cylinder points away from the wire and a normal on the inner cylinder points toward the wire.

     (i) Explain why the flux of $\vec{E}$, the electric field, through the washers is 0.

    (ii) Explain why Gauss' Law implies that the flux through the inner cylinder is the same as the flux through the outer cylinder. [Hint: The charge on the wire is not inside the surface S].

   (iii) Use part (ii) to show that $E_b/E_a = a/b$.

   (iv) Explain why part (iii) shows that the strength of the field due to an infinitely long uniformly charged wire is proportional to $1/r$.

| Figure 19.32 | Figure 19.33 | Figure 19.34 |

   (b) Now consider an infinite flat sheet uniformly covered with charge. As in part (a), symmetry shows that the electric field $\vec{E}$ is perpendicular to the sheet and has the same magnitude at all points that are the same distance from the sheet. Use Gauss' Law to explain why, on any one side of the sheet, the electric field is the same at all points in space off the sheet. [Hint: Consider the flux through the box with sides parallel to the sheet shown in Figure 19.34.]

2. Flux Across a Cylinder Due to a Point Charge: Obtaining Gauss' Law from Coulomb's Law

An electric charge q is placed at the origin in 3-space. The induced electric field $\vec{E}\,(\vec{r}\,)$ at the point with position vector $\vec{r}$ is given by Coulomb's Law, which says

$$\vec{E}\,(\vec{r}\,) = q\frac{\vec{r}}{\|\vec{r}\,\|^3}, \qquad \vec{r} \neq \vec{0}.$$

In this project, Gauss' Law is obtained for a cylinder enclosing a point charge by direct calculation from Coulomb's Law.

(a) Let S be the open cylinder of height $2H$ and radius R given by $x^2 + y^2 = R^2$, $-H \leq z \leq H$, oriented outward.

  (i) Show that the flux of $\vec{E}$, the electric field, through S is given by

$$\int_S \vec{E} \cdot d\vec{A} = 4\pi q\frac{H}{\sqrt{H^2 + R^2}}.$$

  (ii) What are the limits of the flux $\int_S \vec{E} \cdot d\vec{A}$ if
   - $H \to 0$ or $H \to \infty$ when R is fixed?
   - $R \to 0$ or $R \to \infty$ when H is fixed?

(b) Let T be the outward oriented, closed cylinder of height $2H$ and radius R whose curved side is given by $x^2 + y^2 = R^2$, $-H \leq z \leq H$, whose top is given by $z = H$, $x^2 + y^2 \leq R^2$, and bottom by $z = -H$, $x^2 + y^2 \leq R^2$. Use part (a) to show that the flux of the electric field, $\vec{E}$, through T is given by

$$\int_T \vec{E} \cdot d\vec{A} = 4\pi q.$$

Notice that this is Gauss' Law. In particular, the flux is independent of both the height, H, and radius, R, of the cylinder.

Chapter Twenty

CALCULUS OF VECTOR FIELDS

We have seen two ways of integrating vector fields in three dimensions: along curves and over surfaces. Now we look at two ways of differentiating them. If we view the vector field as the velocity field of a fluid flow, then one method of differentiation (the divergence) tells us about the net strength of outflow from a point, and the other method (the curl) tells us about the strength of rotation around a point. Each method fits together with one of the ways of integrating to form a vector analogue of the Fundamental Theorem of Calculus: the Divergence Theorem relating the divergence to flux, and Stokes' theorem relating the curl to circulation around a closed path.

20.1 THE DIVERGENCE OF A VECTOR FIELD

Imagine that the vector fields in Figures 20.1 and 20.2 are velocity vector fields describing the flow of a fluid.[1] Figure 20.1 suggests outflow from the origin; for example, it could represent the expanding cloud of matter in the big bang theory of the origin of the universe. We say that the origin is a *source*. Figure 20.2 suggests flow into the origin; in this case we say that the origin is a *sink*.

In this section we will use the flux out of a closed surface surrounding a point to measure the outflow per unit volume there, also called the *divergence*, or *flux density*.

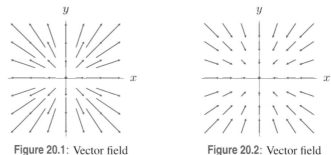

Figure 20.1: Vector field
showing a source

Figure 20.2: Vector field
showing a sink

Definition of Divergence

To measure the outflow per unit volume of a vector field at a point, we calculate the flux out of a small sphere centered at the point, divide by the volume enclosed by the sphere, then take the limit of this flux to volume ratio as the sphere contracts around the point.

Geometric Definition of Divergence

The **divergence**, or **flux density**, of a smooth vector field $\vec{F}$, written **div** $\vec{F}$, is a scalar-valued function defined by

$$\operatorname{div} \vec{F}(x, y, z) = \lim_{\text{Volume} \to 0} \frac{\int_S \vec{F} \cdot d\vec{A}}{\text{Volume of } S}.$$

Here S is a sphere centered at (x, y, z), oriented outward, that contracts down to (x, y, z) in the limit.

The limit can be computed using other shapes as well, such as the cubes in Example 2.

Cartesian Coordinate Definition of Divergence

If $\vec{F} = F_1 \vec{i} + F_2 \vec{j} + F_3 \vec{k}$, then

$$\operatorname{div} \vec{F} = \frac{\partial F_1}{\partial x} + \frac{\partial F_2}{\partial y} + \frac{\partial F_3}{\partial z}.$$

Example 1   Calculate the divergence of $\vec{F}(\vec{r}) = \vec{r}$ at the origin

(a) Using the geometric definition.

(b) Using the Cartesian coordinate definition.

---

[1] Although not all vector fields represent physically realistic fluid flows, it is useful to think of them this way.

Solution
    (a) Using the method of Example 4 on page 920, you can calculate the flux of $\vec{F}$ out of the sphere of radius a, centered at the origin; it is $4\pi a^3$. So we have

$$\operatorname{div}\vec{F}(0,0,0) = \lim_{a \to 0}\frac{\text{Flux}}{\text{Volume}} = \lim_{a \to 0}\frac{4\pi a^3}{\frac{4}{3}\pi a^3} = \lim_{a \to 0}3 = 3.$$

    (b) In coordinates, $\vec{F}(x,y,z) = x\vec{i} + y\vec{j} + z\vec{k}$, so

$$\operatorname{div}\vec{F} = \frac{\partial}{\partial x}(x) + \frac{\partial}{\partial y}(y) + \frac{\partial}{\partial z}(z) = 1 + 1 + 1 = 3.$$

The next example shows that the divergence can be negative if there is net inflow to a point.

Example 2
    (a) Using the geometric definition, find the divergence of $\vec{v} = -x\vec{i}$ at: (i) $(0,0,0)$ (ii) $(2,2,0)$.
    (b) Confirm that the coordinate definition gives the same results.

Solution
    (a) (i) The vector field $\vec{v} = -x\vec{i}$ is parallel to the x axis, as shown in Figure 20.3. To compute the flux density at $(0,0,0)$, we use a cube S_1, centered at the origin with edges parallel to the axes, of length $2c$. Then the flux through the faces perpendicular to the y- and z-axes is zero (because the vector field is parallel to these faces). On the faces perpendicular to the x-axis, the vector field and the outward normal are parallel but point in opposite directions. On the face at $x = c$, we have

$$\vec{v} \cdot \Delta\vec{A} = -c\,\|\Delta\vec{A}\|.$$

On the face at $x = -c$, the dot product is still negative, and

$$\vec{v} \cdot \Delta\vec{A} = -c\,\|\Delta\vec{A}\|.$$

Therefore, the flux through the cube is given by

$$\int_{S_1}\vec{v} \cdot d\vec{A} = \int_{\text{Face }x=-c}\vec{v} \cdot d\vec{A} + \int_{\text{Face }x=c}\vec{v} \cdot d\vec{A}$$
$$= -c \cdot \text{Area of one face} + (-c) \cdot \text{Area of other face} = -2c(2c)^2 = -8c^3.$$

Thus,

$$\operatorname{div}\vec{v}(0,0,0) = \lim_{\text{Volume} \to 0}\frac{\int_S \vec{v} \cdot d\vec{A}}{\text{Volume of cube}} = \lim_{c \to 0}\left(\frac{-8c^3}{(2c)^3}\right) = -1.$$

Since the vector field points inward toward the yz-plane, it makes sense that the divergence is negative at the origin.

    (ii) Take S_2 to be a cube as before, but centered this time at the point $(2,2,0)$. See Figure 20.3. As before, the flux through the faces perpendicular to the y- and z-axes is zero. On the face at $x = 2 + c$,

$$\vec{v} \cdot \Delta\vec{A} = -(2+c)\,\|\Delta\vec{A}\|.$$

On the face at $x = 2 - c$ with outward normal, the dot product is positive, and

$$\vec{v} \cdot \Delta\vec{A} = (2-c)\,\|\Delta\vec{A}\|.$$

Therefore, the flux through the cube is given by

$$\int_{S_2}\vec{v} \cdot d\vec{A} = \int_{\text{Face }x=2-c}\vec{v} \cdot d\vec{A} + \int_{\text{Face }x=2+c}\vec{v} \cdot d\vec{A}$$
$$= (2-c) \cdot \text{Area of one face} - (2+c) \cdot \text{Area of other face} = -2c(2c)^2 = -8c^3.$$

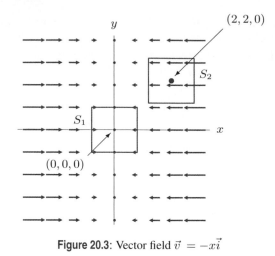

Figure 20.3: Vector field $\vec{v} = -x\vec{i}$

Then, as before,

$$\operatorname{div}\vec{v}\,(2,2,0) = \lim_{\text{Volume}\to 0}\frac{\int_S \vec{v}\cdot d\vec{A}}{\text{Volume of cube}} = \lim_{c\to 0}\left(\frac{-8c^3}{(2c)^3}\right) = -1.$$

Although the vector field is flowing away from the point $(2,2,0)$ on the left, this outflow is smaller in magnitude than the inflow on the right, so the the net outflow is negative.

(b) Since $\vec{v} = -x\vec{i} + 0\vec{j} + 0\vec{k}$, the formula gives

$$\operatorname{div}\vec{v} = \frac{\partial}{\partial x}(-x) + \frac{\partial}{\partial y}(0) + \frac{\partial}{\partial z}(0) = -1 + 0 + 0 = -1.$$

Why Do the Two Definitions of Divergence Give the Same Result?

The geometric definition defines $\operatorname{div}\vec{F}$ as the flux density of $\vec{F}$. To see why the coordinate definition is also the flux density, imagine computing the flux out of a small box-shaped surface S at (x_0, y_0, z_0), with sides of length Δx, Δy, and Δz parallel to the axes. On S_1 (the back face of the box shown in Figure 20.4, where $x = x_0$), the outward normal is in the negative x-direction, so $d\vec{A} = -dy\,dz\,\vec{i}$. Assuming $\vec{F}$ is approximately constant on S_1, we have

$$\int_{S_1}\vec{F}\cdot d\vec{A} = \int_{S_1}\vec{F}\cdot(-\vec{i})\,dy\,dz \approx -F_1(x_0, y_0, z_0)\int_{S_1}dy\,dz$$
$$= -F_1(x_0, y_0, z_0)\cdot\text{Area of }S_1 = -F_1(x_0, y_0, z_0)\,\Delta y\,\Delta z.$$

On S_2, the face where $x = x_0 + \Delta x$, the outward normal points in the positive x-direction, so $d\vec{A} = dy\,dz\,\vec{i}$. Therefore,

$$\int_{S_2}\vec{F}\cdot d\vec{A} = \int_{S_2}\vec{F}\cdot\vec{i}\,dy\,dz \approx F_1(x_0 + \Delta x, y_0, z_0)\int_{S_2}dy\,dz$$
$$= F_1(x_0 + \Delta x, y_0, z_0)\cdot\text{Area of }S_2 = F_1(x_0 + \Delta x, y_0, z_0)\,\Delta y\,\Delta z.$$

Figure 20.4: Box used to find $\operatorname{div}\vec{F}$ at (x_0, y_0, z_0)

Thus

$$\int_{S_1} \vec{F} \cdot d\vec{A} + \int_{S_2} \vec{F} \cdot d\vec{A} \approx F_1(x_0 + \Delta x, y_0, z_0)\Delta y \Delta z - F_1(x_0, y_0, z_0)\Delta y \Delta z$$

$$= \frac{F_1(x_0 + \Delta x, y_0, z_0) - F_1(x_0, y_0, z_0)}{\Delta x}\Delta x \Delta y \Delta z$$

$$\approx \frac{\partial F_1}{\partial x}\Delta x \Delta y \Delta z.$$

By an analogous argument, the contribution to the flux from S_3 and S_4 (the surfaces perpendicular to the y-axis) is approximately

$$\frac{\partial F_2}{\partial y}\Delta x \, \Delta y \, \Delta z,$$

and the contribution to the flux from S_5 and S_6 is approximately

$$\frac{\partial F_3}{\partial z}\Delta x \, \Delta y \, \Delta z.$$

Thus, adding these contributions, we have

$$\text{Total flux through } S \approx \frac{\partial F_1}{\partial x}\Delta x \, \Delta y \, \Delta z + \frac{\partial F_2}{\partial y}\Delta x \, \Delta y \, \Delta z + \frac{\partial F_3}{\partial z}\Delta x \, \Delta y \, \Delta z.$$

Since the volume of the box is $\Delta x \, \Delta y \, \Delta z$, the flux density is

$$\frac{\text{Total flux through } S}{\text{Volume of box}} \approx \frac{\dfrac{\partial F_1}{\partial x}\Delta x \Delta y \Delta z + \dfrac{\partial F_2}{\partial y}\Delta x \Delta y \Delta z + \dfrac{\partial F_3}{\partial z}\Delta x \Delta y \Delta z}{\Delta x \Delta y \Delta z}$$

$$= \frac{\partial F_1}{\partial x} + \frac{\partial F_2}{\partial y} + \frac{\partial F_3}{\partial z}.$$

Divergence Free Vector Fields

A vector field $\vec{F}$ is said to be *divergence free* or *solenoidal* if $\text{div}\vec{F} = 0$ everywhere that $\vec{F}$ is defined.

Example 3   Figure 20.5 shows, for three values of the constant p, the vector field

$$\vec{E} = \frac{\vec{r}}{\|\vec{r}\|^p} \qquad \vec{r} \neq \vec{0}.$$

(a) Find a formula for $\text{div } \vec{E}$.
(b) Is there a value of p for which $\vec{E}$ is divergence free? If so, find it.

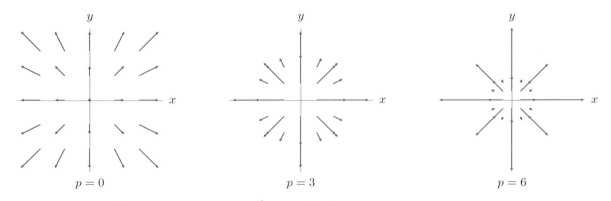

Figure 20.5: The vector field $\vec{E}(\vec{r}) = \vec{r}/\|\vec{r}\|^p$ for $p = 0, 3,$ and 6

Solution

(a) The components of $\vec{E}$ are

$$\vec{E} = \frac{x}{(x^2 + y^2 + z^2)^{p/2}}\vec{i} + \frac{y}{(x^2 + y^2 + z^2)^{p/2}}\vec{j} + \frac{z}{(x^2 + y^2 + z^2)^{p/2}}\vec{k}.$$

We compute the partial derivatives

$$\frac{\partial}{\partial x}\left(\frac{x}{(x^2 + y^2 + z^2)^{p/2}}\right) = \frac{1}{(x^2 + y^2 + z^2)^{p/2}} - \frac{px^2}{(x^2 + y^2 + z^2)^{(p/2)+1}}$$

$$\frac{\partial}{\partial y}\left(\frac{y}{(x^2 + y^2 + z^2)^{p/2}}\right) = \frac{1}{(x^2 + y^2 + z^2)^{p/2}} - \frac{py^2}{(x^2 + y^2 + z^2)^{(p/2)+1}}$$

$$\frac{\partial}{\partial z}\left(\frac{z}{(x^2 + y^2 + z^2)^{p/2}}\right) = \frac{1}{(x^2 + y^2 + z^2)^{p/2}} - \frac{pz^2}{(x^2 + y^2 + z^2)^{(p/2)+1}}.$$

So

$$\text{div } \vec{E} = \frac{3}{(x^2 + y^2 + z^2)^{p/2}} - \frac{p(x^2 + y^2 + z^2)}{(x^2 + y^2 + z^2)^{(p/2)+1}}$$

$$= \frac{3 - p}{(x^2 + y^2 + z^2)^{p/2}} = \frac{3 - p}{\|\vec{r}\|^p}.$$

(b) The divergence is zero when $p = 3$, so $\vec{F}(\vec{r}) = \vec{r}/\|\vec{r}\|^3$ is a divergence free vector field.

Magnetic Fields

An important class of divergence free vector fields is the magnetic fields. One of Maxwell's Laws of Electromagnetism is that the magnetic field $\vec{B}$ satisfies

$$\text{div } \vec{B} = 0.$$

Example 4   An infinitesimal current loop, similar to that shown in Figure 20.6, is called a *magnetic dipole*. Its magnitude is described by a constant vector $\vec{\mu}$, called the dipole moment. The magnetic field due to a magnetic dipole with moment $\vec{\mu}$ is

$$\vec{B} = -\frac{\vec{\mu}}{\|\vec{r}\|^3} + \frac{3(\vec{\mu} \cdot \vec{r})\vec{r}}{\|\vec{r}\|^5}, \qquad \vec{r} \neq \vec{0}.$$

Show that $\text{div } \vec{B} = 0$.

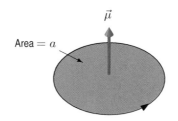

Area = a

$\vec{\mu}$

Figure 20.6: A current loop

Solution

To show that $\text{div } \vec{B} = 0$ we can use the following version of the product rule for the divergence: if g is a scalar function and $\vec{F}$ is a vector field, then

$$\text{div}(g\vec{F}) = (\text{grad } g) \cdot \vec{F} + g \text{ div } \vec{F}.$$

(See Problem 24 on page 946.) Thus, since $\text{div } \vec{\mu} = 0$, we have

$$\text{div}\left(\frac{\vec{\mu}}{\|\vec{r}\|^3}\right) = \text{div}\left(\frac{1}{\|\vec{r}\|^3}\vec{\mu}\right) = \text{grad}\left(\frac{1}{\|\vec{r}\|^3}\right) \cdot \vec{\mu} + \frac{1}{\|\vec{r}\|^3} \cdot 0$$

and

$$\text{div}\left(\frac{(\vec{\mu}\cdot\vec{r})\vec{r}}{\|\vec{r}\|^5}\right) = \text{grad}(\vec{\mu}\cdot\vec{r})\cdot\frac{\vec{r}}{\|\vec{r}\|^5} + (\vec{\mu}\cdot\vec{r})\,\text{div}\left(\frac{\vec{r}}{\|\vec{r}\|^5}\right).$$

From Problems 39 and 40 on page 718 and Example 3 on page 943 we have

$$\text{grad}\left(\frac{1}{\|\vec{r}\|^3}\right) = \frac{-3\vec{r}}{\|\vec{r}\|^5}, \qquad \text{grad}(\vec{\mu}\cdot\vec{r}) = \vec{\mu}, \qquad \text{div}\left(\frac{\vec{r}}{\|\vec{r}\|^5}\right) = \frac{-2}{\|\vec{r}\|^5}.$$

Putting these results together gives

$$\text{div}\,\vec{B} = -\text{grad}\left(\frac{1}{\|\vec{r}\|^3}\right)\cdot\vec{\mu} + 3\,\text{grad}(\vec{\mu}\cdot\vec{r})\cdot\frac{\vec{r}}{\|\vec{r}\|^5} + 3(\vec{\mu}\cdot\vec{r})\,\text{div}\left(\frac{\vec{r}}{\|\vec{r}\|^5}\right)$$

$$= \frac{3\vec{r}\cdot\vec{\mu}}{\|\vec{r}\|^5} + \frac{3\vec{\mu}\cdot\vec{r}}{\|\vec{r}\|^5} - \frac{6\vec{\mu}\cdot\vec{r}}{\|\vec{r}\|^5}$$

$$= 0.$$

Alternative Notation for Divergence

Using $\nabla = \dfrac{\partial}{\partial x}\vec{i} + \dfrac{\partial}{\partial y}\vec{j} + \dfrac{\partial}{\partial z}\vec{k}$, we can write

$$\text{div}\,\vec{F} = \nabla\cdot\vec{F} = \left(\frac{\partial}{\partial x}\vec{i} + \frac{\partial}{\partial y}\vec{j} + \frac{\partial}{\partial z}\vec{k}\right)\cdot(F_1\vec{i} + F_2\vec{j} + F_3\vec{k}) = \frac{\partial F_1}{\partial x} + \frac{\partial F_2}{\partial y} + \frac{\partial F_3}{\partial z}.$$

Exercises and Problems for Section 20.1

Exercises

1. Which of the following two vector fields appears to have the greater divergence at the origin? Assume the scales are the same on each.

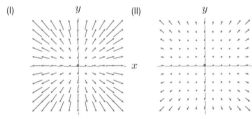

In Exercises 2–8, find the divergence of the given vector field. (Note: $\vec{r} = x\vec{i} + y\vec{j} + z\vec{k}$.)

2. $\vec{F}(x, y) = -x\vec{i} + y\vec{j}$

3. $\vec{F}(x, y) = -y\vec{i} + x\vec{j}$

4. $\vec{F}(x, y) = (x^2 - y^2)\vec{i} + 2xy\vec{j}$

5. $\vec{F}(x, y, z) = (-x + y)\vec{i} + (y + z)\vec{j} + (-z + x)\vec{k}$

6. $\vec{F}(\vec{r}) = \vec{a} \times \vec{r}$

7. $\vec{F}(x, y) = \dfrac{-y\vec{i} + x\vec{j}}{x^2 + y^2}$

8. $\vec{F}(\vec{r}) = \dfrac{\vec{r} - \vec{r}_0}{\|\vec{r} - \vec{r}_0\|}, \quad \vec{r} \neq \vec{r}_0$

9. For each of the following vector fields, decide if the divergence is positive, zero, or negative at the indicated point.

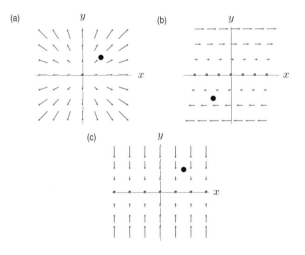

10. Draw two vector fields that have positive divergence everywhere.

11. Draw two vector fields that have negative divergence everywhere.

12. Draw two vector fields that have zero divergence everywhere.

Problems

13. A smooth vector field $\vec{F}$ has div $\vec{F}(1,2,3) = 5$. Estimate the flux of $\vec{F}$ out of a small sphere of radius 0.01 centered at the point $(1, 2, 3)$.

14. A vector field $\vec{F}$ has the property that the flux of $\vec{F}$ out of a small cube of side 0.01 centered around the point $(1, 5, 7)$, is 0.0025. Estimate div $\vec{F}$ at the point $(1, 5, 7)$.

15. (a) Let $\text{div}(\vec{F}) = x^2 + y^2 - z^2$. Estimate the flux out of a small sphere of radius 0.1 centered at each of the following points

    (i) $(2, 1, 1)$ (ii) $(0, 0, 1)$

  (b) What do the signs of your answers to part (a) tell you about the vector field near each of these two points?

16. Let $\vec{F}(x, y, z) = z\vec{k}$.

  (a) Calculate div $\vec{F}$.
  (b) Sketch $\vec{F}$. Does it appear to be diverging? Does this agree with your answer to part (a)?

17. Let $\vec{F}(\vec{r}) = \vec{r}/\|\vec{r}\|^3$ (in 3-space), $\vec{r} \neq \vec{0}$.

  (a) Calculate div $\vec{F}$.
  (b) Sketch $\vec{F}$. Does it appear to be diverging? Does this agree with your answer to part (a)?

For Problems 18–20,

(a) Find the flux of the given vector field through a cube in the first octant with edge length c, one corner at the origin and edges along the axes.

(b) Use your answer to part (a) to find div $\vec{F}$ at the origin using the geometric definition.

(c) Compute div $\vec{F}$ at the origin using partial derivatives.

18. $\vec{F} = x\vec{i}$ **19.** $\vec{F} = 2\vec{i} + y\vec{j} + 3\vec{k}$

20. $\vec{F} = x\vec{i} + y\vec{j}$

21. The divergence of a magnetic vector field $\vec{B}$ must be zero everywhere. Which of the following vector fields cannot be a magnetic vector field?

  (a) $\vec{B}(x, y, z) = -y\vec{i} + x\vec{j} + (x + y)\vec{k}$
  (b) $\vec{B}(x, y, z) = -z\vec{i} + y\vec{j} + x\vec{k}$
  (c) $\vec{B}(x, y, z) = (x^2 - y^2 - x)\vec{i} + (y - 2xy)\vec{j}$

22. (a) Let $f(x, y) = axy + ax^2y + y^3$. Find div grad f.
  (b) If possible, choose a so that div grad $f = 0$ for all x, y.

23. Show that if $\vec{a}$ is a constant vector and $f(x, y, z)$ is a function, then $\text{div}(f\vec{a}) = (\text{grad } f) \cdot \vec{a}$.

24. Show that if $g(x, y, z)$ is a scalar valued function and $\vec{F}(x, y, z)$ is a vector field, then

$$\text{div}(g\vec{F}) = (\text{grad } g) \cdot \vec{F} + g\,\text{div } \vec{F}.$$

25. If $f(x, y, z)$ and $g(x, y, z)$ are functions with continuous second partial derivatives, show that

$$\text{div}(\text{grad } f \times \text{grad } g) = 0.$$

26. Let $r = (x^2 + y^2)^{1/2}$. Figure 20.7 shows the vector field $r^A \cdot (x\vec{i} + y\vec{j})$ for $r \neq 0$ and $A = -1, -2, -3$. The vector fields are shown in the xy-plane; they have no z-component and are independent of z.

  (a) Show that $\text{div}(r^A \cdot (x\vec{i} + y\vec{j})) = (2 + A)r^A$ for any constant A.

  (b) Using your answer to part (a), find the sign of the divergence of the three vector fields for $A = -1, -2, -3$.

  (c) For each value of A, what (if anything) does your answer to part (b) tell you about the sign of the flux of the vector field out of a small sphere centered at $(1, 1, 1)$? At $(0, 0, 0)$?

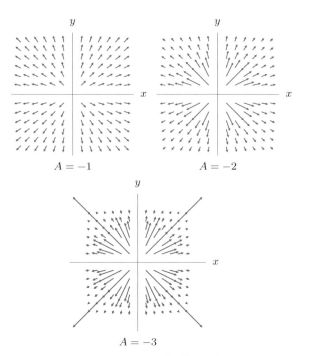

$A = -1$ $A = -2$

$A = -3$

Figure 20.7: The vector field $r^A(x\vec{i} + y\vec{j})$ for the three values of A

27. If $r = (x^2 + y^2)^{1/2}$, show that

$$\text{div}\left(\frac{h(r)}{r^2}(x\vec{i} + y\vec{j})\right) = \frac{h'(r)}{r}.$$

28. Let $\vec{F} = e^{-r}(x\vec{i} + y\vec{j})$, where $r = (x^2 + y^2)^{1/2}$. Use the result of Problem 27 to determine the values of r for which div $\vec{F}$ is positive, negative, or zero.

29. In Problem 38 on page 924 it was shown that the rate of heat loss from a volume V in a region of non-uniform temperature equals $k\int_S(\text{grad } T) \cdot d\vec{A}$, where k is a constant, S is the surface bounding V, and $T(x, y, z)$ is the temperature at the point (x, y, z) in space. By taking the limit as V contracts to a point, show that, at that point,

$$\frac{\partial T}{\partial t} = B \operatorname{div} \operatorname{grad} T$$

where B is a constant with respect to x, y, z, but may depend on time, t.

Problems 30–31 involve electric fields. Electric charge produces a vector field $\vec{E}$, called the electric field, which represents the force on a unit positive charge placed at the point. Two positive or two negative charges repel one another, whereas two charges of opposite sign attract one another. The divergence of $\vec{E}$ is proportional to the density of the electric charge (that is, the charge per unit volume), with a positive constant of proportionality.

30. A certain distribution of electric charge produces the electric field shown in Figure 20.8. Where are the charges that produced this electric field concentrated? Which concentrations are positive and which are negative?

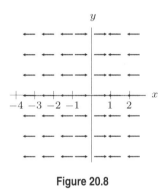

Figure 20.8

31. The electric field at the point $\vec{r}$ as a result of a point charge at the origin is $\vec{E}(\vec{r}) = k\vec{r}/\|\vec{r}\|^3$.

  (a) Calculate div $\vec{E}$ for $\vec{r} \neq \vec{0}$.
  (b) Calculate the limit suggested by the geometric definition of div $\vec{E}$ at the point $(0, 0, 0)$.
  (c) Explain what your answers mean in terms of charge density.

32. A vector field, $\vec{v}$, in the plane is a *point source* at the origin if its direction is away from the origin at every point,

its magnitude depends only on the distance from the origin, and its divergence is zero away from the origin.

  (a) Explain why a point source at the origin must be of the form $\vec{v} = \left[f(x^2 + y^2)\right](x\vec{i} + y\vec{j})$ for some positive function f.
  (b) Show that $\vec{v} = K(x^2 + y^2)^{-1}(x\vec{i} + y\vec{j})$ is a point source at the origin if $K > 0$.
  (c) What is the magnitude $\|\vec{v}\|$ of the source in part (b) as a function of the distance from its center?
  (d) Sketch the vector field $\vec{v} = (x^2 + y^2)^{-1}(x\vec{i} + y\vec{j})$.
  (e) Show that $\phi = \frac{K}{2} \log(x^2 + y^2)$ is a potential function for the source in part (b).

33. A vector field, $\vec{v}$, in the plane is a *point sink* at the origin if its direction is toward the origin at every point, its magnitude depends only on the distance from the origin, and its divergence is zero away from the origin.

  (a) Explain why a point sink at the origin must be of the form $\vec{v} = \left[f(x^2 + y^2)\right](x\vec{i} + y\vec{j})$ for some negative function f.
  (b) Show that $\vec{v} = K(x^2 + y^2)^{-1}(x\vec{i} + y\vec{j})$ is a point sink at the origin if $K < 0$.
  (c) Determine the magnitude $\|\vec{v}\|$ of the sink in part (b) as a function of the distance from its center.
  (d) Sketch $\vec{v} = -(x^2 + y^2)^{-1}(x\vec{i} + y\vec{j})$.
  (e) Show that $\phi = \frac{K}{2} \log(x^2 + y^2)$ is a potential function for the sink in part (b).

34. Let $\vec{F}(x, y) = u(x, y)\vec{i} + v(x, y)\vec{j}$ be a 2-dimensional vector field. Let $F = \|\vec{F}\|$, and let $\phi(x, y)$ be the angle of $\vec{F}$ with the positive x-axis at the point (x, y), so that $u = F \cos \phi$ and $v = F \sin \phi$. Let $\vec{T}$ be the unit vector in the direction of $\vec{F}$, and let $\vec{N}$ be the unit vector in the direction of $\vec{k} \times \vec{F}$. Show that

$$\operatorname{div} \vec{F} = F\phi_{\vec{N}} + F_{\vec{T}}$$

This problem shows that the divergence of the vector field $\vec{F}$ is the sum of two terms. The first term, $F\phi_{\vec{N}}$, is due to changes in the direction of $\vec{F}$ perpendicular to the flow lines, so it reflects the extent to which flow lines fail to be parallel. The second term, $F_{\vec{T}}$, is due to changes in the magnitude of $\vec{F}$ along the flow lines of $\vec{F}$.

20.2 THE DIVERGENCE THEOREM

The Divergence Theorem is a multivariable analogue of the Fundamental Theorem of Calculus; it says that the integral of the flux density over a solid region equals the flux integral through the boundary of the region.

The Boundary of a Solid Region

A solid region is an open region in 3-space. The boundary of a solid region may be thought of as the skin between the interior of the region and the space around it. For example, the boundary of a solid ball is a spherical surface, the boundary of a solid cube is its six faces, and the boundary of a solid cylinder is a tube sealed at both ends by disks. (See Figure 20.9). A surface which is the boundary of a solid region is called a *closed surface*.

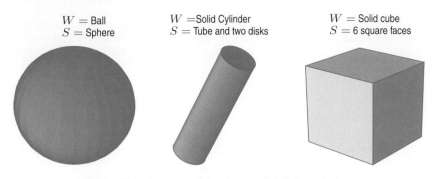

W = Ball
S = Sphere

W = Solid Cylinder
S = Tube and two disks

W = Solid cube
S = 6 square faces

Figure 20.9: Several solid regions and their boundaries

Calculating the Flux from the Flux Density

Consider a solid region W in 3-space whose boundary is the closed surface S. There are two ways to find the total flux of a vector field $\vec{F}$ out of W. One is to calculate the flux of $\vec{F}$ through S:

$$\text{Flux out of } W = \int_S \vec{F} \cdot d\vec{A}\,.$$

Another way is to use $\operatorname{div} \vec{F}$, which gives the flux density at any point in W. We subdivide W into small boxes, as shown in Figure 20.10. Then, for a small box of volume ΔV,

$$\text{Flux out of box} \approx \text{Flux density} \cdot \text{Volume} = \operatorname{div} \vec{F}\ \Delta V.$$

What happens when we add the fluxes out of all the boxes? Consider two adjacent boxes, as shown in Figure 20.11. The flux through the shared wall is counted twice, once out of the box on each side. When we add the fluxes, these two contributions cancel, so we get the flux out of the solid region formed by joining the two boxes. Continuing in this way, we find that

$$\text{Flux out of } W = \sum \text{Flux out of small boxes} \approx \sum \operatorname{div} \vec{F}\ \Delta V.$$

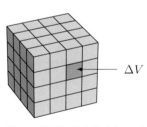

Figure 20.10: Subdivision of region into small boxes

ΔV

Fluxes through inner wall cancel

Figure 20.11: Adding the flux out of adjacent boxes

We have approximated the flux by a Riemann sum. As the subdivision gets finer, the sum approaches an integral, so

$$\text{Flux out of } W = \int_W \operatorname{div} \vec{F}\ dV.$$

We have calculated the flux in two ways, as a flux integral and as a volume integral. Therefore, these two integrals must be equal. This result holds even if W is not a rectangular solid, as shown in Figure 20.10. Thus, we have the following result.[2]

---

[2] A proof of the Divergence Theorem using the coordinate definition of the divergence can be found in the online supplement at www.wiley.com/college/hugheshallett.

Theorem 20.1: The Divergence Theorem

If W is a solid region whose boundary S is a piecewise smooth surface, and if $\vec{F}$ is a smooth vector field on an open region containing W and S, then

$$\int_S \vec{F} \cdot d\vec{A} = \int_W \operatorname{div} \vec{F} \, dV,$$

where S is given the outward orientation.

Example 1   Use the Divergence Theorem to calculate the flux of the vector field $\vec{F}(\vec{r}) = \vec{r}$ through the sphere of radius a centered at the origin.

Solution   In Example 4 on page 920 we computed the flux integral directly:

$$\int_S \vec{r} \cdot d\vec{A} = 4\pi a^3.$$

Now we use $\operatorname{div} \vec{F} = 3$ and the Divergence Theorem:

$$\int_S \vec{r} \cdot d\vec{A} = \int_W \operatorname{div} \vec{F} \, dV = \int_W 3 \, dV = 3(\frac{4}{3}\pi a^3) = 4\pi a^3.$$

Example 2   Use the Divergence Theorem to calculate the flux of the vector field

$$\vec{F}(x, y, z) = (x^2 + y^2)\vec{i} + (y^2 + z^2)\vec{j} + (x^2 + z^2)\vec{k}$$

through the cube in Figure 20.12.

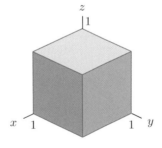

Figure 20.12

Solution   The divergence of $\vec{F}$ is $\operatorname{div} \vec{F} = 2x + 2y + 2z$. Since $\operatorname{div} \vec{F}$ is positive everywhere in the first quadrant, the flux through S is positive. By the Divergence Theorem,

$$\int_S \vec{F} \cdot d\vec{A} = \int_0^1 \int_0^1 \int_0^1 2(x + y + z) \, dx \, dy \, dz = \int_0^1 \int_0^1 x^2 + 2x(y + z) \Big|_0^1 \, dy \, dz$$

$$= \int_0^1 \int_0^1 1 + 2(y + z) \, dy \, dz = \int_0^1 y + y^2 + 2yz \Big|_0^1 \, dz$$

$$= \int_0^1 (2 + 2z) \, dz = 2z + z^2 \Big|_0^1 = 3.$$

The Divergence Theorem and Divergence Free Vector Fields

An important application of the Divergence Theorem is the study of divergence free vector fields.

Example 3   In Example 3 on page 943 we saw that the following vector field is divergence free:

$$\vec{F}(\vec{r}) = \frac{\vec{r}}{\|\vec{r}\|^3}, \qquad \vec{r} \neq \vec{0}.$$

Calculate $\int_S \vec{F} \cdot d\vec{A}$, using the Divergence Theorem if possible, for the following surfaces:

(a) S_1 is the sphere of radius a centered at the origin.
(b) S_2 is the sphere of radius a centered at the point $(2a, 0, 0)$.

Solution   (a) We cannot use the Divergence Theorem directly because $\vec{F}$ is not defined everywhere inside the sphere (it is not defined at the origin). Since $\vec{F}$ points outward everywhere on S_1, the flux out of S_1 is positive. On S_1,

$$\vec{F} \cdot d\vec{A} = \|\vec{F}\| dA = \frac{a}{a^3} dA,$$

so

$$\int_{S_1} \vec{F} \cdot d\vec{A} = \frac{1}{a^2} \int_{S_1} dA = \frac{1}{a^2}(\text{Area of } S_1) = \frac{1}{a^2} 4\pi a^2 = 4\pi.$$

Notice that the flux is not zero, although div $\vec{F}$ is zero everywhere it is defined.

(b) Suppose W is the solid region enclosed by S_2. Since div $\vec{F} = 0$ everywhere in W, we can use the Divergence Theorem in this case, giving

$$\int_{S_2} \vec{F} \cdot d\vec{A} = \int_W \text{div } \vec{F} \, dV = \int_W 0 \, dV = 0.$$

The Divergence Theorem applies to any solid region W and its boundary S, even in cases where the boundary consists of two or more surfaces. For example, if W is the solid region between the sphere S_1 of radius 1 and the sphere S_2 of radius 2, both centered at the same point, then the boundary of W consists of both S_1 and S_2. The Divergence Theorem requires the outward orientation, which on S_2 points away from the center and on S_1 points toward the center. (See Figure 20.13.)

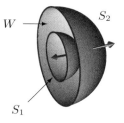

Figure 20.13: Cut-away view of the region W between two spheres, showing orientation vectors

Example 4   Let S_1 be the sphere of radius 1 centered at the origin and let S_2 be the ellipsoid $x^2 + y^2 + 4z^2 = 16$, both oriented outward. For

$$\vec{F}(\vec{r}) = \frac{\vec{r}}{\|\vec{r}\|^3}, \qquad \vec{r} \neq \vec{0},$$

show that

$$\int_{S_1} \vec{F} \cdot d\vec{A} = \int_{S_2} \vec{F} \cdot d\vec{A}.$$

Solution   The ellipsoid contains the sphere; let W be the solid region between them. Since W does not contain the origin, div $\vec{F}$ is defined and equal to zero everywhere in W. Thus, if S is the boundary of W, then

$$\int_S \vec{F} \cdot d\vec{A} = \int_W \text{div } \vec{F} \, dV = 0.$$

But S consists of S_2 oriented outward and S_1 oriented inward, so

$$0 = \int_S \vec{F} \cdot d\vec{A} = \int_{S_2} \vec{F} \cdot d\vec{A} - \int_{S_1} \vec{F} \cdot d\vec{A},$$

and thus

$$\int_{S_2} \vec{F} \cdot d\vec{A} = \int_{S_1} \vec{F} \cdot d\vec{A}.$$

In Example 3 we showed that $\int_{S_1} \vec{F} \cdot d\vec{A} = 4\pi$, so $\int_{S_2} \vec{F} \cdot d\vec{A} = 4\pi$ also. Note that it would have been more difficult to compute the integral over the ellipsoid directly.

Electric Fields

The electric field produced by a positive point charge q placed at the origin is

$$\vec{E} = q\frac{\vec{r}}{\|\vec{r}\|^3}.$$

Using Example 3 we see that the flux of the electric field through any sphere centered at the origin is $4\pi q$. In fact, using the idea of Example 4, we can show that the flux of $\vec{E}$ through any simple closed surface containing the origin is $4\pi q$. See Problems 40 and 41 on page 953. This is a special case of Gauss's Law, which states that the flux of an electric field through any closed surface is proportional to the total charge enclosed by the surface. Carl Friedrich Gauss (1777-1855) also discovered the Divergence Theorem, which is sometimes called Gauss's Theorem.

Exercises and Problems for Section 20.2

Exercises

1. Can you evaluate the flux integral in Exercise 25 on page 935 by application of the Divergence Theorem? Why or why not?

For Exercises 2–5, compute the flux integral $\int_S \vec{F} \cdot d\vec{A}$ in two ways, if possible, directly and using the Divergence Theorem. In each case, S is closed and oriented outward.

2. $\vec{F}(\vec{r}) = \vec{r}$ and S is the cube enclosing the volume $0 \le x \le 2, 0 \le y \le 2$, and $0 \le z \le 2$.

3. $\vec{F} = x^2\vec{i} + 2y^2\vec{j} + 3z^2\vec{k}$ and S is the surface of the box with faces $x = 1, x = 2, y = 0, y = 1, z = 0, z = 1$.

4. $\vec{F} = -z\vec{i} + x\vec{k}$ and S is a square pyramid with height 3 and base on the xy-plane of side length 1.

5. $\vec{F} = y\vec{j}$ and S is a closed vertical cylinder of height 2, with its base a circle of radius 1 on the xy-plane, centered at the origin.

In Exercises 6–7, use the Divergence Theorem to calculate the flux of the vector field through the surface.

6. $\vec{F} = x^2\vec{i} + (y - 2xy)\vec{j} + 10z\vec{k}$ through a sphere of radius 5 centered at the origin, oriented outward.

7. $\vec{F} = -z\vec{i} + x\vec{k}$ through the sphere of radius a centered at the origin. Explain your answer geometrically.

Find the flux of the vector fields in Exercises 8–12 out of the closed box $0 \le x \le 2, 0 \le y \le 3, 0 \le z \le 4$.

8. $\vec{F} = 4\vec{i} + 7\vec{j} - \vec{k}$

9. $\vec{G} = y\vec{i} + z\vec{k}$

10. $\vec{H} = xy\vec{i} + z\vec{j} + y\vec{k}$

11. $\vec{J} = xy^2\vec{j} + x\vec{k}$

12. $\vec{N} = e^z\vec{i} + \sin(xy)\vec{k}$

Problems

13. A vector field $\vec{F}$ satisfies $\text{div}\vec{F} = 0$ everywhere. Show that $\int_S \vec{F} \cdot d\vec{A} = 0$ for every closed surface S.

14. Find the flux of $\vec{F} = z\vec{i} + y\vec{j} + x\vec{k}$ out of a sphere of radius 3 centered at the origin.

15. Find the flux of $\vec{F} = xy\vec{i} + yz\vec{j} + zx\vec{k}$ out of a sphere of radius 1 centered at the origin.

16. Find the flux of $\vec{F} = x^3\vec{i} + y^3\vec{j} + z^3\vec{k}$ through the closed surface bounding the solid region $x^2 + y^2 \le 4$, $0 \le z \le 5$, oriented outward.

17. The region W lies between the spheres $x^2 + y^2 + z^2 = 4$ and $x^2 + y^2 + z^2 = 9$ and within the cone $z = \sqrt{x^2 + y^2}$ with $z \ge 0$; its boundary is the closed surface, S, oriented outward. Find the flux of $\vec{F} = x^3\vec{i} + y^3\vec{j} + z^3\vec{k}$ out of S.

18. Suppose $\vec{F}$ is a vector field with div $\vec{F} = 10$. Find the flux of $\vec{F}$ out of a cylinder of height a and radius a, centered on the z-axis and with base in the xy-plane.

19. A cone has its tip at the point $(0, 0, 5)$ and its base the disk D, $x^2 + y^2 \le 1$, in the xy-plane. The surface of the cone is the curved and slanted face, S, oriented upward, and the flat base, D, oriented downward. The flux of the constant vector field $\vec{F} = a\vec{i} + b\vec{j} + c\vec{k}$ through S is given by

$$\int_S \vec{F} \cdot d\vec{A} = 3.22.$$

Is it possible to calculate $\int_D \vec{F} \cdot d\vec{A}$? If so, give the answer. If not, explain what additional information you would need to be able to make this calculation.

20. The function div $\vec{F}(x, y, z)$ is monotonically increasing in each variable, x, y, z. Use Tables 20.1 and 20.2 to find upper and lower bounds for $\int_S \vec{F} \cdot d\vec{A}$, where S is the closed box $0 \le x \le 1, 0 \le y \le 1, 0 \le z \le 1$.

Table 20.1   div $\vec{F}$
for $z = 0$

$x\backslash y$	0	0.5	1
0	2	3	5
0.5	4	5	8
1	5	6	7

Table 20.2   div $\vec{F}$
for $z = 0$

$x\backslash y$	0	1	2
0	3	4	6
1	4	6	8
2	6	7	9

In Problems 21–25, calculate the flux of $\vec{F}$ through the cylinder $x^2 + y^2 = 2, -3 \le z \le 3$ and its base, oriented outward. The cylinder is open at the top.

21. $\vec{F} = z^2\vec{i} + x^2\vec{j} + 5\vec{k}$

22. $\vec{F} = y^2\vec{i} + z^2\vec{j} + (x^2 + y^2)\vec{k}$

23. $\vec{F} = z\vec{i} + x\vec{j} + y\vec{k}$

24. $\vec{F} = y^2\vec{i} + x^2\vec{j} + 7z\vec{k}$

25. $\vec{F} = x^3\vec{i} + y^3\vec{j} + \vec{k}$

26. In Figure 20.14, the rectangular solids V_1 and V_2 are in the first quadrant. Both have sides of length 1 parallel to the axes; V_1 has one corner at the origin, while V_2 has the corresponding corner at the point $(1, 0, 0)$. The six-faced surfaces of V_1 and V_2 are S_1 and S_2, respectively. The box-shaped volume, V, consisting of V_1 and V_2 together, has outside surface S. Are the following true or false? Give reasons.

(a) If $\vec{F}$ is a constant vector field, then $\int_S \vec{F} \cdot d\vec{A} = 0$.

(b) If S_1, S_2 and S are all oriented outward and $\vec{F}$ is any vector field, then

$$\int_S \vec{F} \cdot d\vec{A} = \int_{S_1} \vec{F} \cdot d\vec{A} + \int_{S_2} \vec{F} \cdot d\vec{A}.$$

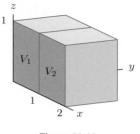

Figure 20.14

27. (a) Let $\vec{F} = \vec{r}/\|\vec{r}\|^3$. Calculate div $\vec{F}$ for $\vec{r} \ne \vec{0}$.
   (b) Calculate the flux of $\vec{F}$ out of a box of side a centered at the origin and with edges parallel to the axes.

28. The gravitational field, $\vec{F}$, of a planet of mass m at the origin is given by

$$\vec{F} = -Gm\frac{\vec{r}}{\|\vec{r}\|^3}.$$

Use the Divergence Theorem to show that the flux of the gravitational field through the sphere of radius a is independent of a. [Hint: Consider the region bounded by two concentric spheres.]

29. Let $\vec{F} = \frac{1}{3}(x^3 + y^3)\vec{i} + \frac{1}{3}(y^3 + z^3)\vec{j} + \left(\frac{1}{3}(z^3 + x^3) - z\right)\vec{k}$. Without calculating the integrals, decide if the flux out of the following regions is positive, negative, or zero. Give reasons for your answers.

(a) $\int_S \vec{F} \cdot d\vec{A}$, where S is the sphere $x^2 + y^2 + z^2 = 1$.
(b) $\int_B \vec{F} \cdot d\vec{A}$, where B is the sphere $(x - 10)^2 + (y - 10)^2 + (z - 10)^2 = 1$.

30. Let $\vec{a} = a_1\vec{i} + a_2\vec{j} + a_3\vec{k}$ be a constant vector and let $\vec{r} = x\vec{i} + y\vec{j} + z\vec{k}$.

(a) Calculate div$(\vec{r} \times \vec{a})$.
(b) Calculate the flux of $\vec{r} \times \vec{a}$ out of a cube of side 5 centered at the origin with edges parallel to the axes.

31. (a) Let $\vec{a} = a_1\vec{i} + a_2\vec{j} + a_3\vec{k}$ be a constant vector and let $\vec{r} = x\vec{i} + y\vec{j} + z\vec{k}$. Calculate div $((\vec{a} \cdot \vec{r})\vec{r})$.
   (b) Find the flux of the vector field $(\vec{a} \cdot \vec{r})\vec{r}$ out of a sphere of radius 1 centered at the origin.

32. Suppose div $\vec{F} = xyz^2$.

(a) Find div $\vec{F}$ at the point $(1, 2, 1)$. [Note: You are given div $\vec{F}$, not $\vec{F}$.]
(b) Using your answer to part (a), but no other information about the vector field $\vec{F}$, estimate the flux out of a small box of side 0.2 centered at the point $(1, 2, 1)$ and with edges parallel to the axes.
(c) Without computing the vector field $\vec{F}$, calculate the exact flux out of the box.

33. Suppose div $\vec{F} = x^2 + y^2 + 3$. Find a surface S such that $\int_S \vec{F} \cdot d\vec{A}$ is negative, or explain why no such surface exists.

34. Suppose $\vec{G}$ is a vector field with the property that $\operatorname{div}\vec{G} = 3$ for $2 \le \|\vec{r}\| \le 7$ and that the flux of $\vec{G}$ through the sphere of radius 3 centered at the origin is 8π. Find the flux of $\vec{G}$ through the sphere of radius 5 centered at the origin.

35. If $2 \le \operatorname{div}\vec{F} \le 5$ on a sphere S of radius 3, find upper and lower bounds for the flux of $\vec{F}$ out of S.

36. The vector field $\vec{G}$ is given by $\vec{G} = 4r^2\vec{r}$ outside the unit sphere centered at the origin, where $r = \|\vec{r}\|$ and $\vec{r} = x\vec{i} + y\vec{j} + z\vec{k}$. However, $\vec{G}$ is not defined inside the unit sphere.

  (a) Calculate the flux of G out of the sphere of radius 2 centered at the origin. Explain your reasoning.

  (b) Consider (but do not calculate) the flux of $\vec{G}$ through the box of side 4 centered at the origin and with sides parallel to the axes. Is this flux larger or smaller than the flux you calculated in part (a)? Justify your answer.

37. A basic property of the electric field $\vec{E}$ is that its divergence is zero at points where there is no charge. Suppose that the only charge is along the z-axis, and that the electric field $\vec{E}$ points radially out from the z-axis and its magnitude depends only on the distance r from the z-axis. Use the Divergence Theorem to show that the magnitude of the field is proportional to $1/r$. [Hint: Consider a solid region consisting of a cylinder of finite length whose axis is the z-axis, and with a smaller concentric cylinder removed.]

38. If a surface S is submerged in an incompressible fluid, a force $\vec{F}$ is exerted on one side of the surface by the pressure in the fluid. If the z-axis is vertical, with the positive direction upward and the fluid level at $z = 0$, then the component of force in the direction of a unit vector $\vec{u}$ is given by the following:

$$\vec{F} \cdot \vec{u} = -\int_S z\delta g\vec{u} \cdot d\vec{A},$$

where δ is the density of the fluid (mass/volume), g is the acceleration due to gravity, and the surface is oriented away from the side on which the force is exerted. In this problem we consider a totally submerged closed surface enclosing a volume V. We are interested in the force of the liquid on the external surface, so S is oriented inward. Use the Divergence Theorem to show that:

  (a) The force in the $\vec{i}$ and $\vec{j}$ directions is zero.

  (b) The force in the $\vec{k}$ direction is δgV, the weight of the volume of fluid with the same volume as V. This is *Archimedes' Principle*.

39. As a result of radioactive decay, heat is generated uniformly throughout the interior of the earth at a rate of 30 watts per cubic kilometer. (A watt is a rate of heat production.) The heat then flows to the earth's surface where it is lost to space. Let $\vec{F}(x, y, z)$ denote the rate of flow of heat measured in watts per square kilometer. By definition, the flux of $\vec{F}$ across a surface is the quantity of heat flowing through the surface per unit of time.

  (a) What is the value of $\operatorname{div}\vec{F}$? Include units.

  (b) Assume the heat flows outward symmetrically. Verify that $\vec{F} = \alpha\vec{r}$, where $\vec{r} = x\vec{i} + y\vec{j} + z\vec{k}$ and α is a suitable constant, satisfies the given conditions. Find α.

  (c) Let $T(x, y, z)$ denote the temperature inside the earth. Heat flows according to the equation $\vec{F} = -k\operatorname{grad}T$, where k is a constant. Explain why this makes sense physically.

  (d) If T is in $°C$, then $k = 30{,}000$ watts/km$°$C. Assuming the earth is a sphere with radius 6400 km and surface temperature 20$°$C, what is the temperature at the center?

40. According to Coulomb's Law, the electrostatic field $\vec{E}$ at the point $\vec{r}$ due to a charge q at the origin is given by

$$\vec{E}(\vec{r}) = q\frac{\vec{r}}{\|\vec{r}\|^3}.$$

  (a) Compute $\operatorname{div}\vec{E}$.

  (b) Let S_a be the sphere of radius a centered at the origin and oriented outward. Show that the flux of $\vec{E}$ through S_a is $4\pi q$.

  (c) Could you have used the Divergence Theorem in part (b)? Explain why or why not.

  (d) Let S be an arbitrary, closed, outward-oriented surface surrounding the origin. Show that the flux of $\vec{E}$ through S is again $4\pi q$. [Hint: Apply the Divergence Theorem to the solid region lying between a small sphere S_a and the surface S.]

41. According to Coulomb's Law, the electric field $\vec{E}$ at the point $\vec{r}$ due to a charge q at the point $\vec{r}_0$ is given by

$$\vec{E}(\vec{r}) = q\frac{(\vec{r} - \vec{r}_0)}{\|\vec{r} - \vec{r}_0\|^3}.$$

Suppose S is a closed, outward-oriented surface and that $\vec{r}_0$ does not lie on S. Use Problem 40 to show that

$$\int_S \vec{E} \cdot d\vec{A} = \begin{cases} 4\pi q & \text{if } q \text{ lies inside } S, \\ 0 & \text{if } q \text{ lies outside } S. \end{cases}$$

42. At the point with position vector $\vec{r}$, the electric field $\vec{E}$ of an ideal electric dipole with moment $\vec{p}$ located at the origin is given by

$$\vec{E}(\vec{r}) = 3\frac{(\vec{r} \cdot \vec{p})\vec{r}}{\|\vec{r}\|^5} - \frac{\vec{p}}{\|\vec{r}\|^3}.$$

  (a) What is $\operatorname{div}\vec{E}$?

  (b) Suppose S is an outward oriented, closed smooth surface surrounding the origin. Compute the flux of $\vec{E}$ through S. Can you use the Divergence Theorem directly to compute the flux? Explain why or why not. [Hint: First compute the flux of the dipole field $\vec{E}$ through an outward-oriented sphere S_a with center at the origin and radius a. Then apply the Divergence Theorem to the region W lying between S and a small sphere S_a.]

20.3   THE CURL OF A VECTOR FIELD

The divergence of a vector field is a sort of scalar derivative which measures its outflow per unit volume. Now we introduce a vector derivative, the curl, which measures the circulation of a vector field. Imagine holding the paddle-wheel in Figure 20.15 in the flow shown by Figure 20.16. The speed at which the paddle-wheel spins measures the strength of circulation. Note that the angular velocity depends on the direction in which the stick is pointing. If the stick is pointing horizontally it does not spin at all, because the velocity field strikes opposite vanes of the paddle with equal force.

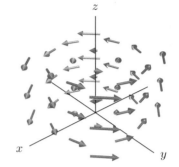

Figure 20.15: A device for measuring circulation

Figure 20.16: A vector field with circulation about the z-axis

Circulation Density

We measure the strength of the circulation using a closed curve. Suppose C is a circle with center (x, y, z) in the plane perpendicular to $\vec{n}$, traversed in the direction determined from $\vec{n}$ by the right-hand rule. (See Figures 20.17 and 20.18.)

Figure 20.17: Direction of C relates to direction of $\vec{n}$ by the right-hand rule

Figure 20.18: When the thumb points in the direction of $\vec{n}$, the fingers curl in the forward direction around C

We make the following definition:

The **circulation density** of a smooth vector field $\vec{F}$ at (x, y, z) around the direction of the unit vector $\vec{n}$ is defined to be

$$\mathrm{circ}_{\vec{n}} \, \vec{F}(x, y, z) = \lim_{\text{Area} \to 0} \frac{\text{Circulation around } C}{\text{Area inside } C} = \lim_{\text{Area} \to 0} \frac{\int_C \vec{F} \cdot d\vec{r}}{\text{Area inside } C},$$

provided the limit exists.

The circulation density determines the angular velocity[3] of the paddle-wheel in Figure 20.15 provided you could make one sufficiently small and light and insert it without disturbing the flow.

---

[3] In fact it is twice the angular velocity. See Example 3 on page 957.

Example 1   Consider the vector field $\vec{F}$ in Figure 20.16. Suppose that $\vec{F}$ is parallel to the xy-plane and that at a distance r from the z-axis it has magnitude $2r$. Calculate $\text{circ}_{\vec{n}}\ \vec{F}$ at the origin for
(a) $\vec{n} = \vec{k}$ (b) $\vec{n} = -\vec{k}$ (c) $\vec{n} = \vec{i}$.

Solution   (a) Take a circle C of radius a in the xy-plane, centered at the origin, traversed in a direction determined from $\vec{k}$ by the right hand rule. Then, since $\vec{F}$ is tangent to C everywhere and points in the forward direction around C, we have

$$\text{Circulation around } C = \int_C \vec{F} \cdot d\vec{r} = \|\vec{F}\| \cdot \text{Circumference of } C = 2a(2\pi a) = 4\pi a^2.$$

Thus, the circulation density is

$$\text{circ}_{\vec{k}}\ \vec{F} = \lim_{a \to 0} \frac{\text{Circulation around } C}{\text{Area inside } C} = \lim_{a \to 0} \frac{4\pi a^2}{\pi a^2} = 4.$$

(b) If $\vec{n} = -\vec{k}$ the circle is traversed in the opposite direction, so the line integral changes sign. Thus

$$\text{circ}_{-\vec{k}}\ \vec{F} = -4.$$

(c) The circulation around $\vec{i}$ is calculated using circles in the yz-plane. Since $\vec{F}$ is everywhere perpendicular to such a circle C,

$$\int_C \vec{F} \cdot d\vec{r} = 0.$$

Thus, we have

$$\text{circ}_{\vec{i}}\ \vec{F} = \lim_{a \to 0} \frac{\int_C \vec{F} \cdot d\vec{r}}{\pi a^2} = \lim_{a \to 0} \frac{0}{\pi a^2} = 0.$$

Definition of the Curl

Example 1 shows that the circulation density of a vector field can be positive, negative, or zero, depending on the direction. We assume that there is one direction in which the circulation density is greatest. Now we define a single vector quantity that incorporates all these different circulation densities.

Geometric Definition of Curl

The curl of a smooth vector field $\vec{F}$, written $\text{curl}\,\vec{F}$, is the vector field with the following properties
- The direction of $\text{curl}\,\vec{F}\,(x, y, z)$ is the direction $\vec{n}$ for which $\text{circ}_{\vec{n}}\ \vec{F}\,(x, y, z)$ is the greatest.
- The magnitude of $\text{curl}\,\vec{F}\,(x, y, z)$ is the circulation density of $\vec{F}$ around that direction.

If the circulation density is zero around every direction, then we define the curl to be $\vec{0}$.

Cartesian Coordinate Definition of Curl

If $\vec{F} = F_1\vec{i} + F_2\vec{j} + F_3\vec{k}$, then

$$\text{curl}\,\vec{F} = \left(\frac{\partial F_3}{\partial y} - \frac{\partial F_2}{\partial z} \right) \vec{i} + \left(\frac{\partial F_1}{\partial z} - \frac{\partial F_3}{\partial x} \right) \vec{j} + \left(\frac{\partial F_2}{\partial x} - \frac{\partial F_1}{\partial y} \right) \vec{k}.$$

Example 2   For each field in Figure 20.19, use the sketch and the geometric definition to decide whether the curl at the origin points up, down, or is the zero vector. Then check your answer using the coordinate definition of curl. Note that the vector fields have no z-components and are independent of z.

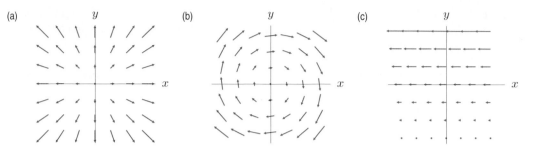

(a)

(b)

(c)

Figure 20.19: Sketches in the xy-plane of (a) $\vec{F} = x\vec{i} + y\vec{j}$ (b) $\vec{F} = y\vec{i} - x\vec{j}$ (c) $\vec{F} = -(y+1)\vec{i}$

Solution   (a) This vector field shows no rotation, and the circulation around any closed curve appears to be zero, so we suspect that the curl is the zero vector. The coordinate definition of curl gives

$$\operatorname{curl} \vec{F} = \left(\frac{\partial(0)}{\partial y} - \frac{\partial y}{\partial z}\right)\vec{i} + \left(\frac{\partial x}{\partial z} - \frac{\partial(0)}{\partial x}\right)\vec{j} + \left(\frac{\partial y}{\partial x} - \frac{\partial x}{\partial y}\right)\vec{k} = \vec{0}.$$

(b) This vector field is rotating around the z-axis. By the right-hand rule, the circulation density around $\vec{k}$ is negative, so we expect the z-component of the curl points down. The coordinate definition gives

$$\operatorname{curl} \vec{F} = \left(\frac{\partial(0)}{\partial y} - \frac{\partial(-x)}{\partial z}\right)\vec{i} + \left(\frac{\partial y}{\partial z} - \frac{\partial(0)}{\partial x}\right)\vec{j} + \left(\frac{\partial(-x)}{\partial x} - \frac{\partial y}{\partial y}\right)\vec{k} = -2\vec{k}.$$

(c) At first glance, you might expect this vector field to have zero curl, as all the vectors are parallel to the x-axis. However, if you find the circulation around the curve C in Figure 20.20, the sides contribute nothing (they are perpendicular to the vector field), the bottom contributes a negative quantity (the curve is in the opposite direction to the vector field), and the top contributes a larger positive quantity (the curve is in the same direction as the vector field and the magnitude of the vector field is larger at the top than at the bottom). Thus, the circulation around C is positive and hence we expect the curl to be nonzero and point up. The coordinate definition gives

$$\operatorname{curl} \vec{F} = \left(\frac{\partial(0)}{\partial y} - \frac{\partial(0)}{\partial z}\right)\vec{i} + \left(\frac{\partial(-(y+1))}{\partial z} - \frac{\partial(0)}{\partial x}\right)\vec{j} + \left(\frac{\partial(0)}{\partial x} - \frac{\partial(-(y+1))}{\partial y}\right)\vec{k} = \vec{k}.$$

Another way to see that the curl is nonzero in this case is to imagine the vector field representing the velocity of moving water. A boat sitting in the water tends to rotate, as the water moves faster on one side than the other.

Figure 20.20

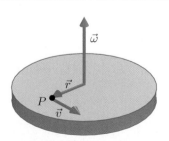

Figure 20.21: Rotating flywheel

Example 3   A flywheel is rotating with angular velocity $\vec{\omega}$ and the velocity of a point P with position vector $\vec{r}$ is given by $\vec{v} = \vec{\omega} \times \vec{r}$. (See Figure 20.21.) Calculate $\operatorname{curl} \vec{v}$.

Solution   If $\vec{\omega} = \omega_1 \vec{i} + \omega_2 \vec{j} + \omega_3 \vec{k}$, we have

$$\vec{v} = \vec{\omega} \times \vec{r} = \begin{vmatrix} \vec{i} & \vec{j} & \vec{k} \\ \omega_1 & \omega_2 & \omega_3 \\ x & y & z \end{vmatrix} = (\omega_2 z - \omega_3 y)\vec{i} + (\omega_3 x - \omega_1 z)\vec{j} + (\omega_1 y - \omega_2 x)\vec{k}.$$

Thus,

$$\operatorname{curl} \vec{v} = \begin{vmatrix} \vec{i} & \vec{j} & \vec{k} \\ \frac{\partial}{\partial x} & \frac{\partial}{\partial y} & \frac{\partial}{\partial z} \\ \omega_2 z - \omega_3 y & \omega_3 x - \omega_1 z & \omega_1 y - \omega_2 x \end{vmatrix}$$

$$= \left(\frac{\partial}{\partial y}(\omega_1 y - \omega_2 x) - \frac{\partial}{\partial z}(\omega_3 x - \omega_1 z) \right)\vec{i} + \left(\frac{\partial}{\partial z}(\omega_2 z - \omega_3 y) - \frac{\partial}{\partial x}(\omega_1 y - \omega_2 x) \right)\vec{j}$$

$$+ \left(\frac{\partial}{\partial x}(\omega_3 x - \omega_1 z) - \frac{\partial}{\partial y}(\omega_2 z - \omega_3 y) \right)\vec{k}$$

$$= 2\omega_1 \vec{i} + 2\omega_2 \vec{j} + 2\omega_3 \vec{k} = 2\vec{\omega}.$$

Thus, as we would expect, $\operatorname{curl} \vec{v}$ is parallel to the axis of rotation of the flywheel (namely, the direction of $\vec{\omega}$) and the magnitude of $\operatorname{curl} \vec{v}$ is larger the faster the flywheel is rotating (that is, the larger the magnitude of $\vec{\omega}$).

Why Do the Two Definitions of Curl Give the Same Result?

Using Green's Theorem in Cartesian coordinates, we can show that for $\operatorname{curl} \vec{F}$ defined in Cartesian coordinates

$$\boxed{\operatorname{curl} \vec{F} \cdot \vec{n} = \operatorname{circ}_{\vec{n}} \vec{F}.}$$

This shows that $\operatorname{curl} \vec{F}$ defined in Cartesian coordinates satisfies the geometric definition, since the left hand side takes its maximum value when $\vec{n}$ points in the same direction as $\operatorname{curl} \vec{F}$, and in that case its value is $\| \operatorname{curl} \vec{F} \|$.

The following example justifies this formula in a specific case.

Example 4   Use the definition of curl in Cartesian coordinates and Green's Theorem to show that

$$\left(\operatorname{curl} \vec{F} \right) \cdot \vec{k} = \operatorname{circ}_{\vec{k}} \vec{F}.$$

Solution   Using the definition of curl in Cartesian coordinates, the left hand side of the formula is

$$\left(\operatorname{curl} \vec{F} \right) \cdot \vec{k} = \frac{\partial F_2}{\partial x} - \frac{\partial F_1}{\partial y}.$$

Now let's look at the right hand side. The circulation density around $\vec{k}$ is calculated using circles perpendicular to $\vec{k}$; hence the $\vec{k}$-component of $\vec{F}$ does not contribute to it, that is, the circulation density of $\vec{F}$ around $\vec{k}$ is the same as the circulation density of $F_1\vec{i} + F_2\vec{j}$ around $\vec{k}$. But in any plane perpendicular to $\vec{k}$, z is constant, so in that plane F_1 and F_2 are functions of x and y alone. Thus $F_1\vec{i} + F_2\vec{j}$ can be thought of as a two-dimensional vector field on the horizontal plane through the point (x, y, z) where the circulation density is being calculated. Let C be a circle in this plane, with radius a and centered at (x, y, z), and let R be the region enclosed by C. Green's Theorem says that

$$\int_C (F_1\vec{i} + F_2\vec{j}) \cdot d\vec{r} = \int_R \left(\frac{\partial F_2}{\partial x} - \frac{\partial F_1}{\partial y} \right) dA.$$

When the circle is small, $\partial F_2/\partial x - \partial F_1/\partial y$ is approximately constant on R, so

$$\int_R \left(\frac{\partial F_2}{\partial x} - \frac{\partial F_1}{\partial y} \right) dA \approx \left(\frac{\partial F_2}{\partial x} - \frac{\partial F_1}{\partial y} \right) \cdot \text{Area of } R = \left(\frac{\partial F_2}{\partial x} - \frac{\partial F_1}{\partial y} \right) \pi a^2.$$

Thus, taking a limit as the radius of the circle goes to zero, we have

$$\text{circ}_{\vec{k}}\, \vec{F}\,(x, y, z) = \lim_{a \to 0} \frac{\int_C (F_1\vec{i} + F_2\vec{j}) \cdot d\vec{r}}{\pi a^2} = \lim_{a \to 0} \frac{\int_R \left(\frac{\partial F_2}{\partial x} - \frac{\partial F_1}{\partial y} \right) dA}{\pi a^2} = \frac{\partial F_2}{\partial x} - \frac{\partial F_1}{\partial y}.$$

Curl Free Vector Fields

A vector field is said to be *curl free* or *irrotational* if curl $\vec{F} = \vec{0}$ everywhere that $\vec{F}$ is defined.

Example 5   Figure 20.22 shows the vector field $\vec{B}$ for three values of the constant p, where $\vec{B}$ is defined on 3-space by

$$\vec{B} = \frac{-y\vec{i} + x\vec{j}}{(x^2 + y^2)^{p/2}}.$$

(a) Find a formula for curl $\vec{B}$.
(b) Is there a value of p for which $\vec{B}$ is curl free? If so, find it.

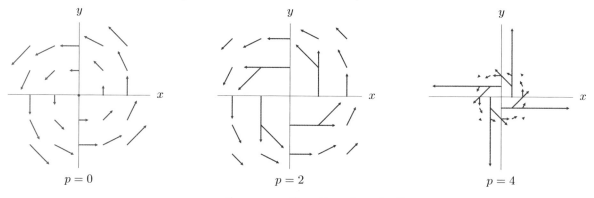

$p = 0$ $p = 2$ $p = 4$

Figure 20.22: The vector field $\vec{B}\,(\vec{r}) = (-y\vec{i} + x\vec{j})/(x^2 + y^2)^{p/2}$ for $p = 0, 2,$ and 4

Solution   (a) We can use the following version of the product rule for curl. If ϕ is a scalar function and $\vec{F}$ is a vector field, then

$$\text{curl}(\phi\vec{F}) = \phi\, \text{curl}\, \vec{F} + (\text{grad}\,\phi) \times \vec{F}.$$

(See Problem 28 on page 961.) We write $\vec{B} = \phi\vec{F} = \dfrac{1}{(x^2 + y^2)^{p/2}}(-y\vec{i} + x\vec{j})$. Then

$$\text{curl}\, \vec{F} = \text{curl}(-y\vec{i} + x\vec{j}) = 2\vec{k}$$

$$\text{grad}\,\phi = \text{grad}\left(\frac{1}{(x^2 + y^2)^{p/2}} \right) = \frac{-p}{(x^2 + y^2)^{(p/2)+1}}(x\vec{i} + y\vec{j}).$$

Thus, we have

$$\text{curl}\,\vec{B} = \frac{1}{(x^2+y^2)^{p/2}}\text{curl}(-y\vec{i}+x\vec{j}) + \text{grad}\left(\frac{1}{(x^2+y^2)^{p/2}}\right)\times(-y\vec{i}+x\vec{j})$$

$$= \frac{1}{(x^2+y^2)^{p/2}}2\vec{k} + \frac{-p}{(x^2+y^2)^{(p/2)+1}}(x\vec{i}+y\vec{j})\times(-y\vec{i}+x\vec{j})$$

$$= \frac{1}{(x^2+y^2)^{p/2}}2\vec{k} + \frac{-p}{(x^2+y^2)^{(p/2)+1}}(x^2+y^2)\vec{k}$$

$$= \frac{2-p}{(x^2+y^2)^{p/2}}\vec{k}.$$

(b) The curl is zero when $p = 2$. Thus, when $p = 2$ the vector field is curl free:

$$\vec{B} = \frac{-y\vec{i}+x\vec{j}}{x^2+y^2}.$$

Alternative Notation for Curl

Using $\nabla = \dfrac{\partial}{\partial x}\vec{i} + \dfrac{\partial}{\partial y}\vec{j} + \dfrac{\partial}{\partial z}\vec{k}$, we can write

$$\text{curl}\,\vec{F} = \nabla\times\vec{F} = \begin{vmatrix} \vec{i} & \vec{j} & \vec{k} \\ \dfrac{\partial}{\partial x} & \dfrac{\partial}{\partial y} & \dfrac{\partial}{\partial z} \\ F_1 & F_2 & F_3 \end{vmatrix}.$$

Exercises and Problems for Section 20.3

Exercises

Decide whether the vector fields in Exercises 1–4 have a nonzero curl at the origin. The vector field is shown in the xy-plane; it has no z-component and is independent of z.

1.

2.

3.

4.

Decide whether the vector fields in Exercises 5–6 have a nonzero curl at the point marked. The vector field is shown in the xy-plane; it has no z-component and is independent of z.

5.

6.

Compute the curl of the vector fields in Exercises 7–13.

7. $\vec{F} = (x^2-y^2)\vec{i} + 2xy\vec{j}$

8. $\vec{F} = x^2\vec{i} + y^3\vec{j} + z^4\vec{k}$

9. $\vec{F} = e^x\vec{i} + \cos y\vec{j} + e^{z^2}\vec{k}$

10. $\vec{F} = 2yz\vec{i} + 3xz\vec{j} + 7xy\vec{k}$

11. $\vec{F} = (-x+y)\vec{i} + (y+z)\vec{j} + (-z+x)\vec{k}$

12. $\vec{F} = (x+yz)\vec{i} + (y^2+xzy)\vec{j} + (zx^3y^2+x^7y^6)\vec{k}$

13. $\vec{F}(\vec{r}) = \vec{r}/\|\vec{r}\|$

14. Let $\vec{F} = (xy+z^2)\vec{i} + x^2\vec{j} + (xz-2)\vec{k}$.

   (a) Find $\text{curl}\,\vec{F}$ at the point $(0,-1,0)$.
   (b) Is this vector field irrotational?

Problems

15. Using your answers to Exercises 8–9, make a conjecture about the value of curl $\vec{F}$ when the vector field $\vec{F}$ has a certain form. (What form?) Show why your conjecture is true.

16. Use the geometric definition to find the curl of the vector field $\vec{F}(\vec{r}) = \vec{r}$. Check your answer using the coordinate definition.

17. The vector field $\vec{F}(x, y)$ sketched in Figure 20.23 has no z-component and is independent of z.

   (a) Do you think that $\vec{F}(x, y)$ is a gradient vector field? Justify your answer using the sketch.
   (b) Do you think curl $\vec{F} \cdot \vec{k}$ is positive, negative, or zero? Explain.
   (c) Give a possible formula for $\vec{F}$.
   (d) Calculate curl $\vec{F}$, using the formula you gave for $\vec{F}$.

Figure 20.23

18. Let $\vec{F}$ be the vector field in Figure 20.16 on page 954. It is rotating counterclockwise around the z-axis when viewed from above. At a distance r from the z-axis, $\vec{F}$ has magnitude $2r$.

   (a) Find a formula for $\vec{F}$.
   (b) Find curl $\vec{F}$ using the coordinate definition and relate your answer to circulation density.

19. A smooth vector field $\vec{G}$ has curl $\vec{G}(0, 0, 0) = 2\vec{i} - 3\vec{j} + 5\vec{k}$. Estimate the circulation around a circle of radius 0.01 in each of the following planes:

   (a) xy-plane, oriented counterclockwise when viewed from the positive z-axis.
   (b) yz-plane, oriented counterclockwise when viewed from the positive x-axis.
   (c) xz-plane, oriented counterclockwise when viewed from the positive y-axis.

20. Three small circles, C_1, C_2, and C_3, each with radius 0.1 and centered at the origin are in the xy-, yz-, and xz-planes, respectively. The circles are oriented counterclockwise when viewed from the positive z-, x-, and y-axes, respectively. A vector field, $\vec{F}$, has circulation around C_1 of 0.02π, around C_2 of 0.5π, and around C_3 of 3π. Estimate curl $\vec{F}$ at the origin.

21. Figure 20.24 gives a sketch of a velocity vector field $\vec{F} = y\vec{i} + x\vec{j}$ in the xy-plane.

   (a) What is the direction of rotation of a thin twig placed at the origin along the x-axis?
   (b) What is the direction of rotation of a thin twig placed at the origin along the y-axis?
   (c) Compute curl $\vec{F}$.

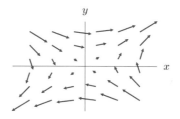

Figure 20.24

22. A tornado is formed when a tube of air circling a horizontal axis is tilted up vertical by the updraft from a thunderstorm. If t is time, this process can be modeled by the wind velocity field

$$\vec{F}(t, x, y, z) = (\cos t\vec{j} + \sin t\vec{k}) \times \vec{r} \quad \text{and} \quad 0 \le t \le \frac{\pi}{2}.$$

Determine the direction of curl $\vec{F}$:

   (a) At $t = 0$ **(b)** At $t = \pi/2$
   (c) For $0 < t < \pi/2$

23. A large fire becomes a fire-storm when the nearby air acquires a circulatory motion. The associated updraft has the effect of bringing more air to the fire, causing it to burn faster. Records show that a fire-storm developed during the Chicago Fire of 1871 and during the Second World War bombing of Hamburg, Germany, but there was no fire-storm during the Great Fire of London in 1666. Explain how a fire-storm could be identified using the curl of a vector field.

24. Show that curl $(\vec{F} + \vec{C}) = $ curl $\vec{F}$ for a constant vector field $\vec{C}$.

25. If $\vec{F}$ is any vector field whose components have continuous second partial derivatives, show div curl $\vec{F} = 0$.

26. We have seen that the Fundamental Theorem of Calculus for Line Integrals implies $\int_C \text{grad } f \cdot d\vec{r} = 0$ for any smooth closed path C and any smooth function f.

   (a) Use the geometric definition of curl to deduce that curl grad $f = \vec{0}$.
   (b) Show that curl grad $f = \vec{0}$ using the coordinate definition.

27. For any constant vector field $\vec{c}$, and any vector field, $\vec{F}$, show that $\operatorname{div}(\vec{F} \times \vec{c}) = \vec{c} \cdot \operatorname{curl}\vec{F}$.

28. Show that $\operatorname{curl}(\phi\vec{F}) = \phi\operatorname{curl}\vec{F} + (\operatorname{grad}\phi) \times \vec{F}$ for a scalar function ϕ and a vector field $\vec{F}$.

29. Show that if $\vec{F} = f\operatorname{grad}g$ for some scalar functions f and g, then $\operatorname{curl}\vec{F}$ is everywhere perpendicular to $\vec{F}$.

30. Let $\vec{F}$ be a smooth vector field and let $\vec{u}$ and $\vec{v}$ be constant vectors. Using the definition of $\operatorname{curl}\vec{F}$ in Cartesian coordinates, show that

$$\operatorname{grad}(\vec{F} \cdot \vec{v}) \cdot \vec{u} - \operatorname{grad}(\vec{F} \cdot \vec{u}) \cdot \vec{v} = (\operatorname{curl}\vec{F}) \cdot \vec{u} \times \vec{v}.$$

31. A vortex that rotates at constant angular velocity ω about the z-axis has velocity vector field $\vec{v} = \omega(-y\vec{i} + x\vec{j})$.

    (a) Sketch the vector field with $\omega = 1$ and the vector field with $\omega = -1$.

    (b) Determine the speed $\|\vec{v}\|$ of the vortex as a function of the distance from its center.

    (c) Compute $\operatorname{div}\vec{v}$ and $\operatorname{curl}\vec{v}$.

    (d) Compute the circulation of $\vec{v}$ counterclockwise about the circle of radius R in the xy-plane, centered at the origin.

32. Let $r = (x^2+y^2)^{1/2}$. Figure 20.25 shows the vector field $r^A \cdot (-y\vec{i} + x\vec{j})$ for $r \neq 0$ and $A = -1, -2$, and -3. The vector fields are shown in the xy-plane; they have no z-component and are independent of z.

    (a) Show that $\operatorname{curl}(r^A \cdot (-y\vec{i} + x\vec{j})) = (2 + A)r^A\vec{k}$ for any constant A.

    (b) Using your answer to part (a), find the direction of the curl of the three vector fields for $A = -1, -2, -3$.

    (c) For each value of A, what (if anything) does your answer to part (b) tell you about the sign of the circulation around a small circle oriented counterclockwise when viewed from above, and centered at $(1, 1, 1)$? Centered at $(0, 0, 0)$?

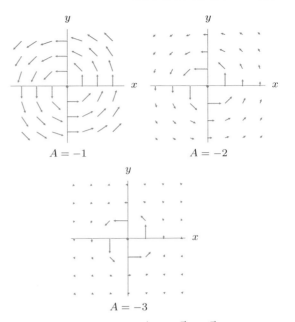

Figure 20.25: The vector field $r^A \cdot (-y\vec{i} + x\vec{j})$ for three values of A

33. Let $\vec{T} = a\vec{i} + b\vec{j}$ be a fixed unit vector, and let $\vec{F} = F(x, y)\vec{T}$ be a vector field everywhere parallel to $\vec{T}$, but of varying magnitude F. Show that $\operatorname{curl}\vec{F}$ equals $\vec{k}$ times the directional derivative of F in the direction of $\vec{F} \times \vec{k}$. Do this in two ways:

    (a) Graphically, using line integrals

    (b) Algebraically

Note that the direction of $\vec{F} \times \vec{k}$ is obtained by rotating $\vec{F}$ through $90°$ clockwise as viewed from above the xy-plane.

20.4   STOKES' THEOREM

The Divergence Theorem says that the integral of the flux density over a solid region is equal to the flux through the surface bounding the region. Similarly, Stokes' Theorem says that the integral of the circulation density over a surface is equal to the circulation around the boundary of the surface.

The Boundary of a Surface

The *boundary* of a surface S is the curve running around the edge of S (like the hem around the edge of a piece of cloth). An orientation of S determines an orientation for its boundary, C, as follows. Pick a positive normal vector $\vec{n}$ on S, near C, and use the right hand rule to determine a direction of travel around $\vec{n}$. This in turn determines a direction of travel around the boundary C. See Figure 20.26. Another way of describing the orientation on C is that someone walking along C in the forward direction, body upright in the direction of the positive normal on S, would have the surface on their left. Notice that the boundary can consist of two or more curves, as the surface on the right in Figure 20.26 shows.

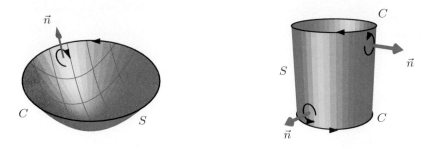

Figure 20.26: Two oriented surfaces and their boundaries

Calculating the Circulation from the Circulation Density

Consider a closed, oriented curve C in 3-space. We can find the circulation of a vector field $\vec{F}$ around C by calculating the line integral:

$$\begin{array}{c}\text{Circulation}\\ \text{around } C\end{array} = \int_C \vec{F} \cdot d\vec{r}.$$

If C is the boundary of an oriented surface S, there is another way to calculate the circulation using $\operatorname{curl} \vec{F}$. We subdivide S into pieces as shown on the surface on the left in Figure 20.26. If $\vec{n}$ is a positive unit normal vector to a piece of surface with area ΔA, then $\Delta \vec{A} = \vec{n}\,\Delta A$. In addition, $\operatorname{circ}_{\vec{n}} \vec{F}$ is the circulation density of $\vec{F}$ around $\vec{n}$, so

$$\begin{array}{c}\text{Circulation of } \vec{F} \text{ around}\\ \text{boundary of the piece}\end{array} \approx \left(\operatorname{circ}_{\vec{n}} \vec{F} \right) \Delta A = ((\operatorname{curl} \vec{F}) \cdot \vec{n})\Delta A = (\operatorname{curl}\vec{F}) \cdot \Delta \vec{A}.$$

Next we add up the circulations around all the small pieces. The line integral along the common edge of a pair of adjacent pieces appears with opposite sign in each piece, so it cancels out. (See Figure 20.27.) When we add up all the pieces the internal edges cancel and we are left with the circulation around C, the boundary of the entire surface. Thus,

$$\begin{array}{c}\text{Circulation}\\ \text{around } C\end{array} = \sum \begin{array}{c}\text{Circulation around}\\ \text{boundary of pieces}\end{array} \approx \sum \operatorname{curl} \vec{F} \cdot \Delta \vec{A}.$$

Taking the limit as $\Delta A \to 0$, we get

$$\begin{array}{c}\text{Circulation}\\ \text{around } C\end{array} = \int_S \operatorname{curl} \vec{F} \cdot d\vec{A}.$$

Figure 20.27: Two adjacent pieces of the surface

We have expressed the circulation as a line integral around C and as a flux integral over S; thus, the two integrals must be equal. Hence we have[4]

---

[4]A proof of Stokes' Theorem using the coordinate definition of curl can be found in the online supplement at www.wiley.com/college/hugheshallett.

Theorem 20.2: Stokes' Theorem

If S is a smooth oriented surface with piecewise smooth, oriented boundary C, and if $\vec{F}$ is a smooth vector field on an open region containing S and C, then

$$\int_C \vec{F} \cdot d\vec{r} = \int_S \text{curl}\, \vec{F} \cdot d\vec{A}.$$

The orientation of C is determined from the orientation of S according to the right hand rule.

Example 1   Let $\vec{F}(x, y, z) = -2y\vec{i} + 2x\vec{j}$. Use Stokes' Theorem to find $\int_C \vec{F} \cdot d\vec{r}$, where C is a circle

(a) Parallel to the yz-plane, of radius a, centered at a point on the x-axis, with either orientation.
(b) Parallel to the xy-plane, of radius a, centered at a point on the z-axis, oriented counterclockwise as viewed from a point on the z-axis above the circle.

Solution   We have $\text{curl}\, \vec{F} = 4\vec{k}$. Figure 20.28 shows sketches of $\vec{F}$ and $\text{curl}\, \vec{F}$.

(a) Let S be the disk enclosed by C. Since S lies in a vertical plane and $\text{curl}\, \vec{F}$ points vertically everywhere, the flux of $\text{curl}\, \vec{F}$ through S is zero. Hence, by Stokes' Theorem,

$$\int_C \vec{F} \cdot d\vec{r} = \int_S \text{curl}\, \vec{F} \cdot d\vec{A} = 0.$$

It makes sense that the line integral is zero. If C is parallel to the yz-plane (even if it is not lying in the plane), the symmetry of the vector field means that the line integral of $\vec{F}$ over the top half of the circle cancels the line integral over the bottom half.

(b) Let S be the horizontal disk enclosed by C. Since $\text{curl}\, \vec{F}$ is a constant vector field pointing in the direction of $\vec{k}$, we have, by Stokes' Theorem,

$$\int_C \vec{F} \cdot d\vec{r} = \int_S \text{curl}\, \vec{F} \cdot d\vec{A} = \|\text{curl}\, \vec{F}\| \cdot \text{Area of } S = 4\pi a^2.$$

Since $\vec{F}$ is circling around the z-axis in the same direction as C, we expect the line integral to be positive. In fact, in Example 1 on page 955, we computed this line integral directly.

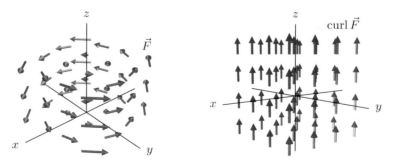

Figure 20.28: The vector fields $\vec{F}$ and $\text{curl}\, \vec{F}$

Curl Free Vector Fields

Stokes' Theorem applies to any oriented surface S and its boundary C, even in cases where the boundary consists of two or more curves. This is useful in studying curl free vector fields.

Example 2   A current I flows along the z-axis in the $\vec{k}$ direction. The induced magnetic field $\vec{B}(x, y, z)$ is

$$\vec{B}(x, y, z) = \frac{2I}{c}\left(\frac{-y\vec{i} + x\vec{j}}{x^2 + y^2}\right),$$

where c is the speed of light. In Example 5 on page 958 we showed that curl $\vec{B} = \vec{0}$.

(a) Compute the circulation of $\vec{B}$ around the circle C_1 in the xy-plane of radius a, centered at the origin, and oriented counterclockwise when viewed from above.
(b) Use part (a) and Stokes' Theorem to compute $\int_{C_2} \vec{B} \cdot d\vec{r}$, where C_2 is the ellipse $x^2 + 9y^2 = 9$ in the plane $z = 2$, oriented counterclockwise when viewed from above.

Solution   (a) On the circle C_1, we have $\|\vec{B}\| = 2I/(ca)$. Since $\vec{B}$ is tangent to C_1 everywhere and points in the forward direction around C_1,

$$\int_{C_1} \vec{B} \cdot d\vec{r} = \int_{C_1} \|\vec{B}\| \, dr = \frac{2I}{ca} \cdot \text{Length of } C_1 = \frac{2I}{ca} \cdot 2\pi a = \frac{4\pi I}{c}.$$

(b) Let S be the conical surface extending from C_1 to C_2 in Figure 20.29 on page 964. The boundary of this surface has two pieces, $-C_2$ and C_1. The orientation of C_1 leads to the outward normal on S, which forces us to choose the clockwise orientation on C_2. By Stokes' Theorem,

$$\int_S \text{curl}\,\vec{B} \cdot d\vec{A} = \int_{-C_2} \vec{B} \cdot d\vec{r} + \int_{C_1} \vec{B} \cdot d\vec{r} = -\int_{C_2} \vec{B} \cdot d\vec{r} + \int_{C_1} \vec{B} \cdot d\vec{r}.$$

Since curl $\vec{B} = \vec{0}$, we have $\int_S \text{curl}\,\vec{B} \cdot d\vec{A} = 0$, so the two line integrals must be equal:

$$\int_{C_2} \vec{B} \cdot d\vec{r} = \int_{C_1} \vec{B} \cdot d\vec{r} = \frac{4\pi I}{c}.$$

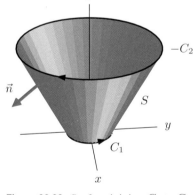

Figure 20.29: Surface joining C_1 to C_2, oriented to satisfy the conditions of Stoke's Theorem

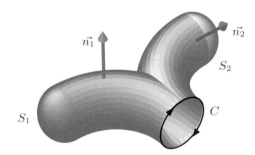

Figure 20.30: The flux of a curl is the same through the two surfaces S_1 and S_2 if they determine the same orientation on the boundary, C

Curl Fields

A vector field $\vec{F}$ is called a *curl field* if $\vec{F} = \text{curl}\,\vec{G}$ for some vector field $\vec{G}$. Recall that if $\vec{F} = \text{grad}\,f$, then f is called a potential function. By analogy, if a vector field $\vec{F} = \text{curl}\,\vec{G}$, then $\vec{G}$ is called a *vector potential* for $\vec{F}$. The following example shows that the flux of a curl field through a surface depends only on the boundary of the surface. This is analogous to the fact that the line integral of a gradient field depends only on the endpoints of the path.

Example 3 Suppose $\vec{F} = \operatorname{curl}\vec{G}$. Suppose that S_1 and S_2 are two oriented surfaces with the same boundary C. Show that, if S_1 and S_2 determine the same orientation on C (as in Figure 20.30), then

$$\int_{S_1}\vec{F}\cdot d\vec{A} = \int_{S_2}\vec{F}\cdot d\vec{A}.$$

If S_1 and S_2 determine opposite orientations on C, then

$$\int_{S_1}\vec{F}\cdot d\vec{A} = -\int_{S_2}\vec{F}\cdot d\vec{A}.$$

Solution Since $\vec{F} = \operatorname{curl}\vec{G}$, by Stokes' Theorem we have

$$\int_{S_1}\vec{F}\cdot d\vec{A} = \int_{S_1}\operatorname{curl}\vec{G}\cdot d\vec{A} = \int_C\vec{G}\cdot d\vec{r}$$

and

$$\int_{S_2}\vec{F}\cdot d\vec{A} = \int_{S_2}\operatorname{curl}\vec{G}\cdot d\vec{A} = \int_C\vec{G}\cdot d\vec{r}.$$

In each case the line integral on the right must be computed using the orientation determined by the surface. Thus, the two flux integrals of $\vec{F}$ are the same if the orientations are the same and they are opposite if the orientations are opposite.

Exercises and Problems for Section 20.4

Exercises

1. Can you use Stokes' Theorem to compute the line integral $\int_C(2x\vec{i}+2y\vec{j}+2z\vec{k})\cdot d\vec{r}$ where C is the straight line from the point $(1,2,3)$ to the point $(4,5,6)$? Why or why not?

2. At all points in 3-space $\operatorname{curl}\vec{F}$ points in the direction of $\vec{i}-\vec{j}-\vec{k}$. Let C be a circle in the yz-plane, oriented clockwise when viewed from the positive x-axis. Is the circulation of $\vec{F}$ around C positive, zero, or negative? Explain.

In Exercises 3–6, use Stokes' Theorem to find the circulation of the vector field around the given paths.

3. $\vec{F} = (z-2y)\vec{i} + (3x-4y)\vec{j} + (z+3y)\vec{k}$ and C is the circle $x^2+y^2=4$, $z=1$, oriented counterclockwise when viewed from above.

4. $\vec{F} = (2x-y)\vec{i} + (x+4y)\vec{j}$ and C is a circle of radius 10, centered at the origin.

   (a) In the xy-plane, oriented clockwise as viewed from the positive z-axis.
   (b) In the yz-plane, oriented clockwise as viewed from the positive x-axis.

5. $\vec{F} = \vec{r}/\|\vec{r}\|^3$ and C is the path consisting of line segments from $(1,0,1)$ to $(1,0,0)$ to $(0,0,1)$ to $(1,0,1)$.

6. $\vec{F} = xz\vec{i} + (x+yz)\vec{j} + x^2\vec{k}$ and C is the circle $x^2+y^2=1$, $z=2$, oriented counterclockwise when viewed from above.

In Exercises 7–11, calculate the circulation, $\int_C\vec{F}\cdot d\vec{r}$, in two ways, directly and using Stokes' Theorem.

7. $\vec{F} = (x-y+z)(\vec{i}+\vec{j})$ and C is the triangle with vertices $(0,0,0)$, $(5,0,0)$, $(5,5,0)$, traversed in that order.

8. $\vec{F} = (x+z)\vec{i} + x\vec{j} + y\vec{k}$ and C is the upper half of the circle $x^2+z^2=9$ in the plane $y=0$, together with the x-axis from $(3,0,0)$ to $(-3,0,0)$, traversed counterclockwise when viewed from the positive y-axis.

9. $\vec{F} = y\vec{i} - x\vec{j}$ and C is the boundary of S, the part of the surface $z=4-x^2-y^2$ above the xy-plane, oriented upward.

10. $\vec{F} = xy\vec{i} + yz\vec{j} + xz\vec{k}$ and C is the boundary of S, the surface $z=1-x^2$ for $0\le x\le 1$ and $-2\le y\le 2$, oriented upward. Sketch S and C.

11. $\vec{F} = y\vec{i} + z\vec{j} + x\vec{k}$ and C is the boundary of S, the paraboloid $z=1-(x^2+y^2)$, $z\ge 0$ oriented upward. [Hint: Use polar coordinates.]

Problems

12. The vector fields $\vec{F}$ and $\vec{G}$ are sketched in Figures 20.31 and 20.32. Each vector field has no z-component and is independent of z. All the axes have the same scales.

(a) What can you say about div $\vec{F}$ and div $\vec{G}$ at the origin?

(b) What can you say about curl $\vec{F}$ and curl $\vec{G}$ at the origin?

(c) Is there a closed surface around the origin such that $\vec{F}$ has a nonzero flux through it?

(d) Repeat part (c) for $\vec{G}$.

(e) Is there a closed curve around the origin such that $\vec{F}$ has a nonzero circulation around it?

(f) Repeat part (e) for $\vec{G}$.

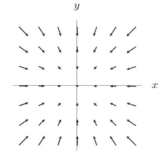

Figure 20.31: Cross-section of $\vec{F}$

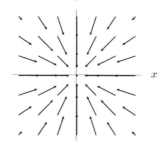

Figure 20.32: Cross-section of $\vec{G}$

13. (a) Find $\operatorname{curl}(y\vec{i} + z\vec{j} + x\vec{k})$.

(b) Find $\int_C (y\vec{i} + z\vec{j} + x\vec{k}) \cdot d\vec{r}$ where C is the boundary of the triangle with vertices $(2, 0, 0)$, $(0, 3, 0)$, $(-2, 0, 0)$, traversed in that order.

14. (a) Let $\vec{F} = y\vec{i} + z\vec{j} + x\vec{k}$. Find curl $\vec{F}$.

(b) Calculate $\int_C \vec{F} \cdot d\vec{r}$ where C is

(i) A circle of radius 2 centered at $(1, 1, 3)$ in the plane $z = 3$, oriented counterclockwise when viewed from above.

(ii) The triangle obtained by tracing out the path $(2, 0, 0)$ to $(2, 0, 5)$ to $(2, 3, 5)$ to $(2, 0, 0)$.

15. (a) Find $\operatorname{curl}(z\vec{i} + x\vec{j} + y\vec{k})$.

(b) Find $\int_C (z\vec{i} + x\vec{j} + y\vec{k}) \cdot d\vec{r}$ where C is a square of side 2 lying in the plane $x + y + z = 5$, oriented counterclockwise when viewed from the origin.

In Problems 16–21, find $\int_C \vec{F} \cdot d\vec{r}$ where C is a circle of radius 2 in the plane $x + y + z = 3$, centered at $(1, 1, 1)$ and oriented clockwise when viewed from the origin.

16. $\vec{F} = \vec{i} + \vec{j} + 3\vec{k}$

17. $\vec{F} = -y\vec{i} + x\vec{j} + z\vec{k}$

18. $\vec{F} = y\vec{i} - x\vec{j} + (y - x)\vec{k}$

19. $\vec{F} = (2y + e^x)\vec{i} + ((\sin y) - x)\vec{j} + (2y - x + \cos z^2)\vec{k}$

20. $\vec{F} = -z\vec{j} + y\vec{k}$

21. $\vec{F} = (z - y)\vec{i} + (x - z)\vec{j} + (y - x)\vec{k}$

22. Evaluate $\int_C (-z\vec{i} + y\vec{j} + x\vec{k}) \cdot d\vec{r}$, where C is a circle of radius 2, parallel to the xz-plane and around the y-axis with the orientation shown in Figure 20.33.

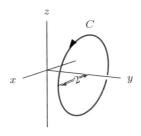

Figure 20.33

23. Evaluate the circulation of $\vec{G} = xy\vec{i} + z\vec{j} + 3y\vec{k}$ around a square of side 6, centered at the origin, lying in the yz-plane, and oriented counterclockwise viewed from the positive x-axis.

24. Suppose that C is a closed curve in the xy-plane, oriented counterclockwise when viewed from above. Show that $\frac{1}{2}\int_C (-y\vec{i} + x\vec{j}) \cdot d\vec{r}$ equals the area of the region R in the xy-plane enclosed by C.

25. (a) Let $\vec{F} = (y - z)\vec{i} + (x + z)\vec{j} + xy\vec{k}$, let C be the circle of radius 3 centered at $(2, 1, 0)$ in the xy-plane, oriented counterclockwise when viewed from above. Compute $\int_C \vec{F} \cdot d\vec{r}$.

(b) Is $\vec{F}$ path-independent (conservative)? Explain.

26. (a) If $\vec{b} = b_1\vec{i} + b_2\vec{j} + b_3\vec{k}$ is a constant vector and $\vec{r} = x\vec{i} + y\vec{j} + z\vec{k}$, show by direct calculation that

$$\operatorname{curl}(\vec{b} \times \vec{r}) = 2\vec{b}.$$

(b) Let C be the circle of radius 5 in the xy-plane, centered at $(3, 3, 0)$ and oriented counterclockwise when seen from above. Use the result of part (a) to calculate

$$\int_C (\vec{b} \times \vec{r}) \cdot d\vec{r}.$$

27. Let $\vec{F} = -y\vec{i} + x\vec{j} + \cos(xy)z\vec{k}$ and let S be the surface of the lower unit hemisphere $x^2 + y^2 + z^2 = 1$, $z \leq 0$, oriented with outward pointing normal. Find $\int_S \operatorname{curl}\vec{F} \cdot d\vec{A}$.

28. Let $\vec{F} = -z\vec{j} + y\vec{k}$, let C be the circle of radius a in the yz-plane oriented clockwise as viewed from the positive x-axis, and let S be the disk in the yz-plane enclosed by C, oriented in the positive x-direction. See Figure 20.34.

   (a) Evaluate directly $\int_C \vec{F} \cdot d\vec{r}$.

   (b) Evaluate directly $\int_S \text{curl}\,\vec{F} \cdot d\vec{A}$.

   (c) The answers in parts (a) and (b) are not equal. Explain why this does not contradict Stokes' Theorem.

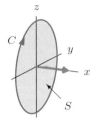

Figure 20.34

29. (a) Show that $\vec{G} = 4yz^2\vec{i} + 3x\vec{j} + xz\vec{k}$ is a vector potential for $\vec{F} = (8yz - z)\vec{j} + (3 - 4z^2)\vec{k}$.

   (b) Evaluate $\int_S \vec{F} \cdot d\vec{A}$ where S is the hemisphere of radius 5 shown in Figure 20.35, oriented upward. [Hint: Use Example 3 on page 965 to simplify the calculation.]

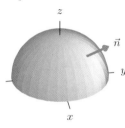

Figure 20.35

30. Use Stokes' Theorem to show that $\int_S \text{curl}\,\vec{F} \cdot d\vec{A} = 0$ for any surface S which is the boundary surface of a solid region W. Use the geometric definition of the divergence to deduce that $\text{div}\,\text{curl}\,\vec{F} = 0$.

31. Show that Green's Theorem is a special case of Stokes' Theorem.

32. A vector field $\vec{F}$ is defined everywhere except on the z-axis and $\text{curl}\,\vec{F} = \vec{0}$ everywhere where $\vec{F}$ is defined. What can you say about $\int_C \vec{F} \cdot d\vec{r}$ if C is a circle of radius 1 in the xy-plane, and if the center of C is at
   (a) The origin?   **(b)** The point $(2, 0, 0)$?

33. For constants a, b, c, m, consider the vector field

$$\vec{F} = (ax + by + 5z)\vec{i} + (x + cz)\vec{j} + (3y + mx)\vec{k}.$$

   (a) Suppose that the flux of $\vec{F}$ through any closed surface is 0. What does this tell you about the values of the constants a, b, c, m?

   (b) Suppose instead that the circulation of $\vec{F}$ around any closed curve is 0. What does this tell you about the values of the constants a, b, c, m?

34. Water in a bathtub has velocity vector field near the drain given, for x, y, z in cm, by

$$\vec{F} = -\frac{y + xz}{(z^2 + 1)^2}\vec{i} - \frac{yz - x}{(z^2 + 1)^2}\vec{j} - \frac{1}{z^2 + 1}\vec{k} \text{ cm/sec.}$$

   (a) Rewriting $\vec{F}$ as follows, describe in words how the water is moving:

$$\vec{F} = \frac{-y\vec{i} + x\vec{j}}{(z^2 + 1)^2} + \frac{-z(x\vec{i} + y\vec{j})}{(z^2 + 1)^2} - \frac{\vec{k}}{z^2 + 1}.$$

   (b) The drain in the bathtub is a disk in the xy-plane with center at the origin and radius 1 cm. Find the rate at which the water is leaving the bathtub. (That is, find the rate at which water is flowing through the disk.) Give units for your answer.

   (c) Find the divergence of $\vec{F}$.

   (d) Find the flux of the water through the hemisphere of radius 1, centered at the origin, lying below the xy-plane and oriented downward.

   (e) Find $\int_C \vec{G} \cdot d\vec{r}$ where C is the edge of the drain, oriented clockwise when viewed from above, and where

$$\vec{G} = \frac{1}{2}\left(\frac{y}{z^2 + 1}\vec{i} - \frac{x}{z^2 + 1}\vec{j} - \frac{x^2 + y^2}{(z^2 + 1)^2}\vec{k} \right).$$

   (f) Calculate $\text{curl}\,\vec{G}$.

   (g) Explain why your answers to parts (d) and (e) are equal.

35. Let $\vec{F} = \dfrac{-y\vec{i} + x\vec{j}}{x^2 + y^2}$.

   (a) Calculate $\text{curl}\,\vec{F}$. What is the domain of $\text{curl}\,\vec{F}$?

   (b) Find the circulation of $\vec{F}$ around the unit circle C_1 in the xy-plane, oriented counterclockwise when viewed from above.

   (c) Find the circulation of $\vec{F}$ around the circle C_2 in the plane $z = 4$ and with equation $(x - 3)^2 + y^2 = 1$, oriented counterclockwise when viewed from above.

   (d) Find the circulation of $\vec{F}$ around the square S with corners $(2, 2, 0)$, $(-2, 2, 0)$, $(-2, -2, 0)$, $(2, -2, 0)$, oriented counterclockwise when viewed from above.

   (e) Using your results to parts (b)–(d) explain how, with no additional calculation, you can find the circulation of $\vec{F}$ around any simple closed curve in the xy-plane, provided it does not intersect the z-axis. (A simple closed curve does not cross itself.)

36. Let $\vec{r} = x\vec{i} + y\vec{j}$. In the xy-plane, let C be the circle of radius a centered at the origin and oriented counterclockwise when viewed from above.

   (a) At the point $(x, y, 0)$, find a unit vector, $\vec{T}$, perpendicular to $\vec{r}$ and pointing counterclockwise around the origin when viewed from above.

   (b) Let $F = ||\vec{r}||^n \vec{T}$ for $n \geq 1$. Find $\int_C \vec{F} \cdot d\vec{r}$ by direct computation.

   (c) Calculate $\text{curl}\,\vec{F}$.

   (d) Find $\int_C \vec{F} \cdot d\vec{r}$ using your answer to part (c).

20.5 THE THREE FUNDAMENTAL THEOREMS

We have now seen three multivariable versions of the Fundamental Theorem of Calculus. In this section we will examine some consequences of these theorems.

Fundamental Theorem of Calculus for Line Integrals

$$\int_C \operatorname{grad} f \cdot d\vec{r} = f(Q) - f(P).$$

Stokes' Theorem

$$\int_S \operatorname{curl} \vec{F} \cdot d\vec{A} = \int_C \vec{F} \cdot d\vec{r}.$$

Divergence Theorem

$$\int_W \operatorname{div} \vec{F}\, dV = \int_S \vec{F} \cdot d\vec{A}.$$

Notice that, in each case, the region of integration on the right is the boundary of the region on the left (except that for the first theorem we simply evaluate f at the boundary points); the integrand on the left is a sort of derivative of the integrand on the right; see Figure 20.36.

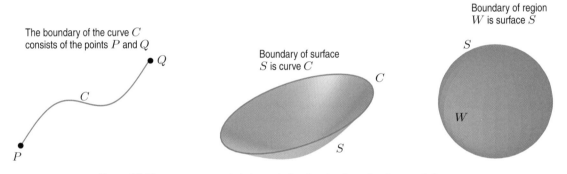

Figure 20.36: Regions and their boundaries for the three fundamental theorems

The Gradient and the Curl

Suppose that $\vec{F}$ is a smooth gradient field, so $\vec{F} = \operatorname{grad} f$ for some function f. Using the Fundamental Theorem for Line Integrals, we saw in Chapter 18 that

$$\int_C \vec{F} \cdot d\vec{r} = 0$$

for any closed curve C. Thus, for any unit vector $\vec{n}$

$$\operatorname{circ}_{\vec{n}} \vec{F} = \lim_{\text{Area} \to 0} \frac{\displaystyle\int_C \vec{F} \cdot d\vec{r}}{\text{Area of } C} = \lim_{\text{Area} \to 0} \frac{0}{\text{Area}} = 0,$$

where the limit is taken over circles C in a plane perpendicular to $\vec{n}$, and oriented by the right hand rule. Thus the circulation density of $\vec{F}$ is zero in every direction, so $\operatorname{curl} \vec{F} = \vec{0}$, that is,

$$\boxed{\operatorname{curl} \operatorname{grad} f = \vec{0}.}$$

(This formula can also be verified using the coordinate definition of curl. See Problem 26 on page 960.)

Is the converse true? Is any vector field whose curl is zero a gradient field? Suppose that $\text{curl}\,\vec{F} = \vec{0}$ and let us consider the line integral $\int_C \vec{F} \cdot d\vec{A}$ for a closed curve C contained in the domain of $\vec{F}$. If C is the boundary curve of an oriented surface S that lies wholly in the domain of $\text{curl}\,\vec{F}$, then Stokes' Theorem asserts that

$$\int_C \vec{F} \cdot d\vec{r} = \int_S \text{curl}\,\vec{F} \cdot d\vec{A} = \int_S \vec{0} \cdot d\vec{A} = 0.$$

If we knew that $\int_C \vec{F} \cdot d\vec{r} = 0$ for every closed curve C, then $\vec{F}$ would be path-independent, and hence a gradient field. Thus we need to know whether every closed curve in the domain of $\vec{F}$ is the boundary of an oriented surface contained in the domain. It can be quite difficult to determine if a given curve is the boundary of a surface (suppose, for example, that the curve is knotted in a complicated way). However, if the curve can be contracted smoothly to a point, remaining all the time in the domain of $\vec{F}$, then it is the boundary of a surface, namely, the surface it sweeps through as it contracts. Thus, we have proved the test for a gradient field that we stated in Chapter 18.

The Curl Test for Vector Fields in 3-Space

Suppose $\vec{F}$ is a smooth vector field on 3-space such that
- The domain of $\vec{F}$ has the property that every closed curve in it can be contracted to a point in a smooth way, staying at all times within the domain.
- $\text{curl}\,\vec{F} = \vec{0}$.

Then $\vec{F}$ is path-independent, and thus is a gradient field.

Example 6 on page 905 shows how the curl test is applied.

The Curl and The Divergence

In this section we will use the second two fundamental theorems to get a test for a vector field to be a curl field, that is, a field of the form $\vec{F} = \text{curl}\,\vec{G}$ for some $\vec{G}$.

Example 1   Suppose that $\vec{F}$ is a smooth curl field. Use Stokes' Theorem to show that for any closed surface, S, contained in the domain of $\vec{F}$

$$\int_S \vec{F} \cdot d\vec{A} = 0.$$

Solution   Suppose $\vec{F} = \text{curl}\,\vec{G}$. Draw a closed curve C on the surface S, thus dividing S into two surfaces S_1 and S_2 as shown in Figure 20.37. Pick the orientation for C corresponding to S_1; then the orientation of C corresponding to S_2 is the opposite. Thus, using Stokes' Theorem,

$$\int_{S_1} \vec{F} \cdot d\vec{A} = \int_{S_1} \text{curl}\,\vec{G} \cdot d\vec{A} = \int_C \vec{G} \cdot d\vec{r} = -\int_{S_2} \text{curl}\,\vec{G} \cdot d\vec{A} = -\int_{S_2} \vec{F} \cdot d\vec{A}.$$

Thus, for any closed surface S, we have

$$\int_S \vec{F} \cdot d\vec{A} = \int_{S_1} \vec{F} \cdot d\vec{A} + \int_{S_2} \vec{F} \cdot d\vec{A} = 0.$$

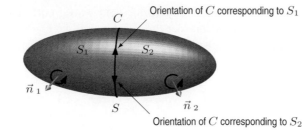

Figure 20.37: The closed surface S divided into two surfaces S_1 and S_2

Thus, if $\vec{F} = \operatorname{curl} \vec{G}$, we use the result of Example 1 to see that

$$\operatorname{div} \vec{F} = \lim_{\text{Volume} \to 0} \frac{\displaystyle\int_S \vec{F} \cdot d\vec{A}}{\text{Volume enclosed by } S} = \lim_{\text{Volume} \to 0} \frac{0}{\text{Volume}} = 0,$$

where the limit is taken over spheres S contracting down to a point. So we conclude that:

$$\boxed{\operatorname{div} \operatorname{curl} \vec{G} = 0.}$$

(This formula can also be verified using coordinates. See Problem 25 on page 960.)

Is every vector field whose divergence is zero a curl field? It turns out that we have the following analogue of the curl test, though we will not prove it.

The Divergence Test for Vector Fields in 3-Space

Suppose $\vec{F}$ is a smooth vector field on 3-space such that
- The domain of $\vec{F}$ has the property that every closed surface in it is the boundary of a solid region completely contained in the domain.
- $\operatorname{div} \vec{F} = 0$.

Then $\vec{F}$ is a curl field.

Example 2   Consider the vector fields $\vec{E} = q\dfrac{\vec{r}}{\|\vec{r}\|^3}$ and $\vec{B} = \dfrac{2I}{c}\left(\dfrac{-y\vec{i} + x\vec{j}}{x^2 + y^2}\right)$.

(a) Calculate $\operatorname{div} \vec{E}$ and $\operatorname{div} \vec{B}$.
(b) Do $\vec{E}$ and $\vec{B}$ satisfy the divergence test?
(c) Is either $\vec{E}$ or $\vec{B}$ a curl field?

Solution   (a) Example 3 on page 943 shows that $\operatorname{div} \vec{E} = 0$. The following calculation shows $\operatorname{div} \vec{B} = 0$ also:

$$\operatorname{div} \vec{B} = \frac{2I}{c}\left(\frac{\partial}{\partial x}\left(\frac{-y}{x^2 + y^2}\right) + \frac{\partial}{\partial y}\left(\frac{x}{x^2 + y^2}\right) + \frac{\partial}{\partial z}(0)\right)$$

$$= \frac{2I}{c}\left(\frac{2xy}{(x^2 + y^2)^2} + \frac{-2yx}{(x^2 + y^2)^2}\right) = 0.$$

(b) The domain of $\vec{E}$ is 3-space minus the origin, so a region is contained in the domain if it misses the origin. Thus the surface of a sphere centered at the origin is contained in the domain of E, but the solid ball inside is not. Hence $\vec{E}$ does not satisfy the divergence test.

The domain of $\vec{B}$ is 3-space minus the z-axis, so a region is contained in the domain if it avoids the z-axis. If S is a surface bounding a solid region W, then the z-axis cannot pierce

W without piercing S as well. Hence, if S avoids the z-axis, so does W. Thus $\vec{B}$ satisfies the divergence test.

(c) In Example 3 on page 950 we computed the flux of $\vec{r}/\|\vec{r}\|^3$ through a sphere centered at the origin, and found it was 4π, so the flux of $\vec{E}$ through this sphere is $4\pi q$. Thus, $\vec{E}$ cannot be a curl field, because by Example 1, the flux of a curl field through a closed surface is zero.

On the other hand, $\vec{B}$ satisfies the divergence test, so it must be a curl field. In fact, Problem 7 below shows that

$$\vec{B} = \operatorname{curl}\left(\frac{-I}{c}\ln(x^2 + y^2)\vec{k}\right).$$

Exercises and Problems for Section 20.5

Exercises

Which of the vector fields in Exercises 1–2 is a gradient field?

1. $\vec{F} = yz\vec{i} + (xz + z^2)\vec{j} + (xy + 2yz)\vec{k}$

2. $\vec{G} = -y\vec{i} + x\vec{j}$

3. Let $\vec{B} = b\vec{k}$, for some constant b. Show that the following are all possible vector potentials for $\vec{B}$:

(a) $\vec{A} = -by\vec{i}$ (b) $\vec{A} = bx\vec{j}$

(c) $\vec{A} = \frac{1}{2}\vec{B}\times\vec{r}$.

Problems

4. Express $(3x + 2y)\vec{i} + (4x + 9y)\vec{j}$ as the sum of a curl free vector field and a divergence free vector field.

5. Find a vector field $\vec{F}$ such that $\operatorname{curl}\vec{F} = 2\vec{i} - 3\vec{j} + 4\vec{k}$. [Hint: Try $\vec{F} = \vec{v}\times\vec{r}$ for some vector $\vec{v}$.]

6. Find a vector potential for the constant vector field $\vec{B}$ whose value at every point is $\vec{b}$.

7. Show that $\vec{A} = \dfrac{-I}{c}\ln(x^2 + y^2)\vec{k}$ is a vector potential for $\vec{B} = \dfrac{2I}{c}\left(\dfrac{-y\vec{i} + x\vec{j}}{x^2 + y^2}\right)$.

8. Is there a vector field $\vec{G}$ such that $\operatorname{curl}\vec{G} = y\vec{i} + x\vec{j}$? How do you know?

For each vector field in Problems 9–10, determine whether a vector potential exists. If so, find one.

9. $\vec{F} = 2x\vec{i} + (3y - z^2)\vec{j} + (x - 5z)\vec{k}$

10. $\vec{G} = x^2\vec{i} + y^2\vec{j} + z^2\vec{k}$

11. An electric charge q at the origin produces an electric field $\vec{E} = q\vec{r}/\|\vec{r}\|^3$.

(a) Does $\operatorname{curl}\vec{E} = \vec{0}$?
(b) Does $\vec{E}$ satisfy the curl test?
(c) Is $\vec{E}$ a gradient field?

12. Suppose c is the speed of light. A thin wire along the z-axis carrying a current I produces a magnetic field

$$\vec{B} = \frac{2I}{c}\left(\frac{-y\vec{i} + x\vec{j}}{x^2 + y^2}\right),$$

(a) Does $\operatorname{curl}\vec{B} = \vec{0}$?
(b) Does $\vec{B}$ satisfy the curl test?
(c) Is $\vec{B}$ a gradient field?

13. For constant p, consider the vector field $\vec{E} = \dfrac{\vec{r}}{\|\vec{r}\|^p}$.

(a) Find $\operatorname{curl}\vec{E}$.
(b) Find the domain of $\vec{E}$.
(c) For which values of p does $\vec{E}$ satisfy the curl test? For those values of p, find a potential function for $\vec{E}$.

14. The magnetic field, $\vec{B}$, due to a magnetic dipole with moment $\vec{\mu}$ satisfies $\operatorname{div}\vec{B} = 0$ and is given by

$$\vec{B} = -\frac{\vec{\mu}}{\|\vec{r}\|^3} + \frac{3(\vec{\mu}\cdot\vec{r})\vec{r}}{\|\vec{r}\|^5}, \qquad \vec{r}\neq\vec{0}.$$

(a) Does $\vec{B}$ satisfy the divergence test?
(b) Show that a vector potential for $\vec{B}$ is given by $\vec{A} = \dfrac{\vec{\mu}\times\vec{r}}{\|\vec{r}\|^3}$.
[Hint: Use Problem 28 on page 961. The identities in Example 3 on page 957, Problem 40 on page 718, and Problem 32 on page 679 might also be useful.]
(c) Does your answer to part (a) contradict your answer to part (b)? Explain.

15. Suppose that $\vec{A}$ is a vector potential for $\vec{B}$.

(a) Show that $\vec{A} + \operatorname{grad}\psi$ is also a vector potential for $\vec{B}$, for any function ψ with continuous second-order partial derivatives. (The vector potentials $\vec{A}$ and $\vec{A} + \operatorname{grad}\psi$ are called *gauge equivalent* and the transformation, for any ψ, from $\vec{A}$ to $\vec{A} + \operatorname{grad}\psi$ is called a *gauge transformation*.)
(b) What is the divergence of $\vec{A} + \operatorname{grad}\psi$? How should ψ be chosen such that $\vec{A} + \operatorname{grad}\psi$ has zero divergence? (If $\operatorname{div}\vec{A} = 0$, the magnetic vector potential $\vec{A}$ is said to be in *Coulomb gauge*.)

CHAPTER SUMMARY (see also Ready Reference at the end of the book)

- **Divergence**
    Geometric and coordinate definition of divergence, calculating divergence, interpretation in terms of outflow per unit volume.
- **Curl**
    Geometric and coordinate definition of curl, calculating curl, interpretation in terms of circulation per unit

area.
- **The Divergence Theorem**
    Statement of the theorem, divergence free fields, harmonic functions.
- **Stokes' Theorem**
    Statement of the theorem, curl free and curl fields.

REVIEW EXERCISES AND PROBLEMS FOR CHAPTER TWENTY

Exercises

Exercises 1–3 concern the vector fields in Figure 20.38. In each case, assume that the cross-section is the same in all other planes parallel to the given cross-section.

1. Three of the vector fields have zero curl at each point shown. Which are they? How do you know?

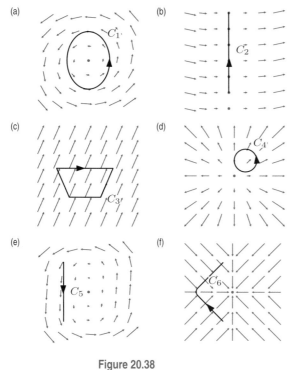

(a) (b) (c) (d) (e) (f)

Figure 20.38

2. Three of the vector fields have zero divergence at each point shown. Which are they? How do you know?

3. Four of the line integrals $\int_{C_i} \vec{F} \cdot d\vec{r}$ are zero. Which are they? How do you know?

In Exercises 4–7, calculate $\text{div} \vec{F}$ and $\text{curl} \vec{F}$. Is $\vec{F}$ solenoidal or irrotational?

4. $\vec{F} = x^2\vec{i} + y^3\vec{j} + z^4\vec{k}$

5. $\vec{F} = xy\vec{i} + yz\vec{j} + zx\vec{k}$

6. $\vec{F} = (\cos x)\,\vec{i} + e^y\vec{j} + (x + y + z)\,\vec{k}$

7. $\vec{F} = e^{y+z}\vec{i} + \sin(x + z)\vec{j} + \left(x^2 + y^2\right)\vec{k}$

8. Compute the flux integral $\int_S (x^3\vec{i} + 2y\vec{j} + 3\vec{k})\cdot d\vec{A}$, where S is the $2 \times 2 \times 2$ rectangular surface centered at the origin, oriented outward. Do this in two ways:

    (a) Directly   **(b)** Using the Divergence Theorem

9. Calculate the flux of $\vec{F} = xy\vec{i} + yz\vec{j} + zx\vec{k}$ out of the closed box $0 \le x \le 1, 0 \le y \le 1, 0 \le z \le 1$

    (a) Directly   **(b)** Using the Divergence Theorem.

10. Compute the line integral $\int_C ((yz^2 - y)\vec{i} + (xz^2 + x)\vec{j} + 2xyz\vec{k})\cdot d\vec{r}$ where C is the circle of radius 3 in the xy-plane, centered at the origin, oriented counterclockwise as viewed from the positive z-axis. Do it two ways:
    (a) Directly   **(b)** Using Stokes' Theorem

In Exercises 11–14, use the Divergence Theorem to calculate the flux of the vector field out of the given surface.

11. $\vec{F} = (2x + y)\vec{i} + (3y + z)\vec{j} + (4z + x)\vec{k}$ and S is the sphere of radius 5 centered at the origin.

12. $\vec{F} = (e^z + x)\,\vec{i} + (2y + \sin x)\vec{j} + \left(e^{x^2} - z\right)\vec{k}$ and S is the sphere of radius 1 centered at $(2, 1, 0)$.

13. $\vec{F} = x^2\vec{i} + y^2\vec{j} + e^{xy}\vec{k}$ and S is the cube of side 3 in the first octant with one corner at the origin, and sides parallel to the axes.

14. $\vec{F} = x^3\vec{i} + y^3\vec{j} + \left(x^2 + y^2\right)\vec{k}$ and S is the cylinder $x^2 + y^2 = 4$ with $0 \le z \le 5$.

In Exercises 15–18, use Stokes' Theorem to calculate the integral.

15. $\int_C \vec{F} \cdot d\vec{r}$ where $\vec{F} = x^2\vec{i} + y^2\vec{j} + z^2\vec{k}$ and C is the unit circle in the xz-plane, oriented counterclockwise when viewed from the positive y-axis.

16. $\int_C \vec{F} \cdot d\vec{r}$ where $\vec{F} = (y-x)\vec{i} + (z-y)\vec{j} + (x-z)\vec{k}$ and C is the circle $x^2 + y^2 = 5$ in the xy-plane, oriented counterclockwise when viewed from above.

17. $\int_S \text{curl } \vec{F} \cdot d\vec{A}$ where $\vec{F} = -y\vec{i} + x\vec{j} + (xy + \cos z)\vec{k}$ and S is the disk $x^2 + y^2 \leq 9$, oriented upward in the xy-plane.

18. $\int_S \text{curl } \vec{F} \cdot d\vec{A}$ where $\vec{F} = (x+7)\vec{j} + e^{x+y+z}\vec{k}$ and S is the rectangle $0 \leq x \leq 3$, $0 \leq y \leq 2$, $z = 0$, oriented counterclockwise when viewed from above.

19. (a) If $\vec{F} = (\cos x)\vec{i} + e^y\vec{j} + (x - y - z)\vec{k}$, find curl $\vec{F}$.

(b) Find $\int_C \vec{F} \cdot d\vec{r}$ where C is the circle of radius 3 in the plane $x + y + z = 1$, centered at $(1, 0, 0)$ oriented counterclockwise when viewed from above.

20. Let $\vec{F} = y\vec{i} - x\vec{j}$ and let C be the unit circle in the xy-plane centered at the origin and oriented counterclockwise when viewed from above.

(a) Calculate $\int_C \vec{F} \cdot d\vec{r}$ by parameterizing the circle.

(b) Calculate curl $\vec{F}$.

(c) Calculate $\int_C \vec{F} \cdot d\vec{r}$ using your result from part (b).

(d) What theorem did you use in part (c)?

Problems

21. Choose a, b, c such that, for all x, y, z,

$$\text{curl}((2z + ax^2y)\vec{i} + (5z + x^3)\vec{j} + (bx + cy + z)\vec{k}) = \vec{0}.$$

22. A central vector field is one of the form $\vec{F} = f(r)\vec{r}$ where f is any function of $r = \|\vec{r}\|$. Show that any central vector field is irrotational.

23. The flux of $\vec{F}$ out of a small sphere of radius 0.1 centered at $(4, 5, 2)$, is 0.0125. Estimate:

(a) div $\vec{F}$ at $(4, 5, 2)$

(b) The flux of $\vec{F}$ out of a sphere of radius 0.2 centered at $(4, 5, 2)$.

24. Use the geometric definition of divergence to find div $\vec{v}$ at the origin, where $\vec{v} = -2\vec{r}$. Check that you get the same result using the definition in Cartesian coordinates.

25. (a) Find the flux of $\vec{F} = 2x\vec{i} - 3y\vec{j} + 5z\vec{k}$ through a box with four of its corners at the points $(a, b, c), (a+w, b, c), (a, b+w, c), (a, b, c+w)$ and edge length w. See Figure 20.39.

(b) Use the geometric definition and part (a) to find div $\vec{F}$ at the point (a, b, c).

(c) Find div $\vec{F}$ using partial derivatives.

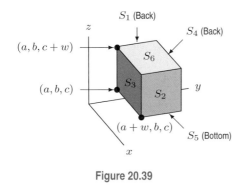

Figure 20.39

26. Suppose $\vec{F} = (3x + 2)\vec{i} + 4x\vec{j} + (5x + 1)\vec{k}$. Use the method of Exercise 25 to find div $\vec{F}$ at the point (a, b, c) by two different methods.

27. The vector field, $\vec{F}$, in Figure 20.40 has no z-component and is independent of z.

(a) Do you think that $\vec{F}$ is a conservative (that is, path independent) vector field? Justify your answer using a sketch.

(b) Do you think curl $\vec{F} \cdot \vec{k}$ is positive, negative, or zero? Explain.

(c) Give a possible formula for $\vec{F}$.

(d) Calculate curl $\vec{F}$ using the formula you gave for $\vec{F}$.

Figure 20.40

In Problems 28–34, calculate the integral.

28. $\int_S \vec{F} \cdot d\vec{A}$, where $\vec{F} = 3x\vec{i} + 4y\vec{j} + xy\vec{k}$ and S is the closed rectangular box whose top face has corners $(0, 0, 0), (3, 0, 0), (3, 5, 0), (0, 5, 0)$, and whose bottom face contains the corner $(0, 0, -2)$.

29. $\int_S \vec{F} \cdot d\vec{A}$ where $\vec{F} = (y^2 + 3x)\vec{i} + (x^2 - y)\vec{j} + 2z\vec{k}$ and S is the unit sphere centered at the origin.

30. $\int_S \text{curl}\vec{F} \cdot d\vec{A}$ where $\vec{F} = (z + y)\vec{i} - (z + x)\vec{j} + (y + x)\vec{k}$ and S is the disk $y^2 + z^2 \leq 3, x = 0$, oriented in the positive x-direction.

31. $\int_S \vec{F} \cdot d\vec{A}$ where $\vec{F} = x^3\vec{i} + y^3\vec{j} + z^3\vec{k}$ and S is the sphere of radius 1 centered at the origin

32. $\int_C \vec{F} \cdot d\vec{r}$ where $\vec{F} = (x + y)\vec{i} + (y + 2z)\vec{j} + (z + 3x)\vec{k}$ and C is a square of side 7 in the xz-plane, oriented counterclockwise when viewed from the positive y-axis.

33. $\int_C \vec{F} \cdot d\vec{r}$ where $\vec{F} = (x - y^3 + z)\vec{i} + (x^3 + y + z)\vec{j} + (x + y + z^3)\vec{k}$ and C is the circle $x^2 + y^2 = 10$, oriented counterclockwise when viewed from above.

34. $\int_S \vec{F} \cdot d\vec{A}$ where $\vec{F} = (y^3 z^3)\vec{i} + y^3 \vec{j} + z^3 \vec{k}$ and S is the cylinder $y^2 + z^2 = 16$, $-1 \le x \le 1$.

In Problems 35–37, find the flux of the vector fields through the upper half of the sphere of radius 5, centered at the origin and oriented upward.

35. $\vec{F} = (x + \cos y)\vec{i} + (y + \sin x)\vec{j} + (z + 3)\vec{k}$

36. $\vec{F} = (\sin z)\vec{i} + e^{xz}\vec{j} + (x^2 + y^2)\vec{k}$

37. $\vec{F} = \text{curl}(-y\vec{i} + x\vec{j})$

38. Are the following vector fields conservative?

(a) $\vec{F}(x, y, z) = y^2 z \vec{i} + 2xyz\vec{j} + xy\vec{k}$
(b) The vector field in Figure 20.41.

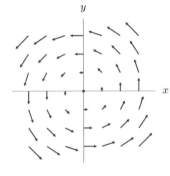

Figure 20.41

39. Figure 20.42 shows the part of a vector field $\vec{E}$ that lies in the xy-plane. Assume that the vector field is independent of z, so that any horizontal cross section looks the same. What can you say about div $\vec{E}$ at the points marked P and Q, assuming it is defined?

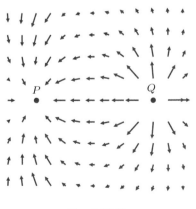

Figure 20.42

40. A box of side 1 in the first octant has faces in the planes $x = 0$, $x = 3$, $y = 0$, $y = 3$, $z = 0$, $z = 3$. Remove the face in the yz-plane to have an open surface, S, with a square boundary C. Let $\vec{F} = (x + y)\vec{i} - z\vec{j} + y\vec{k}$.

(a) Calculate $\int_C \vec{F} \cdot d\vec{r}$, where C is oriented counterclockwise when viewed from the positive x-axis.
(b) Calculate $\int_S \vec{F} \cdot d\vec{A}$, when S is oriented in the direction of the positive x-axis.

41. Let $\vec{r} = x\vec{i} + y\vec{j} + z\vec{k}$ and $\vec{c} = c_1\vec{i} + c_2\vec{j} + c_3\vec{k}$, a constant vector; let S be a sphere of radius R centered at the origin. Find

(a) $\text{div}(\vec{r} \times \vec{c})$
(b) $\int_S (\vec{r} \times \vec{c}) \cdot d\vec{A}$

42. (a) Let $\text{div} \vec{F} = x^2 + y^2 + z^2 + 3$. Calculate $\int_{S_1} \vec{F} \cdot d\vec{A}$ where S_1 is the sphere of radius 1 centered at the origin.
(b) Let S_2 be the sphere of radius 2 centered at the origin; let S_3 be the sphere of radius 3 centered at the origin; let S_4 be the box of side 6 centered at the origin with edges parallel to the axes. Without calculating them, arrange the following integrals in increasing order:

$$\int_{S_2} \vec{F} \cdot d\vec{A}, \quad \int_{S_3} \vec{F} \cdot d\vec{A}, \quad \int_{S_4} \vec{F} \cdot d\vec{A}.$$

43. Assume $\vec{r} \neq \vec{0}$. Let $\vec{F} = \vec{r}/\|\vec{r}\|^3$.

(a) Calculate the flux of $\vec{F}$ out of the unit sphere $x^2 + y^2 + z^2 = 1$ oriented outward.
(b) Calculate div $\vec{F}$. Simplify your answer completely.
(c) Use your answer to parts (a) and (b) to calculate the flux out of a box of side 10 centered at the origin and with sides parallel to the coordinate planes. (The box is also oriented outward.)

44. Let $\vec{r} = x\vec{i} + y\vec{j} + z\vec{k}$. Suppose $\vec{F} = \vec{r}/\|\vec{r}\|^3$. Find $\int_S \vec{F} \cdot d\vec{A}$ where S is the ellipsoid $x^2 + 2y^2 + 3z^2 = 6$. Give reasons for your computation.

45. Let $\vec{r} = x\vec{i} + y\vec{j} + z\vec{k}$. Find $\int_S \vec{F} \cdot d\vec{A}$, where $\vec{F} = \vec{r}/\|\vec{r}\|^3$ and S is the closed cylinder $y^2 + z^2 = 4$, $-2 \le x \le 2$. Give reasons for your computation.

46. For $\vec{r} = x\vec{i} + y\vec{j} + z\vec{k}$, $\vec{r} \neq \vec{0}$, and constant a, let

$$\vec{F} = \frac{\vec{r}}{\|\vec{r}\|^a}.$$

(a) Find div $\vec{F}$. Simplify completely. Your answer will contain a.
(b) Find a value of a such that div $\vec{F} = 0$.
(c) Using a sketch, explain why you would expect curl $\vec{F} = \vec{0}$ for all a.

47. If V is a volume surrounded by a closed surface S, show that $\frac{1}{3}\int_S \vec{r} \cdot d\vec{A} = V$.

48. Use Problem 47 to compute the volume of a sphere of radius R given that its surface area is $4\pi R^2$.

49. Use Problem 47 to compute the volume of a cone of base radius b and height h. [Hint: Stand the cone with its point downward and its axis along the positive z-axis.]

50. Suppose $\text{div } \vec{F}(x,y,z) = 4$ everywhere. Which of the following quantities can be computed from this information? Give the value of those that can be computed.

(a) $\int_S \vec{F} \cdot d\vec{A}$, where S is a sphere of radius 2 centered at the origin and oriented outward.

(b) $\int_C \vec{F} \cdot d\vec{r}$, where C is the unit circle in the xy-plane, oriented counterclockwise viewed from above.

(c) $\int_S \text{curl } \vec{F} \cdot d\vec{A}$, where S is a sphere of radius 2 centered at the origin and oriented outward.

51. Due to roadwork ahead, the traffic on a highway slows linearly from 55 miles/hour to 15 miles/hour over a 2000 foot stretch of road, then crawls along at 15 miles/hour for 5000 feet, then speeds back up linearly to 55 miles/hour in the next 1000 feet, after which it moves steadily at 55 miles/hour.

(a) Sketch a velocity vector field for the traffic flow.

(b) Write a formula for the velocity vector field $\vec{v}$ (miles/hour) as a function of the distance x feet from the initial point of slowdown. (Take the direction of motion to be $\vec{i}$ and consider the various sections of the road separately.)

(c) Compute $\text{div } \vec{v}$ at $x = 1000, 5000, 7500, 10{,}000$. Be sure to include the proper units.

52. The velocity field $\vec{v}$ in Problem 51 does not give a complete description of the traffic flow, for it takes no account of the spacing between vehicles. Let ρ be the density (cars/mile) of highway, where we assume that ρ depends only on x.

(a) Using your highway experience, arrange in ascending order: $\rho(0), \rho(1000), \rho(5000)$.

(b) What are the units and interpretation of the vector field $\rho\vec{v}$?

(c) Would you expect $\rho\vec{v}$ to be constant? Why? What does this mean for $\text{div}(\rho\vec{v})$?

(d) Determine $\rho(x)$ if $\rho(0) = 75$ cars/mile and $\rho\vec{v}$ is constant.

(e) If the highway has two lanes, find the approximate number of feet between cars at $x = 0, 1000$, and 5000.

53. (a) A river flows across the xy-plane in the positive x-direction and around a circular rock of radius 1 centered at the origin. The velocity of the river can be modeled using the potential function $\phi = x + (x/(x^2 + y^2))$. Compute the velocity vector field, $\vec{v} = \text{grad } \phi$.

(b) Show that $\text{div } \vec{v} = 0$.

(c) Show that the flow of $\vec{v}$ is tangent to the circle $x^2 + y^2 = 1$. This means that no water crosses the circle. The water on the outside must therefore all flow around the circle.

(d) Use a computer to sketch the vector field $\vec{v}$ in the region outside the unit circle.

54. A basic property of the magnetic field $\vec{B}$ is that $\text{curl } \vec{B} = \vec{0}$ in a region where there is no current. Consider the magnetic field around a long thin wire carrying a constant current. The magnitude of the magnetic field depends only on the distance from the wire and its direction is always tangent to the circle around the wire traversed in a direction related to the direction of the current by the right hand rule. Use Stokes' Theorem to deduce that the magnitude of the magnetic field is proportional to the reciprocal of the distance from the wire. [Hint: Consider an annulus (ring) around the wire. Its boundary has two pieces: an inner and an outer circle.]

55. A vector field is a *point source* at the origin in 3-space if its direction is away from the origin at every point, its magnitude depends only on the distance from the origin, and its divergence is zero except at the origin. (Such a vector field might be used to model the photon flow out of a star or the neutrino flow out of a supernova.)

(a) Show that $\vec{v} = K(x^2+y^2+z^2)^{-3/2}(x\vec{i} + y\vec{j} + z\vec{k})$ is a point source at the origin if $K > 0$.

(b) Determine the magnitude $\|\vec{v}\|$ of the source in part (a) as a function of the distance from its center.

(c) Compute the flux of $\vec{v}$ through a sphere of radius r centered at the origin.

(d) Compute the flux of $\vec{v}$ through a closed surface that does not contain the origin.

Problems 56–57 use the fact that the electric field, $\vec{E}$, is related to the charge density, $\rho(x,y,z)$, in units of charge/volume, by the equation

$$\text{div } \vec{E} = 4\pi\rho.$$

In addition, there is an electric potential, ϕ, whose gradient gives the electric field:

$$\vec{E} = -\text{grad } \phi.$$

56. Calculate and describe in words the electric field and the charge distribution corresponding to the potential function defined as follows:

$$\phi = \begin{cases} x^2 + y^2 + z^2 & \text{for } x^2+y^2+z^2 \leq \frac{b^2}{4} \\ \frac{b^2}{4} - \frac{b^3}{4(x^2+y^2+z^2)^{1/2}} & \text{for } \frac{b^2}{4} \leq x^2+y^2+z^2 \end{cases}$$

57. A vector field which could possibly represent an electric field is given by

$$\vec{E} = 10xy\vec{i} + (5x^2 - 5y^2)\vec{j}.$$

(a) Calculate the line integral of $\vec{E}$ from the origin to the point (a,b) along the path which runs straight from the origin to the point $(a,0)$ and then straight from $(a,0)$ to the point (a,b).

(b) Calculate the line integral of $\vec{E}$ between the same points as in part (a) but via the point $(0, b)$.

(c) Why do your answers to part (a) and (b) suggest that $\vec{E}$ could indeed be an electric field?

(d) Find the electric potential, ϕ, and calculate $\operatorname{grad} \phi$ to confirm that $\vec{E} = -\operatorname{grad} \phi$.

58. The relations between the electric field, $\vec{E}$, the magnetic field, $\vec{B}$, the charge density, ρ, and the current density, $\vec{J}$, at a point in space are described by the equations

$$\operatorname{div} \vec{E} = 4\pi\rho,$$

$$\operatorname{curl} \vec{B} - \frac{1}{c}\frac{\partial \vec{E}}{\partial t} = \frac{4\pi}{c}\vec{J},$$

where c is a constant (the speed of light).

(a) Using the results of Problem 25 on page 960, show that

$$\frac{\partial \rho}{\partial t} + \operatorname{div} \vec{J} = 0.$$

(b) What does the equation in part (a) say about charge and current density? Explain in intuitive terms why this is reasonable.

(c) Why do you think the equation in part (a) is called the charge conservation equation?

59. The speed of a naturally occurring vortex (tornado, waterspout, whirlpool) is a decreasing function of the distance from its center, so the constant angular velocity model of Problem 31 on page 961 is inappropriate. A *free vortex* circulating about the z-axis has vector field $\vec{v} = K(x^2 + y^2)^{-1}(-y\vec{i} + x\vec{j})$ where K is a constant.

(a) Sketch the vector field with $K = 1$ and the vector field with $K = -1$.

(b) Determine the speed $\|\vec{v}\|$ of the vortex as a function of the distance from its center.

(c) Compute $\operatorname{div} \vec{v}$.

(d) Show that $\operatorname{curl} \vec{v} = \vec{0}$.

(e) Compute the circulation of $\vec{v}$ counterclockwise about the circle of radius R at the origin.

(f) The computations in parts (d) and (e) show that $\vec{v}$ has curl $\vec{0}$, but has nonzero circulation around the closed curve in part (e). Explain why this does not contradict Stokes' Theorem.

CAS Challenge Problems

60. (a) Let $\vec{F} = x^3 y\vec{i} + 2xz^3\vec{j} + (z^3 + 4x^2)\vec{k}$. Compute $\operatorname{curl} \vec{F}(1, 2, 1)$.

(b) Consider the family of curves C_a given, for $0 \leq t \leq 2\pi$, by

$$\vec{r}(t) = \vec{i} + (2 + a\cos t)\vec{j} + (1 + a\sin t)\vec{k}.$$

Evaluate the line integral $\int_{C_a} \vec{F} \cdot d\vec{r}$ and compute the limit

$$\lim_{a \to 0} \frac{\int_{C_a} \vec{F} \cdot d\vec{r}}{\pi a^2}.$$

(c) Repeat part (b) for the family D_a given, for $0 \leq t \leq 2\pi$, by

$$\vec{r}(t) = (1 + a\sin t)\vec{i} + 2\vec{j} + (1 + a\cos t)\vec{k}.$$

(d) Repeat part (b) for the family E_a given, for $0 \leq t \leq 2\pi$, by

$$\vec{r}(t) = (1 + a\cos t)\vec{i} + (2 + a\sin t)\vec{j} + \vec{k}.$$

(e) Compare your answers to parts (b)-(d) with part (a) and explain using the geometric definition of curl.

61. Let S be the sphere of radius R centered at the origin with outward orientation and let

$$\vec{F} = (ax^2 + bxz)\vec{i} + (cy^2 + py)\vec{j} + (qz + rx^3)\vec{k}.$$

(a) Use the Divergence Theorem to express the flux integral $\int_S \vec{F} \cdot d\vec{A}$ as a triple integral. Then use symmetry and the volume formula for a sphere to evaluate the triple integral.

(b) Check your answer in part (a) by computing the flux integral directly.

62. Let $\vec{F}(x, y, z) = x^2 y\vec{i} + 2xz\vec{j} + (z^3 + 4x^2)\vec{k}$ and let S_a be the sphere of radius a centered at $(1, 1, 1)$, oriented outward, parameterized by $\vec{r}(\phi, \theta) = (1 + a\sin\phi\cos\theta)\vec{i} + (1 + a\sin\phi\sin\theta)\vec{j} + (1 + a\cos\phi)\vec{k}$, $0 \leq \phi \leq \pi, 0 \leq \theta \leq 2\pi$.

(a) Compute $\operatorname{div} \vec{F}(1, 1, 1)$.

(b) Use the geometric definition of divergence to estimate $\int_{S_a} \vec{F} \cdot d\vec{A}$ for $a = 0.1$.

(c) Evaluate the flux integral $\int_{S_a} \vec{F} \cdot d\vec{A}$. Compare its value for $a = 0.1$ with your answer to part (b). Then compute the limit

$$\lim_{a \to 0} \frac{\int_{S_a} \vec{F} \cdot d\vec{A}}{\text{Volume inside } S_a}$$

and compare your result with part (a). Explain your answer in terms of the geometric definition of the divergence.

CHECK YOUR UNDERSTANDING

1. If $\vec{F}$ is a smooth vector field, defined everywhere in three-space, with curl $\vec{F}$ everywhere perpendicular to the z-axis, and if C is a circle in the xy-plane, then the circulation of $\vec{F}$ around C is zero. True or false? If true, explain why; if false, give a counterexample.

Are the statements in Problems 2–7 true or false? Assume $\vec{F}$ and $\vec{G}$ are smooth vector fields in 3-space. Explain.

2. $\operatorname{div}(\vec{F} + \vec{G}) = \operatorname{div}\vec{F} + \operatorname{div}\vec{G}$

3. $\operatorname{grad}(\vec{F} \cdot \vec{G}) = \vec{F}(\operatorname{div}\vec{G}) + (\operatorname{div}\vec{F})\vec{G}$

4. $\operatorname{curl}(f\vec{G}) = (\operatorname{grad} f) \times \vec{G} + f(\operatorname{curl}\vec{G})$

5. $\operatorname{div}\vec{F}$ is a scalar whose value can vary from point to point.

6. If $\int_S \vec{F} \cdot d\vec{A} = 12$ and S is a flat disk of area 4π, then $\operatorname{div}\vec{F} = 3/\pi$.

7. If $\vec{F}$ is as shown in Figure 20.43, then $\operatorname{curl}\vec{F} \cdot \vec{j} > 0$.

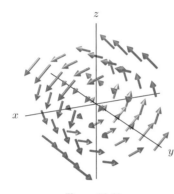

Figure 20.43

Are the statements in Problems 8–41 true or false? Give reasons for your answer.

8. If $\vec{F}$ is a vector field in 3-space, and f is a scalar function, then $\operatorname{div}(f\vec{F}) = f\operatorname{div}\vec{F}$.

9. If $\vec{F}$ is a vector field in 3-space, and $\vec{F} = \operatorname{grad} f$, then $\operatorname{div}\vec{F} = 0$.

10. If $\vec{F}$ is a vector field in 3-space, then $\operatorname{grad}(\operatorname{div}\vec{F}) = \vec{0}$.

11. The field $\vec{F}(\vec{r}) = \vec{r}$ is divergence free.

12. If $f(x, y, z)$ is any given continuous scalar function, then there is at least one vector field $\vec{F}$ such that $\operatorname{div}\vec{F} = f$.

13. If $\vec{F}$ and $\vec{G}$ are vector fields satisfying $\operatorname{div}\vec{F} = \operatorname{div}\vec{G}$ then $\vec{F} = \vec{G}$.

14. If $\vec{F}$ is a vector field in 3-space, and W is a solid region with boundary surface S, then $\int_S \operatorname{div}\vec{F} \cdot d\vec{A} = \int_W \vec{F}\, dV$.

15. If $\vec{F}$ is a divergence free vector field in 3-space, and S is a closed surface oriented inward, then $\int_S \vec{F} \cdot d\vec{A} = 0$.

16. If $\vec{F}$ is a vector field in 3-space satisfying $\operatorname{div}\vec{F} = 1$, and S is a closed surface oriented outward, then $\int_S \vec{F} \cdot d\vec{A}$ is equal to the volume enclosed by S.

17. Let W be the solid region between the sphere S_1 of radius 1 and S_2 of radius 2, both centered at the origin. If $\vec{F}$ is a vector field in 3-space, then $\int_W \operatorname{div}\vec{F}\, dV = \int_{S_2} \vec{F} \cdot d\vec{A} - \int_{S_1} \vec{F} \cdot d\vec{A}$, where both S_1 and S_2 are oriented outward.

18. Let S_1 be the square $0 \le x \le 1, 0 \le y \le 1, z = 0$ oriented downward and let S_2 be the square $0 \le x \le 1, 0 \le y \le 1, z = 1$ oriented upward. If $\vec{F}$ is a vector field, then $\int_W \operatorname{div}\vec{F}\, dV = \int_{S_2} \vec{F} \cdot d\vec{A} + \int_{S_1} \vec{F} \cdot d\vec{A}$, where W is the solid cube $0 \le x \le 1, 0 \le y \le 1, 0 \le z \le 1$.

19. Let S_1 be the square $0 \le x \le 1, 0 \le y \le 1, z = 0$ oriented downward and let S_2 be the square $0 \le x \le 1, 0 \le y \le 1, z = 1$ oriented upward. If $\vec{F} = \cos(xyz)\vec{k}$, then $\int_W \operatorname{div}\vec{F}\, dV = \int_{S_2} \vec{F} \cdot d\vec{A} + \int_{S_1} \vec{F} \cdot d\vec{A}$, where W is the solid cube $0 \le x \le 1, 0 \le y \le 1, 0 \le z \le 1$.

20. If S is a sphere of radius 1, centered at the origin, oriented outward, and $\vec{F}$ is a vector field satisfying $\int_S \vec{F} \cdot d\vec{A} = 0$, then $\operatorname{div}\vec{F} = 0$ at all points inside S.

21. There exist a scalar function f and a vector field $\vec{F}$ satisfying $\operatorname{div}(\operatorname{grad} f) = \operatorname{grad}(\operatorname{div} F)$.

22. Let S_h be the surface consisting of a cylinder of height h, closed at the top. The curved sides are $x^2 + y^2 = 1$, for $0 \le z \le h$, and the top $x^2 + y^2 \le 1$, for $z = h$, oriented outward. If $\vec{F}$ is divergence free, then $\int_{S_h} \vec{F} \cdot d\vec{A}$ is independent of the height h.

23. The circulation density, $\operatorname{circ}_{\vec{n}}\vec{F}(x, y, z)$, is a scalar.

24. If $\vec{F}$ is a vector field with $\operatorname{div}\vec{F} = 0$ and $\operatorname{curl}\vec{F} = \vec{0}$, then $\vec{F} = \vec{0}$.

25. If $\vec{F}$ and $\vec{G}$ are vector fields, then $\operatorname{curl}(\vec{F} + \vec{G}) = \operatorname{curl}\vec{F} + \operatorname{curl}\vec{G}$.

26. If $\vec{F}$ and $\vec{G}$ are vector fields, then $\operatorname{curl}(\vec{F} \cdot \vec{G}) = \operatorname{curl}\vec{F} \cdot \operatorname{curl}\vec{G}$.

27. There exists a vector field $\vec{F}$ with $\operatorname{curl}\vec{F} = \vec{i}$.

28. At all points other than the origin, the vector field $\vec{F} = \frac{-y}{x^2+y^2}\vec{i} + \frac{x}{x^2+y^2}\vec{j}$ has $\operatorname{curl}\vec{F} = \vec{0}$ and $\operatorname{div}\vec{F} = 0$.

29. For any vector field $\vec{F}$, the curl of $\vec{F}$ is perpendicular at every point to $\vec{F}$.

30. If $\vec{F}$ and $\vec{G}$ are vector fields, then $\operatorname{curl}(\vec{F} \times \vec{G}) = (\operatorname{curl}\vec{F}) \times (\operatorname{curl}\vec{G})$.

31. There exists a vector field $\vec{F}$ (whose components have continuous second partial derivatives) satisfying $\operatorname{curl}\vec{F} = x\vec{i}$.

32. If S is the upper unit hemisphere $x^2+y^2+z^2 = 1, z \geq 0$, oriented upward, then the boundary of S used in Stokes' theorem is the circle $x^2+y^2 = 1, z = 0$, with orientation counterclockwise when viewed from the positive z-axis.

33. If C is the boundary of an oriented surface S, oriented by the right hand rule, then $\int_C \text{curl}\, \vec{F} \cdot d\vec{r} = \int_S \vec{F} \cdot d\vec{A}$.

34. Let S_1 be the disk $x^2 + y^2 \leq 1, z = 0$ and let S_2 be the upper unit hemisphere $x^2 + y^2 + z^2 = 1, z \geq 0$, both oriented upward. If $\vec{F}$ is a vector field then $\int_{S_1} \text{curl}\vec{F} \cdot d\vec{A} = \int_{S_2} \text{curl}\vec{F} \cdot d\vec{A}$.

35. Let S be the closed unit sphere $x^2+y^2+z^2 = 1$, oriented outward. If $\vec{F}$ is a vector field then $\int_S \text{curl}\vec{F} \cdot d\vec{A} = 0$.

36. If $\vec{F}$ and $\vec{G}$ are vector fields satisfying $\text{curl}\vec{F} = \text{curl}\vec{G}$, then $\int_C \vec{F} \cdot d\vec{r} = \int_C \vec{G} \cdot d\vec{r}$. where C is any oriented circle in 3-space.

37. If $\vec{F}$ is a vector field satisfying $\text{curl}\vec{F} = \vec{0}$, then $\int_C \vec{F} \cdot d\vec{r} = 0$, where C is any oriented path around a rectangle in 3-space.

38. If $\vec{F}$ is a gradient field, then $\int_S \text{curl}\, \vec{F} \cdot d\vec{r} = 0$, for any smooth oriented surface, S, in 3-space.

39. Let S be the cylinder $x^2 + z^2 = 1, 0 \leq y \leq 2$, oriented with inward pointing normal. Then the boundary of S consists of two circles C_1 $(x^2 + z^2 = 1, y = 0)$ and C_2 $(x^2 + z^2 = 1, y = 2)$, both oriented clockwise when viewed from the positive y-axis.

40. Let S be an oriented surface, with oriented boundary C, and suppose that $\vec{F}$ is a vector field such that $\int_C \vec{F} \cdot d\vec{r} = 0$. Then $\text{curl}\, \vec{F} = \vec{0}$ everywhere on S.

41. Let S be an oriented surface, with oriented boundary C, and suppose that $\vec{F}$ is a vector field such that $\int_S \text{curl}\vec{F} \cdot d\vec{A} = 0$. Then $\vec{F}$ is a gradient field.

PROJECTS FOR CHAPTER TWENTY

1. Divergence of Spherically Symmetric Vector Fields

A vector field is spherically symmetric about the origin if, on every sphere centered at the origin, it has constant magnitude and points either away from or toward the origin. A vector field that is spherically symmetric about the origin can be written in terms of the spherical coordinate $\rho = \|\vec{r}\|$ as $\vec{F} = f(\rho)\vec{e}_\rho$ where f is a function of the distance ρ from the origin, $f(0) = 0$, and $\vec{e}_\rho$ is a unit vector pointing away from the origin.

(a) Show that

$$\text{div}\, \vec{F} = \frac{1}{\rho^2}\frac{d}{d\rho}\left(\rho^2 f(\rho)\right), \quad \rho \neq 0.$$

(b) Use part (a) to show that if $\vec{F}$ is a spherically symmetric vector field such that $\text{div}\, \vec{F} = 0$ away from the origin then, for some constant k,

$$\vec{F} = k\frac{1}{\rho^2}\vec{e}_\rho, \quad \rho \neq 0.$$

(c) Use part (a) to confirm the Divergence Theorem for the flux of a spherically symmetric vector field through a sphere centered at the origin.

(d) A form of Gauss's Law for an electric field $\vec{E}$ states that $\text{div}\, \vec{E}(\vec{r}) = \delta(\vec{r})$, where δ is a scalar-valued function giving the density of electric charge at every point. Use Gauss's Law and part (a) to find the electric field when the charge density is

$$\delta(\vec{r}) = \begin{cases} \delta_0 & \|\vec{r}\| \leq a \\ 0 & \|\vec{r}\| > a, \end{cases}$$

for some nonnegative a. Assume that $\vec{E}$ is spherically symmetric and continuous everywhere.

2. Divergence of Cylindrically Symmetric Vector Fields

We say that a vector field is cylindrically symmetric about a line if, on every cylinder centered along that line, it has constant magnitude and points either away from or toward the line. A vector field that is cylindrically symmetric about the z-axis can be written in terms of the cylindrical coordinate r as $\vec{F} = f(r)\vec{e}_r$ where f is a function of the distance r from the z-axis, $f(0) = 0$, and $\vec{e}_r$ is a unit vector pointing away from the z-axis.

(a) Show that

$$\operatorname{div} \vec{F} = \frac{1}{r}\frac{d}{dr}\left(rf(r)\right), \quad r \neq 0.$$

(b) Use part (a) to show that if $\vec{F}$ is cylindrically symmetric about the z-axis and $\operatorname{div} \vec{F} = 0$ away from the z-axis, then, for some constant k,

$$\vec{F} = k\frac{\vec{e}_r}{r}, \quad r \neq 0.$$

(c) Use part (a) to confirm the Divergence Theorem for the flux of a cylindrically symmetric vector field through a cylinder of radius R and height h centered on the z-axis.

(d) A form of Gauss's Law for an electric field $\vec{E}$ states that $\operatorname{div} \vec{E}(\vec{r}) = \delta(\vec{r})$, where δ is a scalar-valued function giving the density of electric charge at every point. Use Gauss's Law and part (a) to find the electric field when the charge density is given in cylindrical coordinates by

$$\delta(r, \theta, z) = \begin{cases} \delta_0 & r \leq a \\ 0 & r > a, \end{cases}$$

for some nonnegative constant a. Assume that $\vec{E}$ is cylindrically symmetric and continuous everywhere.

APPENDICES

A   ROOTS, ACCURACY, AND BOUNDS

It is often necessary to find the zeros of a polynomial or the points of intersection of two curves. So far, you have probably used algebraic methods, such as the quadratic formula, to solve such problems. Unfortunately, however, mathematicians' search for similar solutions to more complicated equations, has not been all that successful. The formulas for the solutions to third- and fourth-degree equations are so complicated that you'd never want to use them. Early in the nineteenth century, it was proved that there is no algebraic formula for the solutions to equations of degree 5 and higher. Most nonpolynomial equations cannot be solved using a formula either.

However, we can still find roots of equations, provided we use approximation methods, not formulas. In this section we will discuss three ways to find roots: algebraic, graphical, and numerical. Of these, only the algebraic method gives exact solutions.

First, let's get some terminology straight. Given the equation $x^2 = 4$, we call $x = -2$ and $x = 2$ the *roots*, or *solutions of the equation*. If we are given the function $f(x) = x^2 - 4$, then -2 and 2 are called the *zeros of the function*; that is, the zeros of the function f are the roots of the equation $f(x) = 0$.

The Algebraic Viewpoint: Roots by Factoring

If the product of two numbers is zero, then one or the other or both must be zero, that is, if $AB = 0$, then $A = 0$ or $B = 0$. This observation lies behind finding roots by factoring. You may have spent a lot of time factoring polynomials. Here you will also factor expressions involving trigonometric and exponential functions.

Example 1   Find the roots of $x^2 - 7x = 8$.

Solution   Rewrite the equation as $x^2 - 7x - 8 = 0$. Then factor the left side: $(x + 1)(x - 8) = 0$. By our observation about products, either $x + 1 = 0$ or $x - 8 = 0$, so the roots are $x = -1$ and $x = 8$.

Example 2   Find the roots of $\dfrac{1}{x} - \dfrac{x}{(x + 2)} = 0$.

Solution   Rewrite the left side with a common denominator:

$$\frac{x + 2 - x^2}{x(x + 2)} = 0.$$

Whenever a fraction is zero, the numerator must be zero. Therefore we must have

$$x + 2 - x^2 = (-1)(x^2 - x - 2) = (-1)(x - 2)(x + 1) = 0.$$

We conclude that $x - 2 = 0$ or $x + 1 = 0$, so 2 and -1 are the roots. They can be checked by substitution.

Example 3   Find the roots of $e^{-x} \sin x - e^{-x} \cos x = 0$.

Solution   Factor the left side: $e^{-x}(\sin x - \cos x) = 0$. The factor e^{-x} is never zero; it is impossible to raise e to a power and get zero. Therefore, the only possibility is that $\sin x - \cos x = 0$. This equation is equivalent to $\sin x = \cos x$. If we divide both sides by $\cos x$, we get

$$\frac{\sin x}{\cos x} = \frac{\cos x}{\cos x} \quad \text{so} \quad \tan x = 1.$$

The roots of this equation are

$$\dots, \frac{-7\pi}{4}, \frac{-3\pi}{4}, \frac{\pi}{4}, \frac{5\pi}{4}, \frac{9\pi}{4}, \frac{13\pi}{4}, \dots .$$

Warning: Using factoring to solve an equation only works when one side of the equation is 0. It is not true that if, say, $AB = 7$ then $A = 7$ or $B = 7$. For example, you *cannot* solve $x^2 - 4x = 2$ by factoring $x(x - 4) = 2$ and then assuming that either x or $x - 4$ equals 2.

The problem with factoring is that factors are not easy to find. For example, the left side of the quadratic equation $x^2 - 4x - 2 = 0$ does not factor, at least not into "nice" factors with integer coefficients. For the general quadratic equation

$$ax^2 + bx + c = 0,$$

there is the quadratic formula for the roots:

$$x = \frac{-b \pm \sqrt{b^2 - 4ac}}{2a}.$$

Thus the roots of $x^2 - 4x - 2 = 0$ are $(4 \pm \sqrt{24})/2$, or $2 + \sqrt{6}$ and $2 - \sqrt{6}$.

Notice that in each of these examples, we have found the roots exactly.

The Graphical Viewpoint: Roots by Zooming

To find the roots of an equation $f(x) = 0$, it helps to draw the graph of f. The roots of the equation, that is the zeros of f, are *the values of x where the graph of f crosses the x-axis*. Even a very rough sketch of the graph can be useful in determining how many zeros there are and their approximate values. If you have a computer or graphing calculator, then finding solutions by graphing is the easiest method, especially if you use the zoom feature. However, a graph can never tell you the exact value of a root, only an approximate one.

Example 4   Find the roots of $x^3 - 4x - 2 = 0$.

Solution   Attempting to factor the left side with integer coefficients will convince you it cannot be done, so we cannot easily find the roots by algebra. We know the graph of $f(x) = x^3 - 4x - 2$ will have the usual cubic shape; see Figure A.1.

There are clearly three roots: one between $x = -2$ and $x = -1$, another between $x = -1$ and $x = 0$, and a third between $x = 2$ and $x = 3$. Zooming in on the largest root with a graphing calculator or computer shows that it lies in the following interval:

$$2.213 < x < 2.215.$$

Thus, the root is $x = 2.21$, accurate to two decimal places. Zooming in on the other two roots shows them to be $x = -1.68$ and $x = -0.54$, accurate to two decimal places.

Useful trick: Suppose you want to solve the equation $\sin x - \cos x = 0$ graphically. Instead of graphing $f(x) = \sin x - \cos x$ and looking for zeros, you may find it easier to rewrite the equation as $\sin x = \cos x$ and graph $g(x) = \sin x$ and $h(x) = \cos x$. (After all, you already know what these two graphs look like. See Figure A.2.) The roots of the original equation are then precisely the x coordinates of the points of intersection of the graphs of $g(x)$ and $h(x)$.

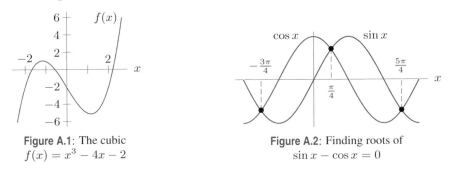

Figure A.1: The cubic $f(x) = x^3 - 4x - 2$

Figure A.2: Finding roots of $\sin x - \cos x = 0$

Example 5   Find the roots of $2 \sin x - x = 0$.

Solution   Rewrite the equation as $2 \sin x = x$, and graph both sides. Since $g(x) = 2 \sin x$ is always between -2 and 2, there are no roots of $2 \sin x = x$ for $x > 2$ or for $x < -2$. We need only consider the graphs between -2 and 2 (or between $-\pi$ and π, which makes graphing the sine function easier). Figure A.3 shows the graphs. There are three points of intersection: one appears to be at $x = 0$, one between $x = \pi/2$ and $x = \pi$, and one between $x = -\pi/2$ and $x = -\pi$. You can tell that $x = 0$ is the exact value of one root because it satisfies the original equation exactly. Zooming in shows that there is a second root $x \approx 1.9$, and the third root is $x \approx -1.9$ by symmetry.

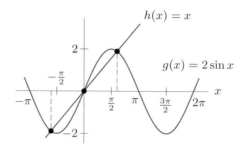

Figure A.3: Finding roots of $2 \sin x - x = 0$

The Numerical Viewpoint: Roots by Bisection

We now look at a numerical method of approximating the solutions to an equation. This method depends on the idea that if the value of a function $f(x)$ changes sign in an interval, and if we believe there is no break in the graph of the function there, then there is a root of the equation $f(x) = 0$ in that interval.

Let's go back to the problem of finding the root of $f(x) = x^3 - 4x - 2 = 0$ between 2 and 3. To locate the root, we close in on it by evaluating the function at the midpoint of the interval, $x = 2.5$. Since $f(2) = -2$, $f(2.5) = 3.625$, and $f(3) = 13$, the function changes sign between $x = 2$ and $x = 2.5$, so the root is between these points. Now we look at $x = 2.25$.

Since $f(2.25) = 0.39$, the function is negative at $x = 2$ and positive at $x = 2.25$, so there is a root between 2 and 2.25. Now we look at 2.125. We find $f(2.125) = -0.90$, so there is a root between 2.125 and 2.25, ... and so on. (You may want to round the decimals as you work.) See Figure A.4. The intervals containing the root are listed in Table A.1 and show that the root is $x = 2.21$ to two decimal places.

Table A.1   *Intervals containing root of* $x^3 - 4x - 2 = 0$ *(Note $[2, 3]$ means $2 \le x \le 3$)*

$[2, 3]$
$[2, 2.5]$
$[2, 2.25]$
$[2.125, 2.25]$
$[2.1875, 2.25]$ So $x = 2.2$ rounded to one decimal place
$[2.1875, 2.21875]$
$[2.203125, 2.21875]$
$[2.2109375, 2.21875]$
$[2.2109375, 2.2148438]$ So $x = 2.21$ rounded to two decimal places

Figure A.4: Locating a root of $x^3 - 4x - 2 = 0$

This method of estimating roots is called the **Bisection Method**:

- To solve an equation $f(x) = 0$ using the bisection method, we need two starting values for x, say, $x = a$ and $x = b$, such that $f(a)$ and $f(b)$ have opposite signs and f is continuous on $[a, b]$.

- Evaluate f at the midpoint of the interval $[a, b]$, and decide in which half-interval the root lies.

- Repeat, using the new half-interval instead of $[a, b]$.

There are some problems with the bisection method:

- The function may not change signs near the root. For example, $f(x) = x^2 - 2x + 1 = 0$ has a root at $x = 1$, but $f(x)$ is never negative because $f(x) = (x - 1)^2$, and a square cannot be negative. (See Figure A.5.)

- The function f must be continuous between the starting values $x = a$ and $x = b$.

- If there is more than one root between the starting values $x = a$ and $x = b$, the method will find only one of the roots. For example, if we had tried to solve $x^3 - 4x - 2 = 0$ starting at $x = -12$ and $x = 10$, the bisection method would zero in on the root between $x = -2$ and $x = -1$, not the root between $x = 2$ and $x = 3$ that we found earlier. (Try it! Then see what happens if you use $x = -10$ instead of $x = -12$.)

- The bisection method is slow and not very efficient. Applying bisection three times in a row only traps the root in an interval $(\frac{1}{2})^3 = \frac{1}{8}$ as large as the starting interval. Thus, if we initially know that a root is between, say, 2 and 3, then we would need to apply the bisection method at least four times to know the first digit after the decimal point.

There are much more powerful methods available for finding roots, such as Newton's method, which are more complicated but which avoid some of these difficulties.

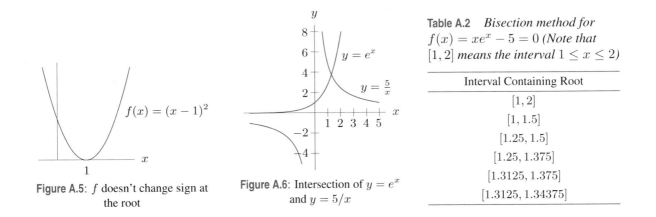

Figure A.5: f doesn't change sign at the root

Figure A.6: Intersection of $y = e^x$ and $y = 5/x$

Table A.2   *Bisection method for* $f(x) = xe^x - 5 = 0$ *(Note that* $[1, 2]$ *means the interval* $1 \le x \le 2$*)*

Interval Containing Root
$[1, 2]$
$[1, 1.5]$
$[1.25, 1.5]$
$[1.25, 1.375]$
$[1.3125, 1.375]$
$[1.3125, 1.34375]$

Example 6   Find all the roots of $xe^x = 5$ to at least one decimal place.

Solution   If we rewrite the equation as $e^x = 5/x$ and graph both sides, as in Figure A.6, it is clear that there is exactly one root, and it is somewhere between 1 and 2. Table A.2 shows the intervals obtained by the bisection method. After five iterations, we have the root trapped between 1.3125 and 1.34375, so we can say the root is $x = 1.3$ to one decimal place.

Iteration

Both zooming in and bisection as discussed here are examples of *iterative* methods, in which a sequence of steps is repeated over and over again, using the results of one step as the input for the next. We can use such methods to locate a root to any degree of accuracy. In bisection, each iteration traps the root in an interval that is half the length of the previous one. Each time you zoom in on a calculator, you trap the root in a smaller interval; how much smaller depends on the settings on the calculator.

Accuracy and Error

In the previous discussion, we used the phrase "accurate to 2 decimal places." For an iterative process where we get closer and closer estimates of some quantity, we take a common-sense approach to accuracy: we watch the numbers carefully, and when a digit stays the same for several iterations, we assume it has stabilized and is correct, especially if the digits to the right of that digit also stay the same. For example, suppose 2.21429 and 2.21431 are two successive estimates for a zero of $f(x) = x^3 - 4x - 2$. Since these two estimates agree to the third digit after the decimal point, we probably have at least 3 decimal places correct.

There is a problem with this, however. Suppose we are estimating a root whose true value is 1, and the estimates are converging to the value from below—say, 0.985, 0.991, 0.997 and so on. In this case, not even the first decimal place is "correct," even though the difference between the estimates and the actual answer is very small—much less than 0.1. To avoid this difficulty, we say that an estimate a for some quantity r is *accurate to p decimal places* if the error, which is the absolute value of the difference between a and r, or $|r - a|$, is as follows:

Accuracy to p decimal places	means	Error less than
$p = 1$		0.05
2		0.005
3		0.0005
$\vdots$		$\vdots$
n		$0.\underbrace{000\ldots0}_{n}5$

This is the same as saying that r must lie in an interval of length twice the maximum error, centered on a. For example, if a is accurate to 1 decimal place, r must lie in the following interval:

$$a - 0.05 \qquad a \qquad a + 0.05$$

Since both the graphing calculator and the bisection method give us an interval in which the root is trapped, this definition of decimal accuracy is a natural one for these processes.

Example 7   Suppose the numbers $\sqrt{10}$, $22/7$, and 3.14 are given as approximations to $\pi = 3.1415\ldots$. To how many decimal places is each approximation accurate?

Solution   Using $\sqrt{10} = 3.1622\ldots$,

$$|\sqrt{10} - \pi| = |3.1622\ldots - 3.1415\ldots| = 0.0206\ldots < 0.05,$$

so $\sqrt{10}$ is accurate to one decimal place. Similarly, using $22/7 = 3.1428\ldots$,

$$\left|\frac{22}{7} - \pi\right| = |3.1428\ldots - 3.1415\ldots| = 0.0013\ldots < 0.005,$$

so 22/7 is accurate to two decimal places. Finally,

$$|3.14 - 3.1415\ldots| = 0.0015\ldots < 0.005,$$

so 3.14 is accurate to two decimal places.

Warning:

- Saying that an approximation is accurate to, say, 2 decimal places does *not* guarantee that its first two decimal places are "correct," that is, that the two digits of the approximation are the same as the corresponding two digits in the true value. For example, an approximate value of 5.997 is accurate to 2 decimal places if the true value is 6.001, but neither of the 9s in the approximation agrees with the 0s in the true value (nor does the digit 5 agree with the digit 6).
- When finding a root r of an equation, the number of decimal places of accuracy refers to the number of digits that have stabilized in the root. It does *not* refer to the number of digits of $f(r)$ that are zero. For example, Table A.1 on page 984 shows that $x = 2.2$ is a root of $f(x) = x^3 - 4x - 2 = 0$, accurate to one decimal place. Yet, $f(2.2) = -0.152$, so $f(2.2)$ does not have one zero after the decimal point. Similarly, $x = 2.21$ is the root accurate to two decimal places, but $f(2.21) = -0.046$ does not have two zeros after the decimal point.

Example 8   Is $x = 2.2143$ a zero of $f(x) = x^3 - 4x - 2$ accurate to four decimal places?

Solution   We want to know whether r, the exact value of the zero, lies in the interval

$$2.2143 - 0.00005 < r < 2.2143 + 0.00005$$

which is the same as

$$2.21425 < r < 2.21435.$$

Since $f(2.21425) < 0$ and $f(2.21435) > 0$, the zero does lie in this interval, and so $r = 2.2143$ is accurate to four decimal places.

How to Write a Decimal Answer

The graphing calculator and bisection method naturally give an interval for a root or a zero. However, other numerical techniques do not give a pair of numbers bounding the true value, but rather a single number near the true value. What should you do if you want a single number, rather than an interval, for an answer? In general, averaging the endpoint of the interval is the best solution.

When giving a single number as an answer and interpreting it, be careful about giving rounded answers. For example, suppose you know a root lies in the interval between 0.81 and 0.87. Averaging gives 0.84 as a single number estimating the root. But it would be wrong to round 0.84 to 0.8 and say that the answer is 0.8 accurate to one decimal place; the true value could be 0.86, which is not within 0.05 of 0.8. The right thing to say is that the answer is 0.84 accurate to one decimal place. Similarly, to give an answer accurate to, say, 2 decimal places, you may have to show 3 decimal places in your answer.

Bounds of a Function

Knowing how big or how small a function gets can sometimes be useful, especially when you can't easily find exact values of the function. You can say, for example, that $\sin x$ always stays between -1 and 1 and that $2\sin x + 10$ always stays between 8 and 12. But 2^x is not confined between any two numbers, because 2^x will exceed any number you can name if x is large enough. We say that $\sin x$ and $2\sin x + 10$ are *bounded* functions, and that 2^x is an *unbounded* function.

A function f is **bounded** on an interval if there are numbers L and U such that

$$L \le f(x) \le U$$

for all x in the interval. Otherwise, f is **unbounded** on the interval.

We say that L is a **lower bound** for f on the interval, and that U is an **upper bound** for f on the interval.

Example 9   Use Figures A.7 and A.8 to decide which of the following functions are bounded.

(a) x^3 on $-\infty < x < \infty$; on $0 \le x \le 100$.

(b) $2/x$ on $0 < x < \infty$; on $1 \le x < \infty$.

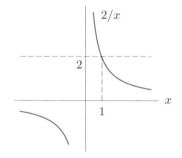

Figure A.7: Is x^3 bounded?   **Figure A.8**: Is $2/x$ bounded?

Solution   (a) The graph of x^3 in Figure A.7 shows that x^3 will exceed any number, no matter how large, if x is big enough, so x^3 does not have an upper bound on $-\infty < x < \infty$. Therefore, x^3 is unbounded on $-\infty < x < \infty$. But on the interval $0 \le x \le 100$, x^3 stays between 0 (a lower bound) and $100^3 = 1{,}000{,}000$ (an upper bound). Therefore, x^3 is bounded on the interval $0 \le x \le 100$. Notice that upper and lower bounds, when they exist, are not unique. For example, -100 is another lower bound and $2{,}000{,}000$ another upper bound for x^3 on $0 \le x \le 100$.

(b) $2/x$ is unbounded on $0 < x < \infty$, since it has no upper bound on that interval. But $0 \le 2/x \le 2$ for $1 \le x < \infty$, so $2/x$ is bounded, with lower bound 0 and upper bound 2, on $1 \le x < \infty$. (See Figure A.8.)

Best Possible Bounds

Consider a group of people whose height in feet, h, ranges from 5 feet to 6 feet. Then 5 feet is a lower bound for the people in the group and 6 feet is an upper bound:

$$5 \le h \le 6.$$

But the people in this group are also all between 4 feet and 7 feet, so it is also true that

$$4 \le h \le 7.$$

So, there are many lower bounds and many upper bounds. However, the 5 and the 6 are considered the best bounds because they are the closest together of all the possible pairs of bounds.

The **best possible bounds** for a function, f, over an interval are numbers A and B such that, for all x in the interval,

$$A \le f(x) \le B$$

and where A and B are as close together as possible. A is called the **greatest lower bound** and B is the **least upper bound**.

What Do Bounds Mean Graphically?

Upper and lower bounds can be represented on a graph by horizontal lines. See Figure A.9.

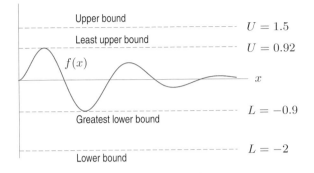

Figure A.9: Upper and lower bounds for the function f

Exercises for Appendix A

1. Use a calculator or computer graph of $f(x) = 13 - 20x - x^2 - 3x^4$ to determine:

   (a) The range of this function;

   (b) The number of zeros of this function.

For Exercises 2–12, determine the roots or points of intersection to an accuracy of one decimal place.

2. (a) The root of $x^3 - 3x + 1 = 0$ between 0 and 1

   (b) The root of $x^3 - 3x + 1 = 0$ between 1 and 2

   (c) The smallest root of $x^3 - 3x + 1 = 0$

3. The root of $x^4 - 5x^3 + 2x - 5 = 0$ between -2 and -1

4. The root of $x^5 + x^2 - 9x - 3 = 0$ between -2 and -1

5. The largest real root of $2x^3 - 4x^2 - 3x + 1 = 0$

6. All real roots of $x^4 - x - 2 = 0$

7. All real roots of $x^5 - 2x^2 + 4 = 0$

8. The smallest positive root of $x \sin x - \cos x = 0$

9. The left-most point of intersection between $y = 2x$ and $y = \cos x$

10. The left-most point of intersection between $y = 1/2^x$ and $y = \sin x$

11. The point of intersection between $y = e^{-x}$ and $y = \ln x$

12. All roots of $\cos t = t^2$

13. Estimate all real zeros of the following polynomials, accurate to 2 decimal places:

   (a) $f(x) = x^3 - 2x^2 - x + 3$

   (b) $f(x) = x^3 - x^2 - 2x + 2$

14. Find the largest zero of

$$f(x) = 10xe^{-x} - 1$$

to two decimal places, using the bisection method. Make sure to demonstrate that your approximation is as good as you claim.

15. (a) Find the smallest positive value of x where the graphs of $f(x) = \sin x$ and $g(x) = 2^{-x}$ intersect.

   (b) Repeat with $f(x) = \sin 2x$ and $g(x) = 2^{-x}$.

16. Use a graphing calculator to sketch $y = 2 \cos x$ and $y = x^3 + x^2 + 1$ on the same set of axes. Find the positive zero of $f(x) = 2 \cos x - x^3 - x^2 - 1$. A friend claims there is one more real zero. Is your friend correct? Explain.

17. Use the table below to investigate the zeros of the function

$$f(\theta) = (\sin 3\theta)(\cos 4\theta) + 0.8$$

in the interval $0 \le \theta \le 1.8$.

θ	0	0.2	0.4	0.6	0.8	1.0	1.2	1.4	1.6	1.8
$f(\theta)$	0.80	1.19	0.77	0.08	0.13	0.71	0.76	0.12	-0.19	0.33

   (a) Decide how many zeros the function has in the interval $0 \le \theta \le 1.8$.

   (b) Locate each zero, or a small interval containing each zero.

   (c) Are you sure you have found all the zeros in the interval $0 \le \theta \le 1.8$? Graph the function on a calculator or computer to decide.

18. (a) Use Table A.3 to locate approximate solution(s) to

$$(\sin 3x)(\cos 4x) = \frac{x^3}{\pi^3}$$

in the interval $1.07 \le x \le 1.15$. Give an interval of length 0.01 in which each solution lies.

Table A.3

x	x^3/π^3	$(\sin 3x)(\cos 4x)$
1.07	0.0395	0.0286
1.08	0.0406	0.0376
1.09	0.0418	0.0442
1.10	0.0429	0.0485
1.11	0.0441	0.0504
1.12	0.0453	0.0499
1.13	0.0465	0.0470
1.14	0.0478	0.0417
1.15	0.0491	0.0340

(b) Make an estimate for each solution accurate to two decimal places.

19. (a) With your calculator in radian mode, take the arctangent of 1 and multiply that number by 4. Now, take the arctangent of the result and multiply it by 4. Continue this process 10 times or so and record each result as in the accompanying table. At each step, you get 4 times the arctangent of the result of the previous step.

1
3.14159...
5.05050...
5.50129...
$\vdots$

(b) Your table allows you to find a solution of the equation

$$4 \arctan x = x.$$

Why? What is that solution?

(c) What does your table in part (a) have to do with Figure A.10?
[Hint: The coordinates of P_0 are $(1, 1)$. Find the coordinates of $P_1, P_2, P_3, \ldots$]

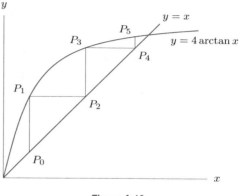

Figure A.10

(d) In part (a), what happens if you start with an initial guess of 10? Of -10? What types of behavior do you observe? (That is, for which initial guesses is the sequence increasing, and for which is it decreasing; does the sequence approach a limit?) Explain your answers graphically, as in part (c).

20. Using radians, apply the iteration method of Problem 19 to the equation

$$\cos x = x.$$

Represent your results graphically, as in Figure A.10.

For Exercises 21–23, draw a graph to decide if the function is bounded on the interval given. Give the best possible upper and lower bounds for any function which is bounded.

21. $f(x) = 4x - x^2$ on $[-1, 4]$

22. $h(\theta) = 5 + 3\sin\theta$ on $[-2\pi, 2\pi]$

23. $f(t) = \dfrac{\sin t}{t^2}$ on $[-10, 10]$

B   COMPLEX NUMBERS

The quadratic equation

$$x^2 - 2x + 2 = 0$$

is not satisfied by any real number x. If you try applying the quadratic formula, you get

$$x = \frac{2 \pm \sqrt{4 - 8}}{2} = 1 \pm \frac{\sqrt{-4}}{2}.$$

Apparently, you need to take a square root of -4. But -4 doesn't have a square root, at least, not one which is a real number. Let's give it a square root.

We define the imaginary number i to be a number such that

$$i^2 = -1.$$

Using this i, we see that $(2i)^2 = -4$, so

$$x = 1 \pm \frac{\sqrt{-4}}{2} = 1 \pm \frac{2i}{2} = 1 \pm i.$$

This solves our quadratic equation. The numbers $1 + i$ and $1 - i$ are examples of complex numbers.

A **complex number** is defined as any number that can be written in the form

$$z = a + bi,$$

where a and b are real numbers and $i^2 = -1$, so we say $i = \sqrt{-1}$.
The *real part* of z is the number a; the *imaginary part* is the number b.

Calling the number i imaginary makes it sound as if i doesn't exist in the same way that real numbers exist. In some cases, it is useful to make such a distinction between real and imaginary numbers. For example, if we measure mass or position, we want our answers to be real numbers. But the imaginary numbers are just as legitimate mathematically as the real numbers are.

As an analogy, consider the distinction between positive and negative numbers. Originally, people thought of numbers only as tools to count with; their concept of "five" or "ten" was not far removed from "five arrows" or "ten stones." They were unaware that negative numbers existed at all. When negative numbers were introduced, they were viewed only as a device for solving equations like $x + 2 = 1$. They were considered "false numbers," or, in Latin, "negative numbers." Thus, even though people started to use negative numbers, they did not view them as existing in the same way that positive numbers did. An early mathematician might have reasoned: "The number 5 exists because I can have 5 dollars in my hand. But how can I have -5 dollars in my hand?" Today we have an answer: "I have -5 dollars" means I owe somebody 5 dollars. We have realized that negative numbers are just as useful as positive ones, and it turns out that complex numbers are useful too. For example, they are used in studying wave motion in electric circuits.

Algebra of Complex Numbers

Numbers such as 0, 1, $\frac{1}{2}$, π, and $\sqrt{2}$ are called *purely real* because they contain no imaginary components. Numbers such as i, $2i$, and $\sqrt{2}i$ are called *purely imaginary* because they contain only the number i multiplied by a nonzero real coefficient.

Two complex numbers are called *conjugates* if their real parts are equal and if their imaginary parts are opposites. The complex conjugate of the complex number $z = a + bi$ is denoted $\overline{z}$, so we have

$$\overline{z} = a - bi.$$

(Note that z is real if and only if $z = \overline{z}$.) Complex conjugates have the following remarkable property: if $f(x)$ is any polynomial with real coefficients ($x^3 + 1$, say) and $f(z) = 0$, then $f(\overline{z}) = 0$. This means that if z is the solution to a polynomial equation with real coefficients, then so is $\overline{z}$.

- Two complex numbers are equal if and only if their real parts are equal and their imaginary parts are equal. Consequently, if $a + bi = c + di$, then $a = c$ and $b = d$.

- Adding two complex numbers is done by adding real and imaginary parts separately:

$$(a + bi) + (c + di) = (a + c) + (b + d)i.$$

- Subtracting is similar:

$$(a + bi) - (c + di) = (a - c) + (b - d)i.$$

- Multiplication works just as for polynomials, using $i^2 = -1$:

$$(a + bi)(c + di) = a(c + di) + bi(c + di)$$
$$= ac + adi + bci + bdi^2$$
$$= ac + adi + bci - bd = (ac - bd) + (ad + bc)i.$$

- Powers of i: We know that $i^2 = -1$; then $i^3 = i \cdot i^2 = -i$, and $i^4 = (i^2)^2 = (-1)^2 = 1$. Then $i^5 = i \cdot i^4 = i$, and so on. Thus we have

$$i^n = \begin{cases} i & \text{for } n = 1, 5, 9, 13, \ldots \\ -1 & \text{for } n = 2, 6, 10, 14, \ldots \\ -i & \text{for } n = 3, 7, 11, 15, \ldots \\ 1 & \text{for } n = 0, 4, 8, 12, 16, \ldots \end{cases}$$

- The product of a number and its conjugate is always real and nonnegative:

$$z \cdot \bar{z} = (a + bi)(a - bi) = a^2 - abi + abi - b^2 i^2 = a^2 + b^2.$$

- Dividing by a nonzero complex number is done by multiplying the denominator by its conjugate, thereby making the denominator real:

$$\frac{a + bi}{c + di} = \frac{a + bi}{c + di} \cdot \frac{c - di}{c - di} = \frac{ac - adi + bci - bdi^2}{c^2 + d^2} = \frac{ac + bd}{c^2 + d^2} + \frac{bc - ad}{c^2 + d^2} i.$$

Example 1   Compute $(2 + 7i)(4 - 6i) - i$.

Solution $(2 + 7i)(4 - 6i) - i = 8 + 28i - 12i - 42i^2 - i = 8 + 15i + 42 = 50 + 15i$.

Example 2   Compute $\dfrac{2 + 7i}{4 - 6i}$.

Solution

$$\frac{2 + 7i}{4 - 6i} = \frac{2 + 7i}{4 - 6i} \cdot \frac{4 + 6i}{4 + 6i} = \frac{8 + 12i + 28i + 42i^2}{4^2 + 6^2} = \frac{-34 + 40i}{52} = \frac{-17}{26} + \frac{10}{13} i.$$

You can check by multiplying out that $(-17/26 + 10i/13)(4 - 6i) = 2 + 7i$.

The Complex Plane and Polar Coordinates

It is often useful to picture a complex number $z = x + iy$ in the plane, with x along the horizontal axis and y along the vertical. The xy-plane is then called the *complex plane*. Figure B.11 shows the complex numbers $-2i$, $1 + i$, and $-2 + 3i$.

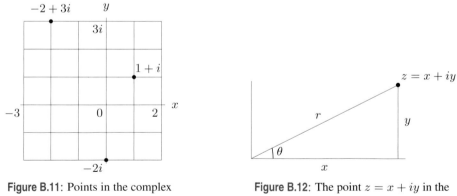

Figure B.11: Points in the complex plane

Figure B.12: The point $z = x + iy$ in the complex plane, showing polar coordinates

The triangle in Figure B.12 shows that a complex number can be written using polar coordinates as follows:

$$z = x + iy = r \cos \theta + ir \sin \theta.$$

Example 3   Express $z = -2i$ and $z = -2 + 3i$ using polar coordinates. (See Figure B.11.)

Solution   For $z = -2i$, the distance of z from the origin is 2, so $r = 2$. Also, one value for θ is $\theta = 3\pi/2$. Using polar coordinates, $-2i = 2\cos(3\pi/2) + i\,2(\sin 3\pi/2)$.
  For $z = -2 + 3i$, we have $x = -2$, $y = 3$. So $r = \sqrt{(-2)^2 + 3^2} \approx 3.61$, and one solution of $\tan\theta = 3/(-2)$ with θ in quadrant II is $\theta \approx 2.16$. So $-2 + 3i \approx 3.61\cos(2.16) + i\,3.61\sin(2.16)$.

Example 4   Consider the point with polar coordinates $r = 5$ and $\theta = 3\pi/4$. What complex number does this point represent?

Solution   Since $x = r\cos\theta$ and $y = r\sin\theta$ we see that $x = 5\cos 3\pi/4 = -5/\sqrt{2}$, and $y = 5\sin 3\pi/4 = 5/\sqrt{2}$, so $z = -5/\sqrt{2} + i\,5/\sqrt{2}$.

Derivatives and Integrals of Complex-Valued Functions

Suppose $z(t) = x(t) + iy(t)$, where t is real, then we define $z'(t)$ and $\int z(t)\,dt$ by treating i like any other constant:

$$z'(t) = x'(t) + iy'(t)$$

$$\int z(t)\,dt = \int x(t)\,dt + i\int y(t)\,dt.$$

With these definitions, all the usual properties of differentiation and integration hold, such as

$$\int z'(t)\,dt = z(t) + C, \qquad \text{for } C \text{ is a complex constant.}$$

Euler's Formula

Consider the complex number z lying on the unit circle in Figure B.13. Writing z in polar coordinates, and using the fact that $r = 1$, we have

$$z = f(\theta) = \cos\theta + i\sin\theta.$$

It turns out that there is a particularly beautiful and compact way of rewriting $f(\theta)$ using complex exponentials. We take the derivative of f using the fact that $i^2 = -1$:

$$f'(\theta) = -\sin\theta + i\cos\theta = i\cos\theta + i^2\sin\theta.$$

Factoring out an i gives

$$f'(\theta) = i(\cos\theta + i\sin\theta) = i \cdot f(\theta).$$

As you know from Chapter 11, page 540, the only real-valued function whose derivative is proportional to the function itself is the exponential function. In other words, we know that if

$$g'(x) = k \cdot g(x), \quad \text{then} \quad g(x) = Ce^{kx}$$

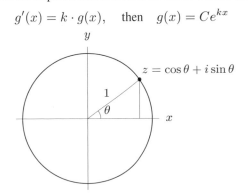

Figure B.13: Complex number represented by a point on the unit circle for some constant C. If we assume that a similar result holds for complex-valued functions, then we have

$$f'(\theta) = i \cdot f(\theta), \quad \text{so} \quad f(\theta) = Ce^{i\theta}$$

for some constant C. To find C we substitute $\theta = 0$. Now $f(0) = Ce^{i \cdot 0} = C$, and since $f(0) = \cos 0 + i \sin 0 = 1$, we must have $C = 1$. Therefore $f(\theta) = e^{i\theta}$. Thus we have

---

Euler's formula

$$e^{i\theta} = \cos \theta + i \sin \theta.$$

This elegant and surprising relationship was discovered by the Swiss mathematician Leonhard Euler in the eighteenth century, and it is particularly useful in solving second-order differential equations. Another way of obtaining Euler's formula (using Taylor series) is given in Problem 45 on page 490. It allows us to write the complex number represented by the point with polar coordinates (r, θ) in the following form:

$$z = r(\cos \theta + i \sin \theta) = re^{i\theta}.$$

Similarly, since $\cos(-\theta) = \cos \theta$ and $\sin(-\theta) = -\sin \theta$, we have

$$re^{-i\theta} = r\left(\cos(-\theta) + i \sin(-\theta)\right) = r(\cos \theta - i \sin \theta).$$

Example 5   Evaluate $e^{i\pi}$.

Solution   Using Euler's formula, $e^{i\pi} = \cos \pi + i \sin \pi = -1$.

Example 6   Express the complex number represented by the point $r = 8$, $\theta = 3\pi/4$ in Cartesian form and polar form, $z = re^{i\theta}$.

Solution   Using Cartesian coordinates, the complex number is

$$z = 8\left(\cos\left(\frac{3\pi}{4}\right) + i \sin\left(\frac{3\pi}{4}\right)\right) = \frac{-8}{\sqrt{2}} + i\frac{8}{\sqrt{2}}.$$

Using polar coordinates, we have

$$z = 8e^{i\,3\pi/4}.$$

The polar form of complex numbers makes finding powers and roots of complex numbers much easier. Writing $z = re^{i\theta}$, we find any power of z as follows:

$$z^p = (re^{i\theta})^p = r^p e^{ip\theta}.$$

To find roots, we let p be a fraction, as in the following example.

Example 7   Find a cube root of the complex number represented by the point with polar coordinates $(8, 3\pi/4)$.

Solution   In Example 6, we saw that this complex number could be written as $z = 8e^{i3\pi/4}$. So,

$$\sqrt[3]{z} = \left(8e^{i\,3\pi/4}\right)^{1/3} = 8^{1/3}e^{i(3\pi/4)\cdot(1/3)} = 2e^{\pi i/4} = 2\left(\cos(\pi/4) + i \sin(\pi/4)\right)$$

$$= 2\left(1/\sqrt{2} + i/\sqrt{2}\right) = \sqrt{2}(1 + i).$$

You can check by multiplying out that $(\sqrt{2}(1 + i))^3 = -(8/\sqrt{2}) + i(8/\sqrt{2}) = z$.

Using Complex Exponentials

Euler's formula, together with the fact that exponential functions are simple to manipulate, allows us to obtain many results about trigonometric functions easily.

The following example uses the fact that for complex z, the function e^z has all the usual algebraic properties of exponents.

Example 8   Use Euler's formula to obtain the double-angle identities

$$\cos 2\theta = \cos^2\theta - \sin^2\theta \qquad \text{and} \qquad \sin 2\theta = 2\cos\theta\sin\theta.$$

Solution   We use the fact that $e^{2i\theta} = e^{i\theta} \cdot e^{i\theta}$. This can be rewritten as

$$\cos 2\theta + i\sin 2\theta = (\cos\theta + i\sin\theta)^2.$$

Multiplying out $(\cos\theta + i\sin\theta)^2$, using the fact that $i^2 = -1$ gives

$$\cos 2\theta + i\sin 2\theta = \cos^2\theta - \sin^2\theta + i(2\cos\theta\sin\theta).$$

Since two complex numbers are equal only if the real and imaginary parts are equal, we must have

$$\cos 2\theta = \cos^2\theta - \sin^2\theta \qquad \text{and} \qquad \sin 2\theta = 2\cos\theta\sin\theta.$$

If we solve $e^{i\theta} = \cos\theta + i\sin\theta$ and $e^{-i\theta} = \cos\theta - i\sin\theta$ for $\sin\theta$ and $\cos\theta$, we obtain

$$\sin\theta = \frac{e^{i\theta} - e^{-i\theta}}{2i} \qquad \text{and} \qquad \cos\theta = \frac{e^{i\theta} + e^{-i\theta}}{2}.$$

By differentiating the formula $e^{ik\theta} = \cos(k\theta) + i\sin(k\theta)$, for θ real and k a real constant, it can be shown that

$$\frac{d}{d\theta}\left(e^{ik\theta}\right) = ike^{ik\theta} \qquad \text{and} \qquad \int e^{ik\theta}\, d\theta = \frac{1}{ik}e^{ik\theta} + C.$$

Thus complex exponentials are differentiated and integrated just like real exponentials.

Example 9   Use $\cos\theta = \left(e^{i\theta} + e^{-i\theta}\right)/2$ to obtain the derivative formula for $\cos\theta$.

Solution   Differentiating gives

$$\frac{d}{d\theta}(\cos\theta) = \frac{d}{d\theta}\left(\frac{e^{i\theta} + e^{-i\theta}}{2}\right) = \frac{ie^{i\theta} - ie^{-i\theta}}{2} = \frac{i(e^{i\theta} - e^{-i\theta})}{2}$$

$$= -\frac{e^{i\theta} - e^{-i\theta}}{2i} = -\sin\theta.$$

The facts that e^z has all the usual properties when z is complex leads to

$$\frac{d}{d\theta}(e^{(a+ib)\theta}) = (a+ib)e^{(a+ib)\theta} \qquad \text{and} \qquad \int e^{(a+ib)\theta}\, d\theta = \frac{1}{a+ib}e^{(a+ib)\theta} + C.$$

Example 10 Use the formula for $\int e^{(a+ib)\theta}\,d\theta$ to obtain formulas for $\int e^{ax}\cos bx\,dx$ and $\int e^{ax}\sin bx\,dx$.

Solution The formula for $\int e^{(a+ib)\theta}\,d\theta$ allows us to write

$$\int e^{ax}e^{ibx}\,dx = \int e^{(a+ib)x}\,dx = \frac{1}{a+ib}e^{(a+ib)x} + C = \frac{a-ib}{a^2+b^2}e^{ax}e^{ibx} + C.$$

The left-hand side of this equation can be rewritten as

$$\int e^{ax}e^{ibx}\,dx = \int e^{ax}\cos bx\,dx + i\int e^{ax}\sin bx\,dx,$$

The right-hand side can be rewritten as

$$\frac{a-ib}{a^2+b^2}e^{ax}e^{ibx} = \frac{e^{ax}}{a^2+b^2}(a-ib)(\cos bx + i\sin bx),$$

$$= \frac{e^{ax}}{a^2+b^2}(a\cos bx + b\sin bx + i(a\sin bx - b\cos bx)).$$

Equating real parts gives

$$\int e^{ax}\cos bx\,dx = \frac{e^{ax}}{a^2+b^2}(a\cos bx + b\sin bx) + C,$$

and equating imaginary parts gives

$$\int e^{ax}\sin bx\,dx = \frac{e^{ax}}{a^2+b^2}(a\sin bx - b\cos bx) + C.$$

These two formulas are usually obtained by integrating by parts twice.

Example 11 Using complex exponentials, find a formula for $\int \sin 2x \sin 3x\,dx$.

Solution Replacing $\sin 2x$ and $\sin 3x$ by their exponential form, we have

$$\int \sin 2x \sin 3x\,dx = \int \frac{(e^{2ix}-e^{-2ix})}{2i}\frac{(e^{3ix}-e^{-3ix})}{2i}\,dx$$

$$= \frac{1}{(2i)^2}\int\left(e^{5ix} - e^{-ix} - e^{ix} + e^{-5ix}\right)dx$$

$$= -\frac{1}{4}\left(\frac{1}{5i}e^{5ix} + \frac{1}{i}e^{-ix} - \frac{1}{i}e^{ix} - \frac{1}{5i}e^{-5ix}\right) + C$$

$$= -\frac{1}{4}\left(\frac{e^{5ix}-e^{-5ix}}{5i} - \frac{e^{ix}-e^{-ix}}{i}\right) + C$$

$$= -\frac{1}{4}\left(\frac{2}{5}\sin 5x - 2\sin x\right) + C$$

$$= -\frac{1}{10}\sin 5x + \frac{1}{2}\sin x + C.$$

This result is usually obtained by using a trigonometric identity.

Exercises for Appendix B

For Exercises 1–8, express the given complex number in polar form, $z = re^{i\theta}$.

For Exercises 9–18, perform the indicated calculations. Give your answer in Cartesian form, $z = x + iy$.

1. $2i$ **2.** -5 **3.** $1+i$ **4.** $-3-4i$ **9.** $(2+3i)+(-5-7i)$ **10.** $(2+3i)(5+7i)$

5. 0 **6.** $-i$ **7.** $-1+3i$ **8.** $5-12i$ **11.** $(2+3i)^2$ **12.** $(1+i)^2 + (1+i)$

13. $(0.5 - i)(1 - i/4)$

14. $(2i)^3 - (2i)^2 + 2i - 1$

15. $(e^{i\pi/3})^2$

16. $\sqrt{e^{i\pi/3}}$

17. $(5e^{i7\pi/6})^3$

18. $\sqrt[4]{10e^{i\pi/2}}$

By writing the complex numbers in polar form, $z = re^{i\theta}$, find a value for the quantities in Exercises 19–28. Give your answer in Cartesian form, $z = x + iy$.

19. $\sqrt{i}$

20. $\sqrt{-i}$

21. $\sqrt[3]{i}$

22. $\sqrt{7i}$

23. $(1+i)^{100}$

24. $(1+i)^{2/3}$

25. $(-4 + 4i)^{2/3}$

26. $(\sqrt{3} + i)^{1/2}$

27. $(\sqrt{3}+i)^{-1/2}$

28. $(\sqrt{5} + 2i)^{\sqrt{2}}$

29. Calculate i^n for $n = -1, -2, -3, -4$. What pattern do you observe? What is the value of i^{-36}? Of i^{-41}?

Solve the simultaneous equations in Exercises 30–31 for A_1 and A_2.

30. $A_1 + A_2 = 2$
$(1 - i)A_1 + (1 + i)A_2 = 3$

31. $A_1 + A_2 = 2$
$(i - 1)A_1 + (1 + i)A_2 = 0$

32. (a) Calculate a and b if $\dfrac{3 - 4i}{1 + 2i} = a + bi$.

(b) Check your answer by calculating $(1 + 2i)(a + bi)$.

33. Check that $z = \dfrac{ac + bd}{c^2 + d^2} + \dfrac{bc - ad}{c^2 + d^2}i$ is the quotient $\dfrac{a + bi}{c + di}$ by showing that the product $z \cdot (c + di)$ is $a + bi$.

34. Let $z_1 = -3 - i\sqrt{3}$ and $z_2 = -1 + i\sqrt{3}$.

(a) Find $z_1 z_2$ and z_1/z_2. Give your answer in Cartesian form, $z = x + iy$.

(b) Put z_1 and z_2 into polar form, $z = re^{i\theta}$. Find $z_1 z_2$ and z_1/z_2 using the polar form, and verify that you get the same answer as in part (a).

35. Let $z_1 = a_1 + b_1 i$ and $z_2 = a_2 + b_2 i$. Show that $\overline{z_1 z_2} = \bar{z_1} \bar{z_2}$.

36. If the roots of the equation $x^2 + 2bx + c = 0$ are the complex numbers $p \pm iq$, find expressions for p and q in terms of b and c.

Are the statements in Exercises 37–42 true or false? Explain your answer.

37. Every nonnegative real number has a real square root.

38. For any complex number z, the product $z \cdot \bar{z}$ is a real number.

39. The square of any complex number is a real number.

40. If f is a polynomial, and $f(z) = i$, then $f(\bar{z}) = i$.

41. Every nonzero complex number z can be written in the form $z = e^w$, where w is another complex number.

42. If $z = x+iy$, where x and y are positive, then $z^2 = a+ib$ has a and b positive.

For Exercises 43–47, use Euler's formula to derive the following relationships. (Note that if a, b, c, d are real numbers, $a + bi = c + di$ means that $a = c$ and $b = d$.)

43. $\sin^2\theta + \cos^2\theta = 1$

44. $\sin 2\theta = 2\sin\theta\cos\theta$

45. $\cos 2\theta = \cos^2\theta - \sin^2\theta$

46. $\dfrac{d}{d\theta}\sin\theta = \cos\theta$

47. $\dfrac{d^2}{d\theta^2}\cos\theta = -\cos\theta$

48. Use complex exponentials to show that
$$\sin(-x) = -\sin x.$$

49. Use complex exponentials to show that
$$\sin(x + y) = \sin x \cos y + \cos x \sin y.$$

50. For real t, show that if $z_1(t) = x_1(t) + iy_1(t)$ and $z_2(t) = x_2(t) + iy_2(t)$ then
$$(z_1 + z_2)' = z_1' + z_2' \quad \text{and} \quad (z_1 z_2)' = z_1' z_2 + z_1 z_2'.$$

C   NEWTON'S METHOD

Many problems in mathematics involve finding the root of an equation. For example, we might have to locate the zeros of a polynomial, or determine the point of intersection of two curves. Here we will see a numerical method for approximating solutions which cannot be calculated exactly.

One such method, bisection, is described in Appendix A. Although it is very simple, the bisection method has two major drawbacks. First, it cannot locate a root where the curve is tangent to, but does not cross, the x-axis. Second, it is relatively slow in the sense that it requires a considerable number of iterations to achieve a desired level of accuracy. Although speed may not be important in solving a single equation, a practical problem may involve solving thousands of equations as a parameter changes. In such a case, any reduction in the number of steps can be important.

Using Newton's Method

We now consider a powerful root-finding method developed by Newton. Suppose we have a function $y = f(x)$. The equation $f(x) = 0$ has a root at $x = r$, as shown in Figure C.14. We begin with an initial estimate, x_0, for this root. (This can be a guess.) We will now obtain a better estimate x_1. To do this, construct the tangent line to the graph of f at the point $x = x_0$, and extend it until it crosses the x-axis, as shown in Figure C.14. The point where it crosses the axis is usually much closer to r, and we use that point as the next estimate, x_1. Having found x_1, we now repeat the process starting with x_1 instead of x_0. We construct a tangent line to the curve at $x = x_1$, extend it until it crosses the x-axis, use that x-intercept as the next approximation, x_2, and so on. The resulting sequence of x-intercepts usually converges rapidly to the root r.

Let's see how this looks algebraically. We know that the slope of the tangent line at the initial estimate x_0 is $f'(x_0)$, and so the equation of the tangent line is

$$y - f(x_0) = f'(x_0)(x - x_0).$$

At the point where this tangent line crosses the x-axis, we have $y = 0$ and $x = x_1$, so that

$$0 - f(x_0) = f'(x_0)(x_1 - x_0).$$

Solving for x_1, we obtain

$$x_1 = x_0 - \frac{f(x_0)}{f'(x_0)}$$

provided that $f'(x_0)$ is not zero. We now repeat this argument and find that the next approximation is

$$x_2 = x_1 - \frac{f(x_1)}{f'(x_1)}.$$

Summarizing, for any $n = 0, 1, 2, \ldots$, we obtain the following result.

Newton's Method to Solve the Equation $f(x) = 0$

Choose x_0 near a solution and compute the sequence $x_1, x_2, x_3 \ldots$ using the rule

$$x_{n+1} = x_n - \frac{f(x_n)}{f'(x_n)}$$

provided that $f'(x_n)$ is not zero. For large n, the solution is well-approximated by x_n.

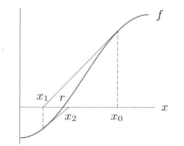

Figure C.14: Newton's method: successive approximations $x_0, x_1, x_2, \ldots$ to the root, r

---

Example 1   Use Newton's method to find the fifth root of 23. (By calculator, this is 1.872171231, correct to nine decimal places.)

Solution   To use Newton's method, we need an equation of the form $f(x) = 0$ having $23^{1/5}$ as a root. Since $23^{1/5}$ is a root of $x^5 = 23$ or $x^5 - 23 = 0$, we take $f(x) = x^5 - 23$. The root of this equation is between 1 and 2 (since $1^5 = 1$ and $2^5 = 32$), so we will choose $x_0 = 2$ as our initial estimate. Now $f'(x) = 5x^4$, so we can set up Newton's method as

$$x_{n+1} = x_n - \frac{x_n^5 - 23}{5x_n^4}.$$

In this case, we can simplify using a common denominator, to obtain

$$x_{n+1} = \frac{4x_n^5 + 23}{5x_n^4}.$$

Therefore, starting with $x_0 = 2$, we find that $x_1 = 1.8875$. This leads to $x_2 = 1.872418193$ and $x_3 = 1.872171296$. These values are in Table C.4. Since $f(1.872171231) > 0$ and $f(1.872171230) < 0$, the root lies between 1.872171230 and 1.872171231. Therefore, in just four iterations of Newton's method, we have achieved eight-decimal accuracy.

Table C.4   *Newton's method:* $x_0 = 2$

n	x_n	$f(x_n)$
0	2	9
1	1.8875	0.957130661
2	1.872418193	0.015173919
3	1.872171296	0.000004020
4	1.872171231	0.000000027

Table C.5   *Newton's method:* $x_0 = 10$

n	x_n	n	x_n
0	10	6	2.679422313
1	8.000460000	7	2.232784753
2	6.401419079	8	1.971312452
3	5.123931891	9	1.881654220
4	4.105818871	10	1.872266333
5	3.300841811	11	1.872171240

As a general guideline for Newton's method, once the first correct decimal place is found, each successive iteration approximately doubles the number of correct digits.

What happens if we select a very poor initial estimate? In the preceding example, suppose x_0 were 10 instead of 2. The results are in Table C.5. Notice that even with $x_0 = 10$, the sequence of values moves reasonably quickly toward the solution: we achieve six-decimal place accuracy by the eleventh iteration.

Example 2   Find the first point of intersection of the curves given by $f(x) = \sin x$ and $g(x) = e^{-x}$.

Solution   The graphs in Figure C.15 make it clear that there are an infinite number of points of intersection, all with $x > 0$. In order to find the first one numerically, we consider the function

$$F(x) = f(x) - g(x) = \sin x - e^{-x}$$

whose derivative is $F'(x) = \cos x + e^{-x}$. From the graph, we see that the point we want is fairly close to $x = 0$, so we start with $x_0 = 0$. The values in Table C.6 are approximations to the root. Since $F(0.588532744) > 0$ and $F(0.588532743) < 0$, the root lies between 0.588532743 and 0.588532744. (Remember, your calculator must be set in radians.)

Table C.6   *Successive approximations to root of* $\sin x = e^{-x}$

n	x_n
0	0
1	0.5
2	0.585643817
3	0.588529413
4	0.588532744
5	0.588532744

Figure C.15: Root of $\sin x = e^{-x}$

When Does Newton's Method Fail?

In most practical situations, Newton's method works well. Occasionally, however, the sequence x_0, x_1, x_2, ... fails to converge or fails to converge to the root you want. Sometimes, for example, the sequence can jump from one root to another. This is particularly likely to happen if the magnitude of the derivative $f'(x_n)$ is small for some x_n. In this case, the tangent line is nearly horizontal and so x_{n+1} will be far from x_n. (See Figure C.16.)

If the equation $f(x) = 0$ has *no* root, then the sequence will not converge. In fact, the sequence obtained by applying Newton's method to $f(x) = 1 + x^2$ is one of the best known examples of *chaotic behavior* and has attracted considerable research interest recently. (See Figure C.17.)

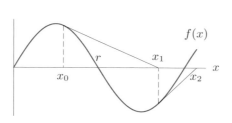

Figure C.16: Problems with Newton's method: Converges to wrong root

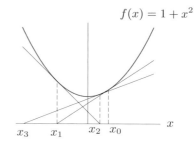

Figure C.17: Problems with Newton's method: Chaotic behavior

Exercises for Appendix C

1. Suppose you want to find a solution of the equation

$$x^3 + 3x^2 + 3x - 6 = 0.$$

Consider $f(x) = x^3 + 3x^2 + 3x - 6$.

(a) Find $f'(x)$, and use it to show that $f(x)$ increases everywhere.
(b) How many roots does the original equation have?
(c) For each root, find an interval which contains it.
(d) Find each root to two decimal places, using Newton's method.

For Exercises 2–4, use Newton's method to find the given quantities to two decimal places:

2. $\sqrt[3]{50}$ **3.** $\sqrt[4]{100}$ **4.** $10^{-1/3}$

For Exercises 5–8, solve each equation and give each answer to two decimal places:

5. $\sin x = 1 - x$ **6.** $\cos x = x$

7. $e^{-x} = \ln x$

8. $e^x \cos x = 1$, for $0 < x < \pi$

9. Find, to two decimal places, all solutions of $\ln x = 1/x$.

10. How many zeros do the following functions have? For each zero, find an upper and a lower bound which differ by no more than 0.1.
(a) $f(x) = x^3 + x - 1$ (b) $f(x) = \sin x - \frac{2}{3}x$
(c) $f(x) = 10xe^{-x} - 1$

11. Find the largest zero of

$$f(x) = x^3 + x - 1$$

to six decimal places, using Newton's method. How do you know your approximation is as good as you claim?

12. For any positive number, a, the problem of calculating the square root, $\sqrt{a}$, is often done by applying Newton's method to the function $f(x) = x^2 - a$. Apply the method to obtain an expression for x_{n+1} in terms of x_n. Use this to approximate $\sqrt{a}$ for $a = 2, 10, 1000$, and π, correct to four decimal places, starting at $x_0 = a/2$ in each case.

D   DETERMINANTS

We introduce the determinant of an array of numbers. Each 2 by 2 array of numbers has another number associated with it, called its determinant, which is given by

$$\begin{vmatrix} a_1 & a_2 \\ b_1 & b_2 \end{vmatrix} = a_1 b_2 - a_2 b_1.$$

For example

$$\begin{vmatrix} 2 & 5 \\ -4 & -6 \end{vmatrix} = 2(-6) - 5(-4) = 8.$$

Each 3 by 3 array of numbers also has a number associated with it, also called a determinant, which is defined in terms of 2 by 2 determinants as follows:

$$\begin{vmatrix} a_1 & a_2 & a_3 \\ b_1 & b_2 & b_3 \\ c_1 & c_2 & c_3 \end{vmatrix} = a_1 \begin{vmatrix} b_2 & b_3 \\ c_2 & c_3 \end{vmatrix} - a_2 \begin{vmatrix} b_1 & b_3 \\ c_1 & c_3 \end{vmatrix} + a_3 \begin{vmatrix} b_1 & b_2 \\ c_1 & c_2 \end{vmatrix}.$$

Notice that the determinant of the 2 by 2 array multiplied by a_i is the determinant of the array found by removing the row and column containing a_i. Also, note the minus sign in the second term. An example is given by

$$\begin{vmatrix} 2 & 1 & -3 \\ 0 & 3 & -1 \\ 4 & 0 & 5 \end{vmatrix} = 2 \begin{vmatrix} 3 & -1 \\ 0 & 5 \end{vmatrix} - 1 \begin{vmatrix} 0 & -1 \\ 4 & 5 \end{vmatrix} + (-3) \begin{vmatrix} 0 & 3 \\ 4 & 0 \end{vmatrix} = 2(15 + 0) - 1(0 - (-4)) + (-3)(0 - 12) = 62.$$

Suppose the vectors $\vec{a}$ and $\vec{b}$ have components $\vec{a} = a_1 \vec{i} + a_2 \vec{j} + a_3 \vec{k}$ and $\vec{b} = b_1 \vec{i} + b_2 \vec{j} + b_3 \vec{k}$. Recall that the cross product $\vec{a} \times \vec{b}$ is given by the expression

$$\vec{a} \times \vec{b} = (a_2 b_3 - a_3 b_2)\vec{i} + (a_3 b_1 - a_1 b_3)\vec{j} + (a_1 b_2 - a_2 b_1)\vec{k}.$$

Notice that if we expand the following determinant, we get the cross product:

$$\begin{vmatrix} \vec{i} & \vec{j} & \vec{k} \\ a_1 & a_2 & a_3 \\ b_1 & b_2 & b_3 \end{vmatrix} = \vec{i}(a_2 b_3 - a_3 b_2) - \vec{j}(a_1 b_3 - a_3 b_1) + \vec{k}(a_1 b_2 - a_2 b_1) = \vec{a} \times \vec{b}.$$

Determinants give a useful way of computing cross products.

READY REFERENCE

This section contains a concise summary of the main definitions, theorems, and ideas presented throughout the book.

READY REFERENCE

Multivariable Functions

Points in 3-space are represented by a system of **Cartesian coordinates** (p. 606). The **distance** between (x, y, z) and (a, b, c) is $\sqrt{(x-a)^2 + (y-b)^2 + (z-c)^2}$ (p. 609).

Functions of two variables can be represented by **graphs** (p. 611), **contour diagrams** (p. 618), **cross-sections** (p. 613), and **tables** (p. 605).

Functions of three variables can be represented by the family of **level surfaces** $f(x, y, z) = c$ for various values of the constant c (p. 634).

A **linear function** $f(x, y)$ has equation

$$f(x, y) = z_0 + m(x - x_0) + n(y - y_0) \text{ (p. 629)}$$
$$= c + mx + ny, \text{ where } c = z_0 - mx_0 - ny_0.$$

Its **graph** is a plane with slope m in the x-direction, slope n in the y-direction, through (x_0, y_0, z_0) (p. 629). Its **table of values** has linear rows (of same slope) and linear columns (of same slope) (p. 630). Its **contour diagram** is equally spaced parallel straight lines (p. 631).

The **limit** of f at the point (a, b), written $\lim_{(x,y) \to (a,b)} f(x, y)$, is the number L, if one exists, such that $f(x, y)$ is as close to L as we please whenever the distance from the point (x, y) to the point (a, b) is sufficiently small, but not zero. (p. 640).

A function f is **continuous at the point** (a, b) if $\lim_{(x,y) \to (a,b)} f(x, y) = f(a, b)$. A function is **continuous on a region** R if it is continuous at each point of R (p. 640).

Vectors

A **vector** $\vec{v}$ has **magnitude** (denoted $\|\vec{v}\|$) and **direction**. Examples are **displacement vectors** (p. 650), **velocity and acceleration vectors** (pp. 658, 660). and **force** (p. 660). We can add vectors, and multiply a vector by a scalar (p. 651). Two non-zero vectors, $\vec{v}$ and $\vec{w}$, are **parallel** if one is a scalar multiple of the other (p. 652).

A **unit vector** has magnitude 1. The vectors $\vec{i}$, $\vec{j}$, and $\vec{k}$ are unit vectors in the directions of the coordinate axes. A unit vector in the direction of any nonzero vector $\vec{v}$ is $\vec{u} = \vec{v}/\|\vec{v}\|$ (p. 656). We **resolve** $\vec{v}$ **into components** by writing $\vec{v} = v_1\vec{i} + v_2\vec{j} + v_3\vec{k}$ (p. 653).

If $\vec{v} = v_1\vec{i} + v_2\vec{j} + v_3\vec{k}$ and $\vec{w} = w_1\vec{i} + w_2\vec{j} + w_3\vec{k}$ then

$$\|\vec{v}\| = \sqrt{v_1^2 + v_2^2 + v_3^2} \text{ (p. 654)}$$
$$\vec{v} + \vec{w} = (v_1 + w_1)\vec{i} + (v_2 + w_2)\vec{j} + (v_3 + w_3)\vec{k} \text{ (p. 655)},$$
$$\lambda\vec{v} = \lambda v_1\vec{i} + \lambda v_2\vec{j} + \lambda v_3\vec{k} \text{ (p. 655)}.$$

The **displacement vector** from $P_1 = (x_1, y_1, z_1)$ to $P_2 = (x_2, y_2, z_2)$ is

$$\overrightarrow{P_1 P_2} = (x_2 - x_1)\vec{i} + (y_2 - y_1)\vec{j} + (z_2 - z_1)\vec{k} \quad \text{(p. 654)}.$$

The **position vector** of $P = (x, y, z)$ is $\overrightarrow{OP}$ (p. 654). A **vector in** n **dimensions** is a string of numbers $\vec{v} = (v_1, v_2, \ldots, v_n)$ (p. 662).

Dot Product (Scalar Product) (p. 664).
Geometric definition: $\vec{v} \cdot \vec{w} = \|\vec{v}\|\|\vec{w}\| \cos\theta$ where θ is the angle between $\vec{v}$ and $\vec{w}$ and $0 \le \theta \le \pi$.
Algebraic definition: $\vec{v} \cdot \vec{w} = v_1 w_1 + v_2 w_2 + v_3 w_3$.

Two nonzero vectors $\vec{v}$ and $\vec{w}$ are **perpendicular** if and only if $\vec{v} \cdot \vec{w} = 0$ (p. 666). Magnitude and dot product are related by $\vec{v} \cdot \vec{v} = \|\vec{v}\|^2$ (p. 666). If $\vec{u} = (u_1, \ldots, u_n)$ and $\vec{v} = (v_1, \ldots, v_n)$ then the dot product of $\vec{u}$ and $\vec{v}$ is $\vec{u} \cdot \vec{v} = u_1 v_1 + \ldots + u_n v_n$ (p. 668).

The **equation of the plane** with **normal vector** $\vec{n} = a\vec{i} + b\vec{j} + c\vec{k}$ and containing the point $P_0 = (x_0, y_0, z_0)$ is $\vec{n} \cdot (\vec{r} - \vec{r}_0) = a(x - x_0) + b(y - y_0) + c(z - z_0) = 0$ or $ax + by + cz = d$, where $d = ax_0 + by_0 + cz_0$ (p. 667).

If $\vec{v}_{\text{parallel}}$ and $\vec{v}_{\text{perp}}$ are components of $\vec{v}$ which are parallel and perpendicular, respectively, to a unit vector $\vec{u}$, then $\vec{v}_{\text{parallel}} = (\vec{v} \cdot \vec{u})\vec{u}$ and $\vec{v}_{\text{perp}} = \vec{v} - \vec{v}_{\text{parallel}}$ (p. 668).

The **work**, W, done by a force $\vec{F}$ acting on an object through a displacement $\vec{d}$ is $W = \vec{F} \cdot \vec{d}$ (p. 670).

Cross Product (Vector Product) (p. 673, 674)
Geometric definition

$$\vec{v} \times \vec{w} = \begin{pmatrix} \text{Area of parallelogram} \\ \text{with edges } \vec{v} \text{ and } \vec{w} \end{pmatrix} \vec{n}$$
$$= (\|\vec{v}\|\|\vec{w}\| \sin\theta)\vec{n},$$

where $0 \le \theta \le \pi$ is the angle between $\vec{v}$ and $\vec{w}$ and $\vec{n}$ is the unit vector perpendicular to $\vec{v}$ and $\vec{w}$ pointing in the direction given by the **right-hand rule**.
Algebraic definition

$$\vec{v} \times \vec{w} = (v_2 w_3 - v_3 w_2)\vec{i} + (v_3 w_1 - v_1 w_3)\vec{j}$$
$$+ (v_1 w_2 - v_2 w_1)\vec{k}$$

$$\vec{v} = v_1\vec{i} + v_2\vec{j} + v_3\vec{k}, \vec{w} = w_1\vec{i} + w_2\vec{j} + w_3\vec{k}.$$

To find the **equation of a plane through three points** that do not lie on a line, determine two vectors in the plane and then find a normal vector using the cross product (p. 676). The **area of a parallelogram** with edges $\vec{v}$ and $\vec{w}$ is $\|\vec{v} \times \vec{w}\|$. The **volume of a parallelepiped** with edges $\vec{a}$, $\vec{b}$, $\vec{c}$ is $|(\vec{b} \times \vec{c}) \cdot \vec{a}|$ (p. 677).

Differentiation of Multivariable Functions

Partial derivatives of f **(p. 687).**

$$f_x(a,b) = \begin{array}{c}\text{Rate of change of } f \text{ with respect to } x\\ \text{at the point } (a,b)\end{array}$$
$$= \lim_{h\to 0}\frac{f(a+h,b)-f(a,b)}{h},$$
$$f_y(a,b) = \begin{array}{c}\text{Rate of change of } f \text{ with respect to } y\\ \text{at the point } (a,b)\end{array}$$
$$= \lim_{h\to 0}\frac{f(a,b+h)-f(a,b)}{h}.$$

On the graph of f, the partial derivatives $f_x(a,b)$ and $f_y(a,b)$ give the slope in the x and y directions, respectively (p. 688). The **tangent plane** to $z = f(x,y)$ at (a,b) is

$$z = f(a,b)+f_x(a,b)(x-a)+f_y(a,b)(y-b)\quad\text{(p. 698).}$$

Partial derivatives can be estimated from a contour diagram or table of values using difference quotients (p. 688), and can be computed algebraically using the same rules of differentiation as for one-variable calculus (p. 693). Partial derivatives for functions of three or more variables are defined and computed in the same way (p 693).

The gradient vector $\operatorname{grad} f$ of f is $\operatorname{grad} f(a,b) = f_x(a,b)\vec{i} + f_y(a,b)\vec{j}$ (2 variables) (p. 706) or $\operatorname{grad} f(a,b,c) = f_x(a,b,c)\vec{i} + f_y(a,b,c)\vec{j} + f_z(a,b,c)\vec{k}$ (3 variables) (p. 713). The gradient vector at P: Points in the direction of increasing f; is perpendicular to the level curve or level surface of f through P; and has magnitude $\|\operatorname{grad} f\|$ equal to the maximum rate of change of f at P (pp. 708, 713). The magnitude is large when the level curves or surfaces are close together and small when they are far apart.

The **directional derivative** of f at P in the direction of a unit vector $\vec{u}$ is (pp. 705, 707)

$$f_{\vec{u}}(P) = \begin{array}{c}\text{Rate of change}\\ \text{of } f \text{ in direction}\\ \text{of } \vec{u} \text{ at } P\end{array} = \operatorname{grad} f(P)\cdot\vec{u}$$

The **tangent plane approximation** to $f(x,y)$ for (x,y) near the point (a,b) is

$$f(x,y)\approx f(a,b)+f_x(a,b)(x-a)+f_y(a,b)(y-b).$$

The right-hand side is the **local linearization** (p. 699). The **differential of** $z = f(x,y)$ at (a,b) is the linear function of dx and dy

$$df = f_x(a,b)\,dx + f_y(a,b)\,dy\quad\text{(p. 701).}$$

Local linearity with three or more variables follows the same pattern as for functions of two variables (p. 700).

The **tangent plane to a level surface** of a function of three-variables f at (a,b,c) is (p. 716)

$$f_x(a,b,c)(x-a)+f_y(a,b,c)(y-b)+f_z(a,b,c)(z-c) = 0.$$

The Chain Rule for the partial derivative of one variable with respect to another in a chain of composed functions (p. 721):

- Draw a diagram expressing the relationship between the variables, and label each link in the diagram with the derivative relating the variables at its ends.
- For each path between the two variables, multiply together the derivatives from each step along the path.
- Add the contributions from each path.

If $z = f(x,y)$, and $x = g(t)$, and $y = h(t)$, then

$$\frac{dz}{dt} = \frac{\partial z}{\partial x}\frac{dx}{dt}+\frac{\partial z}{\partial y}\frac{dy}{dt}\quad\text{(p. 719).}$$

If $z = f(x,y)$, with $x = g(u,v)$ and $y = h(u,v)$, then

$$\frac{\partial z}{\partial u} = \frac{\partial z}{\partial x}\frac{\partial x}{\partial u}+\frac{\partial z}{\partial y}\frac{\partial y}{\partial u},$$
$$\frac{\partial z}{\partial v} = \frac{\partial z}{\partial x}\frac{\partial x}{\partial v}+\frac{\partial z}{\partial y}\frac{\partial y}{\partial v}\quad\text{(p. 721).}$$

Second-order partial derivatives (p. 727)

$$\frac{\partial^2 z}{\partial x^2} = f_{xx} = (f_x)_x,\qquad \frac{\partial^2 z}{\partial x\partial y} = f_{yx} = (f_y)_x,$$
$$\frac{\partial^2 z}{\partial y\partial x} = f_{xy} = (f_x)_y,\qquad \frac{\partial^2 z}{\partial y^2} = f_{yy} = (f_y)_y.$$

Theorem: Equality of Mixed Partial Derivatives. If f_{xy} and f_{yx} are continuous at (a,b), an interior point of their domain, then $f_{xy}(a,b) = f_{yx}(a,b)$ (p. 728).

Taylor Polynomial of Degree 1 **Approximating** $f(x,y)$ **for** (x,y) **near** (a,b) **(p. 731)**

$$f(x,y)\approx L(x,y) = f(a,b)+f_x(a,b)(x-a)+f_y(a,b)(y-b).$$

Taylor Polynomial of Degree 2 **(p. 732)**

$$f(x,y)\approx Q(x,y)$$
$$= f(a,b)+f_x(a,b)(x-a)+f_y(a,b)(y-b)$$
$$+\frac{f_{xx}(a,b)}{2}(x-a)^2+f_{xy}(a,b)(x-a)(y-b)$$
$$+\frac{f_{yy}(a,b)}{2}(y-b)^2.$$

Definition of Differentiability (p. 735). A function $f(x,y)$ is **differentiable at the point** (a,b) if there is a linear function $L(x,y) = f(a,b)+m(x-a)+n(y-b)$ such that if the **error** $E(x,y)$ is defined by

$$f(x,y) = L(x,y)+E(x,y),$$

and if $h = x-a, k = y-b$, then the **relative error** $E(a+h,b+k)/\sqrt{h^2+k^2}$ satisfies

$$\lim_{\substack{h\to 0\\ k\to 0}}\frac{E(a+h,b+k)}{\sqrt{h^2+k^2}} = 0.$$

Theorem: Continuity of Partial Derivatives Implies Differentiability (p. 738). If the partial derivatives, f_x and f_y, of a function f exist and are continuous on a small disk centered at the point (a,b), then f is differentiable at (a,b).

Optimization

A function f has a **local maximum** at the point P_0 if $f(P_0) \geq f(P)$ for all points P near P_0, and a **local minimum** at the point P_0 if $f(P_0) \leq f(P)$ for all points P near P_0 (p. 748). A **critical point** of a function f is a point where grad f is either $\vec{0}$ or undefined. If f has a local maximum or minimum at a point P_0, not on the boundary of its domain, then P_0 is a critical point (p. 748). A **quadratic function** $f(x, y) = ax^2 + bxy + cz^2$ generally has one critical point, which can be a local maximum, a local minimum, or a **saddle point** (p. 751).

Second derivative test for functions of two variables (p. 753). Suppose grad $f(x_0, y_0) = \vec{0}$. Let $D = f_{xx}(x_0, y_0) f_{yy}(x_0, y_0) - (f_{xy}(x_0, y_0))^2$.

- If $D > 0$ and $f_{xx}(x_0, y_0) > 0$, then f has a local minimum at (x_0, y_0).

- If $D > 0$ and $f_{xx}(x_0, y_0) < 0$, then f has a local maximum at (x_0, y_0).

- If $D < 0$, then f has a saddle point at (x_0, y_0).

- If $D = 0$, anything can happen.

Unconstrained optimization

A function f defined on a region R has a **global maximum on** R at the point P_0 if $f(P_0) \geq f(P)$ for all points P in R, and a **global minimum on** R at the point P_0 if $f(P_0) \leq f(P)$ for all points P in R (p. 756). For an **unconstrained optimization problem**, find the critical points and investigate whether the critical points give global maxima or minima (p. 756).

A **closed** region is one which contains its boundary; a **bounded** region is one which does not stretch to infinity in any direction (p. 761).

Extreme Value Theorem for Multivariable Functions. If f is a continuous function on a closed and bounded region R, then f has a global maximum at some point (x_0, y_0) in R and a global minimum at some point (x_1, y_1) in R (p. 761).

Constrained optimization

Suppose P_0 is a point satisfying the constraint $g(x, y) = c$. A function f has a **local maximum** at P_0 **subject to the constraint** if $f(P_0) \geq f(P)$ for all points P near P_0 satisfying the constraint (p. 766). It has a **global maximum** at P_0 **subject to the constraint** if $f(P_0) \geq f(P)$ for all points P satisfying the constraint (p. 766). Local and global minima are defined similarly (p. 766). A local maximum or minimum of $f(x, y)$ subject to a constraint $g(x, y) = c$ occurs at a point where the constraint is tangent to a level curve of f, and thus where grad g is parallel to grad f (p. 766).

To optimize f subject to the constraint $g = c$ (p. 766), find the points satisfying the equations

$$\text{grad } f = \lambda \text{ grad } g \quad \text{and} \quad g = c.$$

Then compare values of f at these points, at points on the constraint where grad $g = \vec{0}$, and at the endpoints of the constraint. The number λ is called the **Lagrange multiplier**.

To optimize f subject to the constraint $g \leq c$ (p. 768), find all points in the interior $g(x, y) < c$ where grad f is zero or undefined; then use Lagrange multipliers to find the local extrema of f on the boundary $g(x, y) = c$. Evaluate f at the points found and compare the values.

The value of λ is the rate of change of the optimum value of f as c increases (where $g(x, y) = c$) (p. 770). The **Lagrangian function** $\mathcal{L}(x, y, \lambda) = f(x, y) - \lambda(g(x, y) - c)$ can be used to convert a constrained optimization problem for f subject the constraint $g = c$ into an unconstrained problem for $\mathcal{L}$ (p. 770).

Multivariable Integration

The **definite integral** of f, a continuous function of two variables, over R, the rectangle $a \leq x \leq b$, $c \leq y \leq d$, is called a **double integral**, and is a limit of **Riemann sums**

$$\int_R f \, dA = \lim_{\Delta x, \Delta y \to 0} \sum_{i,j} f(u_{ij}, v_{ij}) \Delta x \Delta y \quad \text{(p. 783)}.$$

The Riemann sum is constructed by subdividing R into subrectangles of width Δx and height Δy, and choosing a point (u_{ij}, v_{ij}) in the ij-th rectangle.

A **triple integral** of f, a continuous function of three variables, over W, the box $a \leq x \leq b, c \leq y \leq d, p \leq z \leq q$ in 3-space, is defined in a similar way using three-variable Riemann sums (p. 798).

Interpretations

If $f(x, y)$ is positive, $\int_R f \, dA$ is the **volume** under graph of f above the region R (p. 784). If $f(x, y) = 1$ for all x and y, then the area of R is $\int_R 1 \, dA = \int_R dA$ (p. 785). If $f(x, y)$ is a **density**, then $\int_R f \, dA$ is the **total quantity** in the region R (p. 782). The **average value** of $f(x, y)$ on the region R is $\frac{1}{\text{Area of } R} \int_R f \, dA$ (p. 786). In probability, if $p(x, y)$ is a **joint density function** then $\int_a^b \int_c^d p(x, y) \, dy \, dx$ is the fraction of population with $a \leq x \leq b$ and $c \leq y \leq d$ (p. 817).

Iterated integrals

Double and triple integrals can be written as **iterated integrals**

$$\int_R f \, dA = \int_c^d \int_a^b f(x, y) \, dx \, dy \quad \text{(p. 790)}$$

$$\int_W f \, dV = \int_p^q \int_c^d \int_a^b f(x, y, z) \, dx \, dy \, dz \quad \text{(p. 799)}$$

Other orders of integration are possible. For iterated integrals over **non-rectangular regions** (p. 792), limits on outer integral are constants and limits on inner integrals involve only the variables in the integrals further out (pp. 793, 801).

Integrals in other coordinate systems

When computing double integrals in polar coordinates, put $dA = r \, dr \, d\theta$ or $dA = r \, d\theta \, dr$ (p. 803). Cylindrical coordinates are given by $x = r \cos \theta, y = r \sin \theta, z = z$, for $0 \leq r < \infty, 0 \leq \theta \leq 2\pi, -\infty < z < \infty$ (p. 807). Spherical coordinates are given by $x = \rho \sin \phi \cos \theta, y = \rho \sin \phi \sin \theta, z = \rho \cos \phi$, for $0 \leq \rho < \infty, 0 \leq \phi \leq \pi, 0 \leq \theta \leq 2\pi$ (p. 810). When computing triple integrals in cylindrical or spherical coordinates, put $dV = r \, dr \, d\theta \, dz$ for cylindrical coordinates (p. 808), $dV = \rho^2 \sin\phi \, d\rho \, d\phi \, d\theta$ for spherical coordinates (p. 811). Other orders of integration are also possible.

For a **change of variables** $x = x(s,t)$, $y = y(s,t)$, the **Jacobian** is

$$\frac{\partial(x,y)}{\partial(s,t)} = \frac{\partial x}{\partial s} \cdot \frac{\partial y}{\partial t} - \frac{\partial x}{\partial t} \cdot \frac{\partial y}{\partial s} = \begin{vmatrix} \frac{\partial x}{\partial s} & \frac{\partial x}{\partial t} \\[2mm] \frac{\partial y}{\partial s} & \frac{\partial y}{\partial t} \end{vmatrix} \quad \text{(p. 822)}.$$

To convert an integral from x, y to s, t coordinates (p. 823): Substitute for x and y in terms of s and t, change the xy region R into an st region T, and change the area element by making the substitution $dxdy = \left| \frac{\partial(x,y)}{\partial(s,t)} \right| dsdt$. For triple integrals, there is a similar formula (p. 824).

Parameterizations and Vector Fields

Parameterized curves

The motion of a particle is described by **parametric equations** $x = f(t), y = g(t)$ (2-space) or $x = f(t), y = g(t), z = h(t)$ (3-space). The path of the particle is a **parameterized curve** (p. 832). Parameterizations are also written in **vector form** $\vec{r}(t) = f(t)\vec{i} + g(t)\vec{j} + h(t)\vec{k}$ (p. 833). For a **curve segment** we restrict the parameter to to a closed interval $a \le t \le b$ (p. 835). **Parametric equations for the graph** of $y = f(x)$ are $x = t, y = f(t)$.

Parametric equations for a line through (x_0, y_0) in the direction of $\vec{v} = a\vec{i} + b\vec{j}$ are $x = x_0 + at, y = y_0 + bt$. In 3-space, the line through (x_0, y_0, z_0) in the direction of $\vec{v} = a\vec{i} + b\vec{j} + c\vec{k}$ is $x = x_0 + at, y = y_0 + bt, z = z_0 + ct$ (p. 833). In vector form, the equation for a line is $\vec{r}(t) = \vec{r}_0 + t\vec{v}$, where $\vec{r}_0 = x_0\vec{i} + y_0\vec{j} + z_0\vec{k}$ (p. 835).

Parametric equations for a circle of radius R in the plane, centered at the origin are $x = R\cos t$, $y = R\sin t$ (counterclockwise), $x = R\cos t, y = -R\sin t$ (clockwise).

To find the **intersection points** of a curve $\vec{r}(t) = f(t)\vec{i} + g(t)\vec{j} + h(t)\vec{k}$ with a surface $F(x,y,z) = c$, solve $F(f(t), g(t), h(t)) = c$ for t (p. 835). To find the intersection points of two curves $\vec{r}_1(t)$ and $\vec{r}_2(t)$, solve $\vec{r}_1(t_1) = \vec{r}_2(t_2)$ for t_1 and t_2 (p. 835).

The **length of a curve segment** C given parametrically for $a \le t \le b$ with velocity vector $\vec{v}$ is $\int_a^b \|\vec{v}\| dt$ if $\vec{v} \ne \vec{0}$ for $a < t < b$ (p. 844).

The **velocity** and **acceleration** of a moving object with position vector $\vec{r}(t)$ at time t are

$$\vec{v}(t) = \lim_{\Delta t \to 0} \frac{\Delta \vec{r}}{\Delta t} \quad \text{(p. 840)}$$

$$\vec{a}(t) = \lim_{\Delta t \to 0} \frac{\Delta \vec{v}}{\Delta t} \quad \text{(p. 842)}$$

We write $\vec{v} = \frac{d\vec{r}}{dt} = \vec{r}'(t)$ and $\vec{a} = \frac{d\vec{v}}{dt} = \frac{d^2\vec{r}}{dt^2} = \vec{r}''(t)$.

The **components of the velocity and acceleration vectors** are

$$\vec{v}(t) = \frac{dx}{dt}\vec{i} + \frac{dy}{dt}\vec{j} + \frac{dz}{dt}\vec{k} \quad \text{(p. 840)}$$

$$\vec{a}(t) = \frac{d^2 x}{dt^2}\vec{i} + \frac{d^2 y}{dt^2}\vec{j} + \frac{d^2 z}{dt^2}\vec{k} \quad \text{(p. 842)}$$

The **speed** is $\|\vec{v}\| = \sqrt{(dx/dt)^2 + (dy/dt)^2 + (dz/dt)^2}$ (p. 844). Analogous formulas for velocity, speed, and acceleration hold in 2-space.

Uniform Circular Motion (p. 843) For a particle $\vec{r}(t) = R\cos(\omega t)\vec{i} + R\sin(\omega t)\vec{j}$: motion is in a circle of radius R with period $2\pi/\omega$; velocity, $\vec{v}$, is tangent to the circle and speed is constant $\|\vec{v}\| = \omega R$; acceleration, $\vec{a}$, points toward the center of the circle with $\|\vec{a}\| = \|\vec{v}\|^2/R$.

Motion in a Straight Line (p. 844) For a particle $\vec{r}(t) = \vec{r}_0 + f(t)\vec{v}_0$: Motion is along a straight line through the point with position vector $\vec{r}_0$ parallel to $\vec{v}_0$; velocity, $\vec{v}$, and acceleration, $\vec{a}$, are parallel to the line.

Vector fields

A **vector field** in 2-space is a function $\vec{F}(x,y)$ whose value at a point (x,y) is a 2-dimensional vector (p. 847). Similarly, a vector field in 3-space is a function $\vec{F}(x,y,z)$ whose values are 3-dimensional vectors (p. 847). Examples are the **gradient** of a differentiable function f, the **velocity field** of a fluid flow, and **force fields** (p. 847). A **flow line** of a vector field $\vec{v} = \vec{F}(\vec{r})$ is a path $\vec{r}(t)$ whose velocity vector equals $\vec{v}$, thus $\vec{r}'(t) = \vec{v} = \vec{F}(\vec{r}(t))$ (p. 854). The **flow** of a vector field is the family of all of its flow line (p. 854). Flow lines can be approximated numerically using Euler's method (p. 856).

Parameterized surfaces

We **parameterize a surface** with two parameters, $x = f_1(s,t)$, $y = f_2(s,t)$, $z = f_3(s,t)$ (p. 859). We also use the vector form $\vec{r}(s,t) = f_1(s,t)\vec{i} + f_2(s,t)\vec{j} + f_3(s,t)\vec{k}$ (p. 859). **Parametric equations for the graph** of $z = f(x,y)$ are $x = s$, $y = t$, and $z = f(s,t)$ (p. 860). **Parametric equation for a plane** through the point with position vector $\vec{r}_0$ and containing the two nonparallel vectors $\vec{v}_1$ and $\vec{v}_2$ is $\vec{r}(s,t) = \vec{r}_0 + s\vec{v}_1 + t\vec{v}_2$ (p. 860). **Parametric equation for a sphere** of radius R centered at the origin is $\vec{r}(\theta, \phi) = R\sin\phi\cos\theta\,\vec{i} + R\sin\phi\sin\theta\,\vec{j} + \cos\phi\,\vec{k}$, $0 \le \theta \le 2\pi, 0 \le \phi \le \pi$ (p. 861). **Parametric equation for a cylinder** of radius R along the z-axis is $\vec{r}(\theta, z) = R\cos\theta\vec{i} + R\sin\theta\vec{j} + z\vec{k}$, $0 \le \theta \le 2\pi, -\infty < z < \infty$ (p. 858). A **parameter curve** is the curve obtained by holding one of the parameters constant and letting the other vary (p. 864).

Line Integrals

The **line integral** of a vector field $\vec{F}$ along an **oriented curve** C (p. 876) is

$$\int_C \vec{F} \cdot d\vec{r} = \lim_{\|\Delta\vec{r}_i\| \to 0} \sum_{i=0}^{n-1} \vec{F}(\vec{r}_i) \cdot \Delta\vec{r}_i,$$

where the direction of $\Delta\vec{r}_i$ is the direction of the orientation (p. 877).

The line integral measures the extent to which C is going with $\vec{F}$ or against it (p. 877). For oriented curves C,

C_1, and C_2, $\int_{-C} \vec{F} \cdot d\vec{r} = -\int_C \vec{F} \cdot d\vec{r}$, where $-C$ is the curve C parameterized in the opposite direction, and $\int_{C_1+C_2} \vec{F} \cdot d\vec{r} = \int_{C_1} \vec{F} \cdot d\vec{r} + \int_{C_2} \vec{F} \cdot d\vec{r}$, where $C_1 + C_2$ is the curve obtained by joining the endpoint of C_1 to the starting point of C_2 (p. 881).

The **work done by a force** $\vec{F}$ along a curve C is $\int_C \vec{F} \cdot d\vec{r}$ (p. 878). The **circulation** of $\vec{F}$ around an oriented closed curve is $\int_C \vec{F} \cdot d\vec{r}$ (p. 880).

Given a parameterization of C, $\vec{r}(t)$, for $a \leq t \leq b$, the line integral can be calculated as

$$\int_C \vec{F} \cdot d\vec{r} = \int_a^b \vec{F}(\vec{r}(t)) \cdot \vec{r}'(t)\, dt \quad \text{(p. 885)}.$$

Fundamental Theorem for Line Integrals (p. 890):

Suppose C is a piecewise smooth oriented path with starting point P and endpoint Q. If f is a function whose gradient is continuous on the path C, then

$$\int_C \operatorname{grad} f \cdot d\vec{r} = f(Q) - f(P).$$

Path-independent fields and gradient fields

A vector field $\vec{F}$ is said to be **path-independent**, or **conservative**, if for any two points P and Q, the line integral $\int_C \vec{F} \cdot d\vec{r}$ has the same value along any piecewise smooth path C from P to Q lying in the domain of $\vec{F}$ (p. 892). A **gradient field** is a vector field of the form $\vec{F} = \operatorname{grad} f$ for some scalar function f, and f is called a **potential function** for the vector field $\vec{F}$ (p. 893). A vector field $\vec{F}$ is path-independent if and only if $\vec{F}$ is a gradient vector field (p. 893). A vector field $\vec{F}$ is path-independent if and only if $\int_C \vec{F} \cdot d\vec{r} = 0$ for every closed curve C (p. 899). If $\vec{F}$ is a gradient field, then $\frac{\partial F_2}{\partial x} - \frac{\partial F_1}{\partial y} = 0$ (p. 900). The quantity $\frac{\partial F_2}{\partial x} - \frac{\partial F_1}{\partial y}$ is called the 2-dimensional or scalar **curl** of $\vec{F}$.

Green's Theorem (p. 902):

Suppose C is a piecewise smooth simple closed curve that is the boundary of an open region R in the plane and oriented so that the region is on the left as we move around the curve. Suppose $\vec{F} = F_1\vec{i} + F_2\vec{j}$ is a smooth vector field defined at every point of the region R and boundary C. Then

$$\int_C \vec{F} \cdot d\vec{r} = \int_R \left(\frac{\partial F_2}{\partial x} - \frac{\partial F_1}{\partial y} \right) dx\, dy.$$

Curl test for vector fields in 2-space: If $\frac{\partial F_2}{\partial x} - \frac{\partial F_1}{\partial y} = 0$ and the domain of $\vec{F}$ has no holes, then $\vec{F}$ is path-independent, and hence a gradient field (p. 903). The condition that the domain have no holes is important. It is not always true that if the scalar curl of $\vec{F}$ is zero then $\vec{F}$ is a gradient field (p. 903).

Surface Integrals

A surface is **oriented** if a unit normal vector $\vec{n}$ has been chosen at every point on it in a continuous way (p. 916). For a closed surface, we usually choose the outward orientation (p. 916). The **area vector** of a flat, oriented surface is a vector $\vec{A}$ whose magnitude is the area of the surface, and whose direction is the direction of the orientation vector $\vec{n}$ (p. 917). If $\vec{v}$ is the velocity vector of a constant fluid flow and $\vec{A}$ is the area vector of a flat surface, then the total flow through the surface in units of volume per unit time is called the **flux** of $\vec{v}$ through the surface and is given by $\vec{v} \cdot \vec{A}$ (p. 917).

The **surface integral** or **flux integral** of the vector field $\vec{F}$ through the oriented surface S is

$$\int_S \vec{F} \cdot d\vec{A} = \lim_{\|\Delta \vec{A}\| \to 0} \sum \vec{F} \cdot \Delta \vec{A},$$

where the direction of $\Delta \vec{A}$ is the direction of the orientation (p. 918). If $\vec{v}$ is a variable vector field and then $\int_S \vec{v} \cdot d\vec{A}$ is the flux through the surface S (p. 918).

Simple flux integrals can be calculated by putting $d\vec{A} = \vec{n}\, dA$ and using geometry or converting to a double integral (p. 920).

The **flux through a graph** of $z = f(x, y)$ above a region R in the xy-plane, oriented upward, is

$$\int_R \vec{F}(x, y, f(x, y)) \cdot \left(-f_x \vec{i} - f_y \vec{j} + \vec{k} \right) dx\, dy \quad \text{(p. 925)}.$$

The **flux through a cylindrical surface** S of radius R and oriented away from the z-axis is

$$\int_T \vec{F}(R, \theta, z) \cdot \left(\cos \theta \vec{i} + \sin \theta \vec{j} \right) R\, dz\, d\theta \quad \text{(p. 927)},$$

where T is the θz-region corresponding to S.

The **flux through a spherical surface** S of radius R and oriented away from the origin is

$$\int_T \vec{F}(R, \theta, \phi) \cdot \left(\sin \phi \cos \theta \vec{i} + \sin \phi \sin \theta \vec{j} + \cos \phi \vec{k} \right)$$
$$R^2 \sin \phi\, d\phi\, d\theta, \quad \text{(p. 928)}$$

where T is the $\theta\phi$-region corresponding to S.

The **flux through a parameterized surface** S, parameterized by $\vec{r} = \vec{r}(s, t)$, where (s, t) varies in a parameter region R, is

$$\int_R \vec{F}(\vec{r}(s, t)) \cdot \left(\frac{\partial \vec{r}}{\partial s} \times \frac{\partial \vec{r}}{\partial t} \right) ds\, dt \quad \text{(p. 931)}.$$

We choose the parameterization so that $\partial \vec{r}/\partial s \times \partial \vec{r}/\partial t$ is never zero and points in the direction of $\vec{n}$ everywhere.

The **area of a parameterized surface** S, parameterized by $\vec{r} = \vec{r}(s, t)$, where (s, t) varies in a parameter region R, is

$$\int_S dA = \int_R \left\| \frac{\partial \vec{r}}{\partial s} \times \frac{\partial \vec{r}}{\partial t} \right\| ds\, dt \quad \text{(p. 932)}.$$

Divergence and Curl

Divergence

Definition of Divergence (p. 940).
Geometric definition: The **divergence** of $\vec{F}$ is

$$\text{div } \vec{F}(x, y, z) = \lim_{\text{Volume} \to 0} \frac{\int_S \vec{F} \cdot d\vec{A}}{\text{Volume of } S}.$$

Here S is a sphere centered at (x, y, z), oriented outwards, that contracts down to (x, y, z) in the limit.

Cartesian coordinate definition: If $\vec{F} = F_1 \vec{i} + F_2 \vec{j} + F_3 \vec{k}$, then

$$\text{div } \vec{F} = \frac{\partial F_1}{\partial x} + \frac{\partial F_2}{\partial y} + \frac{\partial F_3}{\partial z}.$$

The divergence can be thought of as the outflow per unit volume of the vector field. A vector field $\vec{F}$ is said to be **divergence free** or **solenoidal** if $\text{div} \vec{F} = 0$ everywhere that $\vec{F}$ is defined. Magnetic fields are divergence free (p. 943).

The Divergence Theorem (p. 949). If W is a solid region whose boundary S is a piecewise smooth surface, and if $\vec{F}$ is a smooth vector field which is defined everywhere in W and on S, then

$$\int_S \vec{F} \cdot d\vec{A} = \int_W \text{div } \vec{F} \, dV,$$

where S is given the outward orientation. In words, the Divergence Theorem says that the total flux out of a closed surface is the integral of the flux density over the volume it encloses.

Curl

The **circulation density** of a smooth vector field $\vec{F}$ at (x, y, z) around the direction of the unit vector $\vec{n}$ is defined to be

$$\text{circ}_{\vec{n}} \vec{F}(x, y, z) = \lim_{\text{Area} \to 0} \frac{\text{Circulation around } C}{\text{Area inside } C}$$

$$= \lim_{\text{Area} \to 0} \frac{\int_C \vec{F} \cdot d\vec{r}}{\text{Area inside } C} \quad \text{(p. 954)}.$$

Circulation density is calculated using the **right-hand rule** (p. 954).

Definition of curl (p. 955).
Geometric definition The curl of $\vec{F}$, written curl $\vec{F}$, is the vector field with the following properties

- The direction of curl $\vec{F}(x, y, z)$ is the direction $\vec{n}$ for which $\text{circ}_{\vec{n}}(x, y, z)$ is greatest.

- The magnitude of curl $\vec{F}(x, y, z)$ is the circulation density of $\vec{F}$ around that direction.

Cartesian coordinate definition If $\vec{F} = F_1 \vec{i} + F_2 \vec{j} + F_3 \vec{k}$, then

$$\text{curl } \vec{F} = \left(\frac{\partial F_3}{\partial y} - \frac{\partial F_2}{\partial z} \right) \vec{i} + \left(\frac{\partial F_1}{\partial z} - \frac{\partial F_3}{\partial x} \right) \vec{j}$$

$$+ \left(\frac{\partial F_2}{\partial x} - \frac{\partial F_1}{\partial y} \right) \vec{k}.$$

Curl and circulation density are related by $\text{circ}_{\vec{n}} \vec{F} = \text{curl } \vec{F} \cdot \vec{n}$ (p. 957). A vector field is said to be **curl free** or **irrotational** if $\text{curl } \vec{F} = \vec{0}$ everywhere that $\vec{F}$ is defined (p. 958).

Given an oriented surface S with a boundary curve C we use the right-hand rule to determine the orientation of C (p. 961).

Stokes' Theorem (p. 963). If S is a smooth oriented surface with piecewise smooth, oriented boundary C, and if $\vec{F}$ is a smooth vector field which is defined on S and C, then

$$\int_C \vec{F} \cdot d\vec{r} = \int_S \text{curl } \vec{F} \cdot d\vec{A}.$$

Stokes' Theorem says that the total circulation around C is the integral over S of the circulation density. A **curl field** is a vector field $\vec{F}$ that can be written as $\vec{F} = \text{curl } \vec{G}$ for some vector field $\vec{G}$, called a **vector potential** for $\vec{F}$ (p. 964).

Relation between divergence, gradient, and curl
The curl and gradient are related by $\text{curl grad } f = 0$ (p. 968). Divergence and curl are related by $\text{div curl } \vec{F} = 0$ (p. 970).

The curl test for vector fields in 3-space (p. 969) Suppose that $\text{curl } \vec{F} = \vec{0}$, and that the domain of $\vec{F}$ has the property that every closed curve in it can be contracted to a point in a smooth way, staying at all times within the domain. Then $\vec{F}$ is path-independent, so $\vec{F}$ is a gradient field and has a potential function.

The divergence test for vector fields in 3-space (p. 970) Suppose that $\text{div } \vec{F} = 0$, and that the domain of $\vec{F}$ has the property that every closed surface in it is the boundary of a solid region completely contained in the domain. Then $\vec{F}$ is a curl field.

Section 12.1

1 B, B, B

3 $(1, -1, -3)$; Front, left, below

5 Q

7 $x^2 + y^2 + z^2 = 25$

9

11

13 predicted high temperature

17 Increasing function

19 (a) All increasing
   (b) Deposit of \$1250 gives balance
        \$1276 after 25 mos at 1\% int

21 (a) Decreasing
   (b) Increasing

23 (a) $-19°$F
   (b) 20 mph
   (c) About 17.5 mph
   (d) About 16.7°F

25 $f(x, y) = y^2$

27 Cylinder radius 2 along x-axis

29 $\sqrt{a^2 + c^2}$

31 $(x - 4)^2 + (y - 7)^2 + (z - 1)^2 = 4$

33 h

Section 12.2

1 (a) (IV)
  (b) (II)
  (c) (I)
  (d) (V)
  (e) (III)

3 Horizontal plane 3 units above the xy-plane

5 Bowl opening up, vertex $(0, 0, 4)$

7 Parabolic cylinder extended along x-axis

9 Circular cylinder of radius 2 in the z-direction

11 (a) Bowl
   (b) Neither
   (c) Plate
   (d) Bowl
   (e) Plate

13 (a)

   (b)

15 IV

17 (a) C (mg per liter)

   (b) C (mg per liter)

19 (a) IV
   (b) I
   (c) III

21 (a) $y = 0$
   (b) $x = 0$
   (c) $z = 1$

23 (a)

   (b) Increasing x
   (c)

Section 12.3

1 Contours evenly spaced

3

5

7

11 Price: x; Advertising: y

13

15

17 (a) (III)
   (b) (I)
   (c) (V)
   (d) (II)
   (e) (IV)

19 (a) (IV)
   (b) (II)
   (c) (I)
   (d) (III)

21 (a) A
   (b) B
   (c) A

23 E-W, $N = 50$:

Density of the
fox population P

E-W, $N = 100$:

Density of the
fox population P

N-S, $E = 60$:

Density of the
fox population P

N-S, $E = 120$:

Density of the
fox population P

25 Doubles production

27 $\alpha + \beta > 1$: increasing
    $\alpha + \beta = 1$: constant
    $\alpha + \beta < 1$: decreasing

29 (a)

(b)

(c)

(d)

31 (a) (i)

(ii)

(b) Lines parallel to $y = x$

33 $z = y(x-y)(x+y)$. Other answers possible

Section 12.4

1 (a) Yes
(b) Dollars per month
Dollars per minute
(c) Charge to connect to service
(d) \$120

3 (a) $z = -13 + 5x - 3y$
(b)

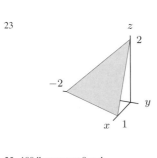

5 Not a linear function

7 Not a linear function

9 $z = 2 - \frac{1}{2}x - \frac{2}{3}y$

11 $z = 4 + 3x + y$

13 Linear

15 Not linear

17 $f(x,y) = -95 + 5x - 5y$

19 $f(x,y) = -0.01x + 0.3y + 1$

21

23

25 180 lb person at 8 mph
120 lb person at 10 mph

27 (a) Linear
(b) Linear
(c) Not linear

29 (a) All $\Delta z = 7$
(b) All $\Delta z = -5$

Section 12.5

1 (a) I

(b) II

3 $f(x,y) = 2x - (y/2) - 3$
$g(x,y,z) = 4x - y - 2z = 6$

5 $f(x,y) = -\sqrt{2(1 - x^2 - y^2)}$
$g(x,y,z) = x^2 + y^2 + z^2/2 = 1$

7 No

9 No

11 $f(x,y,z) = (x-a)^2 + (y-b)^2 + (z-c)^2$

13 $f(x,y) = (1 - x^2 - y)^2$

15 Hyperboloid of two sheets

17 Elliptic paraboloid

19 $f(x,y,z) = x + 2y - z + 1$

21 Elliptical cylinder along y-axis

23 All planes with normal $\vec{i} + \vec{j} + \vec{k}$

25 Parallel planes

27 (a) Graph of f is graph of
$x^2 + y^2 + z^2 = 1, z \geq 0$
(b) $\sqrt{1 - x^2 - y^2} - z = 0$

29 $g(x,y,z) = 1 - \sqrt{x^2 + z^2} - y = 0$

31 Surface of rotation

Section 12.6

1 Not continuous

3 Continuous

5 Not continuous

7 1

9 0

11 1

13

15 $c = 1$

17 (c) No

21 No

Chapter 12 Review

1 Could not be true

3 Might be true

5 True

7 $y = k$

9 $2x^2 + y^2 = k$

11 $h(x,y) = 4 + 2x - y$

13 Vertical line through $(2,1,0)$

15 $(-2,3,-6); 7$

17 $f(x,y) = 2x - y + 2$

19 (a) III

(b) II
(c) I
(d) IV

21 (a)

(b)

23 $72°F; 76°F$

25 Infinite families of parallel planes

27 (a)

(b) $c \geq 0$

29 $z = f(x,y) = 4 - 2x - 4y/3$
$g(x,y,z) = (x/2) + (y/3) + (z/4) = 1$

31 $z = f(x,y) = -\sqrt{4 - (x-3)^2 - y^2}$
$g(x,y,z) = (x-3)^2 + y^2 + z^2 = 4$

37

39

41 (a) $9 + 3x + 4y, 21 + 7x + 8y$
(b) $189 + 63x + 64y$

Ch. 12 Understanding

1 True

3 False

5 False

7 True

9 True

11 False

13 False

15 True

17 True

19 False

21 True

23 True

25 True

27 False

29 False

31 True

33 False

35 False

37 False

39 True

41 False

43 True

45 True

47 True

49 False

51 True

53 True

55 True

57 False

59 False

Section 13.1

1 $\vec{i} + 3\vec{j}$

3 $-4.5\vec{i} + 8\vec{j} + 0.5\vec{k}$

5 $\sqrt{11}$

7 5.6

9 $-15\vec{i} + 25\vec{j} + 20\vec{k}$

11 $21\vec{j}$

13 $\sqrt{65}$

15 (a)

(b) $\|\vec{v}\| = 8.602$
(c) $54.46°$

17 $\vec{a} = -2\vec{j}, \vec{b} = 3\vec{i}, \vec{c} = \vec{i} + \vec{j},$
$\vec{d} = 2\vec{j}, \vec{e} = \vec{i} - 2\vec{j}, \vec{f} = -3\vec{i} - \vec{j}$

19 $\vec{a} = \vec{b} = \vec{c} = 3\vec{k}$
$\vec{d} = 2\vec{i} + 3\vec{k}$
$\vec{e} = \vec{j}$
$\vec{f} = -2\vec{i}$

21 $\vec{u}$ and $\vec{w}$; $\vec{v}$ and $\vec{q}$.

23 (a) $(3/5)\vec{i} + (4/5)\vec{j}$

(b) $6\vec{i} + 8\vec{j}$

25 (a) $-2\vec{i} + 3\vec{j} + 6\vec{k}$, $2\vec{i} + 2\vec{j} - 10\vec{k}$,
$-4\vec{i} + \vec{j} + 16\vec{k}$
(b) Submarine and ship

27 $\vec{k}$

29 $\vec{i} + \vec{k}$

31 $-40\vec{i} - 30\vec{j}$

Section 13.2

1 Scalar

3 Vector

5 Vector

7 $-37.59\vec{i}, -13.68\vec{j}$

9 (a) $50\vec{i}$
(b) $-50\vec{j}$
(c) $25\sqrt{2}\vec{i} - 25\sqrt{2}\vec{j}$
(d) $-25\sqrt{2}\vec{i} + 25\sqrt{2}\vec{j}$

11 $-140.8\vec{i} + 140.8\vec{j} + 18\vec{k}$

13 $4.87°$ north of east
540.63 km/hr

15 $3.4°$ north of east

17

19 (79.00,79.33,89.00,68.33,89.33)

21 $-11\vec{i} + 4\vec{j}$

23 $-98.76\vec{i} + 18.94\vec{j} + 2998.31\vec{k}$
2998.31 newtons directly up

Section 13.3

1 -14

3 14

5 -2

7 $28\vec{j} + 14\vec{k}$

9 185

11 $3\vec{i} + 4\vec{j} - \vec{k}$ (multiples of)

13 $\vec{i} + 3\vec{j} + 2\vec{k}$ (multiples of)

15 Many answers possible
(a) $8\vec{i} + 6\vec{j}$
(b) $-3\vec{i} + 4\vec{j}$

17 Parallel:
$3\vec{i} + \sqrt{3}\vec{j}$ and $\sqrt{3}\vec{i} + \vec{j}$
Perpendicular:
$\sqrt{3}\vec{i} + \vec{j}$ and $\vec{i} - \sqrt{3}\vec{j}$
$3\vec{i} + \sqrt{3}\vec{j}$ and $\vec{i} - \sqrt{3}\vec{j}$

19 $\vec{v}_1, \vec{v}_4, \vec{v}_8$ all parallel
$\vec{v}_3, \vec{v}_5, \vec{v}_7$ all parallel
$\vec{v}_1, \vec{v}_4, \vec{v}_8$ perpendicular to $\vec{v}_3, \vec{v}_5, \vec{v}_7$
$\vec{v}_2$ and $\vec{v}_9$ perpendicular

21 $\vec{a} = -\frac{8}{21}\vec{d} + (\frac{79}{21}\vec{i} + \frac{10}{21}\vec{j} - \frac{118}{21}\vec{k})$

23 $5x + y - 2z = 3$

25 $2x + 4y - 3z = 5$

27 (a) $\lambda = -2.5$
(b) $a = -6.5$

29 $38.7°$

31 $50.91°, 53.13°,$ and $75.96°$

---

33 No

35 Averages; weightings
85.65%; class average

Section 13.4

1 $-\vec{i}$

3 $-\vec{i} + \vec{j} + \vec{k}$

5 $\vec{i} + 3\vec{j} + 7\vec{k}$

7 $\vec{i}$

9 $2\vec{k}$

11 $\vec{v} \times \vec{w} = -6\vec{i} + 7\vec{j} + 8\vec{k}$
$\vec{w} \times \vec{v} = 6\vec{i} - 7\vec{j} - 8\vec{k}$
$\vec{v} \times \vec{w} = -(\vec{w} \times \vec{v})$

13 $x + y + z = 1$

15 (a) 0.6
(b) 0.540

17 Parallel to the z-axis

19 (a) 1.5
(b) $y = 1$

21 $4\vec{i} + 26\vec{j} + 14\vec{k}$

23 $4(x - 4) + 26(y - 5) + 14(z - 6) = 0$

27 (b) $\vec{c} \perp \vec{a}$ and $\|\vec{c}\| = \|\vec{a}\|$
(c) $-a_2 b_1 + a_1 b_2$

Chapter 13 Review

1 $4\vec{i} + \vec{j} + 3\vec{k}$

3 -3

5 $\vec{0}$

7 $-6\vec{i} - 9\vec{j} + 3\vec{k}$

9 0

11 (a) 4
(b) $-4\vec{i} - 11\vec{j} - 17\vec{k}$
(c) $3.64\vec{i} + 2.43\vec{j} - 2.43\vec{k}$
(d) $79.0°$
(e) 0.784.
(f) $2\vec{i} - 2\vec{j} + \vec{k}$ (many answers possible)
(g) $-4\vec{i} - 11\vec{j} - 17\vec{k}$.

13 $\pm(-\vec{i} + \vec{j} - 2\vec{k})/\sqrt{6}$

15 $-3\vec{i} + 4\vec{j}$

17 (a) 1.625
(b) 1.019

19 $4x - 3y - z = 0$

21 $\sqrt{6}/2$

23 (a) $\vec{u}$ and $-\vec{u}$ where
$\vec{u} = \frac{12}{13}\vec{i} - \frac{4}{13}\vec{j} - \frac{3}{13}\vec{k}$
(b) $\theta \approx 49.76°$
(c) 13/2
(d) $13/\sqrt{29}$

25 $9x - 16y + 12z = 5$
0.23

27 (a) $53.1°$ east of south
(b) $37.9°$ west of south; does not move

29 (a) $-500\vec{i}$
(b) $(300, 60, 4)$
(c) $-178.7\vec{i} - 89.4\vec{j} - 7.2\vec{k}$

31 $38.7°$ south of east

33 228.43 newtons
$85.5°$ south of east

39 $0, \vec{0}$

41 (a) $\vec{c} = -\vec{i} - 4\vec{j} + 3\vec{k}$ or $\vec{c} = \vec{i} + 4\vec{j} - 3\vec{k}$
(b) $\vec{i} + 4\vec{j} - 3\vec{k}$

43 (a) $(tv - sw - ty + wy + sz - vz)\vec{i} +$
$(-tu + rw + tx - wx - rz + uz)\vec{j} +$
$(su - rv - sx + vx + ry - uy)\vec{k}$
(b) $((s(u - x) + vx - uy +$
$r(-v + y))/c)(a\vec{i} + b\vec{j} + c\vec{k})$

Ch. 13 Understanding

1 False

3 True

5 False

7 False

9 False

11 False

13 True

15 True

17 False

19 True

21 True

23 False

25 False

27 True

29 True

Section 14.1

1 $f_x(3, 2) \approx 2, f_y(3, 2) \approx -1$

3 (a) Negative
(b) Positive

5 (a) Concentration/distance
Rate of change of concentration with distance
$\partial c / \partial x < 0$
(b) Concentration/time
Rate of change of concentration with time
For small t, $\partial c / \partial t > 0$
For large t, $\partial c / \partial t < 0$

7 (a) Both negative
(b) Both negative

9 Positive, Negative, 10, 2, -4

11 $f_T(5, 20) \approx 1.2°F/°F$

13 $(A)\, 0.06, -0.06$
$(B)\, 0, -0.05$
$(C)\, 0, 0$

15 (a) $2.5, 0.02$
(b) $3.33, 0.02$
(c) $3.33, 0.02$

17 (a) (i) Positive
(ii) Negative
(b) $f_x(P)$: from + to $-$
$f_y(P)$: no change

19 (a) Negative
(b) Negative
(c) Negative

21 -2.5

23

| | | w (gm/m³) | |
T (°C)	0.1	0.2	0.3
10	1300	900	1200
20	800	800	900
30	800	700	800

Section 14.2

1 $f_x = 10xy^3 + 8y^2 - 6x,$
$f_y = 15x^2y^2 + 16xy$

3 $21x^5y^6 - 96x^4y^2 + 5x$

5 $2xy + 10x^4y$

7 $\dfrac{270x^3y^7 - 168x^2y^6 - 15xy^2 + 16y}{(15xy - 8)^2}$

9 $a/(2\sqrt{x})$

11 g

13 $(a + b)/2$

15 $2B/u_0$

17 $2mv/r$

19 Gm_1/r^2

21 $v_0 + at$

23 $\pi/\sqrt{lg}$

25 $f_a = e^a \sin(a + b) + e^a \cos(a + b)$

27 y

29 $\epsilon_0 E$

31 $c\gamma a_1 b_1 K^{b_1 - 1}(a_1 K^{b_1} + a_2 L^{b_2})^{\gamma - 1}$

33 $m_0 v/(c^2(1 - v^2/c^2)^{3/2})$

35 -2.51

37 (a) $3.3, 2.5$
(b) $4.1, 2.1$
(c) $4, 2$

39 (a)

(b) $254.2e^{-0.1t}$ °C/m
(c) $-254.2e^{-0.1t}$ °C/m
(d) $-10e^{-0.1t} \sin(\pi x)$ °C/sec

41 $F = 684$ newtons,
$\partial F/\partial m = 9.77$ newtons/kg,
$\partial F/\partial r = -0.000214$ newtons/meter

43 1.277 m², 0.005 m²/kg, 0.006 m²/cm

45 $f(x, y) = x^4y^2 - 3xy^4 + C$

Section 14.3

1 $z = -4 + 2x + 4y$

3 $z = 6 + 3x + y$

5 $dg = (2u + v)\, du + u\, dv$

7 $dz = -e^{-x}\cos(y)dx - e^{-x}\sin(y)dy$

9 $df = dx - dy$

11 $dg = 4\, dx$

13 $z = 9 + 6(x - 3) + 9(y - 1)$

15 (b) $f(x, y) \approx$
$0.3345 - 0.33(x - 1) - 0.15(y - 2)$
(c) $f(x, y) \approx 0.3345 -$
$0.3345(x - 1) - 0.1531(y - 2)$

17 $P(r, L) \approx$
$80 + 2.5(r - 8) + 0.02(L - 4000)$
$P(r, L) \approx$
$120 + 3.33(r - 8) + 0.02(L - 6000)$
$P(r, L) \approx$
$160 + 3.33(r - 13) + 0.02(L - 7000)$

19 $V \approx 0.0305T - 1.27p + 38.61$

21 (a) 20 hours per day
(b) 18.615 hours per day

23 (a) $d\rho = -\beta\rho\, dT$
(b) $0.00015, \beta \approx 0.0005$

25 8246.68

Section 14.4

1 Negative

3 Approximately zero

5 Negative

7 $\vec{i}$

9 $-\vec{i}$

11 $\vec{i} + \vec{j}$

13 $-\vec{i} + \vec{j}$

15 $(\frac{15}{2}x^4)\vec{i} - (\frac{24}{7}y^5)\vec{j}$

17 $50\vec{i} + 100\vec{j}$

19 $\nabla z = e^y\vec{i} + e^y(1 + x + y)\vec{j}$

21 $(x\vec{i} + y\vec{j})/\sqrt{x^2 + y^2}$

23 $\sin\theta\vec{i} + r\cos\theta\vec{j}$

25 $\nabla z = \dfrac{1}{y}\cos\left(\dfrac{x}{y}\right)\vec{i} - \dfrac{x}{y^2}\cos\left(\dfrac{x}{y}\right)\vec{j}$

27 $\left(\dfrac{-12\beta}{(2\alpha - 3\beta)^2}\right)\vec{i} + \left(\dfrac{12\alpha}{(2\alpha - 3\beta)^2}\right)\vec{j}$

29 $60\vec{i} + 85\vec{j}$

31 $10\pi\vec{i} + 4\pi\vec{j}$

33 $\frac{1}{100}(2\vec{i} - 6\vec{j})$

35 5

37 $-46/5$

39 0.3

41 0.7

43 -0.4

45

47 (a) P
(b) S
(c) Q, S
(d) R

49 Fourth quadrant

51 P

53 (a) -3.268
   (b) -4.919

55 (a) $-16\vec{i} + 12\vec{j}$
   (b) $16\vec{i} - 12\vec{j}$
   (c) $12\vec{i} + 16\vec{j}$; answers may vary

57 $\partial z / \partial x = 1.8$
   $\partial z / \partial y = -0.1$

59 (a) $-5\sqrt{2}/2$
   (b) $4\vec{i} + \vec{j}$

61 $3\vec{i} + 2\vec{j}$; $3(x-2) + 2(y-3) = 0$

63 $-5\vec{i}$; $x = 2$

65 (a) 98.387 ft/mile
   (b) 295.161 ft/hour

67 $5/\sqrt{2}$

69 (a) P, Q
   (b)

   (c) $\| \operatorname{grad} f \|$
      $f_{\vec{u}} = \| \operatorname{grad} f \| \cos \theta$

Section 14.5

1 $2x\vec{i} + 3y^2\vec{j} - 4z^3\vec{k}$

3 $-2(x\vec{i} + y\vec{j} + z\vec{k})/(x^2 + y^2 + z^2)^2$

5 $(e^y + 1/x)\vec{i} + xe^y\vec{j} + (1/z)\vec{k}$

7 $(2x_1 x_2^3 x_3^4)\vec{i} + (3x_1^2 x_2^2 x_3^4)\vec{j} + (4x_1^2 x_2^3 x_3^3)\vec{k}$

9 $6\vec{i} + 3\vec{j} + 2\vec{k}$

11 $9/\sqrt{3}$

13 $-1/\sqrt{2}$

15 $-\sqrt{77/2}$

17 $-2\vec{i} - 2\vec{j} + 4\vec{k}$;
   $-2(x+1) - 2(y-1) + 4(z-2) = 0$

19 $2\vec{i} - 2\vec{j} + \vec{k}$;
   $2(x+1) - 2(y-1) + (z-2) = 0$

21 $-2\vec{i} + \vec{k}$; $-2(x+1) + (z-2) = 0$

23 (a) 0
   (b) $24/\sqrt{19}$

25 (a) $4\vec{i} + 6\vec{j}$ (or any positive multiple)
   (b) $-\sqrt{52}$

27 (a) $-\vec{j}$,
      $\frac{2}{3}\vec{i} - \frac{1}{3}\vec{j} + \frac{2}{3}\vec{k}$
   (b) $y = 0; 2x - y + 2z = 3$
   (c) $x = y = 0$ and $z \neq 0$

29 (a) $6.33\vec{i} + 0.76\vec{j}$
   (b) -34.69

31 $3x + 10y - 5z + 19 = 0$

33 (a) Spheres centered at origin
   (b) $(-2x\vec{i} - 2y\vec{j} - 2z\vec{k})e^{-(x^2+y^2+z^2)}$
   (c) $-3\sqrt{2}e^{-1}$ degrees/sec

35 1.131 atm/sec

Section 14.6

1 $\dfrac{dz}{dt} = e^{-t}\sin(t)(2\cos t - \sin t)$

3 $2e^{1-t^2}(1 - 2t^2)$

5 $2t \sin(\ln t) + 2t \ln(t)\cos(t^2)$
   $+ t\cos(\ln t) + \dfrac{\sin t^2}{t}$

7 $\dfrac{\partial z}{\partial u} = \dfrac{e^v}{u}$
   $\dfrac{\partial z}{\partial v} = e^v \ln u$

9 $\dfrac{\partial z}{\partial u} = 2ue^{(u^2-v^2)}(1 + u^2 + v^2)$
   $\dfrac{\partial z}{\partial v} = 2ve^{(u^2-v^2)}(1 - u^2 - v^2)$

11 $\dfrac{\partial z}{\partial u} = \dfrac{1}{vu}\cos\left(\dfrac{\ln u}{v}\right)$
   $\dfrac{\partial z}{\partial v} = -\dfrac{\ln u}{v^2}\cos\left(\dfrac{\ln u}{v}\right)$

13 $\dfrac{\partial z}{\partial u} =$
   $(e^{-v\cos u} - v(\cos u)e^{-u\sin v})\sin v$
   $- (-u(\sin v)e^{-v\cos u} + e^{-u\sin v})v\sin u$
   $\dfrac{\partial z}{\partial v} =$
   $(e^{-v\cos u} - v(\cos u)e^{-u\sin v})u\cos v$
   $+ (-u(\sin v)e^{-v\cos u} + e^{-u\sin v})\cos u$

15 $\dfrac{\partial w}{\partial u} = \dfrac{\partial w}{\partial x}\dfrac{\partial x}{\partial u} + \dfrac{\partial w}{\partial y}\dfrac{\partial y}{\partial u} + \dfrac{\partial w}{\partial z}\dfrac{\partial z}{\partial u}$
   $\dfrac{\partial w}{\partial v} = \dfrac{\partial w}{\partial x}\dfrac{\partial x}{\partial v} + \dfrac{\partial w}{\partial y}\dfrac{\partial y}{\partial v} + \dfrac{\partial w}{\partial z}\dfrac{\partial z}{\partial v}$

17 -0.6

19 $dV/dt = 0.3812$ volts/sec

21 $b \cdot e + d \cdot p$

23 (a) $F_u(x, y, 3)$
   (b) $F_w(3, y, x)$
   (c) $F_u(x, y, x) + F_w(x, y, x)$
   (d) $F_u(x, y, xy) + yF_w(x, y, xy)$

27 (a) $\dfrac{\partial z}{\partial r} = \cos\theta\dfrac{\partial z}{\partial x} + \sin\theta\dfrac{\partial z}{\partial y}$
      $\dfrac{\partial z}{\partial \theta} = r(\cos\theta\dfrac{\partial z}{\partial y} - \sin\theta\dfrac{\partial z}{\partial x})$
   (b) $\dfrac{\partial z}{\partial y} = \sin\theta\dfrac{\partial z}{\partial r} + \dfrac{\cos\theta}{r}\dfrac{\partial z}{\partial\theta}$
      $\dfrac{\partial z}{\partial x} = \cos\theta\dfrac{\partial z}{\partial r} - \dfrac{\sin\theta}{r}\dfrac{\partial z}{\partial\theta}$

29 $\left(\dfrac{\partial U_3}{\partial P}\right)_V$

31 $\left(\dfrac{\partial U}{\partial T}\right)_V = 7/2$
   $\left(\dfrac{\partial U}{\partial V}\right)_T = 11/4$

33 $\left(\dfrac{\partial U}{\partial P}\right)_T = \left(\dfrac{\partial U}{\partial V}\right)_T\left(\dfrac{\partial V}{\partial P}\right)_T$

37 $F(b, x)$

Section 14.7

1 $f_{xx} = 6y$
   $f_{xy} = 6x + 15y^2$
   $f_{yx} = 6x + 15y^2$
   $f_{yy} = 30xy$

3 $f_{xx} = 6(x + y)$
   $f_{yy} = 6(x + y)$
   $f_{yx} = 6(x + y)$
   $f_{xy} = 6(x + y)$

5 $f_{xx} = 4y^2 e^{2xy}$
   $f_{xy} = 4xye^{2xy} + 2e^{2xy}$
   $f_{yx} = 4xye^{2xy} + 2e^{2xy}$
   $f_{yy} = 4x^2 e^{2xy}$

7 $f_{xx} = \dfrac{y^2}{(x^2+y^2)^{3/2}}$
   $f_{xy} = \dfrac{-xy}{(x^2+y^2)^{3/2}} = f_{yx}$
   $f_{yy} = \dfrac{x^2}{(x^2+y^2)^{3/2}}$

9 $f_{xx} = -(\sin(x^2 + y^2))4x^2$

$+ 2\cos(x^2 + y^2)$
$f_{xy} = -(\sin(x^2 + y^2))4xy = f_{yx}$
$f_{yy} = -(\sin(x^2 + y^2))4y^2$
   $+ 2\cos(x^2 + y^2)$

11 $Q(x, y) = 1 + 2x - 2y + x^2 - 2xy + y^2$

13 $Q(x, y) = 1 - 2x^2 - y^2$

15 $Q(x, y) = 1 + x + x^2/2 - y^2/2$

17 $Q(x, y) = -y + x^2 - y^2/2$

19 (a) Positive
   (b) Zero
   (c) Positive
   (d) Zero
   (e) Zero

21 (a) Negative
   (b) Zero
   (c) Negative
   (d) Zero
   (e) Zero

23 (a) Zero
   (b) Negative
   (c) Zero
   (d) Negative
   (e) Zero

25 (a) Positive
   (b) Positive
   (c) Zero
   (d) Zero
   (e) Zero

27 (a) Positive
   (b) Negative
   (c) Negative
   (d) Zero
   (e) Positive

29 $L(x, y) = y$
   $Q(x, y) = y + 2(x - 1)y$
   $L(0.9, 0.2) = 0.2$
   $Q(0.9, 0.2) = 0.16$
   $f(0.9, 0.2) = 0.162$

31 $L(x, y) = 1 + (x - 1) - y$
   $Q(x, y) = 1 + (x - 1) - y$
   $\quad -(x - 1)y + (1/2)y^2$
   $L(0.9, 0.2) = 0.7$
   $Q(0.9, 0.2) = 0.74$
   $f(0.9, 0.2) \approx 0.737$

33 Positive, negative

35 (a) $z_{yx} = 4y$
   (b) $z_{xyx} = 0$
   (c) $z_{xyy} = 4$

37 (a) Increasing, decreasing
   (b) $f_x > 0, f_y < 0$
   (c) $f_{xx} > 0, f_{yy} < 0$
   (d) y

   (e) P

39 (a) xy
   $1 - \frac{1}{2}(x - \frac{\pi}{2})^2 - \frac{1}{2}(y - \frac{\pi}{2})^2$

(b)

41   f(x, y):

L(x, y):

Q(x, y):

f(x, y):

L(x, y):

Q(x, y):

Section 14.8

1 (0, 0)

3 None

5 y-axis

7 None

9 (1, 2)

11 (a)

(b) No
(c) No
(d) No
(e) Exist, not continuous

13 (a)

(b) Yes
(c) Yes
(d) No
(e) Exist, not continuous

15 (a)

(b) Yes
(d) No
(f) No

17 (a)

(b)

(c) No
(e) No

19 (a) No

Chapter 14 Review

1 $f_x = 2xy + 3x^2 - 7y^6$
   $f_y = x^2 - 42xy^5$

3 $\partial F/\partial L = \frac{3}{2}\sqrt{K/L}$

5 $f_x = \frac{2xy^3}{(x^2+y^2)^2}$, $f_y = \frac{x^4 - x^2y^2}{(x^2+y^2)^2}$

7 $\partial f/\partial p = (1/q)e^{p/q}$
   $\partial f/\partial q = -(p/q^2)e^{p/q}$

9 $f_N = c\alpha N^{\alpha - 1}V^\beta$

11 $e^x \cos(xy) + e^x \cdot (-\sin(xy)) \cdot y$
   $e^x \cdot (-\sin(xy)) \cdot x + 2ay$
   y^2

13 $f_{xx} = (2x^2 - y^2)(x^2 + y^2)^{-5/2}$
   $f_{xy} = 3xy(x^2 + y^2)^{-5/2}$

15 $V_{rr} = 2\pi h$, $V_{rh} = 2\pi r$

17 $(3x^2 - yz)\vec{i} - xz\vec{j} + (3z^2 - xy)\vec{k}$

19 $2(x\vec{i} + y\vec{j})/(x^2 + y^2)$

21 $15/\sqrt{2}$

23 -4

25 $24/\sqrt{19}$

27 $4\vec{i} - 2\vec{j}$

29 $2\vec{i} - 2\vec{j} - 6\vec{k}$

31 $Q(x,y) = 1 - (1/2)x^2 - (9/2)y^2$

33 (a) $f_w(2,2) \approx 2.78$
        $f_z(2,2) \approx 4.01$
   (b) $f_w(2,2) \approx 2.773$
        $f_z(2,2) = 4$

35 -0.5

37 -0.71

39 $84/5$

41 (I); $c = 1/2$

43 (I); $c = 4/9$

45 $\pm 4\sqrt{\frac{2}{11}}\left(\frac{1}{2}, -\frac{1}{2}, \frac{3}{2}\right)$

47 $y = 2x - 7$

49 (a) 35.355
   (b) $35.355v$

53 (a) $-3/8°\text{C/m}, 2/5°\text{C/min}$
   (b) $-1°\text{C/m}, 3/28°\text{C/min}$

57 -0.008544

59 (a) $z = 7 - 3(x - 2) + 4(y - 1)$
   (b) $-3(x - 2) + 4(y - 1) = 0$

61 0.06

63 y

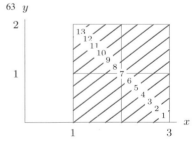

65 $f_r(2,1) = -2/\sqrt{5}, f_\theta(2,1) = 11$

67 (a) $-14, 2.5$
   (b) -2.055
   (c) 14.221 in direction of $-14\vec{i} + 2.5\vec{j}$
   (d) $f(x,y) = f(2,3) = 7.56$
   (e) For example, $\vec{v} = 2.5\vec{i} + 14\vec{j}$
   (f) -0.32

69 (a) $e^{10} - 2e^{10}x - 6e^{10}y$
   (b) $1 + (x - 1)^2 + (y - 3)^2$
   (c) $-2\vec{i} - 6\vec{j}$
   (d) $-2e^{10}\vec{i} - 6e^{10}\vec{j} - \vec{k}$

71 (a) (I)-(E), (II)-(F), (III)-(G), (IV)-(H)
   (b) (I)-(L), (II)-(J), (III)-(M), (IV)-(K)

73 (a) $A_0 + A_1 + 2A_2 + A_3 + 2A_4 + 4A_5 + (A_1 + 2A_3 + 2A_4)(x - 1) + (A_2 + A_4 + 4A_5)(y - 2), 1 + B_1t, 2 + C_1t$
   (b) $A_1B_1 + 2A_3B_1 + 2A_4B_1 + A_2C_1 + A_4C_1 + 4A_5C_1$

75 $\frac{\partial w}{\partial x}\frac{\partial x}{\partial u}\frac{du}{dt} + \frac{\partial w}{\partial y}\frac{\partial y}{\partial u}\frac{du}{dt} + \frac{\partial w}{\partial x}\frac{\partial x}{\partial v}\frac{dv}{dt} + \frac{\partial w}{\partial y}\frac{\partial y}{\partial v}\frac{dv}{dt} + \frac{\partial w}{\partial z}\frac{dz}{dt}$

Ch. 14 Understanding

1 True
3 True
5 True
7 True
9 True
11 True
13 True
15 False
17 True
19 False
21 False
23 True
25 False
27 False
29 True
31 False
33 True
35 False
37 True
39 True
41 True
43 True
45 False
47 True

Section 15.1

1 A: no
   B: yes, max
   C: yes, saddle

3 Local max: $(1,5)$

5 Saddle pts: $(1,-1), (-1,1)$
   local max $(-1,-1)$
   local min $(1,1)$.

7 Local maximum at $(-1,0)$,
   Saddle points at $(1,0)$ and $(-1,4)$,
   Local minimum at $(1,4)$

9 $(0,0)$, $(1,1)$, $(-1,-1)$
   saddle point,
   local maximum,
   local maximum

11 Local minimum at $(0,0)$

13 Saddle point

15 Saddle point

17

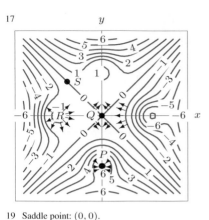

19 Saddle point: $(0,0)$.

21 $y = 0, \pm 2\pi, \pm 4\pi, \dots$ Local minima

23 $A = 10, B = 4, C = -2$

25 (a) (a,b) is a saddle point.
   (b)

27

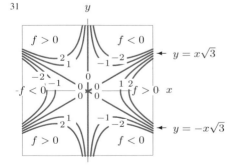

29 (a) $k < 4$
   (b) None
   (c) $k \geq 4$

31

Section 15.2

1. Mississippi:
   $87 - 88$ (max), $83 - 87$ (min)
   Alabama:
   $88 - 89$ (max), $83 - 87$ (min)
   Pennsylvania:
   $89 - 90$ (max), 80 (min)
   New York:
   $81 - 84$ (max), $74 - 76$ (min)
   California:
   $100 - 101$ (max), $65 - 68$ (min)
   Arizona:
   $102 - 107$ (max), $85 - 87$ (min)
   Massachusetts:
   $81 - 84$ (max), 70 (min)

3. Neither

5. Neither

7. Neither

9. Max $= 0$ at $(0, 0)$
   (not on boundary)
   Min $= -2$ at $(1, -1), (-1, -1),$
   $(-1, 1)$ and $(1, 1)$
   (on boundary)

11. Max: 30.5 at $(0, 0)$
    Min: 20.5 at $(2.5, 5)$

13. Max: 11 at $(5.1, 4.9)$
    Min: -1 at $(1, 3.9)$

15. $y = 24x^2/49 - 2/7$

17. $-1/5 + (9/10)x$

19. (b) $p_1 = p_2 = 25$
    Max revenue is 4375

21. (a) $P = p_1^2/8 + p_2^2/8 - 10$
    (b) $\dfrac{\partial}{\partial p_1}(\max\ P) = p_1/4$

23. All edges $(32)^{1/3}$ cm

25. $1/162$

27. $w = 6.35$ cm
    $l = 6.35$ cm
    $h = 12.7$ cm

29. $\sqrt{6}$

31. (a) $L = \left[pA \left(\dfrac{a}{k}\right)^a \left(\dfrac{l}{b}\right)^{(a-1)} \right]^{1/(1-a-b)}$
    $K = \dfrac{la}{kb} L$
    (b) No

35. (c) $h(a) = 2a, M(a) = a^2.$

(c) Max of 5 at $(2, 1)$
    Min of -5 at $(-2, -1)$
(d) Max of 5 at $(2, 1)$
    Min of -5 at $(-2, -1)$

19. $r = \sqrt[3]{50/\pi}$
    $h = 2\sqrt[3]{50/\pi}$

23. (a) $W = 225$
    $K = 37.5$
    (c) $W = 225$
    $K = 37.5$
    $\lambda = 0.29$

25. (a) No
    (b) Yes
    (c) $a + b = 1$

29. (a)–(d)

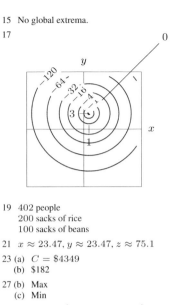

(e) Negative

31. $f(x, y, z) = \dfrac{|Aa + Bb + Cc + D|}{\sqrt{A^2 + B^2 + C^2}}$

35. (a) Cost of producing quantity u when prices are
    p, q
    (b) $2\sqrt{pqu}$

37. (a) $-5\lambda^2 + 10\lambda$
    (b) $1.5, 11.25$
    (c) $11.25, 1.5$
    (d) same

Section 15.3

1. Min $= -\sqrt{2}$, max $= \sqrt{2}$

3. Min $= -2$, max $= 2$

5. Min $= 11.25$; no max

7. No max, min $= 0.866$

9. Max $= 4$, min $= -2$

11. Max: $f(0, 2) = f(0, -2) = 8$
    Min: $f(0, 0) = 0$

13. Max $= \dfrac{\sqrt{2}}{4}$, min $= -\dfrac{\sqrt{2}}{4}$

15. No global extrema

17. (a) and (b)

Chapter 15 Review

1. $(0, 0)$: Local Max
   $(1, 1), (1, -1)$: Saddle point

3. $(0, 0)$: Saddle point

5. $(2, 3)$: Local and global max

7. $(\sqrt{2}, -\sqrt{2}/2)$ saddle point

9. Minimum $f(-3/\sqrt{5}, 4/\sqrt{5}) = -5\sqrt{5}$,
   Maximum $f(3/\sqrt{5}, -4/\sqrt{5}) = 5\sqrt{5}$

11. Maximum $f(10, 12.5) = 250$;
    No minimum

13. Max $= 0$, no min

15. No global extrema.

17.

19. 402 people
    200 sacks of rice
    100 sacks of beans

21. $x \approx 23.47, y \approx 23.47, z \approx 75.1$

23. (a) $C = \$4349$
    (b) $\$182$

27. (b) Max
    (c) Min

29. (a) $f_1 = \dfrac{k_1}{k_1+k_2} mg,\ f_2 = \dfrac{k_2}{k_1+k_2} mg$
    (b) Distance the mass stretches the top spring and compresses the lower spring

31. $x_1 = ((v_1)^{1/2} + (v_2)^{1/2})/(m(v_1)^{1/2})$
    $x_2 = ((v_1)^{1/2} + (v_2)^{1/2})/(m(v_2)^{1/2})$

33. 37.5 m, 4.6 m, 57.9 m

35. (a) $a/(v_1 \cos \theta_1) + b/(v_2 \cos \theta_2)$

37. Student B correct. Local max, not global

Ch. 15 Understanding

1. True

3. False

5. True

7. True

9. False

11. False

13. False

15. False

17. False

19. True

21. True

23. True

25. False

27. False

29. True

31. False

33. False

Section 16.1

1. Lower sum : 0.34
   Upper sum : 0.62

3. Upper sum $= 46.63$
   Lower sum $= 8$
   Average ≈ 27.3

5. Lower Bound $= 25$
   Upper Bound $= 41$

7. $40/3$

9. Positive

11. Positive

13. Zero

15 Negative

17 Negative

19 Zero

21 Positive

23 Zero

25 294 mg

27 $25.2°C$

29 (a) About 148 tornados
   (b) About 56 tornados
   (c) About 2 tornados

31 (b) $(d-c)\displaystyle\int_a^b g(x)\,dx$

33 Decrease

Section 16.2

1 150

3 54

5 $\displaystyle\int_1^4\int_1^2 f\,dy\,dx$ or $\displaystyle\int_1^2\int_1^4 f\,dx\,dy$

7 $\displaystyle\int_{-1}^3\int_{-2}^{(1-3x)/4} f\,dy\,dx$

 or $\displaystyle\int_{-2}^1\int_{-1}^{(1-4y)/3} f\,dx\,dy$

9 $\displaystyle\int_1^3\int_{\frac12(y-1)}^{-\frac12(y-5)} f\,dx\,dy$

11 $(e^4-1)(e^2-1)e$

13 -2.678

15 94.5

17 $\frac{4}{15}(9\sqrt3-4\sqrt2-1)=2.38176$

19 32/9

21 13/6

23 15

25 (a) $\displaystyle\int_0^{1/2}\int_y^{1-y} f(x,y)\,dx\,dy$
    $\displaystyle\int_0^{1/2}\int_0^x f(x,y)\,dy\,dx$
    $\displaystyle\int_{1/2}^1\int_0^{1-x} f(x,y)\,dy\,dx$
   (b) 1/8

27 $(e-1)/2$

29 $\frac29(3\sqrt3-2\sqrt2)$

31 $\frac12(e^2-1)$

33 (a) 180
   (b) 144

35 $\displaystyle\int_{-5}^5\int_{-\sqrt{25-y^2}}^{\sqrt{25-y^2}}(25-x^2-y^2)\,dx\,dy$

37 $\displaystyle\int_0^4\int_{y-4}^{(4-y)/2}(4-2x-y)\,dx\,dy$

39 1/3

41 16/3

43 Volume $= 1/(6abc)$

47 (a) $a+(3/2)b=20$
   (b) $f(x,y)=x+\frac{38}{3}y$:

$f(x,y)=-3x+\frac{46}{3}y$:

49 $k(a^3b+ab^3)/3$

Section 16.3

1 2

3 $(1-e^{-a})(1-e^{-b})(1-e^{-c})$

5

7

9

11

13

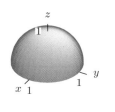

15 $\displaystyle\int_0^r\int_0^{\sqrt{r-x^2}}\int_{-\sqrt{r-x^2-z^2}}^{\sqrt{r-x^2-z^2}} f(x,y,z)\,dy\,dz\,dx$

19 162

21 1/2

23 125

25 (a) $z=\sqrt{1-x^2},\,0\le y\le 10$

(b) $\displaystyle\int_0^{10}\int_{-1}^1\int_0^{\sqrt{1-x^2}} f(x,y,z)\,dz\,dx\,dy$

27 243

29 Positive

31 Zero

33 Positive

35 Zero

37 Positive

39 Positive

41 Zero

43 Zero

45 Negative

47 $\displaystyle\int_{-4}^4\int_{-\sqrt{16-x^2}}^{\sqrt{16-x^2}}\int_3^{\sqrt{25-x^2-y^2}} dz\,dy\,dx$

49 $(4/3)\delta_1+(20/3)\delta_2$ gm

51 $m=1/36;\ (\bar x,\bar y,\bar z)=(1/4,1/8,1/12)$

53 $m(b^2+c^2)/3$

Section 16.4

1 $\displaystyle\int_0^{2\pi}\int_0^{\sqrt2} f\,r\,dr\,d\theta$

3 $\displaystyle\int_{\pi/2}^{3\pi/2}\int_1^2 f\,r\,dr\,d\theta$

5

7

9

11

$\theta = \pi/4$

$r = 2/\sin\theta$
or $r\sin\theta = 2$
or $y = 2$

13 0

15 (a)

(b) $\int_0^1 \int_0^{3y} f(x,y)\, dx\, dy$

(c) $\int_{\tan^{-1}(1/3)}^{\pi/2} \int_0^{1/\sin\theta} f(r\cos\theta, r\sin\theta) r\, dr\, d\theta$

17 6

19 $625\pi/2$

21 (a) $\int_{\pi/2}^{3\pi/2} \int_1^4 \delta(r,\theta)\, r dr\, d\theta$
  (b) (i)
  (c) About 39,000

23 $250\pi/3$ grams

25 Total charge $= 2k\pi R$

Section 16.5

1 (a) $x = 2$
  (b) $r = 3$
  (c) $\rho = \sqrt{3}$
  (d) $\phi = \pi/4$
  (e) $z = -5$
  (f) $\theta = \pi/4$

3 $4\pi(1 - \cos 1) = 5.78$

5 $7\pi(\pi - 2)/12$

7 $\int_0^1 \int_0^{2\pi} \int_0^4 f \cdot r dr\, d\theta\, dz$

9 $\int_0^\pi \int_0^\pi \int_2^3 f \cdot \rho^2 \sin\phi\, d\rho\, d\phi\, d\theta$

11 $\int_0^5 \int_0^2 \int_0^{x/5} f\, dz dy dx$

13

15 $\int_0^{2\pi} \int_{3\pi/4}^\pi \int_0^{\sqrt{2}} f(\rho,\phi,\theta)\rho^2 \sin\phi\, d\rho\, d\phi\, d\theta$

17 $\int_{-\sqrt{2}}^{\sqrt{2}} \int_{-\sqrt{2-x^2}}^{\sqrt{2-x^2}} \int_{-\sqrt{2-x^2-y^2}}^{-\sqrt{x^2+y^2}} h(x,y,z)\, dz\, dy\, dx$

19 (a) $\int_0^1 \int_0^{\sqrt{1-x^2}} \int_{-\sqrt{1-x^2-y^2}}^0 dz\, dy\, dx$
  (b) $\int_0^{\pi/2} \int_0^1 \int_{-\sqrt{1-r^2}}^0 r\, dz\, dr\, d\theta$
  (c) $\int_0^{\pi/2} \int_{\pi/2}^\pi \int_0^1 \rho^2 \sin\phi\, d\rho\, d\phi\, d\theta$

21 (a) $\int_{-1/\sqrt{2}}^{1/\sqrt{2}} \int_{-\sqrt{(1/2)-x^2}}^{\sqrt{(1/2)-x^2}} \int_{\sqrt{x^2+y^2}}^1 dz\, dy\, dx$

(b) $\int_0^{2\pi} \int_0^{1/\sqrt{2}} \int_r^1 r\, dz\, dr\, d\theta$

(c) $\int_0^{2\pi} \int_0^{\pi/4} \int_0^{1/(\sqrt{2}\cos\phi)} \rho^2 \sin\phi\, d\rho\, d\phi\, d\theta$

23 π

25 (a) Positive
  (b) Zero

27 $25\pi/6$

29 (a) $\int_0^{2\pi} \int_0^\pi \int_1^2 \rho^2 \sin\phi\, d\rho\, d\phi\, d\theta$
  (b) $\int_0^{2\pi} \int_0^2 \int_{-\sqrt{4-r^2}}^{\sqrt{4-r^2}} r\, dz\, dr\, d\theta$
    $- \int_0^{2\pi} \int_0^1 \int_{-\sqrt{1-r^2}}^{\sqrt{1-r^2}} r\, dz\, dr\, d\theta$

31 $\int_0^{2\pi} \int_0^{\arccos(3/5)} \int_{3/\cos\phi}^5 \rho^2 \sin\phi\, d\rho\, d\phi\, d\theta$
    $= 52\pi/3 = 54.45 \text{ cm}^3$

33 27π

35 $3/\sqrt{2}$

37 (a) $\int_0^{2\pi} \int_1^5 \int_{-\sqrt{25-r^2}}^{\sqrt{25-r^2}} r\, dz\, dr\, d\theta$
  (b) $64\sqrt{6}\pi = 492.5 \text{ mm}^3$

39 $3/4$

43 $3I = \frac{6}{5}a^2; I = \frac{2}{5}a^2$

45 $2\pi G\delta(H + R - \sqrt{R^2 + H^2})$

47 Total charge $= \frac{k\pi}{2}h^2 R^2$

49 $(q^2/8\pi\epsilon)((1/a) - (1/b))$

Section 16.6

1 0

3 1

5 $7/8$

7 $1/16$

9 (a) 0.3
  (b) 0.9
  (c) 0.9

11 Is a joint density function

13 Is joint density function

15 Is joint density function

17 (a) $k = 8$
  (b) $1/3$

19 $\int_{65}^{100} \int_{0.8}^1 f(x,y)\, dx\, dy$

21 (a) $\lambda/(\lambda + \mu)$

23 (a) 0 if $t \le 0$, $2t^2$ if $0 < t \le 1/2$,
    $1 - 2(1 - t)^2$ if $1/2 < t \le 1$,
    1 if $1 < t$
  (b) 0 if $t \le 0$, $4t$ if $0 < t \le 1/2$,
    $4 - 4t$ if $1/2 < t \le 1$,
    0 if $1 < t$

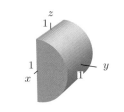

(c) x, y: All equally likely
    z: Near $1/2$

Section 16.7

1 1

3 e^{2s}

5 -13

7 $a = 1/10, b = 1$

9 $a = 1/50, b = 1/10$

11 $a = 1/2, b = 1/3$

13 3

17 $\rho^2 \sin\phi$

19 13.5

21 π

23 72

25 9

Chapter 16 Review

1

$x^2 + y^2 = 1$

3
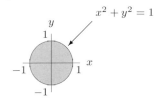

$x = -\sqrt{y}$ $x = \sqrt{y}$

$y = 4$

$y = 1$

5

7 $\int_0^{2\pi} \int_0^2 \int_0^3 fr\, dr dz d\theta$

9 $\int_0^{\pi/2} \int_0^\pi \int_0^5 f\rho^2 \sin\phi\, d\rho d\phi d\theta$

11 $7\pi/3$

13 $\frac{1}{3}(\sin 5 - \sin 2)(\cos 1 - \cos 2)$

15 $-4\cos 4 + 2\sin 4$
    $+ 3\cos 3 - 2\sin 3 - 1$

17 $(\pi/2)(1 - e^{-1})$

19 $\int_0^{2\pi} \int_0^{\arctan(1/2)} \int_0^{4/\cos\phi} g(\rho,\phi,\theta)\rho^2 \sin\theta\, d\rho\, d\phi\, d\theta$

23 $162\pi/5$

25 8π

27 Negative

29 Can't tell

31 Zero

33 Zero

35 Zero

37 Negative

39 Positive

41 108/5

43 ≈ 183

45 $pqr/6$

47 8π gm

49 $\int_{-1}^{1} \int_{-\sqrt{1-x^2}}^{\sqrt{1-x^2}} \int_{-1+\sqrt{2-x^2-y^2}}^{\sqrt{1-x^2-y^2}} dz\,dy\,dx$
$= 2\pi((7/6) - 2\sqrt{2}/3)$

51 $I_x = 25/2$
$I_y = 65/2$
$I_z = 85/2$

53 $2\pi Gm(r_2 - r_1 - \sqrt{r_2^2 + h^2} + \sqrt{r_1^2 + h^2}$

55 $10^7 \pi$ m^3

57 $6\pi/7$

59 $a, b + c, 0$

Ch. 16 Understanding

1 False

3 False

5 True

7 True

9 False

11 False

13 True

15 False

17 False

19 True

21 False

23 False

25 True

27 False

29 False

Section 17.1

1 $x = 3 + t, y = 2t, z = -4 - t$

3 $x = -3 + 2t, y = 4 + 2t, z = -2 - 3t$

5 $x = t, y = 1, z = -t$

7 $x = 1 + 4t, y = 5 - 5t, z = 2 - 3t$

9 $x = -3 + 2t, y = -2 - t, z = 1 - 2t$

11 $x = 3 - 3t, y = 0, z = -5t$

13 $x = 3\cos t, y = 3\sin t, z = 0$

15 $x = 2\cos t, y = 0, z = 2\sin t$

17 $x = 0, y = 3\cos t, z = 2 + 3\sin t$

19 $x = t, y = t^3, z = 0$

21 $x = -3t^2, y = 0, z = t$

23 $x = 3\cos t, y = 2\sin t, z = 0$

25 $x = \frac{3}{2}\cos t, y = 1, z = -2 + \sin t$

27 $\vec{r}(t) = -\vec{i} - 3\vec{j} + t(6\vec{i} + 5\vec{j}), 0 \le t \le 1$,
$x = -1 + 6t, y = -3 + 5t, 0 \le t \le 1$

29 $x = 0, y = 5\cos t, z = -5\sin t$,
$-\pi/2 \le t \le \pi/2$

31 Yes, no

33 No, no

35 $x = 4\cos t, y = -3\sin t, z = 3$,
$0 \le t \le \pi/2$

37 Four possible arcs:
$\vec{r}(t) = 5\vec{i} + 5\sqrt{3}\vec{j} + 10(-\cos t\vec{i} + \sin t\vec{j}), -\pi/3 \le t \le 4\pi/3$, or
$\vec{r}(t) = 5\vec{i} + 5\sqrt{3}\vec{j} + 10(\cos t\vec{i} + \sin t\vec{j}), 4\pi/3 \le t \le 5\pi/3$, or
$\vec{r}(t) = 5\vec{i} - 5\sqrt{3}\vec{j} + 10(-\cos t\vec{i} - $

$\sin t\vec{j}), -\pi/3 \le t \le 4\pi/3$, or
$\vec{r}(t) = 5\vec{i} - 5\sqrt{3}\vec{j} + 10(\cos t\vec{i} - \sin t\vec{j}), 4\pi/3 \le t \le 5\pi/3$

39 $\vec{r}(t) = (2 + (t/5)10)\vec{i} + (5 + (t/5)4)\vec{j}$

41 $\vec{r}(t) = (2 + ((t - 20)/10)10)\vec{i} + (5 + ((t - 20)/10)4)\vec{j}$

43 (b) Reverse of part (a)

45 (a) $2\vec{i} - 5\vec{j} + 3\vec{k}$
(c) $x = 1 + 2t, y = -1 - 5t, z = 1 + 3t$

47 $x = \frac{8}{3}, \quad y = t, \quad z = \frac{1}{3} + t$

49 $x = -4, y = 2 + t, z = 3 + t$

51 Intersection of planes

53 A straight line through the point $(3, 5, 7)$

55 Helix of radius 1 along y-axis

57 Yes

59 (a) $x = 2 + at, y = 1 + bt, z = 3 + ct$
(b) $a/2 = b/1 = c/3$

61 $\vec{r} = 9.65\vec{k} + t(325\vec{i} + 563\vec{j} - 0.84\vec{k})$

63 (a) Yellow
(b) Direction $(-0.4 + t)\vec{i} + (-3.8 + t)\vec{j} + (t - 1)\vec{k}$

Section 17.2

1 $\vec{v} = \vec{i} + (3t^2 - 1)\vec{j}, \quad \vec{a} = 6t\vec{j}$

3 $\vec{v} = -3\sin t\vec{i} + 4\cos t\vec{j}$,
$\vec{a} = -3\cos t\vec{i} - 4\sin t\vec{j}$

5 $\vec{v} = 6t\vec{i} + 3t^2\vec{j}$,
$\|\vec{v}\| = 3|t| \cdot \sqrt{4 + t^2}$,
Stops when $t = 0$

7 $\vec{v} = 6t\cos(t^2)\vec{i} - 6t\sin(t^2)\vec{j}$,
$\|\vec{v}\| = 6|t|$,
Stops when $t = 0$

9 Length $= \sqrt{42}$

11 Length ≈ 24.6

13 $\vec{v} = 6\cos(3t)\vec{j} - 6\sin(3t)\vec{k}$,
$\vec{a} = -18\sin(3t)\vec{j} - 18\cos(3t)\vec{k}$,
$\vec{v} \cdot \vec{a} = 0, \|\vec{v}\| = 6, \|\vec{a}\| = 18$

15 Line through $(1, -5, -2)$ in direction of
$-\vec{i} + 2\vec{j} + 3\vec{k}$,
$\vec{v} = (6t^2 + 3)(-\vec{i} + 2\vec{j} + 3\vec{k})$,
$\vec{a} = 12t(-\vec{i} + 2\vec{j} + 3\vec{k})$

17 Circle radius 3 counterclockwise
$(3, 0), \vec{v} = 6\vec{j}, \vec{a} = -12\vec{i}$

19 (a) Spiral

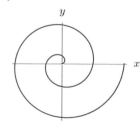

(b) $\vec{v}(2) = -2.24\vec{i} + 0.08\vec{j}$,
$\vec{v}(4) = 2.38\vec{i} - 3.37\vec{j}$,
$\vec{v}(6) = 2.63\vec{i} + 5.48\vec{j}$

(c) $\vec{v}(2) = -2.235\vec{i} + 0.077\vec{j}$,
$\vec{v}(4) = 2.374\vec{i} - 3.371\vec{j}$,
$\vec{v}(6) = 2.637\vec{i} + 5.482\vec{j}$

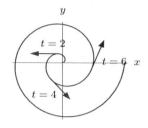

21 Vertical: $t = 3$
Horizontal: $t = \pm 1$
As $t \to \infty, x \to \infty, y \to \infty$
As $t \to -\infty, x \to \infty, y \to -\infty$

23 $x = 5 + 2(t - 4), y = 4 - 3(t - 4)$,
$z = -2 + (t - 4)$.

25 (a) $x = 2 + 0.6t, y = -1 + 0.8t, z = 5 - 1.2t$
(b) $x = 2 + 1.92t, y = -1 + 2.56t, z = 5 - 3.84t$

27 (a) (IV); 4.5 sec; $(0, 8.9 \text{ m}, 0)$
(b) (II); 3.2 sec; base of tower
(c) (V); 10 sec; halfway up

29 $\vec{r}(t) = 22.1t\vec{i} + 66.4t\vec{j} + (442.7t - 4.9t^2)\vec{k}$

31 (a) π m/sec
(b) 2.45 m
(c) 3.01 m

33 (a) $\vec{r}(t) = t\cos(2\pi t)\vec{i} + t\sin(2\pi t)\vec{j}$,
$0 \le t \le 100$
(b) $\vec{v} = \vec{r}'(t) = (\cos(2\pi t) - 2\pi t\sin(2\pi t))\vec{i} + (\sin(2\pi t) + 2\pi t\cos(2\pi t))\vec{j}$
$\|\vec{v}\| = (1 + 4\pi^2 t^2)^{1/2}$ cm/sec
(c) $\vec{a} = \vec{v}'(t) = (-4\pi\sin(2\pi t) - 4\pi^2 t\cos(2\pi t))\vec{i} + (4\pi\cos(2\pi t) - 4\pi^2 t\sin(2\pi t))\vec{j}$
$\|\vec{a}\| = 4\pi(1 + \pi^2 t^2)^{1/2}$ cm/sec^2

35 No

Section 17.3

1 $\vec{V} = x\vec{i}$

3 $\vec{V} = x\vec{i} + y\vec{j} = \vec{r}$

5 $\vec{V} = -y\vec{i} + x\vec{j}$

7

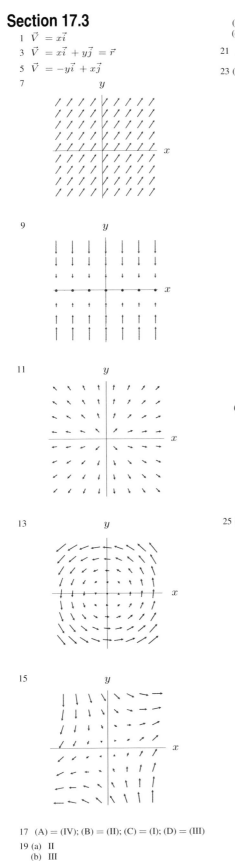

9

11

13

15

17 (A) = (IV); (B) = (II); (C) = (I); (D) = (III)

19 (a) II
   (b) III

(c) IV

(d) I

21 $\vec{F}(x, y) = \frac{-x\vec{i} - y\vec{j}}{\sqrt{x^2 + y^2}}$ (for example)

23 (a) $z = f(x, y)$:

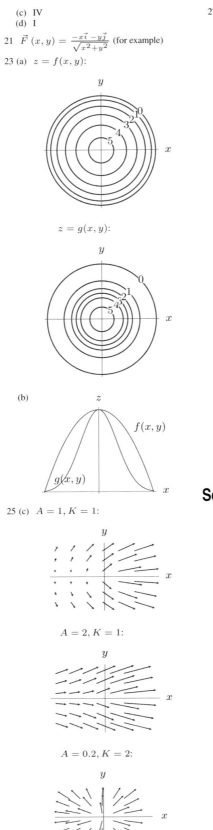

$z = g(x, y)$:

(b)

25 (c) $A = 1, K = 1$:

$A = 2, K = 1$:

$A = 0.2, K = 2$:

27 (a) Radiates out from origin

(b) Spirals outward counterclockwise around origin

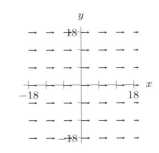

(c) Spirals outward clockwise around origin

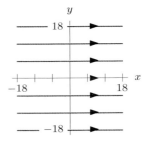

Section 17.4

1 Field:

Flow, $y =$ constant:

3 Field:

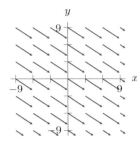

Flow, $y = -(2/3)x + c$:

5 Field:

Flow:

7 Field:

Flow:

9 Field:

Flow:

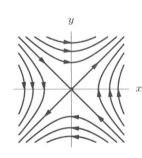

15 (a) Same directions, different magnitudes
   (b) Same curves, different parameterizations
17 (a) $\vec{v} \cdot \text{grad } H = 0$
19 (a) $\vec{v} = \pi(-y\vec{i} + x\vec{j})/12$
   (b) Horizontal circles

Section 17.5

1 Horizontal disk of radius 5 in plane $z = 7$
3 Cylinder radius 5 around z-axis, $0 \leq z \leq 7$
5 Cone of height and radius 5
7 Cylinder, elliptical cross-section
9 Top hemisphere
11 Vertical segment
13 $\vec{r} = \vec{j} + 2\cos\theta\vec{u} + 2\sin\theta\vec{v}$
15 (a) Yes
   (b) No
17 $s = 4, t = 2$
   $(x, y, z) = (x_0 + 10, y_0 - 4, z_0 + 18)$
19 Horizontal circle
21 $x = a\cos\theta, y = a\sin\theta, z = z$

23 $x = a + d\sin\phi\cos\theta, y = b + d\sin\phi\sin\theta,$
   $z = c + d\cos\phi$ for $0 \leq \phi \leq \pi$ and
   $0 \leq \theta \leq 2\pi$.
25 $x = u\cos v, y = u\sin v,$
   $z = u$ for $0 \leq v \leq 2\pi$
27 $x = r\cos\theta,$
   $y = r\sin\theta,$
   $z = r^2,$
   $r \geq 0,$
   $0 \leq \theta \leq 2\pi$
29 $\vec{r}(t) =$
   $x_0\vec{i} + y_0\vec{j} + z_0\vec{k} + a\cos t\vec{u} + a\sin t\vec{v}$
31 (a) $x^2 + y^2 + z^2 = 1, x, y, z \geq 0$.
   (b)

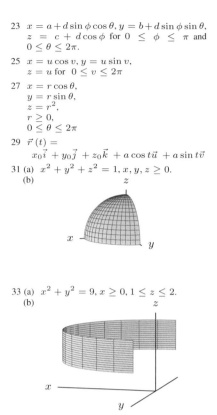

33 (a) $x^2 + y^2 = 9, x \geq 0, 1 \leq z \leq 2$.
   (b)

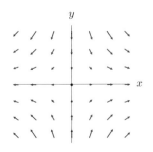

35 $x = tu, y = tu^2, z = tu - tu^2 + 5t,$
   $0 \leq u \leq 3, 0 \leq t \leq 1$
37 (a) $\vec{r} = b\cos\theta\vec{i} + b\sin\theta\vec{j}$
   (b) $\vec{s} =$
      $a\cos\phi(\cos\theta\vec{i} + \sin\theta\vec{j}) + a\sin\phi\vec{k}$
   (c) $x\vec{i} + y\vec{j} + z\vec{k} = \vec{r} + \vec{s} =$
      $(b\cos\theta + a\cos\phi\cos\theta)\vec{i} +$
      $(b\sin\theta + a\cos\phi\sin\theta)\vec{j} + a\sin\phi\vec{k}$

Chapter 17 Review

1 $x = t, y = 5$
3 $x = 4 + 4\sin t, y = 4 - 4\cos t$
5 $x = 2 - t, y = -1 + 3t, z = 4 + t$.
7 $x = 1 + 2t, y = 1 - 3t, z = 1 + 5t$.
9 $x = 3\cos t$
   $y = 5$
   $z = -3\sin t$
11 No
13 (a) The part of the line with
      $x, y < 0$ and $z < 10$
   (b) The line segment between
      $(0, 0, 10)$ and $(1, 2, 13)$
15

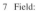

17 (b) (i) Yes
    (ii) No
    (iii) Yes
    (iv) No

19 (a) $(I) = C_4, (II) = C_1,$
    $(III) = C_2, (IV) = C_6$
  (b) $C_3 : 0.5\cos t\vec{i} - 0.5\sin t\vec{j},$
    $C_5 : -2\cos(\frac{t}{2})\vec{i} - 2\sin(\frac{t}{2})\vec{j}$

21 $\vec{E} : (IV); \vec{F} : (I);$
  $\vec{G} : (II); \vec{H} : (III)$

23 $\vec{v} = 0.2\vec{i} - 0.4\vec{j} + 0.4\vec{k}$ m/sec
  $\vec{a} = \vec{0}$

25 (a) Max rate change temp with distance; $^\circ$C/cm
  (b) $\sqrt{(g'(t))^2 + (k'(t))^2}$; cm/sec
  (c) $f_x \cdot g'(t) + f_y \cdot k'(t)$; $^\circ$C/cm

27 (a) In the direction given by the vector: $\vec{i} - \vec{j}$
  (b) Directions given by unit vectors:
    $\frac{1}{\sqrt{2}}\vec{i} + \frac{1}{\sqrt{2}}\vec{j}$
    $-\frac{1}{\sqrt{2}}\vec{i} - \frac{1}{\sqrt{2}}\vec{j}$
  (c) -4

29 (a) $(2t, 0)$ when $0 \le t \le \frac{1}{2}$
    $(\cos\frac{3\pi}{2}(t-\frac{1}{2}), \sin\frac{3\pi}{2}(t-\frac{1}{2}))$
    when $\frac{1}{2} < t \le \frac{5}{6}$
    $(0, -2(t-\frac{4}{3}))$
    when $\frac{5}{6} < t \le \frac{4}{3}$

  (b) $(0, 2t)$ when $0 \le t \le \frac{1}{2}$
    $(\sin\frac{3\pi}{2}(t-\frac{1}{2}), \cos\frac{3\pi}{2}(t-\frac{1}{2}))$
    when $\frac{1}{2} < t \le \frac{5}{6}$
    $(-2(t-\frac{4}{3}), 0)$
    when $\frac{5}{6} < t \le \frac{4}{3}$

31 No, since the point $(0, 1)$ is not on the curve

33 (b) $\vec{i} + \vec{j} + \sqrt{2}\vec{k}$
  (c) $-\vec{i} + \frac{1}{\sqrt{2}}\vec{k}$

35 ω: Rate of change of polar angle θ of particle, a: Rate of change of particle's distance from origin

37 (a) Yes
  (b) No

39 $x = 2 + 5\sin\phi\cos\theta$
  $y = -1 + 5\sin\phi\sin\theta$
  $z = 3 + 5\cos\phi$

41 (a) Cylinder
  (b) Helices

43 $x = a\sin\phi\cos\theta$ $0 \le \phi \le \pi$
  $y = b\sin\phi\sin\theta$ $0 \le \theta \le 2\pi$
  $z = c\cos\phi$

45 No

47 (a) Circles centered at origin
  (c) Parameterized at different speeds

49 (a) $x = \frac{20}{13} - \frac{6t}{13}, y = \frac{-1}{13} - \frac{t}{13}, z = t$
  (b) $x = t, y = \frac{1}{6}(-2 - 2t + 3t^2),$
    $z = \frac{1}{6}(20 - 10t - 3t^2).$
  (c) $x = \sqrt{2 - t^2}, y = t,$
    $z = 5 + 5t - 3\sqrt{2 - t^2}$ and
    $x = -\sqrt{2 - t^2}, y = t,$
    $z = 5 + 5t + 3\sqrt{2 - t^2}$

Ch. 17 Understanding

1 False
3 False
5 False
7 True
9 True

11 True
13 False
15 False
17 False
19 False
21 True
23 False
25 True
27 True
29 True
31 False
33 False
35 False
37 True
39 True
41 False

Section 18.1

1 Positive
3 Positive
5 16
7 10
9 0
11 24
13 -48
15 $\int_{C_3}\vec{F}\cdot d\vec{r} < \int_{C_1}\vec{F}\cdot d\vec{r} < \int_{C_2}\vec{F}\cdot d\vec{r}$
17 C_1 is pos; C_2, C_3 are zero
19 C_1 is pos; C_2 is neg; C_3 is zero
21 Positive
23 0
25 (a) (i)

  (ii)

  (iii)

(iv)

(b) (i), (iii)

27

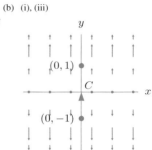

31 $-2.5 \cdot 10^{-5} GMm$
33 $-\int_C \vec{E}\cdot d\vec{r}$
35 Spheres centered at origin

Section 18.2

1 -6
3 9
5 π
7 116.28
9 12
11 $e^2 - e$
13 85.32
15 24π
17 (a) 11/6
  (b) 7/6
19 $560g$
21 (a) 3/2
  (b) 3/2

Section 18.3

3 Yes
5 No
7 Indep
9 Indep
11 12
13 $\cos(2) - \cos(54)$
15 $(e^7 - 1)/7 - 3/5$
17 (a) e
  (b) e
19 (a) $a^2/2 + ab + b^2/2$
  (b) $a^2/2 + 2ab + b^2/2$
  (c) $a^2/2 + 3ab/2 + b^2/2$
21 9/2
23 $\frac{3}{\sqrt{2}}\ln(\frac{3}{\sqrt{2}} + 1)$
25 1859
27 -2
29 (b) No
31 (a) $2\pi mg$
  (b) Yes
33 (a) $\vec{a}$
  (b) $\vec{r}_0 \cdot \vec{a}$
  (c) $10\|\vec{a}\|$

Section 18.4

1 $f(x, y) = x^2y + 2y^4 + K$
   $K = $ constant

3 No

5 Yes, $f = x^2y^3 + xy + C$

7 No

9 -2π

11 -6

13

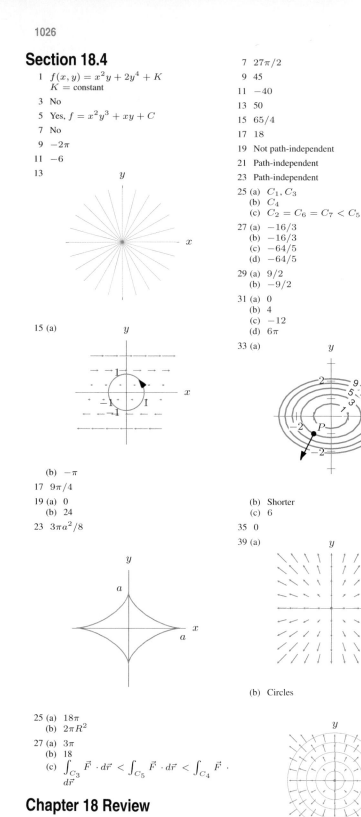

15 (a)

   (b) $-\pi$

17 $9\pi/4$

19 (a) 0
   (b) 24

23 $3\pi a^2/8$

25 (a) 18π
   (b) $2\pi R^2$

27 (a) 3π
   (b) 18
   (c) $\int_{C_3} \vec{F} \cdot d\vec{r} < \int_{C_5} \vec{F} \cdot d\vec{r} < \int_{C_4} \vec{F} \cdot d\vec{r}$

Chapter 18 Review

1 Positive

3 (a) Zero
   (b) C_1, C_3: Zero
        C_2: Negative
        C_4: Positive
   (c) Zero

5 350

7 $27\pi/2$

9 45

11 -40

13 50

15 $65/4$

17 18

19 Not path-independent

21 Path-independent

23 Path-independent

25 (a) C_1, C_3
   (b) C_4
   (c) $C_2 = C_6 = C_7 < C_5$

27 (a) $-16/3$
   (b) $-16/3$
   (c) $-64/5$
   (d) $-64/5$

29 (a) $9/2$
   (b) $-9/2$

31 (a) 0
   (b) 4
   (c) -12
   (d) 6π

33 (a)

   (b) Shorter
   (c) 6

35 0

39 (a)

   (b) Circles

   (c) No

41 (a) $-(\pi/2)(-6a^2 + a^4)$, $a = \sqrt{3}$
   (b) Integrand $3 - x^2 - y^2$ positive inside disk of radius $\sqrt{3}$, negative outside

43 $18a + 18b + 36c + (81d/2)$, $-18a - 18b - 36c - (81d/2)$, curves go in opposite directions

Ch. 18 Understanding

5 False

7 False

9 True

11 False

13 False

15 False

17 False

19 False

21 False

23 True

25 True

27 False

29 True

31 True

33 False

35 False

37 False

39 True

41 False

Section 19.1

1 (a) Positive
  (b) Negative
  (c) Zero
  (d) Zero
  (e) Zero

3 (a) Zero
  (b) Zero
  (c) Zero
  (d) Negative
  (e) Zero

5 4

7 9

9 80

11 75π

13 27/2

15 64/3

17 756π

19 -96π

21 $4\sqrt{3}$

23 $\pi(e^4 - 1)$

25 36π

27 (a) Zero
   (b) Zero

29 $a = 6$
    Cannot say anything about b and c

31 (a) Flux through S_3
   (b) 13π

37 (a) Maximum speed
   (b) 0
   (c) $\pi u a^2/2$

Section 19.2

1 6

3 $\pi/2$

5 7/3

7 $\pi \sin 25$

9 1296π

11 $81\pi/4$

13 12π

15 $625\pi/2$

17 $11\pi/2$

19 (a) Constant inside sphere of radius a

(b) $\vec{E} = \begin{cases} k\dfrac{\delta_0}{3}\rho\vec{e}_\rho & \rho \le a \\ k\dfrac{\delta_0 a^3}{3r^3}\vec{e}_\rho & \rho > a \end{cases}$

Section 19.3

1 $4/3$

3 195

5 $-\pi R^7/28$

7 $2\pi c(a^2 + b^2)$

Chapter 19 Review

1 0

3 3

5 2π

7 $4\pi^4$

9 45π

11 -98

13 1

15 100

17 28π

19 0

21 114

23 $3(8 + \sin 4)$

25 20

27 2π

29 Flux through S_1 = Flux through S_2 < Flux through S_3 < Flux through S_4

31 Q

33 $(\rho/\epsilon_0)\pi R^2 h$. No

37 (a)

(b) 0

(c) $Ih \ln|b/a|/2\pi$

39 (a) 0.0958

(b) 0.0958

(c) 0.0958

Ch. 19 Understanding

1 True

3 False

5 True

7 True

9 True

11 False

13 False

15 True

Section 20.1

1 (I)

3 0

5 -1

7 0

9 (a) Positive

(b) Zero

(c) Negative

11

13 $0.00002\pi/3$

15 (a) (i) 0.0168

(ii) -0.004

(i) Pointing outward

(ii) Pointing inward

17 (a) 0

(b)

19 (a) Flux = c^3

(b) 1

(c) 1

21 (b)

31 (a) 0

(b) Undefined

33 (d)

Section 20.2

1 No

3 8

5 2π

9 24

11 72

13 $\int_S \vec{F} \cdot d\vec{A} = \int_W \operatorname{div}\vec{F}\, dV = 0$

15 0

17 $633(2 - \sqrt{2})\pi/5 = 232.98$

19 Yes; -3.22

21 -10π

23 0

25 34π

27 (a) 0

(b) 4π

29 (a) Negative

(b) Positive

31 (a) $4\vec{a} \cdot \vec{r}$

(b) 0

33 S closed, oriented inward

35 $72\pi \le$ Flux $\le 180\pi$

39 (a) 30 watts/km^3

(b) $\alpha = 10$ watts/km^3

(d) $6847°$C

Section 20.3

1 Zero curl

3 Nonzero curl

5 Zero

7 $4y\vec{k}$

9 $\vec{0}$

11 $-\vec{i} - \vec{j} - \vec{k}$

13 $\vec{0}$

15 $\operatorname{curl}(F_1(x)\vec{i} + F_2(y)\vec{j} + F_3(z)\vec{k}) = 0$

17 (a) No

(b) Positive

(c) $x\vec{j}$

(d) $\vec{k}$

19 (a) 0.0005π

(b) 0.0002π

(c) -0.0003π

21 (a) Counterclockwise

(b) Clockwise

(c) $\vec{0}$

31 (a) $w = 1$

$w = -1$

(b) $|\omega| \cdot \sqrt{x^2 + y^2}$
(c) div $\vec{v} = 0$
curl $\vec{v} = 2\omega \vec{k}$
(d) $2\pi\omega R^2$

Section 20.4

1 No

3 20π

5 0

7 25

9 -8π

11 $-\pi$

13 (a) $-\vec{i} - \vec{j} - \vec{k}$
(b) -6

15 (a) $\vec{i} + \vec{j} + \vec{k}$
(b) $-4\sqrt{3}$

17 $8\pi/\sqrt{3}$

19 0

21 $8\sqrt{3}\pi$

23 72

25 (a) 0
(b) No

27 -2π

29 (b) 75π

33 (a) $a = 0$
(b) $b = 1, c = 3, m = 5$

35 (a) $\vec{0}$; all except z-axis
(b) 2π
(c) 0
(d) 2π
(e) $\pm 2\pi$ if encircles z-axis
0 otherwise

Section 20.5

1 Yes

5 $\vec{F} = (-\frac{3}{2}z - 2y)\vec{i}$
$+ (2x - z)\vec{j} + (y + \frac{3}{2}x)\vec{k}$

9 Yes
$(-xy + 5yz)\vec{i} + (2xy + xz^2)\vec{k}$.

11 (a) Yes
(b) Yes
(c) Yes

13 (a) curl $\vec{E} = \vec{0}$
(b) 3-space minus a point if $p > 0$
3-space if $p \le 0$.
(c) Satisfies test for all p.
$\phi(r) = r^{2-p}$ if $p \ne 2$.
$\phi(r) = \ln r$ if $p = 2$.

Chapter 20 Review

1 $(c), (d), (f)$

3 $C_2, C_3, C_4, C_6,$

5 div $\vec{F} = y + z + x$
curl $\vec{F} = -y\vec{i} - z\vec{j} - x\vec{k}$
$\vec{F}$ not solenoidal; not irrotational

7 div $\vec{F} = 0$
curl $\vec{F} = (2y - \cos(x + z))\vec{i}$
$- \left(2x - e^{y+z}\right)\vec{j} + \left(\cos(x + z) - e^{y+z}\right)\vec{k}$
$\vec{F}$ is solenoidal; not irrotational

9 (a) 3/2
(b) 3/2

11 1500π

13 162

15 0

17 18π

19 $-6\sqrt{3}\pi$

21 $a = 3, b = 2, c = 5$

23 (a) 2.984
(b) 0.100

25 (a) $4w^3$
(b) 4
(c) 4

27 (a) Conservative.
(b) 0
(c) $-y\vec{j}$
(d) $\vec{0}$

29 $16\pi/3$

31 $12\pi/5$

33 150π

35 325π

37 50π

39 div $\vec{E}(P) \le 0$, div $\vec{E}(Q) \ge 0$.

41 (a) 0
(b) 0

43 (a) 4π
(b) 0
(c) 4π

45 4π

49 $\pi b^2 h/3$

51 (a)

(b) $\vec{v}(x) = (55 - x/50)\vec{i}$ mph
if $0 \le x < 2000$
$\vec{v}(x) = 15\vec{i}$ mph
if $2000 \le x < 7000$
$\vec{v}(x) = (15 + (x - 7000)/25)\vec{i}$ mph
if $7000 \le x < 8000$
$\vec{v}(x) = 55\vec{i}$ mph
if $x \ge 8000$

(c) div $\vec{v}(1000) = -1/50$ mph/ft
div $\vec{v}(5000) = 0$ mph/ft
div $\vec{v}(7500) = 1/25$ mph/ft
div $\vec{v}(10,000) = 0$ mph/ft

53 (a) $\left(1 + \frac{y^2 - x^2}{(x^2 + y^2)^2}\right)\vec{i} + \frac{-2xy}{(x^2 + y^2)^2}\vec{j}$
(d)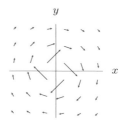

55 (b) $\|\vec{v}\| = K/r^2$

(c) Flux $= 4\pi K/3$
(d) Zero

57 (a) $5a^2 b - 5b^3/3$
(b) $5a^2 b - 5b^3/3$
(c) Line integrals are equal
(d) $\frac{5}{3}y^3 - 5x^2 y$

59 (a) $K = 1$:

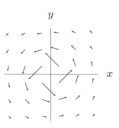

$K = -1$:

(b) $|K|/r$
(c) For$(x, y, z) \ne (0, 0, z)$, div $\vec{v} = 0$
(e) $2\pi K$
(f) $\vec{v}$ undefined on z-axis

61 (a) $(4(p + q)\pi R^3/3)$
(b) $(4(p + q)\pi R^3/3)$

Ch. 20 Understanding

1 True
3 False
5 True
7 True
9 False
11 False
13 False
15 True
17 True
19 True
21 False
23 True
25 True
27 True
29 False
31 False
33 False
35 True
37 True
39 False

41 False

Appendix A

1 (a) $y \leq 30$
  (b) two zeros

3 -1.05

5 2.5

7 $x = -1.1$

9 0.45

11 1.3

13 (a) $x = -1.15$
  (b) $x = 1, x = 1.41$,
    and $x = -1.41$

15 (a) $x \approx 0.7$
  (b) $x \approx 0.4$

17 (a) 4 zeros
  (b) $[0.65, 0.66], [0.72, 0.73]$,
    $[1.43, 1.44], [1.7, 1.71]$

19 (b) $x \approx 5.573$

21 Bounded $-5 \leq f(x) \leq 4$

23 Not bounded

Appendix B

1 $2e^{i\pi/2}$

3 $\sqrt{2}e^{i\pi/4}$

5 $0e^{i\theta}$, for any θ.

7 $\sqrt{10}e^{i\arctan(-3)}$

9 $-3 - 4i$

11 $-5 + 12i$

13 $1/4 - 9i/8$

15 $-1/2 + i\sqrt{3}/2$

17 $-125i$

19 $\sqrt{2}/2 + i\sqrt{2}/2$

21 $\sqrt{3}/2 + i/2$

23 -2^{50}

25 $2i\sqrt[3]{4}$

27 $(1/\sqrt{2})\cos(-\pi/12) + (i/\sqrt{2})\sin(-\pi/12)$

29 $-i, -1, i, 1$
  $i^{-36} = 1, i^{-41} = -i$

31 $A_1 = 1 + i$
  $A_2 = 1 - i$

37 True

39 False

41 True

Appendix C

1 (a) $f'(x) = 3x^2 + 6x + 3$
  (b) At most one
  (c) $[0, 1]$
  (d) $x \approx 0.913$

3 $\sqrt[4]{100} \approx 3.162$

5 $x \approx 0.511$

7 $x \approx 1.310$

9 $x \approx 1.763$

11 $x \approx 0.682328$

INDEX

Differentiation Formulas

1. $(f(x) \pm g(x))' = f'(x) \pm g'(x)$

2. $(kf(x))' = kf'(x)$

3. $(f(x)g(x))' = f(x)g'(x) + g(x)f'(x)$

4. $\left(\dfrac{f(x)}{g(x)}\right)' = \dfrac{g(x)f'(x) - f(x)g'(x)}{(g(x))^2}$

5. $(f(g(x)))' = f'(g(x)) \cdot g'(x)$

6. $\dfrac{d}{dx}(x^n) = nx^{n-1}$

7. $\dfrac{d}{dx}(e^x) = e^x$

8. $\dfrac{d}{dx}(a^x) = a^x \ln a \quad (a > 0)$

9. $\dfrac{d}{dx}(\ln x) = \dfrac{1}{x}$

10. $\dfrac{d}{dx}(\sin x) = \cos x$

11. $\dfrac{d}{dx}(\cos x) = -\sin x$

12. $\dfrac{d}{dx}(\tan x) = \dfrac{1}{\cos^2 x}$

A Short Table of Indefinite Integrals

I. Basic Functions

1. $\displaystyle \int x^n \, dx = \dfrac{1}{n+1} x^{n+1} + C, \quad n \neq -1$

2. $\displaystyle \int \dfrac{1}{x} \, dx = \ln|x| + C$

3. $\displaystyle \int a^x \, dx = \dfrac{1}{\ln a} a^x + C, \quad a > 0$

4. $\displaystyle \int \ln x \, dx = x \ln x - x + C$

5. $\displaystyle \int \sin x \, dx = -\cos x + C$

6. $\displaystyle \int \cos x \, dx = \sin x + C$

7. $\displaystyle \int \tan x \, dx = -\ln|\cos x| + C$

II. Products of e^x, $\cos x$, and $\sin x$

8. $\displaystyle \int e^{ax} \sin(bx) \, dx = \dfrac{1}{a^2 + b^2} e^{ax} [a \sin(bx) - b \cos(bx)] + C$

9. $\displaystyle \int e^{ax} \cos(bx) \, dx = \dfrac{1}{a^2 + b^2} e^{ax} [a \cos(bx) + b \sin(bx)] + C$

10. $\displaystyle \int \sin(ax) \sin(bx) \, dx = \dfrac{1}{b^2 - a^2} [a \cos(ax) \sin(bx) - b \sin(ax) \cos(bx)] + C, \quad a \neq b$

11. $\displaystyle \int \cos(ax) \cos(bx) \, dx = \dfrac{1}{b^2 - a^2} [b \cos(ax) \sin(bx) - a \sin(ax) \cos(bx)] + C, \quad a \neq b$

12. $\displaystyle \int \sin(ax) \cos(bx) \, dx = \dfrac{1}{b^2 - a^2} [b \sin(ax) \sin(bx) + a \cos(ax) \cos(bx)] + C, \quad a \neq b$

III. Product of Polynomial $p(x)$ with $\ln x$, e^x, $\cos x$, $\sin x$

13. $\displaystyle \int x^n \ln x \, dx = \dfrac{1}{n+1} x^{n+1} \ln x - \dfrac{1}{(n+1)^2} x^{n+1} + C, \quad n \neq -1$

14. $\displaystyle \int p(x)e^{ax} \, dx = \dfrac{1}{a} p(x)e^{ax} - \dfrac{1}{a} \int p'(x)e^{ax} \, dx$

$$= \dfrac{1}{a} p(x)e^{ax} - \dfrac{1}{a^2} p'(x)e^{ax} + \dfrac{1}{a^3} p''(x)e^{ax} - \cdots$$
$$(+ - + - \ldots)$$

(signs alternate)